AF786132

The Mycota

Edited by
K. Esser and P.A. Lemke

Springer
Berlin
Heidelberg
New York
Barcelona
Budapest
Hong Kong
London
Milan
Paris
Santa Clara
Singapore
Tokyo

The Mycota

I *Growth, Differentiation and Sexuality*
Ed. by J.G.H. Wessels and F. Meinhardt

II *Genetics and Biotechnology*
Ed. by U. Kück

III *Biochemistry and Molecular Biology*
Ed. by R. Brambl and G. Marzluf

IV *Environmental and Microbial Relationships*
Ed. by D. Wicklow and B. Söderström

V *Plant Relationships*
Ed. by G. Carroll and P. Tudzynski

VI *Human and Animal Relationships*
Ed. by D.H. Howard and J.D. Miller

VII *Systematics and Evolution*
Ed. by D.J. McLaughlin, E.G. McLaughlin, and P.A. Lemke

VIII *Cell Structure and Function*
Ed. by C.E. Bracker and L. Dunkle

IX *Fungal Associations*
Ed. by B. Hock

The Mycota

A Comprehensive Treatise on Fungi as Experimental Systems for Basic and Applied Research

Edited by K. Esser and P.A. Lemke†

V *Plant Relationships Part B*

Volume Editors:
G.C. Carroll and P. Tudzynski

With 81 Figures and 13 Tables

Springer

Series Editors

Professor Dr. Dr. h.c. mult. KARL ESSER
Allgemeine Botanik
Ruhr-Universität
D-44780 Bochum
Germany

Professor Dr. PAUL A. LEMKE†, Auburn, USA

Volume Editors

Professor Dr. George C. Carroll
Department of Biology
University of Oregon
Eugene, OR 97403
USA

Professor Dr. Paul Tudzynski
Institut für Botanik
Westfälische Wilhelms-Universität
Schloßgarten 3
D-48149 Münster
Germany

ISBN-13: 978-3-642-64491-7 e-ISBN-13: 978-3-642-60647-2
DOI: 10.1007/978-3-642-60647-2

Library of Congress Cataloging-in-Publication Data.

The Mycota. Includes bibliographical references and index. Contents: 1. Growth, differentiation, and sexuality/editors, J.G.H. Wessels and F. Meinhardt – 2. Genetics and biotechnology. 1. Mycology. 2. Fungi. 3. Mycology – Research. 4. Research. I. Esser, Karl, 1924– . II. Lemke, Paul A., 1937– . QK603.M87 1994 589.2

This work is subject to copyright. All rights are reserved, whether the whole or part of the material is concerned, specifically the rights of translation, reprinting, reuse of illustrations, recitation, broadcasting, reproduction on microfilm or in any other way, and storage in data banks. Duplication of this publication or parts thereof is permitted only under the provisions of the German Copyright Law of September 9, 1965, in its current version, and permission for use must always be obtained from Springer-Verlag. Violations are liable for prosecution under the German Copyright Law.

© Springer-Verlag Berlin Heidelberg 1997

Softcover reprint of the hardcover 1st edition 1997

The use of general descriptive names, registered names, trademarks, etc, in this publication does not imply, even in the absence of a specific statement, that such names are exempt from the relevant protective laws and regulations and therefore free for general use.

Production Editor: P. Venkateswara Rao

Cover design: Springer-Verlag, E. Kirchner

Typesetting by Best-set Typesetter Ltd., Hong Kong

SPIN: 10546618 31/3137 – 5 4 3 2 1 0 – Printed on acid-free paper

Series Preface

Mycology, the study of fungi, originated as a subdiscipline of botany and was a descriptive discipline, largely neglected as an experimental science until the early years of this century. A seminal paper by Blakeslee in 1904 provided evidence for self-incompatibility, termed "heterothallism", and stimulated interest in studies related to the control of sexual reproduction in fungi by mating-type specificities. Soon to follow was the demonstration that sexually reproducing fungi exhibit Mendelian inheritance and that it was possible to conduct formal genetic analysis with fungi. The names Burgeff, Kniep and Lindegren are all associated with this early period of fungal genetics research.

These studies and the discovery of penicillin by Fleming, who shared a Nobel Prize in 1945, provided further impetus for experimental research with fungi. Thus began a period of interest in mutation induction and analysis of mutants for biochemical traits. Such fundamental research, conducted largely with *Neurospora crassa*, led to the one gene: one enzyme hypothesis and to a second Nobel Prize for fungal research awarded to Beadle and Tatum in 1958. Fundamental research in biochemical genetics was extended to other fungi, especially to *Saccharomyces cerevisiae*, and by the mid-1960s fungal systems were much favored for studies in eukaryotic molecular biology and were soon able to compete with bacterial systems in the molecular arena.

The experimental achievements in research on the genetics and molecular biology of fungi have benefited more generally studies in the related fields of fungal biochemistry, plant pathology, medical mycology, and systematics. Today, there is much interest in the genetic manipulation of fungi for applied research. This current interest in biotechnical genetics has been augmented by the development of DNA-mediated transformation systems in fungi and by an understanding of gene expression and regulation at the molecular level. Applied research initiatives involving fungi extend broadly to areas of interest not only to industry but to agricultural and environmental sciences as well.

It is this burgeoning interest in fungi as experimental systems for applied as well as basic research that has prompted publication of this series of books under the title *The Mycota*. This title knowingly relegates fungi into a separate realm, distinct from that of either plants, animals, or protozoa. For consistency throughout this Series of Volumes the names adopted for major groups of fungi (representative genera in parentheses) are as follows:

Pseudomycota

Division: Oomycota (*Achlya, Phytophthora, Pythium*)
Division: Hyphochytriomycota

Eumycota

Division: Chytridiomycota (*Allomyces*)
Division: Zygomycota (*Mucor, Phycomyces, Blakeslea*)

Division: Dikaryomycota
  Subdivision: Ascomycotina
    Class: Saccharomycetes (*Saccharomyces, Schizosaccharomyces*)
    Class: Ascomycetes (*Neurospora, Podospora, Aspergillus*)
  Subdivision: Basidiomycotina
    Class: Heterobasidiomycetes (*Ustilago, Tremella*)
    Class: Homobasidiomycetes (*Schizophyllum, Coprinus*)

We have made the decision to exclude from *The Mycota* the slime molds which, although they have traditional and strong ties to mycology, truly represent nonfungal forms insofar as they ingest nutrients by phagocytosis, lack a cell wall during the assimilative phase, and clearly show affinities with certain protozoan taxa.

The Series throughout will address three basic questions: what are the fungi, what do they do, and what is their relevance to human affairs? Such a focused and comprehensive treatment of the fungi is long overdue in the opinion of the editors.

A volume devoted to systematics would ordinarily have been the first to appear in this Series. However, the scope of such a volume, coupled with the need to give serious and sustained consideration to any reclassification of major fungal groups, has delayed early publication. We wish, however, to provide a preamble on the nature of fungi, to acquaint readers who are unfamiliar with fungi with certain characteristics that are representative of these organisms and which make them attractive subjects for experimentation.

The fungi represent a heterogeneous assemblage of eukaryotic microorganisms. Fungal metabolism is characteristically heterotrophic or assimilative for organic carbon and some nonelemental source of nitrogen. Fungal cells characteristically imbibe or absorb, rather than ingest, nutrients and they have rigid cell walls. The vast majority of fungi are haploid organisms reproducing either sexually or asexually through spores. The spore forms and details on their method of production have been used to delineate most fungal taxa. Although there is a multitude of spore forms, fungal spores are basically only of two types: (i) asexual spores are formed following mitosis (mitospores) and culminate vegetative growth, and (ii) sexual spores are formed following meiosis (meiospores) and are borne in or upon specialized generative structures, the latter frequently clustered in a fruit body. The vegetative forms of fungi are either unicellular, yeasts are an example, or hyphal; the latter may be branched to form an extensive mycelium.

Regardless of these details, it is the accessibility of spores, especially the direct recovery of meiospores coupled with extended vegetative haploidy, that have made fungi especially attractive as objects for experimental research.

The ability of fungi, especially the saprobic fungi, to absorb and grow on rather simple and defined substrates and to convert these substances, not only into essential metabolites but into important secondary metabolites, is also noteworthy. The metabolic capacities of fungi have attracted much interest in natural products chemistry and in the production of antibiotics and other bioactive compounds. Fungi, especially yeasts, are important in fermentation processes. Other fungi are important in the production of enzymes, citric acid and other organic compounds as well as in the fermentation of foods.

Fungi have invaded every conceivable ecological niche. Saprobic forms abound, especially in the decay of organic debris. Pathogenic forms exist with both plant and animal hosts. Fungi even grow on other fungi. They are found in aquatic as well as soil environments, and their spores may pollute the air. Some are edible; others are poisonous. Many are variously associated with plants as copartners in the formation of lichens and mycorrhizae, as symbiotic endophytes or as overt pathogens. Association with animal systems varies; examples include the predaceous fungi that trap nematodes, the microfungi that grow in the anaerobic environment of the rumen, the many insect-associated fungi and the medically important pathogens afflicting humans. Yes, fungi are ubiquitous and important.

There are many fungi, conservative estimates are in the order of 100000 species, and there are many ways to study them, from descriptive accounts of organisms found in nature to laboratory experimentation at the cellular and molecular level. All such studies expand our knowledge of fungi and of fungal processes and improve our ability to utilize and to control fungi for the benefit of humankind.

We have invited leading research specialists in the field of mycology to contribute to this Series. We are especially indebted and grateful for the initiative and leadership shown by the Volume Editors in selecting topics and assembling the experts. We have all been a bit ambitious in producing these Volumes on a timely basis and therein lies the possibility of mistakes and oversights in this first edition. We encourage the readership to draw our attention to any error, omission or inconsistency in this Series in order that improvements can be made in any subsequent edition.

Finally, we wish to acknowledge the willingness of Springer-Verlag to host this project, which is envisioned to require more than 5 years of effort and the publication of at least nine Volumes.

Bochum, Germany
Auburn, AL, USA
April 1994

Karl Esser
Paul A. Lemke
Series Editors

Addendum to the Series Preface

In early 1989, encouraged by Dieter Czeschlik, Springer-Verlag, Paul A. Lemke and I began to plan *The Mycota*. The first volume was released in 1994, three other volumes followed in the years 1995 and 1996. Also on behalf of Paul A. Lemke, I would like to take this opportunity to thank Dieter Czeschlik, his colleague Andrea Schlitzberger, and Springer-Verlag for their help in realizing the enterprise and for their excellent cooperation for many years.

Unfortunately, after a long and serious illness, *Paul A. Lemke* died in November 1995. Without his expertise, his talent for organization and his capability to grasp the essentials, we would not have been able to work out a concept for the volumes of the series and to acquire the current team of competent volume editors. He also knew how to cope with unexpected problems which occurred after the completion of the manuscripts. His particular concern was directed at Volume VII; in this volume, a posthumous publication of his will be included.

Paul A. Lemke was an outstanding scientist interested in many fields. He was extremely wise, dedicated to his profession and a preeminent teacher and researcher. Together with the volume editors, authors, and Springer-Verlag, I mourn the loss of a very good and reliable friend and colleague.

Bochum, Germany
January 1997

KARL ESSER

Volume Preface

The number of fungal species has been loosely estimated to be on the order of 1 million, while the number of vascular plants is known with considerably greater certainty to lie between 300000 and 350000. Clearly, any volume which purports to deal with interactions between these two vast assemblages of organisms must do so concisely and selectively. In the chapters to follow, we have made no attempt to be all-inclusive, but rather have chosen examples from which general conclusions about fungus/plant interactions might be drawn. The materials presented here come from the core literature on plant pathology, research on fungal mutualisms, and evolutionary biology. A variety of approaches are evident: biochemistry, molecular biology, cellular fine structure, genetics, epidemiology, population biology, ecology, and computer modeling. The frequent overlap of such approaches within single reviews has resulted in a rich array of insights into the factors which regulate fungus/plant interactions. In these chapters, such interactions have also been considered on a variety of scales, both geographic and temporal, from single plant cells to ecosystems, from interactions which occur within minutes of contact to mechanisms which have presumably evolved during the course of several hundred million years.

Volume V consists of two parts: Volume V, Part A, and Volume V, Part B. While section headings provide signposts, we wish to make the rationale for the organization of these volumes absolutely clear. Part A begins with a brief introduction to both volumes. A series of reviews follows (Chaps. 1–6) which deal with the temporal sequence of events from the time fungal spores make contact with a host plant until the point where fungal hyphae are either are either firmly ensconced within a host or the attempted infections have been repulsed. Chapters 7–12 deal with metabolic interactions between host and fungus within the host plant after infection and particularly with the roles played by low molecular weight fungal metabolites such as toxins and phytohormones in pathogenic as well as mutualistic associations.

Chapters 1–8 of Part B are grouped in a section labeled, "Profiles in Pathogenesis and Mutualism"; here, interactions between fungi and host plants are explored in a variety of important model systems. These reviews focus less on processes per se and more on the specific fungi or groups of fungi as examples of pathogens or mutualists on plants. Chapters 9–12 of Part B move from discussions of physiological interactions between individuals to considerations of interactions at an expanded geographic scale, within populations of plants. Here, Chapter 9 provides a treatment of classical plant epidemiology, while Chapter 11 provides the same focus for mutualistic mycorrhizal associations. Chapter 10 covers the fuzzy area between population biology and micro-evolution in a genus of ubiquitous and pleurivorous pathogens; Chapter 12 offers much the same approach for mutualistic endophytes of grasses.

Chapters 13–16 of Part B offer a view of an expanded temporal scale and consider the evolution of plant/fungus interactions. Chapter 13 considers the flexibility of the fungal genome, the ultimate substrate on which evolutionary forces must act. Chapter 14 discusses the evolutionary relationships between pathogenic and mutualistic fungi in one situation which has been particularly well worked out, the clavicipitaceous endophytes of grasses. Chapter 15 considers the evolutionary interplay between fungi and plants as illuminated through the use of mathematical and computer-driven models. The final chapter in the volume (Chap. 16) deals with the overall evolution of fungal parasitism and plant resistance and provides an appropriate coda of this series of essays.

Who is the audience for these volumes? Who might and will read them with profit? Basic literacy in mycology, in particular, and in modern biology, in general, has been assumed as a background for these chapters, and they clearly are not intended for the biological novice. However, we do expect that these volumes will be appreciated by a wide variety of professional biologists including, for example: teachers of upper division courses in general mycology engaged in the valiant (but often futile) attempt to keep their lectures up-to-date; graduate students contemplating literature reviews in connection with a thesis project; nonmycologists who wish to know what the fungi might have to offer in the way of model systems for the study of some fundamental aspect of host/parasite interactions; evolutionary biologists who have just become aware that fungi offer advantages in studying the evolutionary consequences of asexual reproduction. These, and many others, will read these chapters with pleasure. On the whole we are very pleased with the contributions presented here and believe they will prove informative and useful as entrées into the literature on fungus/plant interactions for some years to come.

Eugene, Oregon, USA
Münster, Germany
March 1997

George Carroll
Paul Tudzynski
Volume Editors

Contents Part B

Profiles in Pathogenesis and Mutualism

1 The Gene-for-Gene Concept in Plant-Pathogen Interactions: Tomato – *Cladosporium fulvum*
MATTHIEU H.A.J. JOOSTEN, GUY HONÉE, JAN A.L. VAN KAN, and PIERRE J.G.M. DE WIT (With 3 Figures) 3

2 The Potato Late Blight Pathogen and Other Pathogenic Oomycota *Phytophthora infestans*
FRANCINE GOVERS, ANDRÉ DRENTH, and CORNÉ M.J. PIETERSE (With 4 Figures) 17

3 The Rice Blast Fungus, *Magnaporthe grisea*
BARBARA VALENT (With 9 Figures) 37

4 *Erysiphe graminis*
HENRIETTE GIESE, SIGRUN HIPPE-SANWALD, SHAUNA SOMERVILLE, and JENNIFER WALLER (With 8 Figures) 55

5 The Uredinales
KURT MENDGEN (With 4 Figures) 79

6 Genetics and Molecular Biology of the Fungal Partner in the Ectomycorrhizal Symbiosis *Hebeloma cylindrosporum* × *Pinus pinaster*
JEAN-CLAUDE DEBAUD, GILLES GAY, and ROLAND MARMEISSE (With 12 Figures) 95

7 Vesicular-Arbuscular Mycorrhizal (VAM) Fungi
ROBERT G. LINDERMAN (With 4 Figures) 117

8 Endophytes of Forest Trees: A Model for Fungus-Plant Interactions
JEFFREY K. STONE and ORLANDO PETRINI (With 3 Figures) 129

The Consequences of Fungal Associations in Plant Populations

9 Plant Pathogenic Fungi in Agroecosystems
JAMES T. ENGLISH and JAMES J. MAROIS (With 5 Figures) 143

10 Speciation and Population Biology in *Colletotrichum*
DAVID. O. TEBEEST, J.C. CORRELL, and G.J. WEIDEMANN (With 2 Figures) 157

11 The Epidemiology of Mycorrhizal Fungal Infection During Plant Succession
MICHAEL F. ALLEN, JOHN N. KLIRONOMOS, and SHARON HARNEY (With 4 Figures) 169

12 The Population Biology of Grass Endophytes
ADRIAN LEUCHTMANN and KEITH CLAY (With 4 Figures) 185

The Evolution of Fungus-Plant Associations

13 Genomic Structure and Genetic Flexibility and Microevolution in Pathogenic Fungi
JOHANNES WÖSTEMEYER (With 2 Figures) 205

14 Evolution of Mutualistic Endophytes from Plant Pathogens
CHRISTOPHER L. SCHARDL and KEITH CLAY (With 5 Figures) 221

15 Mathematical Models of Plant/Fungus Interactions
IRMA SALONIEMI (With 6 Figures) 239

16 Evolution of Plant Resistance and Susceptibility to Fungal Parasites
MICHÈLE C. HEATH (With 6 Figures) 257

Biosystematic Index 277

Subject Index 283

Contents Part A

Introduction to Part A and Part B: Fungus/Plant Interactions – An Over view
G.C. CARROLL

External Interactions

1 Adhesion of Spores and Hyphae to Plant Surfaces
LYNN EPSTEIN and RALPH L. NICHOLSON (With 6 Figures)

2 Physical and Chemical and Cues for Spore Germination and Appressorium Formation by Fungal Pathogens
RICHARD C. STAPLES and HARVEY C. HOCH (With 2 Figures)

Fungal Invasion and Plant Responses

3 Breaching the Outer Barriers – Cuticle and Cell Wall Penetration
RICHARD J. HOWARD (With 10 Figures)

4 Fungal Invasion Enzymes and Their Inhibition
G. DE LORENZO, R. CASTORIA, D. BELLINCAMPI, and FELICE CERVONE (With 6 Figures)

5 Signals in Host-Parasite Interactions
JÜRGEN EBEL and DIERK SCHEEL (With 1 Figure)

6 Pathogenesis-Related Proteins and Plant Defense
ERICH KOMBRINK and IMRE E. SOMSSICH (With 2 Figures)

7 Fungal Phytotoxins: Biosynthesis and Activity
THOMAS H. HOHN (With 7 Figures)

8 *Cochliobolus* spp. and Their Host-Specific Toxins
OLEN C. YODER, VLADO MACKO, THOMAS WOLPERT, and B. GILLIAN TURGEON (With 4 Figures)

9 Fungal Phytohormones in Pathogenic and Mutualistic Associations
BETTINA TUDZYNSKI (With 10 Figures)

10 Toxin Production in Grass/Endophyte Associations
MALCOLM R. SIEGEL and LOWELL P. BUSH (With 6 Figures)

11 Metabolic Interactions at the Mycobiont-Photobiont Interface in Lichens
ROSMARIE HONEGGER (With 9 Figures)

12 Altered Gene Expression During Ectomycorrhizal Development
FRANCIS MARTIN, FRÉDÉRIC LAPEYRIE, and DENIS TAGU
(With 11 Figures)

Biosystematic Index

Subject Index

List of Contributors

ALLEN, M.F., Department of Biology, Soil Ecology and Restoration Group, San Diego State University, San Diego, California 92182-4614 USA

CLAY, K., Department of Biology, Jordan Hall, Indiana University, Bloomington, Indiana 47405, USA

CORRELL, J., Department of Plant Pathology, 217 Plant Sciences, University of Arkansas, Fayetteville, Arkansas 72701, USA

DE WIT, P.J.G.M., Department of Phytopathology, Wageningen Agricultural University, P.O. Box 8025, 6700 EE Wageningen, The Netherlands

DEBAUD, J.C., Laboratoire d'Ecologie Microbienne du Sol, Université Lyon 1, Bât. 405, 43 Bd 11 Novembre 1918, F-69622 Villeurbanne Cedex, France

DRENTH, A., Cooperative Research Centre for Tropical Plant Pathology, The University of Queensland, John Hines Buildings, Brisbane QLD 4072, Australia

ENGLISH, J.T., Department of Plant Pathology, University of Missouri, Columbia, Missouri 65211, USA

GAY, G., Laboratoire d'Ecologie Microbienne du Sol, Université Lyon 1, Bât. 405, 43 Bd 11 Novembre 1918, F-69622 Villeurbanne Cedex, France

GIESE, H., Division of Plant Biology, Risø National Laboratory, DK-4000 Roskilde, Denmark

GOVERS, F., Department of Phytopathology, Wageningen Agricultural University, Binnenhaven 9, 6709 PD Wageningen, The Netherlands

HARNEY, S., Department of Biology, Soil Ecology and Restoration Group, San Diego State University, San Diego, California 92182-4614 USA

HEATH, M.C., Department of Botany, University of Toronto, 25 Willcocks Street, Toronto, Ontario, Canada M5S 3B2

HIPPE-SANWALD, S., Institute of Phytopathology, University of Kiel, D-24098 Kiel, Germany

HONÉE, G., Department of Phytopathology, Wageningen Agricultural University, P.O. Box 8025, 6700 EE Wageningen, The Netherlands

JOOSTEN, M.H.A.J., Department of Phytopathology, Wageningen Agricultural University, P.O. Box 8025, 6700 EE Wageningen, The Netherlands

KLIRONOMOS, J.N., Department of Botany, University of Guelph, Guelph, Ontario, N1G 2W1, Canada

LEUCHTMANN, A., Geobotanisches Institut, ETH-Zürich, Zollikerstrasse 107, CH-8008 Zürich, Switzerland

LINDERMAN, R.G., USDA-ARS Horticultural Crops Research Laboratory, 3420 NW Orchard Avenue, Corvallis, Oregon 97330, USA

MARMEISSE, R., Laboratoire d'Ecologie Microbienne du Sol, Université Lyon 1, Bât. 405, 43 Bd 11 Novembre 1918, F-69622 Villeurbanne Cedex, France

MAROIS, J.J., Institute of Food and Agricultural Sciences, University of Florida, Quincy, Florida 32351, USA

MENDGEN, K., Fakultät für Biologie, Lehrstuhl Phytopathologie, Universität Konstanz, Universitätsstr. 10, D-78464 Konstanz, Germany

PETRINI, O., Tera d'Sott 5, CH-6949 Comano, Switzerland

PIETERSE, C.M.J., Department of Plant Ecology and Evolutionary Biology, Section Phytopathology, Utrecht University, P.O. Box 800.84, 3508 TB Utrecht, The Netherlands

SALONIEMI, I., Department of Biology and Kevo Subarctic Research Institute, University of Turku, Turku, FIN-20014 Turku, Finland

SCHARDL, C.L., Department of Plant Pathology, S-305 Agricultural Science Building-North, University of Kentucky, Lexington, Kentucky 40546, USA

SOMERVILLE, S.C., Department of Plant Biology, Carnegie Institution of Washington, 290 Panama Street, Stanford, California 94305, USA

STONE, J.K., Department of Botany and Plant Pathology, Oregon State University, Corvallis, Oregon 97331-2902, USA

TEBEEST, D.O., Department of Plant Pathology, 217 Plant Sciences, University of Arkansas, Fayetteville, Arkansas 72701, USA

VALENT, B., DuPont Central Research and Development, P.O. Box 80402, Experimental Station, Wilmington, Delaware 19880-0402, USA

VAN KAN, J.A.L., Department of Phytopathology, Wageningen Agricultural University, P.O. Box 8025, 6700 EE Wageningen, The Netherlands

WEIDEMANN, G.J., Department of Plant Pathology, 217 Plant Sciences, University of Arkansas, Fayetteville, Arkansas 72701, USA

WELLER, J., PE Zoogen Lab., 1756 Picasso Drive, Davis, California 95616, USA

WÖSTEMEYER, J., Institut für Mikrobiologie, Lehrstuhl für Allgemeine Mikrobiologie und Mikrobengenetik, Friedrich-Schiller-Universität, Neugasse 24, 07743 Jena, Germany

Profiles in Pathogenesis and Mutualism

1 The Gene-for-Gene Concept in Plant-Pathogen Interactions: Tomato-*Cladosporium fulvum*

M.H.A.J. Joosten, G. Honée, J.A.L. Van Kan, and P.J.G.M. De Wit

CONTENTS

I. Introduction ... 3
II. The Gene-for-Gene Concept ... 3
III. Avirulence Genes of *Cladosporium fulvum* ... 5
A. The Model System Tomato-*Cladosporium fulvum* ... 5
B. Race-Specific Elicitors ... 6
C. Cloning and Characterization of Avirulence Genes ... 7
1. Isolation and Characterization of *avr9* and *avr4* cDNA ... 7
2. Demonstration that *avr9* and *avr4* Are Avirulence Genes ... 7
3. Structure of the *avr4* Allele Present in Races of *C. fulvum* Virulent on Tomato Genotype Cf4 ... 8
4. Regulation of Expression of the *avr4* and *avr9* Genes ... 9
5. Possible Involvement of the *avr4* and *avr9* Genes in Basic Pathogenicity ... 9
IV. Elicitor Perception and Signal Tansduction ... 10
A. Biochemical Responses Resulting from Elicitor Perception ... 10
B. Key Enzymes in Signal Transduction ... 10
V. Towards Cloning of *Cf* Resistance Genes ... 12
A. Identification of Genes Involved in Signal Perception and Transduction ... 12
B. Map-Based Cloning ... 12
C. Transposon Tagging ... 12
D. Cloning of Genes Encoding Receptors of Race-Specific Elicitors ... 13
VI. Conclusions and Future Prospects ... 13
References ... 14

I. Introduction

The phenomenon of host-genotype specificity in interactions between plants and fungal pathogens has intrigued plant pathologists for more than a century. The discovery that resistance to infection by the rust fungus *Puccinia graminis* f. sp. *tritici* in wheat is genetically determined (Farrer 1898) and obeys Mendel's laws (Biffen 1905) provided the possibility to breed for resistance against pathogens. However, as a result of selection pressure, the pathogen is able to develop new variants that regain virulence. Already at the beginning of this century, Stakman and coworkers showed that newly introduced genes for resistance against the stem rust fungus of wheat were rapidly overcome by new "biological forms" of the pathogen (Stakman 1917; Stakman et al. 1918). This competition between plant and pathogen resulted in the so-called boom and bust cycle; farmers rapidly abandoned the use of cultivars that had lost resistance and started to grow new resistant varieties which, however, again gave rise to the emergence of new virulent biological forms (also called races) of the pathogen (Crute 1985). This boom and bust cycle, which resulted from resistance breeding based on single genes and took place in modern, monoculture-based agriculture over a period of only a few decades, illustrates the continuous adaptation of a pathogen to its host and vice versa, during co-evolution in nature.

The apparent ease by which biotrophic fungal pathogens are able to circumvent host plant resistance has initiated various studies aimed at the elucidation of the molecular basis of the interaction between plants and their pathogens. In this chapter we will first briefly discuss the gene-for-gene concept and then focus on the model system *Cladosporium fulvum*-tomato.

II. The Gene-for-Gene Concept

According to the model introduced by Heath (Heath 1982, 1987, 1991; Chap. 16, this Vol.), all plants are potentially able to defend themselves against fungal invaders. A general, nonspecific, defense mechanism allows plants to prevent parasites from establishing a pathogenic relationship. During evolution, however, some fungi have

Department of Phytopathology, Wageningen Agricultural University, P.O. Box 8025, NL-6700 EE Wageningen, The Netherlands

The Mycota V Part B
Plant Relationships
Carroll/Tudzynski (Eds.)
© Springer-Verlag Berlin Heidelberg 1997

evolved into pathogens by developing tools (pathogenicity factors) that enabled them to inactivate, suppress, or circumvent defense reactions and to colonize the plant tissue and reproduce in the host. The establishment of this so-called basic compatibility by a pathogen results in host species susceptibility, and subsequently imposes selection pressure on the host. This, in turn, may lead to the development of resistance in certain members of the host species, which is specific for certain races of the pathogen. Plant genes responsible for perception of the presence of the invading pathogen can be considered resistance genes, whereas fungal genes, which encode products that trigger activation of host defense responses are defined as avirulence genes. In this way, a host-pathogen interaction evolves, the outcome of which is determined by the combination of resistance genes in the plant and avirulence genes in the fungus (the gene-for-gene concept). If one or more matching resistance and avirulence genes are present, the interaction will be incompatible (host resistant, pathogen avirulent). If the genes do not match, the fungus can successfully colonize its host (compatible interaction, host susceptible, pathogen virulent). According to this model, it is proposed that a gene-for-gene relationship develops as a result of a mutual selection pressure on host and pathogen after the pathogen has established basic compatibility.

The first report describing the genetic basis of a gene-for-gene relationship between host and pathogen was published by Flor (1942), who reported on the interaction between flax and the obligate fungal pathogen *Melampsora lini*, the causal agent of flax rust. He showed that resistance of the host to the fungus was inherited as a dominant monogenic trait, whereas virulence of the pathogen was a recessive monogenic trait (Flor 1942, 1946, 1955). By crossing races of the pathogen that differed in their virulence towards flax and checking the pathogenicity of the offspring on a differential set of flax varieties, he found that the number of resistance genes present in the various flax lines determined the maximum number of (a)virulence genes that could be identified in the parental races. Thus, for every gene determining resistance in the host, there is a corresponding gene conditioning avirulence in the pathogen. By carrying out several crosses and assessing the inheritance of (a)virulence, Flor was able to identify the genotype of the different biotypes or physiologic races of the rust fungus.

In subsequent years, several plant-pathogen interactions were found to fit a pattern complying with the gene-for-gene concept, and various models have been proposed to explain the molecular basis of the relationship (reviewed by De Wit 1986). The current view holds that resistance is associated with the so-called hypersensitive-response (HR), which is triggered by specific recognition of the pathogen by the host (recently reviewed by Gabriel and Rolfe 1990; Keen 1990; De Wit 1992; Atkinson 1993). HR can be considered as a rapid, programmed death of a limited number of host cells at the site of penetration and is accompanied by transcriptional activation of various defense genes of the plant. The combination of a rapid collapse of the cells in the vicinity of the invading pathogen, and activation of various defense responses in the area surrounding the site of penetration, prevents further spreading of the pathogen through the plant tissue. Recognition of the pathogen by the host is thought to be the result of a specific interaction between products encoded by avirulence genes of the pathogen (the so-called race-specific elicitors) and resistance genes of the host. Genetic and biochemical data obtained from various host-pathogen interactions for which a gene-for-gene relationship has been proposed support the specific elicitor-specific receptor model (Gabriel and Rolfe 1990; Keen 1990; De Wit 1992). This model (Fig. 1) implies that specific elicitors produced by the pathogen bind to receptors in the host plant. As a result of elicitor binding, a signal transduction cascade is activated, which finally results in the induction of HR and activation of host defense genes (Scheel 1990; Lamb 1994). It is assumed that resistance gene products are either specific receptors that are directly involved in elicitor perception or proteins that are associated with these receptors and are involved in one of the first steps of the signal transduction pathway that connects the perception event with the activation of defense responses.

In order to prove whether the proposed elicitor-receptor model holds for plant-fungus interactions that comply with the gene-for-gene concept, isolation of resistance and avirulence genes and their products is required. In recent years, research on several pathosystems has provided a better understanding of the biochemical and molecular basis of the gene-for-gene model. Besides the studies on the interaction between rice and *Magnaporthe grisea* (Chap. 3, this Vol.), the analysis of the interaction between barley and *Rhyn-*

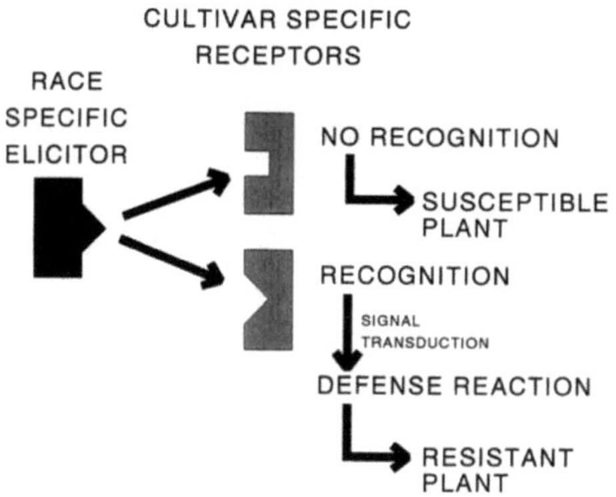

Fig. 1. The specific elicitor-specific receptor model. See text for details

chosporium secalis (Wevelsiep et al. 1991, 1993; Hahn et al. 1993) and tomato and *Cladosporium fulvum* has provided a significant contribution to our knowledge of the molecular basis of the gene-for-gene interaction. In the following paragraphs we will discuss the results that have been obtained with studies on the interaction between tomato and *C. fulvum*.

III. Avirulence Genes of *Cladosporium fulvum*

A. The Model System Tomato-*Cladosporium fulvum*

The interaction between the biotrophic, imperfect fungus *Cladosporium fulvum* and its only host, tomato (*Lycopersicon esculentum*), is a well-established model system to study plant-fungus gene-for-gene relationships (De Wit 1992). An advantage of this particular interaction is the strict apoplastic localization of the pathogen in the tomato leaves (De Wit 1977; Lazarovits and Higgins 1976a,b). The fungus penetrates the leaf through stomata and invades the intercellular spaces between the mesophyll cells without formation of haustoria. In compatible interactions there is no clear response of the plant upon penetration, allowing the fungus to grow in close contact with the mesophyll cells and the vascular tissue (Van den Ackerveken et al. 1994). In this way, a large interface is created where communication between plant and fungus takes place.

The fungus obtains its nutrients from the apoplast. Sucrose probably is a carbon source for the fungus; the amount of this carbohydrate in the apoplast decreases significantly during the course of colonization of the leaf mesophyll (Joosten et al. 1990). Sucrose is the main product of photosynthesis and is loaded into the phloem via the apoplast, probably by specific sucrose carriers (Dickinson et al. 1991). Invertases that are secreted by *C. fulvum* convert the apoplastic sucrose into glucose and fructose, which is subsequently taken up and metabolized by the fungus. The end product of this pathway is mannitol, a polyol that cannot be metabolized by the host (Joosten et al. 1990). Nitrogen-containing compounds and other nutrients that are imported in the leaves via the xylem are probably directly taken up by the fungus from the tissue surrounding the vascular bundles.

Different near-isogenic lines originating from tomato cultivar Moneymaker, carrying single genes for resistance (*Cf2*, *Cf4*, *Cf5*, or *Cf9*) give a clear differential response (resistant or susceptible) to the various races of *C. fulvum* that have been isolated. The extracellular confinement of the fungus allows the isolation of apoplastic fluid (AF) containing components that originate from the plant-fungus interphase. AF, which is obtained by in vacuo infiltration of tomato leaves with water, followed by low-speed centrifugation (De Wit and Spikman 1982), contains proteins that are constitutively produced by the plant or the fungus and proteins whose synthesis is induced as a result of the interaction between host and pathogen (De Wit et al. 1986, 1988). In search for proteinaceous factors that enable *C. fulvum* to penetrate, colonize, and reproduce in the leaves of tomato (factors that establish or maintain basic compatibility), the electrophoretic profiles of AF isolated from compatible and incompatible interactions between tomato and *C. fulvum* were compared. In this way, various compatible interaction-specific proteins of *C. fulvum* were identified (De Wit et al. 1986).

Two of these extracellular proteins (ECPs), ECP1 and ECP2, were purified (Joosten and De Wit 1988; J.P. Wubben et al., in prep.) and the encoding genes, *ecp*1 and *ecp*2, were isolated (Van den Ackerveken et al. 1993a). A way to investigate whether *ecp*1 and *ecp*2 are required for pathogenicity of *C. fulvum*, is disruption of the genes by homologous recombination. The *ecp*2 gene appeared not to be essential for pathogenicity of *C. fulvum* on tomato seedlings (Marmeisse et al. 1994). Currently, experiments are being performed to obtain disruptants that have lost *ecp*1 or both *ecp*1 and *ecp*2 genes (G.F.J.M. Van den Ackerveken, pers. comm.).

In incompatible *C. fulvum*-tomato interactions growth of the fungus is arrested immediately after penetration of the leaf. One of the initial defense responses of the host consists of a characteristic HR, which is accompanied by callose deposition (De Wit 1977; Lazarovits and Higgins 1976a,b). The fast accumulation of various host-encoded pathogenesis-related (PR) proteins in the apoplast is very characteristic for the incompatible interaction (De Wit and Van der Meer 1986; De Wit et al. 1986; Joosten and De Wit 1989; Chap. 6, Vol. V, Part A). Biochemical characterization revealed that several of these proteins are 1,3-β-glucanases and chitinases, hydrolytic enzymes potentially able to degrade hyphal walls that contain 1,3-β-glucans and chitin (Joosten and De Wit 1989). The genes encoding the various basic and acidic 1,3-β-glucanases and chitinases have been cloned (Van Kan et al. 1992; Danhash et al. 1993) and the expression of the genes (Van Kan et al. 1992; Danhash et al. 1993; Wubben et al. 1994) and localization of the encoded proteins (Wubben et al. 1992, 1994) have been studied. Although the early accumulation of the hydrolytic enzymes in the incompatible interaction coincides with the expression of HR and arrest of fungal growth, it is not clear whether the induced PR proteins indeed play a significant role in the defense of tomato against *C. fulvum*. Incubation of germinated conidia of *C. fulvum* with protein fractions containing a complete set of intracellular and extracellular 1,3-β-glucanases and chitinases in vitro did not result in inhibition of fungal growth (M.H.A.J. Joosten et al., in prep.).

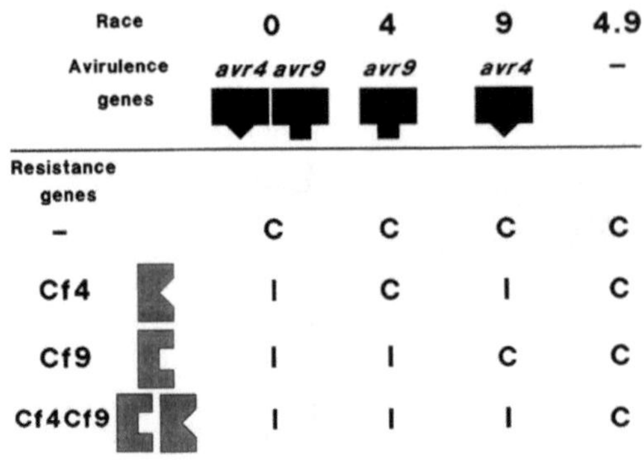

Fig. 2. Outcome of the interaction between races of *C. fulvum* and near-isogenic lines, carrying different resistance genes, of tomato. The combination of the set of avirulence gene products of the fungus and receptors in the host plant determines the outcome of the interaction. If one matching elicitor-receptor combination is present, recognition will take place, and the interaction is incompatible (*I*). If none of the combinations is matching, no recognition will take place, and the interaction is compatible (*C*). Races of *C. fulvum* are named according to their virulence on the different tomato genotypes, the various tomato lines according to the *Cf* genes that they contain

B. Race-Specific Elicitors

Recognition of *C. fulvum*-specific components allows tomato genotypes that contain the appropriate resistance gene(s) to detect the penetrating fungus and to rapidly activate defense responses (Fig. 2). The first evidence for the presence of race-specific elicitors in AF isolated from compatible interactions between tomato and *C. fulvum* was obtained by De Wit and Spikman (1982); AF isolated from a compatible interaction induces chlorosis and/or necrosis when injected into leaflets of tomato plants resistant to the race that was involved in the compatible interaction. In general, the induction of chlorosis and/or necrosis was visible 2 to 7 days after injection. When, however, leaflets of a tomato variety carrying resistance gene *Cf9* were injected with AF containing the product of avirulence gene *avr*9, extensive necrosis and wilting were observed within 1 day. The specificity of the AF was determined by the genotype of the race of *C. fulvum* and not by the genotype of the susceptible cultivar from which the AF had been obtained (De Wit et al. 1984). As tomato cultivar Sonatine (containing resistance genes *Cf2*, *Cf4*, and *Cf9*) reacted very heavily upon injection of AF that contained the putative product of avirulence gene *avr*9, an efficient bioassay was available to identify and purify the AVR9 elicitor. The necrosis-inducing activity was correlated with the presence of a small peptide, visible in a protein preparation from AF separated on native, low pH polyacrylamide gels. In compatible interactions, the peptide was produced in high amounts by all races containing the *avr*9 gene (De Wit et al. 1985).

Detailed comparison of *Cf*-specific responses revealed that, in contrast to the severe response of Cf9 plants upon recognition of the AVR9 peptide, the response of leaflets of tomato genotype Cf4 upon injection of AF containing the *avr*4 gene product is less pronounced. Instead of heavy necrosis and wilting, the *Cf4*-specific response consists of a collapse of the mesophyll of the leaf area that has been infiltrated with AF. This collapse occurs at about 1 to 2 days after injection and is visible as a necrotic sector. *Cf2*- and *Cf5*-specific responses consisted of chlorosis only of the area injected with AF containing the product of *avr*2 and *avr*5, respectively, at about 5 days after treatment (De Wit and Spikman 1982). Whether these different responses are caused by differences in concentration of the responsible elicitors or by physiologically distinct *Cf*-dependent responses remains to be demonstrated.

C. Cloning and Characterization of Avirulence Genes

1. Isolation and Characterization of *avr*9 and *avr*4 cDNA

Purification of the AVR9 elicitor peptide from appropriate AF allowed determination of the complete amino acid sequence of the putative product of avirulence gene *avr*9 (Scholtens-Toma and De Wit 1988). The peptide appeared to consist of 28 amino acids, including 6 cysteine residues. Based on the amino acid sequence of the elicitor, degenerated oligonucleotide probes were designed and used to screen a cDNA library (Van Kan et al. 1991). The primary structure of mRNA encoding the AVR9 elicitor revealed that the peptide is encoded as a pre-pro-protein; the mRNA contains an open reading frame (ORF) encoding a protein of 63 amino acids, which contains the sequence of the mature elicitor at the carboxy-terminus. The pre-pro-protein contains a signal peptide of 23 amino acids, which is cleaved off upon extracellular targeting. The resulting extracellular peptide of 40 amino acids appeared to be N-terminally processed by fungal proteases into peptides of 32, 33, or 34 amino acids, which accumulated in culture filtrates of transformants of *C. fulvum* that constitutively produce the AVR9 elicitor (Van den Ackerveken et al. 1993b). When the purified peptide of 34 amino acids was incubated with AF isolated from uninfected tomato leaflets, it was further processed into the mature elicitor of 28 amino acids, indicating that plant proteases are necessary for the final processing (Van den Ackerveken et al. 1993b). It is, however, not excluded that plant proteases are able to directly process the pro-protein of 40 amino acids to the mature elicitor peptide of 28 amino acids. It is also not clear whether the proteolytic processing steps that take place after extracellular targeting of the peptide are essential for necrosis-inducing activity of the AVR9 elicitor. Preliminary nuclear magnetic resonance (NMR) studies indicated that in the 28 amino acid peptides three disulphide bridges are present (J. Vervoort et al., unpubl.). Treatment of the elicitor with reducing agents destroyed biological activity, suggesting that the disulphide bridges are important for structure and activity of the peptide or are required for its stability (unpubl.).

The race-specific AVR4 elicitor, which specifically induces necrosis in tomato genotypes that carry resistance gene *Cf4*, consists of a 12-kDa protein. For the isolation of *avr*4 cDNA, an approach similar to the method described for cloning of *avr*9 cDNA was followed (Joosten et al. 1994). Analysis of *avr*4 cDNA revealed that, similarly to the AVR9 peptide, the AVR4 protein is encoded as a pre-pro-protein, with a putative amino-terminal signal peptide of 18 amino acids (Fig. 3B). Analogous to AVR9, a stretch of 11 amino acids in front of the mature protein is probably cleaved off by plant and/or fungal proteases after extracellular targeting. The mature AVR4 protein of 106 amino acids contains 8 cysteine residues and shares no significant homology with AVR9 or other proteins present in various databases.

2. Demonstration that *avr*9 and *avr*4 Are Avirulence Genes

Southern blot analysis of a large number of races of *C. fulvum* revealed that presence of the *avr*9 gene was always correlated with avirulence on tomato genotype Cf9, whereas virulent races appeared to circumvent *Cf9*-mediated resistance by complete deletion of the *avr*9 gene (Fig. 3A; Van Kan et al. 1991). In order to prove that this gene indeed determines avirulence of *C. fulvum* towards tomato varieties that carry resistance gene *Cf9*, and therefore can be considered as a genuine avirulence gene, the *avr*9 gene was cloned and transferred to race 2.4.5.9.11, a race of *C. fulvum* which does not contain the *avr*9 gene and is virulent on tomato genotype Cf9 (Van den Ackerveken et al. 1992). One of the *avr*9$^+$ transformants was checked for its virulence on tomato cultivar Moneymaker (MM, a cultivar that lacks resistance genes against *C. fulvum*) and genotype Cf9.

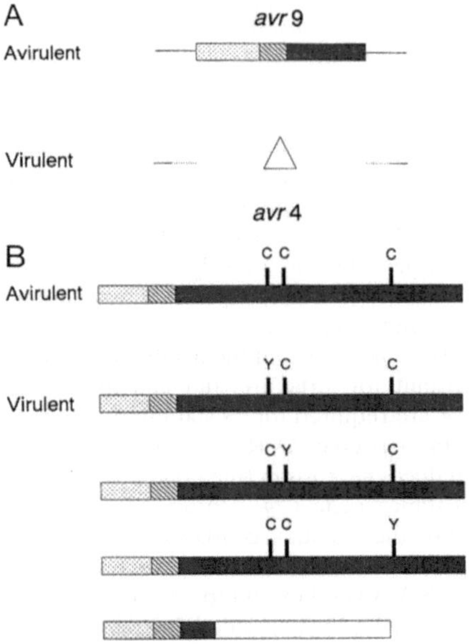

Fig. 3A,B. Circumvention of *Cf* gene-specific recognition by *Cladosporium fulvum*. Avoidance of recognition is achieved either by complete deletion (*avr9*, **A**) or point-mutation (*avr4*, **B**) of an avirulence gene. See text for details. *Dotted box* Putative signal peptide; *single hatched box* stretch of amino acids removed by proteolytic cleavage; *double hatched box* mature elicitor protein as purified from apoplastic fluid; *open box* part of the protein not homologous to AVR4 encoded by an *avr4* frame-shift mutant; *C* cysteine residue; *Y* tyrosine residue

Successful colonization of leaflets of seedlings of cultivar MM indicated that the transformant had retained its pathogenicity, whereas it had become avirulent on tomato genotype Cf9. Thus, integration of the *avr9* gene had converted race 2.4.5.9.11 into race 2.4.5.11. Conversely, disruption of the *avr9* coding sequence in races 4 and 5 of *C. fulvum*, races that do produce the AVR9 elicitor, rendered these races virulent on tomato genotypes that contain resistance gene *Cf9* (Marmeisse et al. 1993). In this way, two new, transgenic races (4.9 and 5.9) were obtained, with a genotype that has not been found in nature thus far.

These results confirmed that the cloned *avr9* gene is the only factor determining avirulence of *C. fulvum* on tomato varieties that carry the *Cf9* resistance gene, and demonstrate that this gene indeed complies with the definitions of an avirulence gene.

Similarly, isolation of the *avr4* gene and subsequent integration into the genome of race 4 (a race virulent on tomato genotype Cf4) by transformation, revealed that this gene is the only factor determining *Cf4*-specific avirulence in *C. fulvum*. Transformants of race 4 which had obtained the *avr4* gene, produced the AVR4 elicitor when inoculated on tomato cultivar MM, and had become avirulent on tomato genotype Cf4 (Joosten et al. 1994).

3. Structure of the *avr4* Allele Present in Races of *C. fulvum* Virulent on Tomato Genotype Cf4

Southern analysis did not reveal any difference between races of *C. fulvum* avirulent or virulent on tomato genotype Cf4. All races were found to contain a homologous, single copy gene, not displaying any restriction fragment length polymorphism with the restriction enzymes that were used (Joosten et al. 1994). Although race 4 of *C. fulvum* does not produce biologically active AVR4 elicitor, RNA isolated from a compatible interaction between this race and tomato genotype Cf4 contains transcripts hybridizing to an *avr4* cDNA probe. The hybridization to an *avr4*-specific probe in DNA and RNA of races of *C. fulvum* virulent on tomato genotype Cf4 prompted us to determine the nucleotide sequence of the ORFs present in virulent alleles of *avr4* and to compare these sequences with the ORFs present in the wild-type *avr4* gene of race 5 of *C. fulvum* and other races avirulent on tomato genotype Cf4. It appeared that all four avirulent races that were analyzed contain an ORF identical to the one present in the *avr4* gene of race 5, the race from which the gene was originally isolated (Fig. 3B), whereas all races that are virulent on tomato genotype Cf4 show a single base pair change in the part of the ORF encoding the mature AVR4 protein. In these races a point mutation, replacing a cysteine residue (codon TGT) by a tyrosine residue (codon TAT), was present in the codon of either the 4th, 5th, or 8th cysteine residue (Fig. 3B; Joosten et al. 1994). The consequence of this codon change for biological activity of the AVR4 protein suggests involvement of the particular cysteine residue in disulphide bridge formation in the AVR4 elicitor protein. Replacement of this amino acid by a tyrosine residue will interfere with folding of the protein and the altered secondary and tertiary structure of the AVR4 elicitor protein might affect binding of the protein to the complementary receptor. Alternatively, the structural change in the AVR4 protein might render the polypeptide more sensitive to proteolytic breakdown. As a result of this decrease in stability, the protein might immediately be degraded by plant and/or fungal proteases upon secretion into the apoplast, and consequently not even come into contact with the receptor. Thus far, we cannot discriminate between either possibility.

Analysis of additional isolates that circumvent *Cf4*-mediated resistance revealed that in one isolate a deletion of a nucleotide, instead of a replacement, had taken place (unpubl.). The deletion, causing a frame-shift, results in an ORF encoding a protein from which only the first 13 amino acids are identical to the mature AVR4 protein (Fig. 3B).

4. Regulation of Expression of the *avr4* and *avr9* Genes

Northern blot analysis indicated that expression of the *avr4* and *avr9* genes of *C. fulvum* is specifically induced in planta; the accumulation of mRNA encoding the race-specific elicitors correlates with the increase in fungal biomass in the tomato leaves during pathogenesis in compatible interactions (Van Kan et al. 1991; Joosten et al. 1994).

Analysis of the promoter of the *avr4* gene for the presence of specific motifs did not reveal any significant homology to sequences known for binding of regulatory proteins. Analysis of transformants of *C. fulvum* carrying *avr4* promoter-GUS fusions in planta should reveal at which particular stage(s) of pathogenesis expression of the gene is induced.

When grown on liquid medium, expression of the *avr9* gene of *C. fulvum* could be induced by limitation of the amount of nitrogen (present as nitrate, ammonium, glutamate, or glutamine; Van den Ackerveken et al. 1994). Limitation of other macronutrients or addition of plant factors (such as homogenized tomato suspension cells, culture medium in which the suspension cells were grown, or AF isolated from healthy tomato plants) to the medium in which *C. fulvum* was grown had no effect on expression of the gene. Analysis of the *avr9* promoter sequence revealed several potential regulatory elements in two large direct repeats. The promoter region contains six copies of the hexanucleotide sequence TAGATA (Van den Ackerveken et al. 1994), which was identified as the recognition site of the NIT2 protein, a transcription factor positively regulating gene expression under nitrogen-limiting conditions in *Neurospora crassa* (Fu and Marzluf 1990). Possibly, expression of *avr9* is regulated in a similar way, by a *C. fulvum*-homologue of the NIT2 protein.

5. Possible Involvement of the *avr4* and *avr9* Genes in Basic Pathogenicity

In contrast to the *avr9* gene, which is absent in races of *C. fulvum* virulent on tomato genotype Cf9, the presence of mutated *avr4* alleles in several races of *C. fulvum* might suggest an essential role for its product in basic pathogenicity. However, the isolate that circumvents *Cf4*-specific resistance by a frame-shift mutation and produces an AVR4 homolog of which only the first 13 amino acids are identical to the AVR4 protein (Fig. 3B) did not show impaired pathogenicity, indicating that the AVR4 protein is not indispensable (unpubl.).

Cytological studies on the expression of the *avr9* gene in planta, using transformants of *C. fulvum* that carry *avr9* promoter-GUS fusions revealed that the promoter is activated in hyphae penetrating the substomatal cavities and reaches maximal expression levels in mycelium growing in the vicinity of the vascular tissue of the tomato leaves (Van den Ackerveken et al. 1994). Possibly the uptake of nutrients from the apoplastic region around mesophyll cells, xylem, and phloem requires high metabolic activity of the fungus, resulting in limited availability of the various nutrients in the immediate vicinity of the growing hyphae. In this way, nitrogen concentrations in the apoplast might be continuously limiting and induce regulatory genes that subsequently activate the *avr9* gene. The AVR9 peptide itself might be involved in the uptake of nitrogen from the apoplast by the fungus or could in some way interfere with nutrient transport in the host. Alternatively, the peptide could modify the plasma membrane of the host cells in such a way that nutrient release is stimulated. The observations that transformants of *C. fulvum* in which the *avr9* gene was disrupted did not show impaired pathogenicity in a monocyclic infection assay, and the fact that, in nature, circumvention of *Cf9*-specific resistance is achieved by complete deletion of the *avr9* gene, suggest that the *avr9* gene is dispensable for growth and pathogenicity of *C. fulvum*. The *Cf9* resistance gene, however, which has been present in tomato breeding lines since 1979, still renders complete protection of commercial tomato crops to *C. fulvum*, indicating that there is only a low frequency of mutation at the *avr9* locus. Possibly, deletion of the *avr9* locus (and possibly linked genes as well) leads to reduced fitness of the pathogen under field conditions. Additional research, using near-isogenic races of *C. fulvum* that are *avr9*$^-$ or *avr9*$^+$ in polycyclic competition experiments, should reveal whether this explanation is plausible.

IV. Elicitor Perception and Signal Transduction

A. Biochemical Responses Resulting from Elicitor Perception

Upon penetration of leaves of a resistant tomato genotype by hyphae of *C. fulvum*, the fungus is recognized by the plant and HR takes place. As a result of this immediate response, fungal growth is limited to only a few mesophyll cells. HR, which in general refers to the rapid cell death which takes place, is accompanied by several biochemical responses common to different plant species that show active resistance to viral, bacterial, or fungal pathogens. Therefore, it is assumed that hypersensitivity has a similar physiological basis in different plant-pathogen interactions (Atkinson 1993).

For the interaction between tomato and *C. fulvum*, some general responses were investigated by Peever and Higgins (1989), who injected a crude preparation of a race-specific elicitor of *C. fulvum* in leaflets of a resistant tomato genotype. They observed that injection of AF containing the AVR9 elicitor induced a specific elicitor-dependent increase in electrolyte leakage in leaflets of tomato genotype *Cf9*. The electrolyte leakage coincided with elevated levels of lipoxygenases and lipid peroxidation in the injected tissue. Vera-Estrella et al. (1992) extended studies on specific biochemical responses to cell suspension cultures. Cell cultures, initiated from callus obtained from tomato genotypes Cf4 and Cf5, retained the specificity of the intact plants from which they originated. Immediately upon addition of AF containing matching elicitor activity, a marked increase in the extracellular production of superoxide (O_2^-) and other active oxygen (AO) species (H_2O_2 and OH·) took place. Besides this oxidative burst, the cells showed increased lipid peroxidation and increases in extracellular peroxidases and phenolic compounds. Similar results were obtained by Hammond-Kosack et al. (1993), who used cotyledons of tomato seedlings. Upon injection of AF containing the AVR9 elicitor in cotyledons of Cf9 plants, an oxidative burst was observed within 3h of injection. By about 9–10h, ethylene production and loss of membrane integrity became evident. Within 24h, extensive electrolyte leakage and alkalization of the apoplastic region of the tissue was detected. Besides these responses, the concentration of salicylic acid, an endogenous signal involved in the induction of systemic plant defense responses (Gaffney et al. 1993), started to increase in the cotyledons about 12h after injection, and reached a maximum level at about 24h after elicitor treatment.

Research on various other host-pathogen interactions has also provided substantial support for the involvement of AO species in the early events following recognition of the pathogen (Sutherland 1991; Atkinson 1993; Tzeng and De Vay 1993; Lamb 1994; Chap. 5, Vol. V, Part A). The general picture that emerges from this research is that the AO species that are generated extracellularly might damage the invading pathogen directly and could initiate host cell wall lignification reactions, whereas intracellular AO will cause the oxidation of lipids present in the membranes of the reacting host cells. This oxidative membrane damage might result in leakage of electrolytes from the host cells and could initiate the process of host cell necrosis (HR). Recent experiments by Chen and coworkers revealed that the activation of genes encoding PR proteins in tobacco can be the result of an increase in the endogenous H_2O_2 concentration, caused by inhibition of H_2O_2-catalases (Chen et al. 1993). Injection of purified race-specific elicitors AVR4 and AVR9 of *C. fulvum* into leaflets of resistant tomato genotypes induced activation of several genes encoding PR proteins within a few hours (J.P. Wubben et al., in prep.). Possibly, an increase in AO species is generated upon such a treatment (Peever and Higgins 1989; Vera-Estrella et al. 1992), which results in the induction of PR protein genes in tomato plants as well.

B. Key Enzymes in Signal Transduction

Histological studies of Hammond-Kosack and Jones (1994) on incompatible tomato-*C. fulvum* combinations indicated that the *Cf* genes of tomato may display semidominance. Assessment of the extent of fungal growth in plants homozygous or heterozygous for the various *Cf* resistance genes suggested that in heterozygous plants defense responses are delayed and more fungal development takes place than in plants homozygous for the *Cf* genes. This gene-dose effect was confirmed by injection of AF prepared from a compatible MM/race 0 interaction (containing all known race-specific elicitors), in a dilution series of 1 to 1/64th of the original concentration. Plants homozygous for each *Cf* gene responded to a twofold lower titer of AF compared to heterozygotes, indicating that the amount of injected race-specific elicitor, as well as the *Cf* gene zygosity, determine the response of the plant. This apparent gene-dosage effect indicates that the amount of *Cf* gene product may be a rate-limiting factor in HR induction.

In order to investigate whether plasma membrane-bound receptors are involved in the process of elicitor perception, Vera-Estrella and coworkers isolated plasma membranes from a suspension culture of Cf5 cells and incubated those with a crude AF preparation containing the AVR5 elicitor (Vera-Estrella et al. 1994). Immediately upon exposure of the membranes to the elicitor preparation, ATPase activity increased fourfold, whereas other AF preparations that did not contain this elicitor had no effect on the ATPase activity. The increase in ATPase activity was due to a stimulation of the proton pump (H^+-ATPase), since incubation of the Cf5 cells with the AVR5 elicitor preparation induced an immediate acidification of the culture medium. Activation of the H^+-ATPase was caused by dephosphorylation of the enzyme, as shown by radiolabeling experiments and studies with inhibitors of protein phosphatases. Inhibition of protein kinase activity did not interfere with the elicitor-induced increase in H^+-ATPase activity.

These findings imply that, as a result of binding of the AVR5 elicitor to a receptor in the plasma membrane, the H^+-ATPase is activated by dephosphorylation via a direct signal transduction pathway. Extensive research on signal transduction chains in mammals has resulted in the identification of a family of heterotrimeric guanine nucleotide (GDP or GTP)-binding regulatory proteins (G proteins) that are involved in transduction across the plasma membrane of stimuli generated by binding of a signal molecule to its corresponding receptor (reviewed by Kaziro et al. 1991; Hepler and Gilman 1992). Incubation of Cf5 plasma membranes with the AVR5 elicitor preparation and guanidine nucleotide analogs that inactivate or activate the Â-subunit of a G protein indicated that *Cf5*-mediated signaling responses also take place via G proteins (Vera-Estrella et al. 1994).

The activation of the H^+-ATPase and subsequent acidification of the cell wall might decrease its rigidity and change the distribution of different ions across the plasma membrane. These processes might, amongst others, stimulate the production of callose, which is deposited at penetration sites in incompatible tomato-*C. fulvum* interactions (De Wit 1977; Lazarovits and Higgins 1976b). Upon addition of the AVR5 elicitor preparation to Cf5 tomato suspension cells, callose synthesis was observed immediately, indicating that this is indeed an early response on elicitor recognition (Vera-Estrella et al. 1994).

The plasma membrane H^+-ATPase is generally considered to be a key enzyme that controls various fundamental cellular processes (Serrano 1989) and interference with its activity might result in the physiological responses described above. More pronounced changes in the H^+-ATPase activity might also directly lead to cell collapse. This is thought to be the case for virulence factors that are secreted by *Rhynchosporium secalis*, the leaf scald fungus of barley. Wevelsiep and coworkers (1993) isolated phytotoxic peptides (necrosis-inducing peptides, NIPs) from liquid shake cultures of *R. secalis* and obtained evidence that their toxic effect on host cells could be due to a stimulation of the plasma membrane H^+-ATPase. Incubation of barley plasma membrane vesicles with NIP1 or NIP3 resulted in a cultivar-aspecific stimulation of the H^+-ATPase activity, in a way similar to the host-nonselective toxin fusicoccin of *Fusicoccum amygdali* (Marrè 1979). Besides its effect on the H^+-ATPase, the NIP1 protein can also be considered as a race-specific elicitor, as it specifically induces defense reactions in barley cultivars carrying resistance gene *Rrs1* (Hahn et al. 1993). The NIP1 peptide therefore could be the product of avirulence gene *avrRrs1*.

It is assumed that *R. secalis* secretes the phytotoxic peptides during colonization to kill the host tissue and stimulate the release of nutrients from the cells. Like fusicoccin, it appeared that the NIPs do not interact directly with the H^+-ATPase. Preliminary data suggest that there is a protein present in the plasma membrane that specifically binds the peptides and therefore could be considered as a receptor. Possibly, according to a similar mechanism as proposed for the AVR5 race-specific elicitor of *C. fulvum*, binding of the peptides to a receptor activates a cascade of events which finally result in stimulation of the H^+-ATPase.

The observed acidification of the culture medium of suspension cells of tomato genotype Cf5 upon addition of the AVR5 elicitor, as found by Vera-Estrella et al. (1994), contrasts with the findings of Felix et al. (1993), who showed that chitin fragments, as aspecific elicitors, induced a rapid alkalization of the medium, when added to a tomato cell suspension culture. This increase in pH was also found in the extracellular space of the mesophyll of cotyledons of tomato genotype Cf9 upon injection of AF containing the AVR9 elicitor (Hammond-Kosack et al. 1993) and suggests that several independent pathways involved in *Cf*-gene-mediated resistance might be present.

V. Towards Cloning of *Cf* Resistance Genes

A. Identification of Genes Involved in Signal Perception and Transduction

A strategy to identify genes that are involved in the activation of plant defense responses involves the characterization of mutants that show reduced resistance to *C. fulvum*. In this way, theoretically, the resistance locus itself or a gene encoding a protein responsible for one of the steps in the signal transduction cascade which ultimately activates defense genes, could be identified. Hammond-Kosack et al. (1994) chemically mutagenized homozygous seeds of tomato genotype Cf9, selfed the mutants, and screened the M_2 families for loss of resistance against a race of *C. fulvum* that produces the AVR9 elicitor. Eight independent families were found that showed reduced resistance to *C. fulvum*. When crossed to tomato cultivar MM (Cf0), the F_1 progeny of six families was also still susceptible to the race of *C. fulvum* carrying the *avr9* gene, indicating that the mutation had occurred at the *Cf9* locus. Two partially susceptible mutants that were obtained gave a resistant F_1 progeny when crossed to MM, indicating that the mutation was not present in the *Cf9* gene itself. Additional crosses with these mutants revealed that both mutant loci inherited independently, indicating that two distinct loci were mutated, and were not linked to the *Cf9* gene. The mutagenized loci in the two different lines were designated *Rcr-1* and *Rcr-2* (**R**equired for ***Cla-dosporium*** **r**esistance) and could represent genes involved in the signal transduction pathway leading to *Cf9*-mediated resistance.

B. Map-Based Cloning

To allow the isolation of *Cf* genes by map-based cloning, a thorough knowledge of the chromosomal location of these genes is necessary. In order to obtain accurate information on the location of the *Cf2*, *Cf4*, *Cf5*, and *Cf9* genes, detailed linkage studies were performed which consisted of classical mapping using phenotypical markers, and analysis of restriction fragment length polymorphism (RFLP), random amplified polymorphic DNA (RAPD), and amplified fragment length polymorphism (AFLP) markers in crosses of tomato plants segregating for *Cf* genes in the offspring (Van der Beek et al. 1992; Dickinson et al. 1993; Jones et al. 1993a). Linkage analysis revealed that in tomato two multiple resistance loci are present. The *Cf2* and *Cf5* genes are very closely linked or even allelic and were assigned to chromosome 6, close to the *Mi* gene, which confers resistance to root knot nematodes (*Meloidogyne* spp.; Dickinson et al. 1993). The *Cf4* gene is very closely linked to the *Cf9* gene and located on the short arm of chromosome 1 (Van der Beek et al. 1992; Jones et al. 1993a). The generation of additional molecular markers in the vicinity of the various resistance genes, combined with the identification of phenotypic markers closely linked to the *Cf* genes, allows rapid screening of recombinants and makes genomic walking to the *Cf* gene clusters and subsequent cloning of individual *Cf* genes possible. This approach is currently being followed by Jones and co-workers, who have generated many markers on either side of the two gene clusters (D.A. Jones, pers. comm.).

C. Transposon Tagging

A more direct approach to identify the *Cf9* gene involves a heterologous transposon tagging system (Hammond-Kosack et al. 1993; Jones et al. 1993b). For this strategy, homozygous Cf9 plants are used that carry the stabilized maize activator

(*Ac*) and dissociator (*Ds*) elements. The *Ac* element is able to *trans*-activate the *Ds* element, which is excised and integrated elsewhere in the genome. The *Ds* element mainly transposes to linked sites, so for a targeted tagging of the *Cf9* gene, experiments should be initiated from a stock of plants that contains the *Ds* element closely linked to the *Cf9* gene. If mutant plants are obtained that show loss of *Cf9*-specific responses, this could mean that the *Ds* element has integrated into the *Cf9* locus. When it has been confirmed that in these plants the *Ds* element is absolutely linked to the *Cf9* gene, further information on this locus can be obtained by sequencing the regions flanking the *Ds* element. An efficient way to identify *Cf9* inactivation consists of crossing Cf9 lines in which transposition events might have occurred, to Cf0 plants that express the *avr9* gene of *C. fulvum*. Constitutive production of biologically active AVR9 elicitor in tomato MM (Cf0) and tobacco plants was achieved by transformation with a construct that contains a synthetic ORF of the truncated *avr9* gene, fused to the sequence of a signal peptide that is responsible for extracellular targeting (Hammond-Kosack et al. 1993; G. Honée, unpubl.). The fact that expression of the *avr9* gene is lethal in a *Cf9* background provides a strong selection for mutants in the *Cf9* gene that have been induced by transposon insertion. When the transposon has inserted into the *Cf9* gene, or a gene important for *Cf9*-mediated defense responses, the F_1 should be normally viable or might develop necrotic sectors (revertants due to re-excision of *Ds*) as they grow up. In this way, it is possible to efficiently screen large numbers of F_1 plants on *Cf9*-mediated responses and to tag and eventually clone the *Cf9* resistance gene.

D. Cloning of Genes Encoding Receptors of Race-Specific Elicitors

The receptors in the host plant that interact with race-specific elicitors of *C. fulvum* are thought to be encoded by *Cf* resistance genes, and the results of Vera-Estrella et al. (1994) indicate that specific receptors for *C. fulvum*-encoded elicitors are present in the plasma membrane of the host cells. Studies by Van den Ackerveken et al. (1993b) showed that labeling of the AVR9 elicitor with ^{125}I does not affect its specific necrosis-inducing activity. Presently, binding studies using ^{125}I-labeled AVR9 elicitor and plasma membrane fractions of different tomato cultivars are being carried out in our laboratory. The establishment of conditions that allow specific binding of a race-specific elicitor to its corresponding receptor might enable isolation of the receptor by affinity chromatography using a column in which the elicitor is present in an immobilized state. The corresponding receptor, present in plasma membrane protein preparations that are applied to the column, will be retained, and is subsequently eluted. In this way, sufficient receptor protein might be isolated for N-terminal and internal sequencing. Amino acid sequence information on the receptor proteins allows the design of degenerated oligonucleotide probes to screen a cDNA or genomic library of tomato and, eventually, cloning of the encoding gene. Complementation analysis, by transferring the isolated gene to the genome of a susceptible tomato genotype, followed by inoculation of the transformants with the appropriate fungal race, will reveal whether the gene functions as a resistance gene.

Genes encoding plasma membrane-bound receptors of elicitors of *C. fulvum* might also be isolated by exploiting the possible homology that consists between these receptors and receptors that have already been identified in mammalian systems. Conserved stretches of amino acids, identified in certain classes of receptor proteins, can form the basis for degenerated primers that are used in PCR reactions on genomic DNA of the various tomato genotypes. Generated fragments can be analyzed for the presence of sequences encoding parts of a putative plasma membrane-bound receptor and can be used to screen a genomic or cDNA library of tomato.

VI. Conclusions and Future Prospects

The interaction between tomato and *C. fulvum* has proven to be an excellent model system for studies on the molecular dissection of a host-pathogen interaction that complies with the gene-for-gene model. As far as the fungus is concerned, substantial progress has been made. We have shown that two entirely different proteins of *C. fulvum*, AVR4 and AVR9, can be recognized by tomato *Cf* genes as cues that fungal invasion is in progress. Moreover, we have found that *C. fulvum* utilizes two entirely different mechanisms (deletion versus point mutation) to circumvent this recognition. On the plant side, however, our knowledge is still very limited; the information that is available on functioning of resistance genes is mainly based on working models and circumstantial evidence. By map-based cloning, Martin and coworkers (1993) were able to isolate the *Pto* resistance gene of tomato. This gene confers resistance to races of the bacterial speck pathogen (*Pseudomonas syringae* pv. *tomato*) that contain avirulence gene *avrPto*. It was found to encode a hydrophilic protein with significant homology to serine/threonine protein kinases, suggesting that the protein functions in a signal transduction cascade where (de)phosphorylation events take place. As the protein contains no membrane-spanning or extracellular domain, it does not show the characteristic features of a plasma membrane-bound receptor. The protein might, however, be associated with the plasma membrane in the vicinity of a receptor, and be responsible for one of the first steps of the signaling cascade. This mechanism would suggest that in tomato varieties that are susceptible to the bacterium, a receptor for the product of *avrPto* is also present; however, only when the *Pto* gene product is present is the perceived signal further channeled into the signal transduction pathway. Cloning and characterization of the various *Cf* resistance genes might answer the question whether these genes encode receptors for elicitors of *C. fulvum* or proteins that are associated with these receptors and are involved in the signal transduction chain that leads to the activation of host defense responses.

Engineering of disease-resistant tomato plants, and eventually also other plant species, by introduction of a cassette encoding a two-component sensor system into the genome (De Wit 1992), requires detailed information on the processes that establish resistance. The two-component sensor system would combine a resistance gene (the effector) and the complementary avirulence gene, placed under control of a pathogen-inducible promoter (the sensor). Specific activation of the avirulence gene, which should occur both quickly and locally, results in production of an elicitor that interacts with the resistance gene product. As a result of this interaction, localized HR will take place and further spread of the invading pathogen will be prevented. With this system it should be possible, at least in theory, to use a highly specific resistance gene/avirulence gene combination (for example *Cf9*/*avr9* or *Cf4*/*avr4*) to obtain plants resistant to a wide variety of pathogens. For general applications, besides the avirulence gene product, also the complementary resistance gene must be cloned. Since this has not yet been accomplished, studies on the feasibility of this approach are currently being performed in Cf9 tomato plants. These plants will be transformed with a construct that consists of the *avr9* coding region fused to a specific, pathogen-inducible promoter and will be tested for resistance against various (fungal) pathogens (G. Honée et al., unpubl.). If resistance is obtained, the construct can be tested in combination with the eventually cloned *Cf9* gene in tomato genotypes that do not carry this gene and in other plant species.

Detailed information on the mode of action of specific elicitors might be obtained by structure-function analysis of the elicitor proteins that specifically interact with receptors, an approach currently followed in our laboratory. By PCR-mediated site-directed mutagenesis, specific mutations will be introduced in the ORFs of the *avr4* and *avr9* genes and the resulting modified elicitors will be tested for their ability to induce necrosis in tomato genotype Cf4 and Cf9, respectively.

More information on the mechanism that underlies the specific interaction between elicitors of *C. fulvum* and host-encoded receptors might form the basis of the elucidation of the signal perception and transduction pathways that connect the recognition event with the induction of HR and expression of host defense genes.

Acknowledgments. We thank Dr. Jonathan D.G. Jones for sending us papers prior to publication. The contents of this chapter has been updated up to April 1, 1994.

References

Atkinson MM (1993) Molecular mechanisms of pathogen recognition by plants. In: Andrews JH, Tommerup IC (eds) Advances in plant pathology, vol 10. Academic Press, London, pp 35–64

Biffen RH (1905) Mendel's laws of inheritance and wheat breeding. J Agric Sci 1:4–48

Chen Z, Silva H, Klessig DF (1993) Active oxygen species in the induction of plant systemic aquired resistance by salicylic acid. Science 262:1883–1886

Crute IR (1985) The genetic bases of relationships between microbial parasites and their hosts. In: Fraser RSS (ed) Mechanisms of resistance to plant diseases. Nijhoff/Junk, Dordrecht, pp 80–142

Danhash N, Wagemakers CAM, Van Kan JAL, De Wit PJGM (1993) Molecular characterization of four chiti-

nase cDNAs obtained from *Cladosporium fulvum*-infected tomato. Plant Mol Biol 22:1017–1029
De Wit PJGM (1977) A light and scanning electron-microscopic study of infection of tomato plants by virulent and avirulent races of *Cladosporium fulvum*. Neth J Plant Pathol 83:109–122
De Wit PJGM (1986) Elicitation of active resistance mechanisms. In: Bailey J (ed) Biology and molecular biology of plant-pathogen interactions. NATO ASI series, vol H1. Springer, Berlin Heidelberg New York, pp 149–169
De Wit PJGM (1992) Molecular characterization of gene-for-gene systems in plant-fungus interactions and the application of avirulence genes in control of plant pathogens. Annu Rev Phytopathol 30:391–418
De Wit PJGM, Spikman G (1982) Evidence for the occurrence of race- and cultivar-specific elicitors of necrosis in intercellular fluids of compatible interactions between *Cladosporium fulvum* and tomato. Physiol Plant Pathol 21:1–11
De Wit PJGM, Van der Meer FE (1986) Accumulation of the pathogenesis-related tomato leaf protein P14 as an early indicator of incompatibility in the interaction between *Cladosporium fulvum* (syn. *Fulvia fulva*) and tomato. Physiol Mol Plant Pathol 28:203–214
De Wit PJGM, Hofman JE, Aarts JMMJG (1984) Origin of specific elicitors of chlorosis and necrosis occurring in intercellular fluids of compatible interactions of *Cladosporium fulvum* (syn. *Fulvia fulva*) and tomato. Physiol Plant Pathol 24:17–23
De Wit PJGM, Hofman JE, Velthuis GCM, Kuc JA (1985) Isolation and characterization of an elicitor of necrosis isolated from intercellular fluids of compatible interactions of *Cladosporium fulvum* (syn. *Fulvia fulva*) and tomato. Plant Physiol 77:642–647
De Wit PJGM, Buurlage MB, Hammond KE (1986) The occurrence of host, pathogen and interaction-specific proteins in the apoplast of *Cladosporium fulvum* (syn. *Fulvia fulva*) infected tomato leaves. Physiol Mol Plant Pathol 29:159–172
De Wit PJGM, Toma IMJ, Joosten MHAJ (1988) Race-specific elicitors and pathogenicity factors in the *Cladosporium fulvum*-tomato interaction. In: Keen NT, Kosuge T, Walling LL (eds) Physiology and biochemistry of plant-microbial interactions. The American Society of Plant Physiologists, Bethesda, pp 111–119
Dickinson CD, Altabella T, Chripeels MJ (1991) Slow-growth phenotype of transgenic tomato expressing apoplastic invertase. Plant Physiol 95:420–425
Dickinson MJ, Jones DA, Jones JDG (1993) Close linkage between the *Cf-2*/*Cf-5* and *Mi* resistance loci in tomato. Mol Plant-Microbe Interact 6:341–347
Farrer W (1898) The making and improvement of wheats for Australian conditions. Agric Gaz NSW 9:131–168
Felix G, Regenass M, Boller B (1993) Specific perception of subnanomolar concentrations of chitin fragments by tomato cells: induction of extracellular alkalinization, changes in protein phosphorylation, and establishment of a refractory state. Plant J 4:307–316
Flor HH (1942) Inheritance of pathogenicity in *Melampsora lini*. Phytopathology 32:653–669
Flor HH (1946) Genetics of pathogenicity in *Melampsora lini*. J Agric Res 73:335–357
Flor HH (1955) Host-parasite interaction in flax rust. Its genetics and other implications. Phytopathology 45: 680–685
Fu YH, Marzluf GA (1990) *nit-2*, the major positive-acting nitrogen regulatory gene of *Neurospora crassa*, encodes a sequence-specific DNA-binding protein. Proc Natl Acad Sci USA 87:5331–5335
Gabriel DW, Rolfe BG (1990) Working models of specific recognition in plant-microbe interactions. Annu Rev Phytopathol 28:365–391
Gaffney T, Friedrich L, Vernooij B, Negrotto D, Nye G, Uknes S, Ward E, Kessmann H, Ryals J (1993) Requirement of salicylic acid for the induction of systemic acquired resistance. Science 261:754–756
Hahn M, Jüngling S, Knogge W (1993) Cultivar-specific elicitation of barley defense reactions by the phytotoxic peptide NIP1 from *Rhynchosporium secalis*. Mol Plant-Microbe Interact 6:745–754
Hammond-Kosack KE, Jones JDG (1994) Incomplete dominance of tomato *Cf* genes for resistance to *Cladosporium fulvum*. Mol Plant-Microbe Interact 7: 58–70
Hammond-Kosack KE, Oliver R, Harrison K, Ashfield T, Jones J (1993) Studies on the mechanism by which tomato *Cf* (*Cladosporium fulvum*) resistance genes activate plant defence. In: Nester EW, Verma DPS (eds) Advances in molecular genetics of plant-microbe interactions. Kluwer, Dordrecht, pp 457–461
Hammond-Kosack KE, Jones DA, Jones JDG (1994) Identification of two genes required in tomato for full *Cf-9*-dependent resistance to *Cladosporium fulvum*. Plant Cell 6:361–374
Heath MC (1982) The absence of active defense mechanisms in compatible host-pathogen interactions. In: Wood RKS (ed) Active defense mechanisms in plants. Plenum Press, New York, pp 143–156
Heath MC (1987) Evolution of plant resistance and susceptibility to fungal invaders. Can J Plant Pathol 9:389–397
Heath MC (1991) The role of gene-for-gene interactions in the determination of host species specificity. Phytopathology 81:127–130
Hepler JR, Gilman AG (1992) G proteins. Trends Biol Sci 17:383–387
Jones DA, Dikinson MJ, Balint-Kurti PJ, Dixon MS, Jones JDG (1993a) Two complex resistance loci revealed in tomato by classical and RFLP mapping of the *Cf-2*, *Cf-4*, *Cf-5*, and *Cf-9* genes for resistance to *Cladosporium fulvum*. Mol Plant-Microbe Interact 6:348–357
Jones JDG, Jones DA, Balint-Kurti PJ, Scofield S, Dickinson MJ, Thomas CM, Harrison K, Bishop G, Ashfield T, Hammond-Kosack K, Carroll B (1993b) Towards understanding of the function of tomato genes for resistance to *Cladosporium fulvum*. In: Yoder JI (ed), Molecular biology of tomato: fundamental advances and crop improvement. Technomic Publishing, Basel, pp 263–273
Joosten MHAJ, De Wit PJGM (1988) Isolation, purification and preliminary characterization of a protein specific for compatible *Cladosporium fulvum* (syn. *Fulvia fulva*)-tomato interactions. Physiol Mol Plant Pathol 33:241–253
Joosten MHAJ, De Wit PJGM (1989) Identification of several pathogenesis-related proteins in tomato leaves inoculated with *Cladosporium fulvum* (syn. *Fulvia fulva*) as 1,3-β-glucanases and chitinases. Plant Physiol 89:945–951
Joosten MHAJ, Hendrickx LJM, De Wit PJGM (1990) Carbohydrate composition of apoplastic fluids isolated from tomato leaves inoculated with virulent or avirulent races of *Cladosporium fulvum* (syn. *Fulvia fulva*). Neth J Plant Pathol 96:103–112

Joosten MHAJ, Cozijnsen TJ, De Wit PJGM (1994) Host resistance to a fungal tomato pathogen lost by a single base-pair change in an avirulence gene. Nature 367:384–387

Kaziro Y, Itoh H, Kozasa T, Nakafuku M, Satoh T (1991) Structure and function of signal-transducing GTP-binding proteins. Annu Rev Biochem 60:349–400

Keen NT (1990) Gene-for-gene complementarity in plant-pathogen interactions. Annu Rev Genet 24:447–463

Lamb CJ (1994) Plant disease resistance genes in signal perception and transduction. Cell 76:419–422

Lazarovits G, Higgins VJ (1976a) Histological comparison of *Cladosporium fulvum* race 1 on immune, resistant, and susceptible tomato varieties. Can J Bot 54:224–234

Lazarovits G, Higgins VJ (1976b) Ultrastructure of susceptible, resistant, and immune reactions of tomato to races of *Cladosporium fulvum*. Can J Bot 54:235–247

Marmeisse R, Van den Ackerveken GFJM, Goosen T, De Wit PJGM, Van den Broek HWJ (1993) Disruption of the avirulence gene *avr9* in two races of the tomato pathogen *Cladosporium fulvum* causes virulence on tomato genotypes with the complementary resistance gene *Cf9*. Mol Plant-Microbe Interact 6:412–417

Marmeisse R, Van den Ackerveken GFJM, Goosen T, De Wit PJGM, Van den Broek HWJ (1994) The *in planta* induced *ecp*2 gene of the tomato pathogen *Cladosporium fulvum* is not essential for pathogenicity. Curr Genet 26:245–250

Marrè E (1979) Fusicoccin: a tool in plant physiology. Annu Rev Plant Physiol 30:273–288

Martin GB, Brommonschenkel SH, Chunwongse J, Frary A, Ganal MW, Spivey R, Wu T, Earle ED, Tanksley SD (1993) Map-based cloning of a protein kinase gene conferring disease resistance in tomato. Science 262:1432–1436

Peever TL, Higgins VJ (1989) Electrolyte leakage, lipoxygenase, and lipid peroxidation induced in tomato leaf tissue by specific and nonspecific elicitors from *Cladosporium fulvum*. Plant Physiol 90:867–875

Scheel D (1990) Elicitor recognition and signal transduction in plant defense gene activation. Z Naturforsch 45c:569–575

Scholtens-Toma IMJ, De Wit PJGM (1988) Purification and primary structure of a necrosis-inducing peptide from the apoplastic fluids of tomato infected with *Cladosporium fulvum* (syn. *Fulvia fulva*). Physiol Mol Plant Pathol 33:59–67

Serrano R (1989) Structure and function of plasma membrane ATPase. Annu Rev Plant Physiol 40:61–94

Stakman EC (1917) Biologic forms of *Puccinia graminis* on cereals and grasses. J Agric Res 10:429–495

Stakman EC, Parker JH, Piemeisel FJ (1918) Can biologic forms of stemrust on wheat change rapidly enough to interfere with breeding for rust resistance? J Agric Res 14:111–123

Sutherland MW (1991) The generation of oxygen radicals during host plant responses to infection. Physiol Mol Plant Pathol 39:79–93

Tzeng DD, DeVay JE (1993) Role of oxygen radicals in plant disease development. In: Andrews JH, Tommerup IC (eds) Advances in plant pathology, vol 10. Academic Press, London, pp 1–33

Van den Ackerveken GFJM, Van Kan JAL, De Wit PJGM (1992) Molecular analysis of the avirulence gene *avr9* of the fungal tomato pathogen *Cladosporium fulvum* fully supports the gene-for-gene hypothesis. Plant J 2:359–366

Van den Ackerveken GFJM, Van Kan JAL, Joosten MHAJ, Muisers JM, Verbakel HM, De Wit PJGM (1993a) Characterization of two putative pathogenicity genes of the fungal tomato pathogen *Cladosporium fulvum*. Mol Plant-Microbe Interact 6:210–215

Van den Ackerveken GFJM, Vossen P, De Wit PJGM (1993b) The AVR9 race-specific elicitor of *Cladosporium fulvum* is processed by endogenous and plant proteases. Plant Physiol 103:91–96

Van den Ackerveken GFJM, Dunn RM, Cozijnsen AJ, Vossen JPMJ, Van den Broek HWJ, De Wit PJGM (1994) Nitrogen limitation induces expression of the avirulence gene *avr9* in the tomato pathogen *Cladosporium fulvum* Mol Gen Genet 243:277–285

Van der Beek JG, Verkerk R, Zabel P, Lindhout P (1992) Mapping strategy for resistance genes in tomato based on RFLPs between cultivars: *Cf9* (resistance to *Cladosporium fulvum*) on chromosome 1. Theor Appl Genet 84:106–112

Van Kan JAL, Van den Ackerveken GFJM, De Wit PJGM (1991) Cloning and characterization of cDNA of avirulence gene *avr9* of the fungal tomato pathogen *Cladosporium fulvum*, causal agent of tomato leaf mold. Mol Plant-Microbe Interact 4:52–59

Van Kan JAL, Joosten MHAJ, Wagemakers CAM, Van den Berg-Velthuis GCM, De Wit PJGM (1992) Differential accumulation of mRNAs encoding extracellular and intracellular PR proteins in tomato induced by virulent and avirulent races of *Cladosporium fulvum*. Plant Mol Biol 20:513–527

Vera-Estrella R, Blumwald E, Higgins VJ (1992) Effect of specific elicitors of *Cladosporium fulvum* on tomato suspension cells. Plant Physiol 99:1208–1215

Vera-Estrella R, Barkla BJ, Higgins VJ, Blumwald E (1994) Plant defense response to fungal pathogens. Activation of host-plasma membrane H^+-ATPase by elicitor induced enzyme dephosphorylation. Plant Physiol 104:209–215

Wevelsiep L, Kogel KH, Knogge W (1991) Purification and characterization of peptides from *Rhynchosporium secalis* inducing necrosis in barley. Physiol Mol Plant Pathol 39:471–482

Wevelsiep L, Rüpping E, Knogge W (1993) Stimulation of barley plasmalemma H^+-ATPase by phytotoxic peptides from the fungal pathogen *Rhynchosporium secalis*. Plant Physiol 101:297–301

Wubben JP, Joosten MHAJ, Van Kan JAL, De Wit PJGM (1992) Subcellular localization of plant chitinases and 1,3-β-glucanases in *Cladosporium fulvum* (syn. *Fulvia fulva*)-infected tomato leaves. Physiol Mol Plant Pathol 41:23–32

Wubben JP, Eijkelboom CA, De Wit PJGM (1994) Accumulation of pathogenesis-related proteins in the epidermis of tomato leaves infected by *Cladosporium fulvum*. Neth J Plant pathol 99:231–239

2 The Potato Late Blight Pathogen *Phytophthora infestans* and Other Pathogenic Oomycota

F. GOVERS[1], A. DRENTH[2], and C.M.J. PIETERSE[3]

CONTENTS

I.	Introduction	17
II.	Plant Pathogenic Oomycota	18
III.	Disease Cycle of *Phytophthora infestans*	18
A.	Asexual Life Cycle	18
B.	Sexual Life Cycle	19
C.	Incompatible Interactions	20
IV.	Population Genetics of *Phytophthora infestans*	21
A.	Markers for Population Genetic Analysis	21
B.	Central Mexico: the Center of Origin	22
C.	Migrations	22
D.	Population Displacement	24
E.	The Acquisition of Sexual Reproduction	25
V.	Molecular Genetics of Plant Pathogenic Oomycota	25
A.	DNA Transformation	26
B.	Structure of Oomycetous Genes	26
VI.	Elicitors Produced by *Phytophthora* Species	27
VII.	Pathogenicity of *Phytophthora infestans*	29
A.	Differential Gene Expression During Pathogenesis	29
B.	Differential Gene Expression During Starvation	33
VIII.	Perspectives and Conclusions	33
	References	33

I. Introduction

Potato late blight, caused by a member of the Oomycota, *Phytophthora infestans* (Mont.) De Bary, is one of the most important and devastating diseases of potato (*Solanum tuberosum*). The pathogen attacks both foliage and tubers, and spreads rapidly through host tissues, thereby causing a destructive necrosis. *P. infestans* is a hemibiotrophic pathogen with a rather narrow range of hosts, all of them members of the Solanaceae. The crop plants, potato and tomato (*Lycopersicon esculentum*), are the economically most important hosts. The first late blight epidemic in Europe in 1845 had disastrous effects on potato production. *P. infestans* spread over Europe within 1 year (Bourke 1964) and was found in most potato-growing areas of the world soon thereafter. In Ireland the potato crop was destroyed in two successive years, leading to a famine. As a consequence, about one and a half million people died and another million emigrated, mainly to the United States (Woodham-Smith 1962). De Bary (1876) described the life cycle of the potato late blight pathogen and named it *Phytophthora* ("plant destroyer") *infestans*.

The extremely destructive nature of the disease and its worldwide distribution made potato late blight and the potato-*P. infestans* interaction favorite topics of research. Various aspects have been studied extensively (reviewed in Ingram and Williams 1991). In recent years, the new tools of molecular biology and molecular genetics have been applied to study the interaction in further detail.

In this chapter, we will first present some information on pathogenic Oomycota in general. This is followed by a description of the disease cycle of *P. infestans* and the mode of reproduction. We will then focus on the population genetics of the pathogen, and discuss the significance and consequences of the worldwide population displacement which resulted in the introduction of a new mating type in Europe around 1980. Subsequently, the current status of molecular genetic research on plant pathogenic Oomycota is summarized. The diverse nature of elicitors produced by various *Phytophthora* species is discussed and, finally, we focus on putative pathogenicity factors of *P. infestans*.

[1] Department of Phytopathology, Wageningen Agricultural University, Binnenhaven 9, 6709 PD Wageningen, The Netherlands
[2] CRC for Tropical Plant Pathology, The University of Queensland, Brisbane QLD 4072, Australia
[3] Department of Plant Ecology and Evolutionary Biology, Section Phytopathology, Utrecht University, P.O. Box 800.84, 3508 TB Utrecht, The Netherlands

The Mycota V Part B
Plant Relationships
Carroll/Tudzynski (Eds.)
© Springer-Verlag Berlin Heidelberg 1997

II. Plant Pathogenic Oomycota

Phytophthora infestans belongs to the class Oomycetes, order Peronosporales, family Pythiaceae. Traditionally, the Oomycota were placed within the kingdom Fungi. However, structurally, biochemically, and genetically, they differ from all other taxa in the kingdom Fungi and are now placed in a separate monophyletic clade, the Stramenopila or chromista, together with the Hyphochytriomycota, Labyrinthulomycota, and several groups of algae with heterokont zoospores (Patterson and Sogin 1992; Hawksworth et al. 1995; Alexopoulos et al. 1996).

The most important plant-pathogenic Oomycota belong to the Pythiaceae and the Peronosporaceae, two families in the order Peronosporales. The best-known genera of the Pythiaceae are *Phytophthora* and *Pythium*. The genus *Pythium* includes some notorious soil-borne plant pathogens which cause seedling damping-off, root rot, and soft rot in a large variety of plants. All *Phytophthora* species, about 60 in total, cause a wide range of diseases on a large variety of plants. Some cause root or stem rots, others cause rots of buds or fruits and some cause foliar blight. Some species have a wide range of host plants, whereas others are restricted to one or a few hosts. *P. cinnamomi*, for example, is pathogenic on almost 1000 different species of trees and shrubs. In contrast, *P. sojae* (syn. *P. megasperma* f. sp. *glycinea*) can infect only one host, i.e., soybean.

The family Peronosporaceae comprises the very destructive organisms commonly known as the downy mildew pathogens. From a historical point of view, the best known is *Plasmopara viticola*, the causal agent of downy mildew on grapes, which destroyed vineyards throughout Europe in the 1870s–1880s. In terms of genetics the best known is *Bremia lactucae*. Here, the main focus has been the inheritance of avirulence genes that interact in a gene-for-gene manner with major resistance genes in the host, lettuce (Crute 1992). Unfortunately, the biotrophic nature of *Bremia lactucae* and the inability to transform this organism hamper further progress of these studies. A downy mildew pathogen which is receiving a lot of attention lately is *Peronospora parasitica*. It attacks crucifers, among which is *Arabidopsis thaliana*, the model plant for genetic and molecular studies. Recently, it has been demonstrated that *P. parasitica* and *A. thaliana* have a gene-for-gene relationship (Holub et al. 1994) and genetic analysis of this interaction has already resulted in the identification of genes from *A. thaliana* involved in race-specific resistance. It is likely that the isolation of a gene that confers resistance against an oomycetous pathogen will be accomplished in the near future. Another oomycetous fungus pathogenic on *A. thaliana* is *Albugo candida* (Holub et al. 1993). *Albugo* is the only genus in the third family of the Peronosporales, the Albuginaceae. The genus comprises the white rust pathogens which are economically of minor importance.

III. Disease Cycle of *Phytophthora infestans*

A. Asexual Life Cycle

In spring, when potato tubers are planted, diseased sprouts may arise from tubers which are infested with mycelium of *P. infestans* (Fig. 1). Under favorable conditions, sporangiophores emerge from the stomata and release numerous airborne sporangia causing a rapid spread of the disease. At temperatures above 12–15 °C sporangia may germinate directly. Below 12 °C, sporangia may differentiate into numerous motile zoospores which germinate after encystment. On leaf surfaces and stems, germinated sporangia and cysts form germ tubes with appressoria from which penetration hyphae arise. From the sporangia and cysts that germinate on the leaf surface only a small proportion actually penetrates the epidermal cells. Characteristically, the anticlinal walls of epidermal cells are penetrated by the penetration hyphae and often the cells next to stomatal guard cells are the target sites for penetration. Hyphal structures are formed in the epidermal cell from which the mycelium grows, initially intercellularly, while intracellular haustoria are formed in the mesophyll cell layer (Coffey and Gees 1991; Freytag et al. 1994).

The first macroscopic sign of lesion development is the appearance of water-soaked areas. While the pathogen grows, the center of the spreading lesion becomes necrotic. Sporulation is visible at the advancing edges of the developing lesion where sporangiophores emerge predominantly from the stomata. The expanding necrosis eventually leads to complete destruction of foliage and, consequently, in reduction of tuber yields.

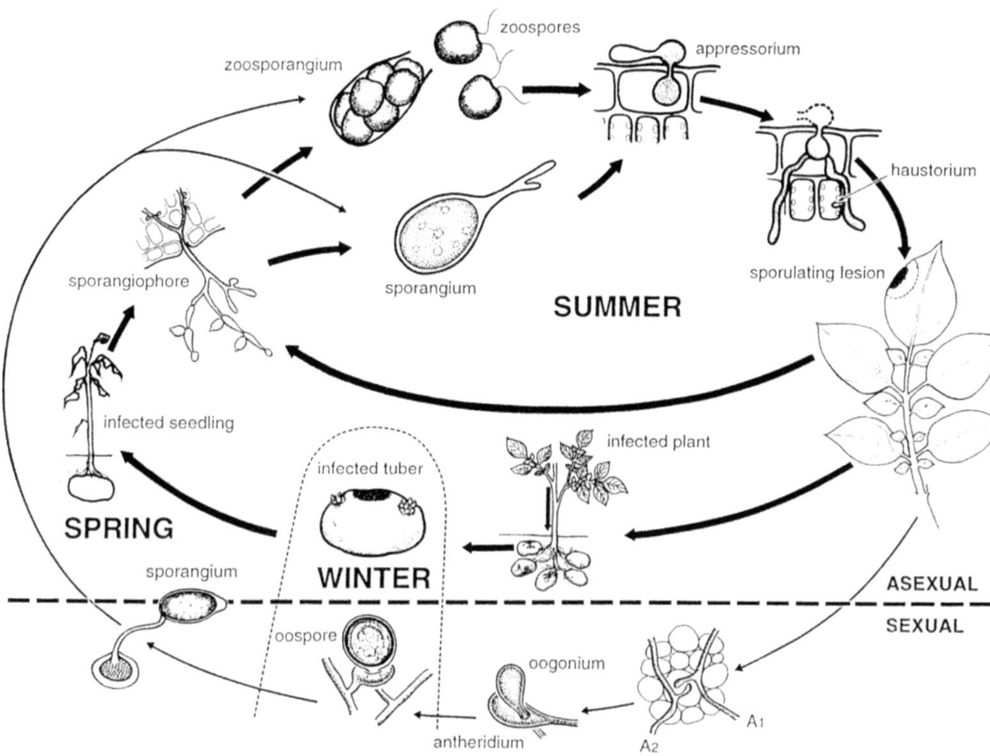

Fig. 1. Disease cycle of potato late blight and life cycle of *Phytophthora infestans*

Under favorable conditions, such a rapidly expanding lesion, carrying numerous sporangiophores with abundant sporangia, is formed within 4 to 5 days after infection. Many asexual generations may be produced in one growing season, which explains the enormous capacity for spread and epidemic development of the disease. The airborne sporangia can spread over distances up to several hundred kilometers. In wet weather conditions, sporangia or zoospores are washed down from the leaves and carried into the soil. Here the spores germinate and the germ tubes may penetrate the tubers at lenticels, wounds, and eyes, or at sites where the surface is not completely suberized. Most of the blighted tubers will rot in the soil or during storage. However, a few will survive the winter and they are the source for new infections when they sprout in the following season.

B. Sexual Life Cycle

P. infestans is a heterothallic species with two known mating types, A1 and A2. Both mating types are required for sexual reproduction, and in nature, oospores are produced only when hyphae of opposite mating type grow in each other's vicinity in infected host tissue. When the A1 and A2 hyphae make contact, antheridia and oogonia are formed. The oogonium grows through the antheridium in an amphigynous configuration in which the antheridium surrounds the oogonial stalk (Fig. 1). There is no exchange of cytoplasm between antheridium and oogonium. The oogonium expands rapidly due to the flow of cytoplasm through the oogonial stalk from its own thallus. Meiosis occurs in the multinucleate gametangium (Shaw 1983). All nuclei in the oogonium, except one, migrate to the periphery, where they disintegrate. A fertilization tube grows from the antheridium through the oogonial wall to deposit an antheridial nucleus in the oogonium. The nuclei fuse and while the oospore matures a thick oospore wall develops. When blighted potato plants containing oospores remain in the field, they decompose and, as a result, oospores are liberated. In general, oospores of Oomycota are highly persistent structures. For some *Phytophythora* species it has been shown that oospores can survive in soil for several years. When the surviving oospores germinate, they can infect tubers and stolons of newly planted potatoes, as well as stems and leaves which come into contact with the soil (Schöber and Turkensteen 1992; Drenth et al. 1995).

Since only A1 mating type isolates were present in most potato-growing areas of the world, the sexual life cycle was not very important for the disease cycle of *P. infestans*. With the population displacement in the late 1970s, A1 and A2 mating types were introduced in Europe (see Sect. IV.C; Fry et al. 1993), and there is strong evidence that here sexual reproduction is now integrated in the disease cycle (Drenth et al. 1994, 1995; Sujkowski et al. 1994).

C. Incompatible Interactions

In nature, many different races of *P. infestans* occur. They vary in their capacity to infect different host cultivars. When a virulent race attacks a susceptible host cultivar, a compatible interaction develops and the host plant is readily destroyed, as described above. However, when the host cultivar has an appropriate major resistance (R) gene which confers race-specific resistance, the pathogen may be recognized by the host in an early stage and the interaction turns out to be incompatible. Supposedly, race-specific resistance in the *P. infestans*-potato interaction is based on a gene-for-gene relationship (Black et al. 1953). For several pathosystems a gene-for-gene relationship has been demonstrated (Thompson and Burdon 1992; Chap. 1, this Vol.). It is suggested that in such a relationship, the products of avirulence genes of the pathogen, the so-called race-specific elicitors, interact with the products of the matching major resistance genes of the host. This interaction results in triggering of a hypersensitive response (HR) which is recognized by rapid cell death of invaded host cells and a limited number of cells surrounding the infection site (Tomiyama 1983). In this way, the biotrophic pathogen is arrested in growth and remains confined to the focus of infection. Little is known about the genetic and molecular basis of virulence/avirulence in *P. infestans*. Spielman et al. (1989, 1990) found that avirulence towards the potato resistance gene R3 and the tomato resistance gene Ph1 is dominant whereas avirulence towards R2 and R4 is recessive. Doke et al. (1993) have indications that certain *P. infestans* races produce race-specific suppressors which are thought to suppress the activity of general elicitors, thereby making an interaction compatible.

Comparison of the behavior of different races of *P. infestans* on various potato cultivars has demonstrated that at the stage of spore germination and penetration, virulent and avirulent races behave similarly (Coffey and Gees 1991; Freytag et al. 1994). However, in an incompatible interaction, further growth of hyphal structures that succeeded in penetrating the epidermis is effectively inhibited within the epidermis or the mesophyll. Further spread of the lesion is prevented by the hypersensitive response which is observed in all invaded host cells. Even in compatible interactions, mesophyll cells may react with hypersensitive cell death. In the potato cultivar Datura, which has the R1 resistance gene, it has been shown that infection with a compatible virulent race leads to varying degrees of hypersensitive responses. At some sites, only a few mesophyll cells respond hypersensitively whereas at other sites many cells do. The final phenotypic outcome of compatible or incompatible interactions, which are qualitatively clearly distinguished as either complete destruction or largely undisturbed survival of the plant, is very likely to be the consequence of comparatively small quantitative effects: a slightly large number of reacting cells appears to be sufficient to prevent spread and sporulation of *P. infestans* in the incompatible interaction (Freytag et al. 1994). Whether this small quantitative difference is related to differences in the amount of race-specific elicitor or suppressor produced at a certain site in the infected tissues in unknown. As long as the nature of race-specific elicitors or suppressors is unclear, this remains an open question.

IV. Population Genetics of *Phytophthora infestans*

Central Mexico is considered to be the center of origin of *P. infestans*. In the mid-19th century, the first appearance of potato late blight outside Mexico was reported. The disease was found in the USA and Europe, and since then has appeared wherever potatoes are grown. Of the two mating types that are required for sexual reproduction of *P. infestans*, only one, A1, was known before the early 1950s. At that time the second mating type, A2, was discovered in isolates collected in the Toluca Valley in Central Mexico, although they were then known from nowhere else in the world. This changed in the early 1980s when A2 mating type isolates appeared in Europe. This was the first indication that the populations were changing dramatically and it stimulated plant pathologists all over the world to analyze local populations. Currently, A2 mating type isolates are present all over northern Europe, in Asia, in several African countries, in South America, and in North America (summarized in Drenth et al. 1993b). Thus, the changes in the populations appear to have occurred on a worldwide scale, and most certainly result from migration.

A. Markers for Population Genetic Analyses

Characterizations of *P. infestans* populations have relied on a series of markers which have contributed significantly to our understanding of the population genetics of the pathogen. The markers available for *P. infestans* can be classified in three categories: biologically significant markers, cytoplasmic markers, and neutral markers. They are briefly summarized below. From the available markers, neurtral polymorphic DNA markers appear to be the most powerful for population genetic analyses.

Biologically significant markers include mating type, fungicide resistance, and virulence. The mating type of an isolate is readily determined by scoring oospore formation upon cocultivation with A1 and A2 tester isolates. Fungicide resistance is easily detectable in mycelium grown in vitro on an appropriate medium. Virulence can be determined by assaying compatibility and incompatibility in a differential set of potato lines with the 11 known resistance genes (R-genes). Before the 1980s, virulence was the only marker available to assess variation in *P. infestans* populations. A problem for using virulence as an unambiguous marker is the lack of a universal differential set of potato lines and a universal virulence assessment method. This seriously limits the interpretation of data in the numerous papers concerning variation for virulence. Moreover, the stability of virulence phenotypes has been the subject of much debate. Virulence and fungicide resistance are traits on which an enormous selection pressure can be exerted through the use of resistant potato cultivars or the application of fungicides. Strong selection on particular characteristics can dramatically alter the frequency of certain genotypes in the fungal population, thus requiring careful interpretation of data.

Cytoplasmic markers are mitochondrial DNA (mt-DNA) and double-stranded RNA (ds-RNA). Their main characteristic is the extrachromosomal inheritance leading to intact transmission from parent to progeny. The origin and the function of ds-RNA, which has been found in only a limited number of isolates, is unknown (Newhouse et al. 1992). Cytoplasmic markers are very useful to determine the global movement of plant pathogens, to infer phylogenetic relationships among different *Phytophthora* species, and to follow anastomosis and cytoplasmic inheritance. However, they are not suited for unambiguous identification of different isolates. The number of polymorphisms for mt-DNA and ds-RNA in *P. infestans* is rather limited. Because they are transmitted without recombination from parent to progeny they cannot reflect the true diversity in a population.

Neutral markers are called neutral because they are not related to biologically important characteristics. Neutral markers can be subdivided in allozyme markers and polymorphic DNA markers. More than 50 different enzymes were tested for their usefulness as allozyme marker for *P. infestans*, but only two appeared to show polymorphism (Tooley et al. 1985; Spielman et al. 1990). A number of different techniques are currently being used to detect DNA polymorphisms. Hybridization of digested genomic DNA with moderately repetitive DNA probes or short oligonucleotide repeats allows simultaneous detection of restriction fragment length polymorphism (RFLP) at numerous loci in the genome. A prerequisite for this so-called DNA fingerprinting is

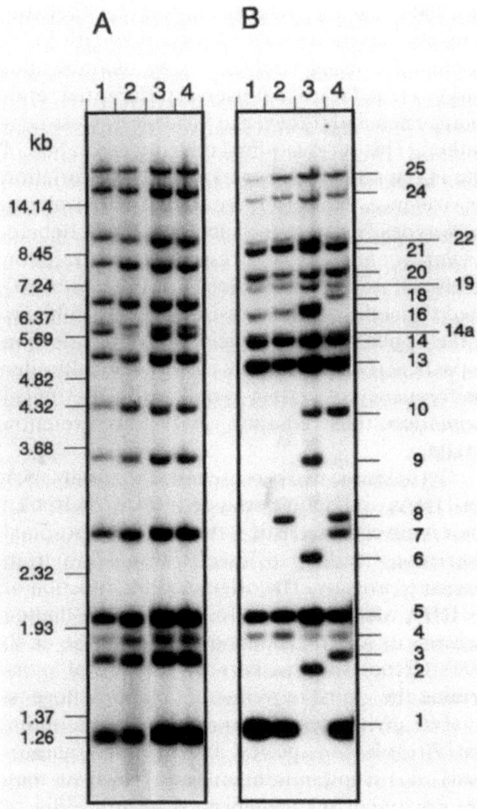

Fig. 2A,B. RG-57 DNA fingerprinting patterns from "old" and "new" *P. infestans* isolates. The "old" isolates shown in **A** were collected in 1966 (*1*), 1968 (*2*), 1974 (*3*), and 1978 (*4*). The "new" isolates shown in **B** were collected in 1986 (*1*) and in 1988 (*2*, *3*, and *4*). Size markers in kilobases (kb) are indicated on the *left*. RG-57 fragments numbers are indicated on the *right*. (Drenth et al. 1994)

that the probe hybridizes to loci which are dispersed all over the genome and which are inherited independently. A moderately repetitive DNA probe that has been used routinely for DNA fingerprinting of *P. infestans* is RG-57, a random genomic DNA fragment of *P. infestans*. It hybridizes to more than 25 nonallelic polymorphic fragments of EcoRI-digested *P. infestans* DNA (Fig. 2). The banding pattern is stable through asexual reproduction and each hybridizing fragment appears to represent a unique genetic locus. Probe RG-57 allows virtually unambiguous identification of different isolates (Goodwin et al. 1992a).

B. Central Mexico: the Center of Origin

It has been known for a long time that many of the wild tuber-bearing *Solanum* species native to central Mexico have resistance to *P. infestans*. Several resistance genes that have been introduced by breeding into domesticated potatoes (*Solanum tuberosum* subsp. *tuberosum*) are derived from wild Mexican *Solanum* species. The natural occurrence of many different resistant *Solanum* species along with a diverse pathogen population suggested that the *Solanum-P. infestans* pathosystem evolved in central Mexico (Niederhauser and Mills 1953). The observation made in the early 1950s that the two mating types of *P. infestans*, A1 and A2, both occur in Mexico was a strong indication that, indeed, the pathogen has its origin in the central highlands of Mexico. In Mexico, the frequency of A1 and A2 mating type isolates is equal and the isolates are extremely diverse for both virulence factors and neutral markers. Many different and complex races have been found there, and all known allozyme alleles and RG-57 hybridizing fragments are represented in Mexican isolates (Goodwin et al. 1992b). Genotypic characterization of isolates collected during an epidemic in Toluca Valley in 1988 revealed an extremely large number of unique genotypes, and this diversity occurred despite the predominance of asexual reproduction during the epidemic. The large number of distinct genotypes strongly indicates that sexual reproduction occurs frequently in *P. infestans* populations in central Mexico (Fry et al. 1993).

C. Migrations

The first reported escape of *P. infestans* from Mexico occurred in 1843, when a new potato disease appeared in the northeastern part of the United States. Two years later the pathogen had reached Europe, and from there it has probably been distributed to the rest of the world via the international trade in seed potatoes. Since there is no evidence for new migrations from Mexico during the interval between 1840 and the mid 1970s, it is now generally assumed that for more than 130 years only a single genotype dominated the *P. infestans* populations outside Mexico. This historical clonal lineage, which was recently designated US-1 (Goodwin et al. 1994), has the A1 mating type.

In 1984, Hohl and Iselin reported the discovery of isolates with the A2 mating type among their *P. infestans* cultures which had been collected in Switzerland in 1981. By the mid 1980s, many other culture collections and field isolates had been screened for mating type and it turned out that the A2 mating type had been present in several European countries since the late 1970s. With the appearance of A2 mating type isolates, the diversity for virulence factors increased significantly (Drenth et al. 1994). The virulence diversity in European isolates was found to be as high as in isolates from central Mexico (Mills and Niederhauser 1953; Goodwin et al. 1992b) and this strongly suggested that the appearance of new isolates was the result of a new introduction. Evidence for this was gathered from the genotypic characterization of large numbers of *P. infestans* isolates (Spielman et al. 1991; Drenth et al. 1994). These studies revealed that in Europe not only new A2 mating type isolates were introduced, but also new A1 mating type isolates, and that this new population has now completely displaced the original historical clonal lineage, the US-1 type isolate. An extensive comparative study of isolates collected in the Netherlands before and after 1980 (Drenth et al. 1994) showed that before 1980 all old Dutch isolates had the US-1 genotype (Figs. 2, 3b). Among 33 isolates only 4 different races, consisting of combinations of virulence factors, 1, 3, 4, and 10, were identified (Fig. 3a). Thus, there was scarcely any diversity in the population, and the limited diversity found for virulence matched the R-genes present in commonly used potato cultivars. Among isolates collected after 1980, the A2 mating type appeared along with the A1 mating type, and new allozyme alleles and RG-57 genotypes were found. In 179 isolates collected between 1980 and 1991, 134 different RG-57 genotypes occurred, of which 121 were found

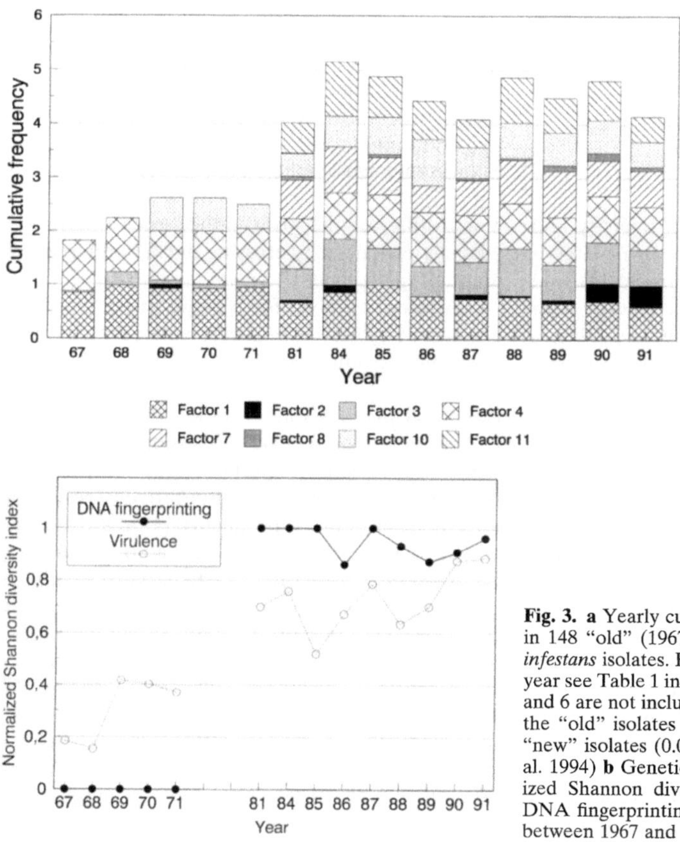

Fig. 3. **a** Yearly cumulative frequency of virulence factors in 148 "old" (1967–1971) and 253 "new" (1981–1991) *P. infestans* isolates. For the number of isolates and races per year see Table 1 in Drenth et al. (1994). Virulence factors 5 and 6 are not included in the bars. They were not found in the "old" isolates and only at very low frequency in the "new" isolates (0.011 and 0.023, respectively). (Drenth et al. 1994) **b** Genetic diversity as calculated by the normalized Shannon diversity index for virulence and RG-57 DNA fingerprinting for *P. infestans* populations collected between 1967 and 1991. (Drenth et al. 1994)

only once (Fig. 3b). Diversity for virulence and complexity of races increased tremendously (Fig. 3a). Among 253 isolates, 73 different races were identified. These races contained many unnecessary virulence alleles (e.g., 5, 6, 7, 8, and 11) for which there are no corresponding resistance genes in the array of potato cultivars grown in The Netherlands.

During the 1980s, the new population that first appeared in western Europe around 1980 and completely displaced the old population, migrated rapidly eastward and is now present in Asia and Africa. More recently, in the late 1980s, the first A2 mating type isolates were found in the United States and Canada. Now A2 mating type isolates are widely distributed over northern America. Genotypic characterization of isolates collected in various regions indicated that there was a massive immigration of new genotypes directly from northwestern Mexico and not via Europe (Goodwin et al. 1994). The appearance of several unique genotypes in British Columbia (Canada) in 1992 and 1993 is the first evidence for the probable occurrence of sexual reproduction of *P. infestans* in America north of Mexico (S.B. Goodwin and W.E. Fry, pers. comm.).

D. Population Displacement

The observation that the virulence diversity and genetic diversity in the new European isolates is as high as in the Mexican isolates indicates that again, as in 1845, *P. infestans* escaped from Mexico and was able to establish in Europe. In 1976, a severe drought reduced the potato harvest in Europe and in the spring of 1977 vast quantities of potatoes were transported from Mexico to Europe. Possibly this potato import was accompanied by new *P. infestans* isolates, either in the form of mycelium in potato tubers or as oospores attached to potato tubers. In 1980 and 1981, genetically different A1 and A2 mating type isolates were found in a number of countries nearly simultaneously whereas the old US-1 genotype started to disappear rapidly. In The Netherlands, no US-1 genotype has been found since 1980. That the discovery of these new isolates took at least 3 years probably can be explained by the widespread use of fungicides, which may have efficiently restricted the spread of the pathogen.

In 1845, a single year sufficed for *P. infestans* to spread over vast parts of Europe (Bourke 1964); at that time, however, no control measures were available and potato fields were heavily blighted, producing immense quantities of spores. The fact that the population displacement in The Netherlands occurred relatively fast, within 1 or 2 years, indicates that the current *P. infestans* isolates possess a higher fitness than the isolates which were present before 1980. We have found that sexual progeny produced by in vitro pairing of old A1 and new A2 isolates have a very low fitness and are not very pathogenic (A. Drenth, unpubl.). This reduced fitness may be explained by the buildup of a huge mutational load in the old A1 population. As the A1 mating type isolates present before the introduction of the new population are all US-1 types, they originate from the initial introduction of *P. infestans* to Europe in 1845. They were thus restricted to asexual reproduction for about 130 years. With an average of 20 generations per year and a mutation rate of 11.2×10^{-6} per gene per generation (Schlager and Dickie 1971), the probability of a mutation over 2600 generations will be 0.029 per gene. The unfavorable effects of many mutations will be masked as long as there are undamaged copies of the genes available. This is especially true for the old population, which was predominantly tetraploid. However, the mutational load in the old A1 mating type isolates may have reduced their suitability for sexual reproduction to a great extent, and it might be responsible for the poor pathogenicity of sexual progeny produced by crossing old A1 and new A2 mating type isolates.

Another catalyst for a rapid population displacement by new *P. infestans* isolates might have been the introduction of metalaxyl in 1979. Very soon after its introduction, in 1980, resistance to this fungicide was found among *P. infestans* isolates (Davidse et al. 1981). Resistance to metalaxyl may have been present in the newly introduced isolates from Mexico, or alternatively, with the advent of sexually reproducing populations, the new isolates might have been able to acquire the resistance more rapidly. The large-scale application of metalaxyl in 1979 and in 1980 provided an intense selection pressure for resistant isolates, leading to a tremendous increase in their frequency in the population (Davidse et al. 1981). No metalaxyl-resistant isolates were found in the old, tetraploid *P. infestans* isolates collected before 1980. In the new *P. infestans* population, resistance to metalaxyl was found in various A1 and A2 mating type isolates with different genetic backgrounds (Drenth et al. 1993a), suggesting many independent mutations leading to metalaxyl resistance and/or the inheritance of resistance through sexual reproduction.

E. The Acquisition of Sexual Reproduction

The observation that the population displacement in 1980 with new A1 and A2 mating type isolates of *P. infestans* coincided with a change from little, if any, diversity in the old population to a very high level of diversity in the new population strongly suggested that the *P. infestans* population has acquired the ability to reproduce sexually which, in turn, is the driving force behind the high level of genetic diversity. To support this suggestion, we searched for the occurrence of oospores in the field and investigated whether the Dutch climate allows survival of oospores in soil between growing seasons. In 1992 we found oospores in blighted potato and tomato plants in the field. In addition, we used well-characterized parental isolates for survival studies. Oospores were allowed to develop in leaves, and the infested leaves were mixed with soil and exposed to natural weather conditions during the winter of 1992–1993. After 8 months, the soil still contained infectious material. DNA fingerprinting confirmed that infections on leaves, which contacted the soil containing the oospores, indeed resulted from oospores and were not caused by mycelium or vegetative spores from any of the parental isolates. Moreover, we found that soil samples taken in June 1993 from potato fields, which had been inoculated the year before, were still infectious. DNA fingerprinting showed that these infections were caused by progeny from the two parental isolates used to inoculate the field plots in 1992, demonstrating that oospores can overwinter in soil (Drenth et al. 1995).

Whether or not sexual reproduction has advantages over asexual reproduction has been a point of much debate (Charlesworth 1989). In the case of *P. infestans* the advantages are obvious. Firstly, *P. infestans* is a hemibiotrophic pathogen whose asexual spores, in contrast to oospores, cannot survive for long periods outside the host plant. When oospores are part of the life cycle, *P. infestans* can survive in soil between growing seasons, thereby increasing its ability to colonize new host plants and to propagate. Secondly, sexual reproduction results in a wider variety of genotypes and this, in turn, leads to an increased ability to respond to changes in the environment and to selection pressure. This is illustrated by the speed at which the new *P. infestans* population developed resistance to the fungicide metalaxyl (Davidse et al. 1981). In some countries, where the new sexually propagating populations have not yet appeared, metalaxyl is still used successfully to control late blight. Thirdly, the increased diversity of virulence patterns and the appearance of new virulence factors, through sexual recombination, enables the pathogen to avoid recognition by any specific resistance gene or combinations of resistance genes in the potato population.

Established disease management strategies are based on the biological characteristics of the old asexual *P. infestans* population. With an asexually propagating population, mycelium retained in seed potatoes or in potato refuse piles produces infectious spores, and this results in typical focal epidemics. Hence, control measures involve disease-free seed potatoes and removal of potato refuse piles. With the acquisition of sexual reproduction resulting in the presence of oospores in soil, late blight epidemics can start anywhere in the field as soon as potatoes have emerged. Thus, a general epidemic, originating from oospores, may appear despite sufficient removal of infected tubers containing mycelium. To anticipate epidemic development, detailed studies on oospore production, survival, germination, and infectivity are needed to determine the parameters of late blight epidemics caused by the new *P. infestans* population. These parameters are likely to be different from those in epidemics caused by old isolates, and we expect that the current disease management strategies will have to be adjusted accordingly.

V. Molecular Genetics of Plant-Pathogenic Oomycota

In the late 1980s, molecular genetic approaches were introduced in the research on *P. infestans* and other plant-pathogenic Oomycota. From various *Phytophthora* and *Pythium* species and from *Bremia lactucae*, random DNA fragments or ribosomal genes have been cloned and they are used for identification of strains, for population analyses and for genome mapping. Currently, PCR-based techniques (e.g., RAPDs) are frequently employed for the same purposes. To elucidate the nature of factors that determine pathogenicity or elicitors which induce resistance reactions in the host plant, putative pathogenicity genes and genes encoding elicitors have been cloned and characterized (see below). In addition, some oomycetous genes encoding housekeeping enzymes and structural proteins have been isolated. After numerous unsuccessful attempts to transform representative Oomycota, stable DNA transformation of several *Phytophthora* species has now been accomplished. The observation that DNA transformation of *P. infestans* is successful only when expression of marker genes or antibiotic resistance genes is regulated by oomycete regulatory sequences suggests that oomycetous gene promoters have special features which are required for high level expression.

A. DNA Transformation

For functional analysis of genes and manipulation of gene products, DNA transformation is a crucial tool. Initial attempts to transform *P. infestans* were reported by Kinghorn et al. (1991), who used various transformation vectors of which several had been proven to be successful in a diverse array of filamentous fungi. However, no stable transformants were generated. Judelson and Michelmore (1991) assayed transient expression of the GUS reporter gene in *P. infestans* regulated by promoters of oomycetous genes. By using optimized transformation conditions deduced from the transient expression assays and vectors containing chimeric genes consisting of transcriptional regulatory sequences from the *Bremia lactucae ham34* or *hsp70* genes and the coding region of the bacterial HPG or NPTII gene conferring resistance to hygromycin B or G-418, respectively, Judelson et al. (1991) were the first to obtain stable transformants of *P. infestans*. In the same year, transformation of *P. parasitica* and *P. capsici* with a stably maintained self-replicating plasmid was reported (Bailey et al. 1991). In these two *Phytophthora* species, as well as in *P. citricola* and *P. cinnamomi*, the promoter of the *hsp70* gene of the heterobasidiomycete *Ustilago maydis* seems to be able to drive the expression of the hygromycin resistance gene (Bailey et al. 1993) whereas in *P. infestans* no activity of the *U. maydis hsp70* promoter, when fused to GUS, was detectable (Judelson et al. 1992).

The most direct way to study the function of a gene and its gene product is by disrupting the gene and examining the effect on biological properties which are thought to be governed by that particular gene. In several plant-pathogenic fungi, e.g., *Cladosporium fulvum* (Marmeisse et al. 1993; Chap. 1, this Vol.), *Fusarium solani* f. sp. *pisi* (Stahl and Schäfer 1992), and *Magnaporthe grisea* (Sweigard et al. 1992; Chap. 3, this Vol.), the functions of avirulence genes or putative pathogenicity genes have been studied by gene disruption or gene replacement. However, this approach requires a highly efficient DNA transformation system and although in recent years the transformation efficiency of *P. infestans* has improved, it is still relatively low when compared to efficiencies obtained with other organisms. Moreover, *Phytophthora* species are diploid organisms; some isolates are even tetraploid. Therefore, inactivation of a target gene requires simultaneous disruption of at least two alleles. Sequential disruption of multiple alleles by integrative transformation has been achieved in the diploid fungus *Candida albicans* with the so-called *ura*-blaster protocol (Alani et al. 1987). However, to apply this protocol to *Phytophthora*, new marker genes and stable auxotrophic mutants must be developed. Disruption of the two alleles can also be obtained by sexual crosses. In the case of a homothallic species this is a straightforward procedure but with heterothallic species like *P. infestans*, forced self-fertilization is necessary.

An alternative approach to study gene function involves reducing or blocking the synthesis of gene products by means of anti-sense RNA. The anti-sense technique has been tested in *P. infestans* using sense and anti-sense GUS constructs, and was shown to be effective in inhibiting the accumulation of GUS up to 98% (Judelson et al. 1993).

B. Structure of Oomycetous Genes

In the past few years, the number of cloned oomycotan genes has been increasing steadily. The picture emerging from gene structure analyses is that most of these genes lack introns. So far, only two oomycetous genes are known which may have introns: the *P. infestans niaA* gene (Pieterse et al. 1995) and the *trp1* gene of *P. parasitica* encoding an enzyme of the tryptophan biosynthetic pathway (Karlovsky and Prell 1991). Transformation experiments have demonstrated that gene promoters isolated from higher fungi, from plants, or from mammalian and plant viruses (SV40 and

Table 1. Conserved sequence motif in promoters of oomycetous genes. Alignment of the DNA regions surrounding the conserved sequence motif GCTCATTYYNCA(A/T)TTT which is present within the first 100 nt upstream of the ATG start codon of ten oomycetous genes. (After Pieterse et al. 1994b)

Gene	DNA sequence surrounding the conserved GCTCATTYYNCA(A/T)TTT motif[a]				
P. infestans ipiB[b]	−94	GCCGTCCAGA	GCTCATTTCCCATTCT ************** *	CCTCCCT	−62
P. infestans ipiO1[c]	−40	TCCGCTAAGA	GCTCATTTGTGAATTC ******** * ****	ATTTCTT	−8
P. infestans ubi3R[d]	−66	CGCCTCCTTT	GCTCATTTTCCATTTT ****************	GAGCGGA	−34
P. infestans calA[e]	−62	TTTTGGATGG	GATCATTGTTGGATTT * ***** ** ****	CCCTCGA	−30
P. infestans actA[f,g]	−84	TCCCTCTTTG	GCTCATTTCCC/TTTT *********** ****	CTTCCAG	−53
P. infestans actB[f]	−60	GTGTCAAAGT	TCTCATTCTGCATTTT ***************	GTCTCGA	−28
P. parasitica parA1[h]	−69	GCCTCCAACT	GC/CATTGTGCAATTT ** ****** ******	GCTCTCA	−38
P. sojae actin[f]	−71	GGACCTTGCT	CGTCATTCCGCAATTT **************	GCTGCCA	−39
P. sojae 42-kDa elicitor[i]	−71		ccattCTCCAATCT *********** *	GCTCGTC	−39
B. lactucae ham34[f]	−85	CGATCGGAAG	GCTCATTCTCC/TTTT *********** ****	CACTCTC	−54
B. lactucae hsp70[f]	−78	TCTCAAGTTT	GCTCACTTTGAAATTT ***** **** *****	TCCATCT	−46

[a] Numbers refer to the position of the first and last nt of the depicted sequence relative to the ATG start codon of the respective gene. Nucleotides marked with an asterisk match with the consensus sequence GCTCATTYYNCA(A/T)TTT. In the *P. infestans actA* sequence, the *P. parasitica parA1* sequence and the *B. lactucae ham34* sequence, a one-nt gap, indicated by a slash (/), is introduced in the conserved motif to obtain optimal alignment. The position of the transcription start point is underlined.
[b] Pieterse et al. (1994b). The nt sequence shown is present in the promoter region of the *P. infestans ipiB2* and *ipiB3* gene. The nt sequence at positions −62 to −94 relative to the ATG start codon of the *ipiB1* gene was not determined. The transcription start point has not been determined.
[c] Pieterse et al. (1994b). The nt sequence at position −8 to −40 relative to the ATG start codon of *ipiO1* and *ipiO2* is nearly identical. Only at position −34, the A in *ipiO1* is a G in *ipiO2*.
[d] Pieterse et al. (1991).
[e] Pieterse et al. (1993b).
[f] For references see Pieterse et al. (1994b).
[g] In the *actB* promoter region five transcription start points have been found which are all located more upstream of the depicted sequence between positions −120 and −164 relative to the ATG start codon.
[h] Kamoun et al. (1993a).
[i] Sacks et al. (1995). Sequence is derived from cDNA clones.

CaMV, respectively) are inactive in *P. infestans*, in *P. sojae*, and in the oomycetous water mold *Achlya ambisexualis* (Judelson et al. 1992). On the other hand, expression of marker genes regulated by oomycete gene promoters is not detectable in higher fungi such as *Aspergillus nidulans*, *Saccharomyces cerevisiae*, *Neurospora crassa*, and *Ustilago maydis* (Judelson et al. 1992; Pieterse et al. 1995). Apparently, the regulatory mechanisms involved in gene activation are not conserved between fungi and chromista. One remarkable feature found in promoters of several *Phytophthora* genes and two *Bremia lactucae* genes is a conserved DNA sequence motif surrounding the transcription start site (Table 1). This motif is not found in gene promoters of higher fungi, plants, or animals, but whether or not it is important for regulation or initiation of transcription is unknown.

VI. Elicitors Produced by *Phytophthora* Species

Elicitors constitute a large class of chemically diverse molecules derived from plant pathogens which stimulate defense responses in intact plants or cultured plant cells. Several different types of elicitor molecules produced by various *Phytophthora* spp. have been identified. For example, the root and stem rot pathogen *P. sojae* produces at least three distinct elicitor molecules. One is a cell wall-bound oligoglucoside that induces phytoalexin synthesis in cotyledons of the host of *P. sojae*, soybean. Specific binding of the oligoglucoside elicitor to soybean plasma membranes has been demonstrated, suggesting that elicitor perception and signal transduction involves a transmembrane receptor-mediated process (Ebel and Cosio 1994; Chap. 5, Vol. V, Part A).

The second elicitor identified in *P. sojae* is an extracellular glycoprotein that induces defense responses, oxidative burst, and ion fluxes in the nonhost parsley (Nürnberger et al. 1994). Characterization of corresponding cDNA clones showed that the entire gene encodes a 57-kDa precursor protein. After cleavage of the signal peptide the N-terminal part of the protein is removed, most likely by proteolytic processing, generating the 43-kDa mature elicitor that is present in culture filtrates. Elicitor activity was found to reside in the protein moiety of the glycoprotein (Sacks et al. 1995). Protease treatment released an oligopeptide consisting of 13 amino acids that stimulated the same responses as the intact elicitor, whereas other fragments of the protein were devoid of elicitor activity (Nürnberger et al. 1994). Also for this elicitor, specific binding to plasma membranes has been demonstrated.

The third elicitor is a 10-kDa protein which belongs to a family of proteins collectively termed elicitins. Elicitins are produced by many *Phytophthora* species (Pernollet 1993). They are synthesized as preproteins with a signal peptide which is removed cotranslationally before secretion in the culture medium. The 98-amino-acid mature proteins are not glycosylated. Elicitins induce a vigorous defense response in tobacco, resulting in resistance to the black shank agent, *Phytophthora parasitica*. *P. parasitica* isolates which are nonpathogenic in tobacco produce the elicitin called parasiticein, whereas most isolates pathogenic on tobacco do not produce elicitins (Ricci et al. 1992). Therefore, these proteins have been proposed to act as avirulence factors in the *P. parasitica*-tobacco interaction. It has been demonstrated that parasiticein is responsible for activation of both localized and distal hypersensitive responses in tobacco (Kamoun et al. 1993b). Recent studies on the three-dimensional structure of the *P. capsici* elicitin capsicein revealed that capsicein has structural features typical for phospholipase A2. Purified capsicein was found to have phosholipase A2 activity, suggesting that the defense signaling pathway triggered by capsicein involves the jasmonic acid route, fostered by the release of linolenic acid (J.-C. Pernollet, pers. comm.).

Neither the 42-kDa glycoprotein nor the 10-kDa *P. sojae* elicitin, which is an active elicitor on tobacco, induces defense responses in soybean, the natural host of *P. sojae*. They seem to serve as signals of nonself-recognition by nonhost plants. However, the biological function of these two elicitors in *P. sojae* remains unknown. The gene encoding the 42-kDa elicitor is part of a gene family. The presence of this gene family appears to be a property common to *Phytophthora* species, but not all species produce a protein that cross-reacts with antiserum raised against the elicitor (Sacks et al. 1995). Also elicitins are coded by members of gene families and elicitin genes are conserved in many tested *Phytophthora* species. Even some *P. parasitica* isolates that do not produce elicitins, and are virulent on tobacco, have elicitin-coding sequences in their genomes (Kamoun et al. 1993a).

In addition to the three different types of elicitor molecules described above, *P. sojae* most likely produces race-specific elicitors. *P. sojae* interacts with soybean in a gene-for-gene manner. Several dominant avirulence genes, which may have a function in the synthesis of race-specific elicitors, have been characterized genetically (Whisson et al. 1994; B.M. Tyler, pers. comm.). However, the biochemical nature of race-specific elicitors produced by *P. sojae* or any other *Phytophthora* species is so far unknown. In view of the observation that from the three *P. sojae* elicitors mentioned above, only the oligoglucoside elicitor is able to induce defense responses in soybean, it might well be that modifications of oligoglucosides determine compatibility or incompatibility, but as yet there are no indications pointing in that direction.

A completely different class of elicitor molecules found in all species of *Phytophthora* are the polyunsaturated fatty acids, eicosapentaenoic acid (EPA) and arachidonic acid (AA). EPA and AA are active elicitors of defense responses in potato. Incompatible as well as compatible races of *P. infestans* release AA while invading potato leaves (Ricker and Bostock 1992). This finding is in agreement with the cytological studies of Freytag et al. (1994), who observed hypersensitive reacting mesophyll cells not only in incompatible but also compatible interactions and suggested that the difference between compatibility and incompatibility is due to comparatively small quantitative difference in the number of mesophyll cells that respond hypersensitively. How these quantitative differences are determined and what the function of putative race-specific elicitors or suppressors is in this process is unclear.

VII. Pathogenicity of *Phytophthora infestans*

P. infestans is hemibiotrophic, which means that in nature the pathogen does not exist in an independent mycelial form. Growth and reproduction require colonization of the host. In initial stages of the interaction encysted zoospores germinate on the leaf surface and form appressoria. When an appropriate artificial surface is provided, germination and appressorium formation proceed in water in the absence of the host. Apparently, the zoospores carry sufficient food reserves for the initial stages, but once these reserves are exhausted, the host must be penetrated. Hence, the pathogen must possess the machinery required for growth and development in the host environment.

Pathogenesis in plants is often associated with the production of extracellular hydrolytic enzymes by the pathogen. These proteins may facilitate penetration of the pathogen and colonization of the host by damaging host tissue. Fungi typically exhibit absorptive nutrition (Hohl 1991a). Therefore, hydrolytic extracellular enzymes are important prerequisites for growth on complex, high-molecular-weight substrates. The first barrier for the pathogen, the plant cuticle, can be degraded by cutinase. For some pathogens cutinase seems to be important for penetration (Chap. 3, Vol. V, Part A; Chap. 3, this Vol.), but other plant pathogens can still penetrate when the cutinase gene is inactivated (Stuhl and Schäfer 1992). To our knowledge, the involvement of cutinase in the potato-*P. infestans* interaction has never been studied. Since *P. infestans* forms an appressorium, it might be able to penetrate host cells directly with physical force in a fashion similar to *M. grisea* (Chap. 3, Vol. V, Part A; Chap. 3, this Vol.) The biotrophic nature of *P. infestans* suggests that the cell wall-degrading enzyme system of the pathogen is under tight control. Cell walls may be degraded to some extent to facilitate the fungal hyphae to grow through the tissue, but the host cells should not be destroyed. During growth of *P. infestans* in vitro, several hydrolytic enzymes are secreted in the culture medium (Moreau and Seibles 1985). Endocellulases, 1,3-β-glucanases, β-glucosidases, pectin-esterases, polygalacturonases, and galactanases of *P. infestans* have all been isolated (Friend 1991), but individually none of these shows massive cell wall-degrading activity. Infection of tubers with *P. infestans* causes dry rot, suggesting that cell wall decomposition does not contribute substantially to nutrition of the pathogen (Hohl 1991a). Probably extracellular hydrolytic enzymes do not play a decisive role in pathogenesis of *P. infestans*.

Phytotoxins are a diverse group of low-molecular-weight secondary metabolites which have been shown to be pathogenicity factors of several plant pathogens. Phytotoxic metabolites have been isolated from culture fluids of *P. infestans* (Möllers et al. 1992), but their involvement in disease development still remains to be demonstrated.

A. Differential Gene Expression During Pathogenesis

It is clear that the importance of the few potential pathogenicity factors studied so far in the potato-*P. infestans* interaction has not been established convincingly; further, little is known about the molecular basis underlying pathogenicity of *P. infestans*. The specificity of *P. infestans* for solanaceous hosts cannot be explained simply by the action of general pathogenicity factors described above. This specificity implies a mechanism for recognition which involves the exchange of signals between host and pathogen. Successful colonization probably requires the expression of particular pathogenicity genes, i.e., genes of the pathogen which are essential for the establishment and maintenance of a compatible interaction. It is possible that, as in other symbiotic or pathogenic plant-microbe interactions, signal molecules originating from the host, or environmental stimuli present in the host, induce the expression of genes in the micro-organism which are necessary for the establishment of a successful interaction (Peters and Verma 1990). Hence, one approach to gain more insight into the molecular processes involved in pathogenesis of *P. infestans* is based on the characterization of genes of the pathogen whose expression is induced during a compatible interaction with the host plant. Once the gene products of these so-called in planta-induced genes are characterized, their role in pathogenicity can be studied.

A powerful technique to select genes induced in planta involves differential screening of a genomic library of the pathogen with probes derived from mRNA isolated from infected plant tissues and mRNA isolated from the in vitro-cultured pathogen. The selection is nonbiased, enabling the identification of genes whose role in pathogenicity cannot be predicted by other methods. Such a differential screening procedure does not exclusively select pathogenicity genes. Also genes induced in planta which are not directly involved in pathogenicity may be selected, e.g., fungal genes encoding products involved in adaptations to a change in nutrient environment encountered upon invasion of the host. Nevertheless, by identifying this type of differentially expressed genes, one can acquire a better understanding of the cellular processes which change during pathogenesis.

By means of differential screening of a genomic library of *P. infestans*, nine distinct genes have been isolated (Pieterse et al. 1993a; Fig. 4; Table 2). A detailed characterization of two of these genes, *ubi3R* and *calA*, showed that they encode the ubiquitous proteins ubiquitin and calmodulin, respectively. *Ubi3R* and *calA* are expressed during growth of *P. infestans* in vitro but the expression levels increase and are continuously fivefold higher during pathogenesis on potato (Pieterse et al. 1991, 1993b). Ubiquitin and calmodulin are highly conserved proteins in all eukaryotic organisms, indicating that they play important roles in basic cellular processes. This is the first time that increased expression of genes encoding ubiquitin and calmodulin has been found in association with fungal pathogenicity. Although these proteins cannot be considered as primary pathogenicity factors, they may have important functions in cellular processes which change during pathogenesis.

Ubiquitin has been shown to play a key role in selective degradation of abnormal proteins which arise during stress, in maintenance of chromatin structure, in modification of cell surface receptors, and in regulation of gene expression (Monia et al. 1990), all processes which are likely to proceed during pathogenesis. Failure to regulate these processes correctly, due to insufficient ubiquitin concentrations, may counteract the pathogenic process. Calmodulin is one of the major intracellular Ca^{2+} receptors and is responsible for mediating

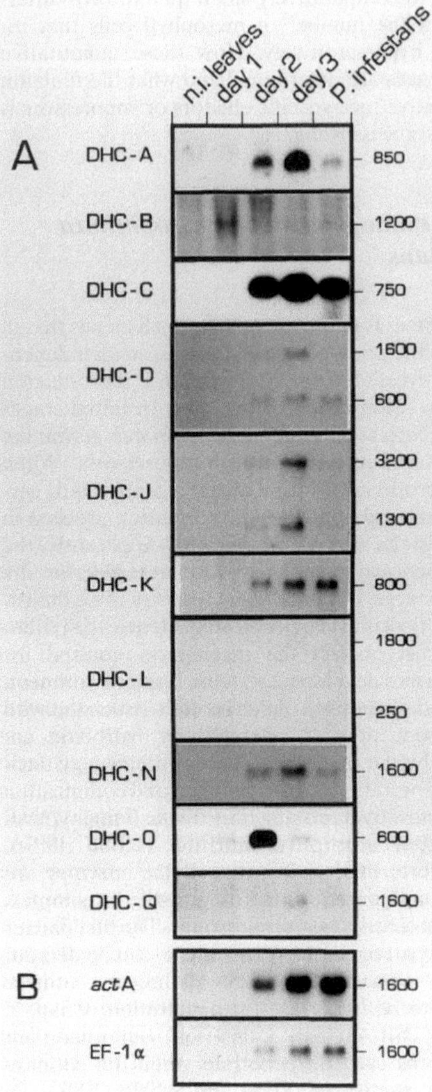

Fig. 4. Exprssion of *Phytophthora infestans* genes during colonization of potato leaves. Autoradiographs of northern blots containing total RNA isolated from non-inoculated potato leaves (*n.i. leaves*), from leaves colonized by *P. infestans* at different time points after inoculation (day 1, day 2, and day 3), and from *P. infestans* mycelium grown in vitro (*P. infestans*). The blots were hybridized with (**A**) probes derived from DNA fragments located on various differentially hybridizing clones (*DHC*) selected from a *P. infestans* genomic library by differential screening and (**B**) as a control, probes derived from constitutively expressed genes. The DNA fragments from the *DHC* clones represent *P. infestans ipi* and *pig* genes as shown in Table 2. (Pieterse et al. 1993a)

Table 2. Normalized mRNA levels of *Phytophthora infestans ipi* and *pig* genes during colonization of potato leaves. (Pieterse et al. 1993a)

Gene	Location	Transcript length (nt)	Normalized mRNA levels[a]				
			n.i.[b]	Day 1[b]	Day 2[b]	Day 3[b]	*P. infestans*[b]
ipiA (*ubi3R*)	DHC-A	850	–	–	10	10	1
ipiB	DHC-B	1200	–	>100	1	1	1
ipiC (*calA*)	DHC-C	750	–	–	5	4	1
ipiD	DHC-D	1600	–	–	10	10	1
pigD	DHC-D	600	–	–	1	1	1
ipiJ1	DHC-J	3200	–	–	10	8	1
ipiJ2	DHC-J	1300	–	–	9	9	1
pigK	DHC-K	800	–	–	1	1	1
pigL	DHC-L	1800	–	–	–	1	1
		250	#	#	–	–	–
ipiN	DHC-N	1600	–	–	5	5	1
ipiO	DHC-O	600	–	35[c]	35[c]	1[c]	–
ipiQ	DHC-Q	1600	–	–	5	9	1

–, No mRNA detectable; #, mRNA of plant origin.
[a] mRNA levels were measured by densitometric scanning of autoradiographs and normalized to actual fungal RNA levels using the signals obtained with probes derived from the constitutively expressed marker genes *actA* and EF-1α. Densitometric scanning of the *actA* and EF-1α signals shown in Fig. 4 revealed that interaction RNAs isolated at day 1, day 2, and day 3 postinoculation contained 5, 25, and 70% fungal RNA, respectively. All values obtained by densitometric scanning of the hybridization signals shown in Fig. 4 were adjusted to the equivalent of 100% fungal RNA and subsequently compared to those values obtained with RNA isolated from *P. infestans* grown in vitro. Numbers indicate the increase in actual mRNA levels in comparison to the mRNA levels in *P. infestans* grown in vitro which is set at 1.
[b] See legend Fig. 4.
[c] In *P. infestans* grown in vitro, *ipiO* mRNA is not detectable. Therefore, signals obtained in lanes containing interaction RNA are compared with each other.

cellular responses to Ca^{2+} fluxes which are generated by a variety of processes. It has been shown in several organisms that, via Ca^{2+}-calmodulin complexes, calmodulin is involved in the transduction of signals from extracellular stimuli into intracellular events (Roberts et al. 1986). Therefore, it is tempting to speculate that calmodulin is involved in the communication between plant and pathogen. Unfortunately, the involvement of both ubiquitin and calmodulin in various basic cellular processes makes it difficult to study the significance of increased *ubi3R* and *calA* gene expression during pathogenesis. Manipulating the synthesis or activity of ubiquitin or calmodulin will influence all other cellular processes which are dependent on these proteins. As a consequence, a straightforward interpretation of such studies will be impossible.

Two other genes, *ipiB* and *ipiO*, show a transient expression pattern, with the higher expression levels in early stages of infection (Fig. 4; Table 2; Pieterse et al. 1994a). They both belong to small gene clusters encoding novel proteins with yet unknown functions. *IpiB* belongs to a gene family of which three highly homologous members, *ipiB1*, *ipiB2*, and *ipiB3*, are closely linked on the genome. The primary structure of the IPI-B proteins consists of an amino terminal signal peptide for secretion followed by a highly repetitive, glycine-rich domain which comprises most of the protein (Pieterse et al. 1994b). The IPI-B proteins share these characteristics with plant glycine-rich proteins, some of which have been shown to be associated with the cell wall (Showalter 1993). In germinating cysts and during early stages of the interaction, *ipiB* mRNA accumulates to very high levels. The possible association of the *ipiB* gene products with the cell wall suggests that they are involved in the development of infection structures like appressoria or primary hyphae. In more advanced stages of the infection process and during growth of *P. infestans* in vitro, *ipiB* mRNA is hardly detectable, indicating that for mycelial growth, high levels of IPI-B proteins are not required. Immunological studies with antisera raised against the IPI-B proteins could provide more insight into this matter.

P. infestans has two almost identical *ipiO* genes which are located within a 5-kb genomic region (Pieterse et al. 1994b). Like the IPI-B proteins, the IPI-O proteins contain an amino terminal signal peptide, suggesting that the proteins are secreted. The mature proteins have a putative N-glycosylation site and contain the tripeptide sequence Arg-Gly-Asp (RGD), a conserved motif found in several mammalian extracellular matrix proteins in which it functions as a cell-attachment sequence (D'Souza et al. 1991). Accumulation of *ipiO* mRNA follows a transient pattern and coincides with colonization of the host cells. Once colonization of host tissue is completed, *ipiO* mRNA level decreases dramatically to low or undetectable levels. Also during growth of the pathogen in vitro, *ipiO* mRNA cannot be detected, indicating that IPI-O proteins are synthesized exclusively during the biotrophic phase of the pathogenic interaction. Based on the assumption that the IPI-O proteins are extracellular, we speculate that they might have a function in the extracellular matrix where components produced by the pathogen interact directly with plant cells. In view of this, the RGD sequence motif in the IPI-O proteins might play a role in the pathogenicity of *P. infestans*. We hypothesize that through the RGD sequence, IPI-O can be involved either in attachment of *P. infestans* to the plant cells or disturbance of the normal RGD-mediated signaling, which supposedly occurs in healthy plant tissues.

The RGD tripeptide motif was first identified in several mammalian adhesion proteins, such as fibronectin, vitronectin, and the Von Willebrand coagulation factor, and it was demonstrated to be a cell attachment sequence crucial for cell-substratum adhesion (D'Souza et al. 1991). The RGD tripeptide is responsible for the interaction of extracellular adhesion proteins with cell surface receptors, the so-called integrins (Hynes 1992). These specific ligand-receptor interactions mediate adhesion of cells to the extracellular matrix and play an essential role in intercellular communication and development of mammalian cells. Also in lower eukaryotes, cell adhesion molecules with a functional RGD-tripeptide are found. An example is discoidin I, which plays a role in the aggregation of the slime mold *Dictyostelium discoideum* (Gabius et al. 1985). In addition, pathogenicity factors of some bacteria pathogenic for humans contain a functional RGD-tripeptide which interacts with integrins on the surface of host cells, thereby mediating uptake into the host cells (Finlay 1990). Furthermore, the RGD-containing VP1 protein of the foot-and-mouth disease virus appears to interact with integrin-like molecules on the surface of the host cells in an RGD-dependent manner, suggesting that also viruses utilize RGD-dependent interactions for host invasion (Fox et al. 1989).

If the IPI-O proteins truly function as adhesion molecules in a RGD-dependent fashion and attach to plant cells, then receptor-ligand interactions, i.e., plasma membrane-cell wall interactions, involving RGD-containing proteins and integrin-like molecules, should occur in plants. Although the walled cells of plants differ from animal cells, there are several observations which suggest that RGD-dependent adhesion processes similar to those found in mammals occur at the surface of the plasma membrane of plant cells. First, proteins immunologically related to human vitronectin and its integrin receptor have been found in plants and green algae (Schindler et al. 1989; Wagner et al. 1992; Wang et al. 1994). Secondly, Schindler et al. (1989) showed that the RGD-containing hexapeptide Gly-Arg-Gly-Asp-Ser-Pro (G**RGD**SP), specifically disrupted cell membrane-cell wall attachment in soybean suspension cells whereas the control hexapeptides Gly-Arg-Gly-Glu-Ser-Pro (G**RGE**SP) and Gly-Asp-Gly-Arg-Ser-Pro (G**DGR**SP) did not. Thirdly, studies of Zhu et al. (1993) demonstrated that protoplasts prepared from tobacco cells which are adapted to grow in high salt concentrations adhere tightly to each other and that this typical tight adhesion is specifically blocked by the G**RGD**SP hexapeptide. They also found that vitronectin-like proteins are enriched in these high salt-adapted cells. Finally, Sanders et al. (1991) found that vitronectin-like proteins were enriched on the surfaces of cells lining the stylar duct through which pollen tubes grow to deliver the pollen nuclei to the ovule.

Using plasmolysis techniques, Nozue et al. (1979) demonstrated tight adherence of the potato plasmalemma to infecting hyphae of *P. infestans* from the earliest time of penetration. In addition, Hohl (1991b) and Gross et al. (1993) demonstrated adhesion of hyphae of *Phytophthora sojae* and *P. infestans* to protoplasts of soybean and parsley, respectively. Based on these findings it can be postulated that hyphae produce extracellular adhesion proteins which facilitate attachment to host cells. When these adhesion molecules function via RGD-dependent ligand-receptor interactions, it is also possible that they compete with RGD-containing proteins produced by the host, thereby disturbing the regular RGD-mediated ligand-receptor interactions and creating beneficial conditions for colonization of the host tissue. Adhesion molecules may also reside in the extrahaustorial matrix, a very thin layer of unknown material between the haustorium and the plasmalemma of the host cell. The extrahaustorial matrix is thought to be of fungal origin and may have a function in recognition or in nutrient uptake, either through suppressing a host resistance response, or by activating physiological changes in the host which are beneficial for the maintenance of a compatible interaction between host and pathogen (Coffey and Gees 1991). Whether the *P. infestans* IPI-O proteins indeed have cell adhesion properties and operate as hypothesized above, and whether RGD-mediated adhesion interactions in general play a role in pathogenesis, are interesting questions to address in the future.

B. Differential Gene Expression During Starvation

In *P. infestans* expression of the *ipiB* and *ipiO* genes is specifically induced in planta, whereas during growth in a rich culture medium expression is not detectable. However, during growth in vitro, the expression of both the *ipiB* and *ipiO* genes can be activated by nutrient deprivation (Pieterse et al. 1994a). Interestingly, transcriptional activation by nutrient deprivation has been reported for two other genes of plant pathogens, i.e., the avirulence gene *avr9* of *Cladosporium fulvum*, which is specifically expressed when the nitrogen source in the growth medium is limited (van den Ackerveken et al. 1994; Chap. 1, this Vol.) and the putative pathogenicity gene *MPG1* of *Magnaporthe grisea*, whose expression can be induced in vitro by glucose deprivation (Talbot et al. 1993; Chap. 3, this Vol.). It is tempting to speculate that upon infection of a host plant, nutrient deprivation or changes in nutrient conditions trigger the expression of a set of pathogenicity genes whose gene products facilitate establishment of a beneficial environment for the pathogen. If transcriptional activation of pathogenicity genes by nutrient deprivation is a general phenomenon, selecting for genes which are specifically expressed in starvation-stressed cells could be a useful approach for the isolation of putative pathogenicity genes of a wide range of plant pathogens.

VIII. Perspectives and Conclusions

To establish effective control strategies for potato late blight, increased investigation and education are needed. The population displacement in Europe in the early 1980s, and the acquired ability of *P. infestans* to reproduce sexually and to overwinter in soil, independent of living host tissue, have dramatically changed the behavior of the pathogen population. Therefore, one must realize that a continuous evaluation of the pathogen population is required. For practical disease management, the danger of the emergence of new virulence factors, fungicide-resistant isolates, and more aggressive isolates should not be underestimated.

The search for alternative methods of disease control should be intensified. This certainly requires more knowledge about pathogenicity factors, elicitors, and other compounds produced by the pathogen, and about defense responses elicited in the host. The signaling processes between plant and pathogen have to be unraveled so that vital links in the signaling pathways, which lead to the destruction of the host, are recognized. Such knowledge should be instrumental in the development of environmentally friendly control strategies which are based on interference in the signaling pathways, either through genetic modification of the host or by application of very specifically targeted inhibitory compounds.

Note added in proof

This review was written in August 1994. Publications that have appeared from that time on are not cited. Recently, an excellent review by Howard S. Judelson covering recent advances in the genetics of oomycete plant pathogens was published in the journal, Molecular Plant-Microbe Interactions (1996, volume 9, pp. 443–449). For updated literature on population genetics, we refer to the article, The Population Genetics of *Phytophthora*, by Stephen B. Goodwin in Phytopathology. (1997, volume 87, pp. 462–473)

Acknowledgments. We thank our colleagues working with *Phytophthora* for allowing us to cite their unpublished results, and Carlos Alfonso, Anke de Jong, and Pieter van West for their critical comments on the manuscript.

References

Alani E, Cao L, Kleckner N (1987) A method for gene disruption that allows repeated use of URA3 selection in construction of multiple disrupted yeast strains. Genetics 116:541–545

Alexopoulos CJ, Mims CW, Blackwell M (1996) Introductory mycology. John Wiley, New York
Bailey AM, Mena GL, Herrera-Estrella L (1991) Genetic transformation of the plant pathogens *Phytophthora capsici* and *Phytophthora parasitica*. Nucleic Acids Res 19:4273–4278
Bailey AM, Mena GL, Herrera-Estrella L (1993) Transformation of four pathogenic *Phytophthora* spp. by microprojectile bombardment on intact mycelia. Curr Genet 23:42–46
Black W, Mastenbroek C, Mills WR, Peterson LC (1953) A proposal for an international momenclature of races of *Phytophthora infestans* and of genes controlling immunity in *Solanum demissum* derivatives. Euphytica 2:173–179
Bourke PMA (1964) Emergence of potato blight, 1843–46. Nature 203:805–808
Charlesworth B (1989) The evolution of sex and recombination. Trends Ecol Evol 9:264–267
Coffey MD, Gees R (1991) The cytology of development. In: Ingram DS, Williams PH (eds) *Phytophthora infestans*, the cause of late blight of potato. Advances in plant pathology 7. Academic Press, London, pp 31–51
Crute I (1992) From breeding to cloning (and back again?): a case study with lettuce downy mildew. Annu Rev Phytopathol 30:485–506
Davidse LC, Looijen D, Turkensteen LJ, Van der Wal D (1981) Occurrence of metalaxyl-resistant strains of *Phytophthora infestans* in The Netherlands. Eur Plant Prot Organ Bull 15:403–409
de Bary A (1876) Researches into the nature of the potato fungus *Phytophthora infestans*. J R Agric Soc Engl 12:239–269
Doke N, Miura Y, Sanchez LM, Yoshioka H, Kawakita K (1993) Signal transduction and hypersensitive reaction in *Phytophthora*-Solanaceae plant pathosystem: superoxide generating reaction. In: Kohmoto K, Yoder OC (eds) Host-specific toxin: biosynthesis, receptor and molecular biology. Tottori University, Tottori, Japan, pp 153–167
Drenth A, Goodwin SB, Fry WE, Davidse LC (1993a) Genotypic diversity of *Phytophthora infestans* in The Netherlands revealed by DNA polymorphisms. Phytopathology 83:1087–1092
Drenth A, Turkensteen LJ, Govers F (1993b) The occurrence of the A2 mating type of *Phytophthora infestans* in The Netherlands: signifiance and consequences. Neth J Plant Pathol 99 (Suppl 3):57–67
Drenth A, Tas ICQ, Govers F (1994) DNA fingerprinting uncovers a new sexually reproducing population of *Phytophthora infestans* in The Netherlands. Eur J Plant Pathol 100:97–107
Drenth A, Janssen EM, Govers F (1995) Formation and survival of oospores of *Phytophthora infestans* under natural conditions. Plant Pathol 44:86–94
D'Souza SE, Ginsberg MH, Plow EF (1991) Arginyl-glycyl-aspartic acid (RGD): a cell adhesion motif. TIBS 16:246–250
Ebel J, Cosio EG (1994) Elicitors of plant defense responses. Int Rev Cytol 148:1–36
Finlay BB (1990) Cell adhesion and invasion mechanisms in microbial pathogenesis. Curr Opin Cell Biol 2:815–820
Fox G, Parry NR, Varnett PV, McGinn B, Rowlands DJ, Brown F (1989) The cell attachment site on foot-and-mouth disease virus includes the amino acid sequence RGD (arginine-glycine-aspartic acid). J Gen Virol 70: 625–637
Freytag S, Arabatzis N, Hahlbrock K, Schmelzer E (1994) Reversible cytoplasmic rearrangements precede wall apposition, hypersensitive cell death and defense-related gene activation in potato/*Phytophthora infestans* interactions. Planta 194:123–135
Friend J (1991) The biochemistry and cell biology of interaction. In: Ingram DS, Williams PH (eds) *Phytophthora infestans*, the cause of late blight of potato. Advances in plant pathology 7. Academic Press, London, pp 85–129
Fry WE, Goodwin SB, Dyer AT, Matuszak JM, Drenth A, Tooley PW, Sujkowski LS, Koh YJ, Cohen BA, Spielman LJ, Deahl KL, Inglis DA, Sandlan KP (1993) Historical and recent migrations of *Phytophthora infestans*: chronology, pathways and implications. Plant Dis 77: 653–661
Gabius H-J, Springer WR, Barondes SH (1985) Receptor for the cell binding site of Discoidin I. Cell 42:449–456
Goodwin SB, Drenth A, Fry WE (1992a) Cloning and genetic analyses of two highly polymorphic, moderately repetitive nuclear DNAs from *Phytophthora infestans*. Curr Genet 22:107–115
Goodwin SB, Spielman LJ, Matuszak JM, Bergeron SN, Fry WE (1992b) Clonal diversity and genetic differentiation among *Phytophthora infestans* populations in northern Mexico. Phytopathology 82:955–961
Goodwin SB, Cohen BA, Deahl KL, Fry WE (1994) Migration from northern Mexico as the probable cause of recent genetic changes in populations of *Phytophthora infestans* in the United States and Canada. Phytopathology 84:553–558
Gross P, Julius C, Schmelzer E, Hahlbrock K (1993) Translocation of cytoplasm and nucleus to fungal penetration sites is associated with depolymerization of microtubules and defence gene activation in infected, cultured parsley cells. EMBO J 12:1735–1744
Hawksworth DL, Kirk PM, Sutton BC, Pegler DN (eds) (1995) Ainsworth and Bisby's Dictionary of the Fungi. International Mycological Institute. CAB International, Oxon UK
Hohl HR (1991a) Nutrition. In: Ingram DS, Williams PH (eds) *Phytophthora infestans*, the cause of the late blight of potato. Advances in plant pathology 7. Academic Press, London, pp 53–83
Hohl HR (1991b) Host-pathogen interactions in *Phytophthora*. In: Lucas JA, Shattock RC, Shaw DS, Cooke LR (eds) *Phytophthora*. Cambridge University Press, Cambridge, pp 295–311
Hohl HR, Iselin K (1984) Strains of *Phytophthora infestans* with A2 mating type behaviour. Trans Br Mycol Soc 83:529–530
Holub E, Crute I, Brose E, Beynon J (1993) Identification and mapping of loci in *Arabidopsis* for resistance to downy mildew and white blister. In: Davis KR, Hammerschmidt R (eds) *Arabidopsis thaliana* as a model for plant-pathogen interactions. APS Press, St Paul, pp 21–35
Holub EB, Beynon JL, Crute IR (1994) Phenotypic and genotypic characterization of interactions between isolates of *Peronospora parasitica* and accessions of *Arabidopsis thaliana*. Mol Plant-Microbe Interact 7:223–239
Hynes RO (1992) Integrins: versatility, modulation, and signalling in cell adhesion. Cell 69:11–25
Ingram DS, Williams PH (eds) (1991) Advances in plant

pathology 7. *Phytophthora infestans*, the cause of late blight of potato. Academic Press, London
Judelson HS, Michelmore RW (1991) Transient expression of genes in the oomycete *Phytophthora infestans* using *Bremia lactucae* regulatory sequences. Curr Genet 19:453–459
Judelson HS, Tyler GM, Michelmore RW (1991) Transformation of the oomycete pathogen *Phytophthora infestans*. Mol Plant-Microbe Interact 4:602–607
Judelson HS, Tyler BM, Michelmore RW (1992) Regulatory sequences for expressing genes in oomycete fungi. Mol Gen Genet 234:138–146
Judelson HS, Dudler R, Pieterse CMJ, Unkles SE, Michelmore RW (1993) Expression and antisense inhibition of transgenes in *Phytophthora infestans* is modulated by choice of promotor and position effects. Gene 133:63–69
Kamoun S, Klucher KM, Coffey MD, Tyler Bm (1993a) A gene encoding a host-specific elicitor protein of *Phytophthora parasitica*. Mol Plant-Microbe Interact 6:573–581
Kamoun S, Young M, Glascock CB, Tyler BM (1993b) Extracellular protein elicitors from *Phytophthora*: host-specificity and induction of resistance to bacterial and fungal phytopathogens. Mol Plant-Microbe Interact 6:15–25
Karlovsky P, Prell HH (1991) The TRP1 gene of *Phytophthora parasitica* encoding indole-3-glycerolphosphate synthase-N-(5′-phosphoribosyl) anthranilate isomerase: structure and evolutionary distance from homologous fungal genes. Gene 109:161–165
Kinghorn JR, Moon RP, Unkles SE, Duncan JM (1991) Gene structure and expression in *Phytophthora infestans* and the development of gene-mediated transformation. In: Lucas JA, Shattock RC, Shaw DS, Cooke LR (eds) Phytophthora. Cambridge University Press, Cambridge, pp 295–311
Marmeisse R, van den Ackerveken GFJM, Goosen T, de Wit PJGM, van den Broek HWJ (1993) Disruption of the avirulence gene avr9 in two races of the tomato pathogen *Cladosporium fulvum* causes virulence on tomato genotypes with the complementary resistance gene Cf9. Mol Plant-Microbe Interact 6:412–417
Mills WR, Niederhauser JS (1953) Observations on races of *Phytophthora infestans* in Mexico. Phytopathology 43:454–455
Möllers C, Zitzlsperger J, Wenzel G (1992) The effects of a toxin preparation from *Phytophthora infestans* on potato protoplasts and microspores. Physiol Mol Plant Pathol 41:427–435
Monia BP, Ecker DJ, Crooke ST (1990) New perspectives on the structure and function of ubiquitin. Bio/Technology 8:209–215
Moreau RA, Seibles TS (1985) Production of extracellular enzymes by germinating cysts of *Phytophthora infestans*. Can J Bot 63:1811–1816
Newhouse JR, Tooley PW, Smith OP, Fishel RA (1992) Characterization of double-stranded RNA in isolates of *Phytophthora infestans* from Mexico, The Netherlands, and Peru. Phytopathology 82:164–169
Niederhauser JS, Mills WR (1953) Resistance of *Solanum* species to *Phytophthora infestans* in Mexico. Phytopathology 43:456–457
Nozue M, Tomiyama K, Doke N (1979) Evidence for adherence of host plasmalemma to infecting hyphae of both compatible and incompatible races of *Phytophthora infestans*. Physiol Plant Pathol 15:111–115
Nürnberger T, Nennstiel D, Jabs T, Sacks WR, Hahlbrock K, Scheel D (1994) High affinity binding of a fungal oligopeptide elicitor to parsley plasma membranes triggers multiple defense responses. Cell 78:449–460
Patterson DJ, Sogin ML (1992) Eukaryote origins and protistan diversity. In: Hartman H, Matsuno K (eds) The origin and evolution of prokaryotic and eukaryotic cells. World Scientific, Singapore, pp 13–46
Pernollet J-C (1993) Structural and biological properties of elicitins, toxic proteins secreted by *Phytophthora* fungi, which induce the plant defence responses. Proc Symp on Plant growth and environment, Suvon, Korea, 16 Oct 1993, pp 45–71
Peters NK, Verma DPS (1990) Phenolic compounds as regulators of gene expression in plant-microbe interactions. Mol Plant-Microbe Interact 3:4–8
Pieterse CMJ, Risseeuw EP, Davidse LC (1991) An in planta-induced gene of *Phytophthora infestans* codes for ubiquitin. Plant Mol Biol 17:799–811
Pieterse CMJ, Riach MBR, Bleker T, van den Berg-Velthuis GCM, Govers F (1993a) Isolation of putative pathogenicity genes of the potato late blight fungus *Phytophthora infestans* by differential hybridization of a genomic library. Physiol Mol Plant Pathol 43:69–79
Pieterse CMJ, Verbakel HM, Hoek Spaans J, Davidse LC, Govers F (1993b) Increased expression of the calmodulin gene of the late blight fungus *Phytophthora infestans* during pathogenesis on potato. Mol Plant-Microbe Interact 6:164–172
Pieterse CMJ, Derksen A-MCE, Folders J, Govers F (1994a) Expression of the *Phytophthora infestans* ipiB and ipiO genes in planta and in vitro. Mol Gen Genet 244:269–277
Pieterse CMJ, van West P, Verbakel HM, Brassé PWHM, van de Berg-Velthuis, GCM, Govers F (1994b) Structure and genomic organization of the ipiB and ipiO gene clusters of *Phytophthora infestans*. Gene 138:67–77
Pieterse CMJ, van 't Klooster J, van den Berg-Velthuis GCM, Govers F (1995) NiaA, the structural nitrate reductase gene of *Phytophthora infestans*: isolation, characterization and expression analysis in *Aspergillus nidulans*. Curr Genet 27:359–366
Ricci P, Trentin T, Bonnet P, Venard P (1992) Differential production of parasiticein, an elicitor of *Phytophthora parasitica*. Plant Pathol 41:298–307
Ricker KE, Bostock RM (1992) Evidence for release of the arachidonic aicd and its metabolites from sporangia of *Phytophthora infestans* during infection of potato. Physiol Mol Plant Pathol 41:61–72
Roberts DM, Lukas TJ, Watterson DM (1986) Structure, function, and mechanism of action of calmodulin. CRC Crit Rev Plant Sci 4:311–339
Sacks W, Nürnberger T, Hahlbrock K, Scheel D (1995) Molecular characterization of nucleotide sequences encoding the extracellular glycoprotein elicitor from *Phytophthora megasperma*. Mol Gen Genet 246:45–55
Sanders LC, Wang C-S, Walling LL, Lord EM (1991) A homolog of the substrate adhesion molecule vitronectin occurs in four species of flowering plants. Plant Cell 3:629–635
Schindler M, Meiners S, Cheresh DA (1989) RGD-dependent linkage between plant cell wall and plasma membrane: consequences for growth. J Cell Biol 108:1955–1965
Schlager G, Dickie MM (1971) Natural mutation rates in

the house mouse. Estimates for five specific loci and dominant mutations. Mutat Res 11:89–96

Schöber B, Turkensteen LJ (1992) Recent and future developments in potato fungal pathology. Neth J Plant Pathol 98 (Suppl 2):73–83

Shaw DS (1983) The Peronosporales. A fungal geneticist's nightmare. In: Buczacki ST (eds) Oosporic plant pathogens a modern perspective. Academic Press, London, pp 85–121

Showalter AW (1993) Structure and function of plant cell wall proteins. Plant Cell 5:9–23

Spielman LJ, McMaster BJ, Fry WE (1989) Dominance and recessiveness at loci for virulence against potato and tomato in *Phytophthora infestans*. Theor Appl Genet 77:832–838

Spielman LJ, Sweigard JA, Shattock RC, Fry WE (1990) The genetics of *Phytophthora infestans*: segregation of allozyme markers in F2 and backcross progeny and the inheritance of virulence against potato resistance R2 and R4 in F1 progeny. Exp Mycol 14:57–69

Spielman LJ, Drenth A, Davidse LC, Sujkowski LS, Gu W, Tooley PW, Fry WE (1991) A second world-wide migration and population displacement of *Phytophthora infestans*? Plant Pathol 40:422–430

Stahl DJ, Schäfer W (1992) Cutinase is not required for fungal pathogenicity on pea. Plant Cell 4:621–629

Sujkowski LS, Goodwin SB, Dyer AT, Fry WE (1994) Increased genotypic diversity via migration and possible occurrence of sexual reproduction of *Phytophthora infestans* in Poland. Phytopathology 84:201–207

Sweigard JA, Chumley FG, Valent B (1992) Disruption of a *Magnaporthe grisea* cutinase gene. Mol Gen Genet 232:183–190

Talbot NJ, Ebbole DJ, Hamer JE (1993) Identification and characterization of MPG1, a gene involved in pathogenicity from the rice blast fungus *Magnaporthe grisea*. Plant Cell 5:1575–1590

Thompson JN, Burdon JJ (1992) Gene-for-gene coevolution between plants and parasites. Nature 360:121–125

Tomiyama K (1983) Research on the hypersensitive response. Annu Rev Phytopathol 21:1–12

Tooley PW, Fry WE, Villarreal Gonzalez MJ (1985) Isozyme characterization of sexual and asexual *Phytophthora infestans* populations. J Hered 76:431–435

Van den Ackerveken GFJM, Dunn RM, Cozijnsen AJ, Vossen JPHJ, van den Broek HWJ, de Wit PJGM (1994) Nitrogen limitation induces expression of the avirulence gene *avr9* in *Cladosporium fulvum*. Mol Gen Genet 243:277–285

Wagner VT, Brian L, Quatrano RS (1992) Role of a vitronectin-like molecule in embryo adhesion of the brown alga *Fucus*. Proc Natl Acad Sci USA 89:3644–3648

Wang C-S, Walling LL, Gu YQ, Ware CF, Lord EM (1994) Two classes of proteins and mRNAs in *Lilium longiflorum* L. identified by human vitroectin probes. Plant Physiol 104:711–717

Whisson SC, Drenth A, Maclean DJ, Irwin JAG (1994) Evidence for outcrossing in *Phytophthora sojae* and linkage of a DNA marker to two avirulence genes. Curr Genet 27:77–82

Woodham-Smith C (1962) The Great Hunger. Harper & Row, New York

Zhu J-K, SHi J, Singh U, Wyatt SE, Bressan RA, Hasegawa PM, Carpita NC (1993) Enrichment of vitronectin- and fibronectin-like proteins in NaCl-adapted plant cells and evidence for their involvement in plasma membrane-cell wall adhesion. Plant J 3:637–646

3 The Rice Blast Fungus, *Magnaporthe grisea*

B. VALENT

CONTENTS

I. Introduction ... 37
A. Historical Perspective on Rice Blast Disease ... 37
B. New Tools for Modern Challenges ... 38
II. *M. grisea* Populations ... 41
A. Host Species Specificity Within *M. grisea* ... 41
B. Clonal Lineages in the Rice Blast Pathogen Population ... 42
III. Molecular Mechanisms of Pathogenesis ... 43
A. Penetration ... 44
B. Proliferation ... 46
C. Sporulation ... 46
IV. Genetic Control of Host Specificity ... 47
A. The *PWL* Host Species Specificity Gene Family ... 47
B. The *AVR2-YAMO* Avirulence Gene Family ... 47
V. New Insights into Genetic Instability in *M. grisea* ... 48
A. Instability Associated with Chromosomal Location ... 49
B. Transposable Elements ... 50
VI. Conclusions ... 51
References ... 51

I. Introduction

A. Historical Perspective on Rice Blast Disease

Magnaporthe grisea (Hebert) Barr (Barr 1977; Yaegashi and Udagawa 1978) is a filamentous, heterothallic Ascomycotina that collectively causes disease on many species of the grass (Poaceae) family. *M. grisea* is the teleomorph corresponding to the previously distinct anamorphs *Pyricularia oryzae*, infecting rice (*Oryza sativa*), and *P. grisea*, infecting other grasses. However, *P. oryzae* and *P. grisea* have now been synonymized, with the earlier name, *P. grisea*, having priority (Rossman et al. 1990), because strains with diverse host specificities are interfertile and morphologically indistinguishable. Although the collective host range of *M. grisea* is broad, individual strains infect a limited number of grass species (Ou 1985; Borromeo et al. 1993; Urashima et al. 1993). A comprehensive description of the economically important rice blast disease has recently been published (Zeigler et al. 1994).

The fungus infects all aboveground parts of the plant, producing multiple rounds of conidiating lesions (Fig. 1). The hyaline to pale-colored, pyriform conidia have three cells with a single nucleus per cell (Fig. 2a). *M. grisea* is a Pyrenomycete, producing fusiform, curved ascospores in unordered asci within perithecia (Fig. 2b). Fully fertile strains are self-sterile hermaphrodites, with compatibility for mating governed by alternate alleles of the mating type locus, *Mat1* (Kang et al. 1994). Crescent-shaped microconidia (average size 6 μm long and 0.7 μm wide) produced from phialides have recently been reported (Kato et al. 1994).

Rice pathogens are divided into numerous races, or pathotypes, defined according to virulence spectrums on differential rice cultivars (Zeigler et al. 1994). Sasaki first distinguished differences in rice cultivar specificity between field isolates in Japan in 1922, and he reported the first genetic analysis of resistance in rice (Ou 1985). Since that time, major effort has been directed toward identifying dominant resistance genes (Kiyosawa 1976), and into developing blast-resistant rice cultivars (Latterell 1975; Yamada et al. 1976; Ou 1985). However, resistance has been unstable, quickly failing or breaking down under field conditions as new races of the pathogen appear with the ability to attack the previously resistant rice cultivars.

Reports on the degree of variability associated with races of the rice pathogen have been the subject of much debate. Kiyosawa (1976) reported genetic instabilities in fungal determinants of cultivar specificities corresponding to specific blast resistance genes. Ou and his colleagues (Ou 1980),

Central Research and Development, The DuPont Company, P.O. Box 80402, Experimental Station, Wilmington, Delaware 19880-0402, USA

The Mycota V Part B
Plant Relationships
Carroll/Tudzynski (Eds.)
© Springer-Verlag Berlin Heidelberg 1997

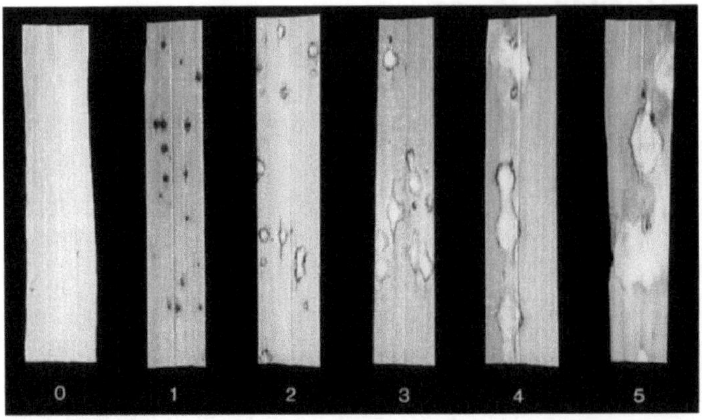

Fig. 1, 0–5. Typical symptoms of rice blast disease. Leaf segments of cultivar CO39 showing a range of possible lesion types, ranging from no visible evidence of infection (type **0**) to the maximum-sized, large eyespot lesions (type **5**, approximately 5 mm in length). Types **0** and **1** are considered nonpathogenic interactions because affected tissue does not produce conidia, and therefore, the fungus does not complete the infection cycle. Types **2** through **5** are pathogenic, but differ in aggressiveness, or extent of colonization within the tissue

working in the Philippines, reported such extreme instability that vegetative cultures derived from a single conidium produced conidia of so many different races that it was impossible to define a race of the fungus. Latterell (Latterell 1975; Latterell and Rossi 1986) presented evidence that races were stable, although mutants with altered rice cultivar specificity were detected in rare cases. Bonman et al. (1987), again working in the Philippines, failed to observe the continuous variation reported by Ou. Genetic studies have now identified both stable and unstable avirulence (AVR) genes, the pathogen genes that control cultivar specificity, although the degree of instability described in these studies does not appear as extreme as reported by Ou (Valent and Chumley 1994).

B. New Tools for Modern Challenges

Hebert (1971) discovered the sexual state corresponding to *P. grisea* with strains that infect crabgrass (*Digitaria sanguinalis*). The degree of sexual fertility exhibited by fungal isolates directly from the field ranges from highly fertile hermaphrodites, to female-sterile strains that mate only with hermaphrodites, to totally infertile strains. Host range is generally predictive of fertility. For example, most rice pathogens isolated from the field are poorly fertile. Early genetic studies were conducted using strains that parasitize finger millet (*Eleusine coracana*) or weeping love-grass (*Eragrostis curvula*) because field isolates infecting these grasses are often hermaphrodites that cross to produce viable ascospores (Valent and Chumley 1991; Notteghem et al. 1994). Breeding for improved fertility, as well as identification of rare hermaphroditic field isolates that infect rice, have made genetic analysis possible with rice pathogens (Zeigler et al. 1994).

M. grisea is haploid with a genome size of 37 to 40 Megabases (Mbp) (Hamer et al. 1989a; Skinner et al. 1993). The mitochondrial genome ranges in size from 32–39 kb (Borromeo et al. 1993). Electrophoretic separation of chromosome-size DNAs by contour-clamped homogeneous electric field (CHEF) gel electrophoresis showed that all *M. grisea* strains have chromosomes ranging in size from 3 to 12 Mbp (Fig. 3; Hamer et al. 1989a; Orbach et al. 1996). Infertile field isolates, including most rice pathogens, have a relatively high number of chromosome length polymorphisms (Talbot et al. 1993a; Orbach et al. 1996). In contrast, some *M. grisea* strains from diverse hosts around the world are interfertile and have a relatively uniform karyotype, although translocations are present even among these fertile strains (Skinner et al. 1993; Orbach et al. 1996). Many strains, including most rice pathogens from the field, contain one to four small, apparently nonessential, chromosomes ranging in size from 0.5 to 3 Mbp (Fig. 3; Hamer et al. 1989a; Talbot et al. 1993a). These minichromosomes resemble the B chromo-

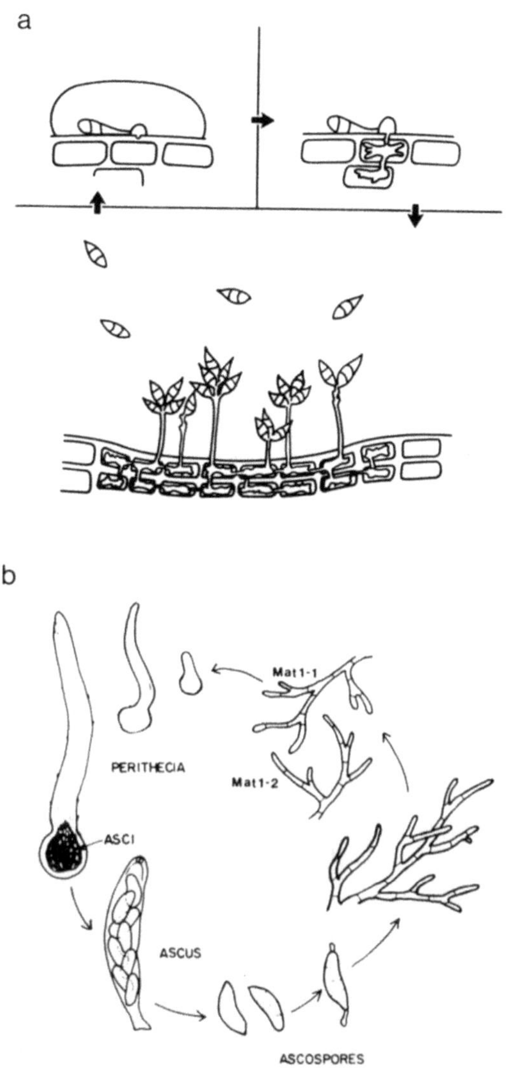

Fig. 2. **a** Asexual cycle of the pathogen. Conidia (25–30 × 9–12 μm) infect host tissue in the presence of water from dew or rain, using an appressorium to penetrate directly through the outer surface of the plant. The fungus grows intracellularly, and produces sporulating lesions within 5 to 7 days. **b** Strains of opposite mating type mate to form pigmented perithecia (500–1200 μm in length) with spherical bases (80–260 μm in diameter) and long cylindrical necks. Hyaline, crescent-shaped ascospores (16–25 × 4–8 μm) have four cells with a single nucleus per cell

somes of higher eukaryotes. There appears to be a correlation between the presence of minichromosomes and low levels of sexual fertility, although it is not yet clear if minichromosomes are a result or a cause of the low fertility (Orbach et al. 1996; see also Chap. 13, this Vol.).

Several laboratories have undertaken construction of restriction fragment length polymorphism (RFLP) maps for cloning of pathogen genes accessible only by map position (Fig. 4; Romao and Hamer 1992; Hayashi and Naito 1993; Skinner et al. 1993; Sweigard et al. 1993; Notteghem et al. 1994; Farman and Leong 1995). Although previous cytological studies indicated that *M. grisea* has six chromosomes (Leung and Williams 1987), mapping and karyotype studies have identified seven linkage groups. Current efforts toward integration of the maps from various laboratories will produce a valuable resource for blast researchers (N. Nitta and S.A. Leong, pers. comm.). Several genes controlling host specificity have been mapped (Fig. 4; Hayashi and Naito 1993; Sweigard et al. 1993; Notteghem et al. 1994; Smith and Leong 1994), and two host specificity genes have been cloned based on map positions (Sweigard et al. 1995; also see below).

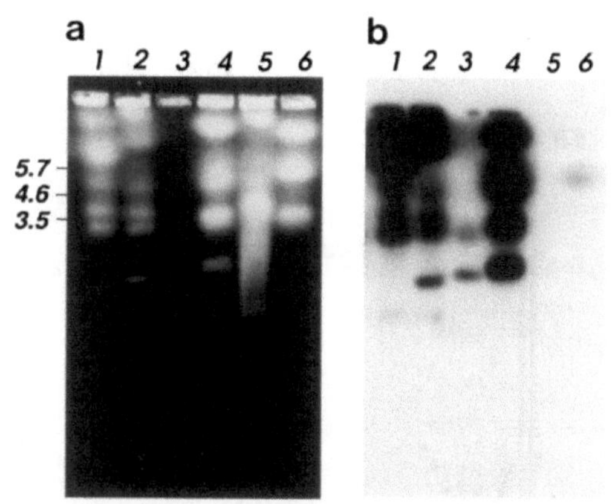

Fig. 3a,b. Electrophoretic separation of chromosome-size DNAs and hybridization analysis with middle repetitive DNA sequences. **a** Ethidium bromide-stained CHEF gel of chromosome-size DNAs from four rice pathogens (*lanes 1–4*; *lane 3* poor preparation), a *Neurospora crassa* strain (*lane 5*), and a weeping lovegrass pathogen that does not infect rice (*lane 6*). Positions of migration of *Schizosaccharomyces pombe* chromosome-size DNAs (sized in Mbp) are indicated. **b** Autoradiograph of these chromosomal DNAs after blotting and hybridizing with the M. *grisea* repeat (MGR) subclone, pCB586 (Hamer et al. 1989a). MGR586 hybridizes to all rice pathogen chromosomes and to only one chromosome of the weeping lovegrass pathogen. Chromosomal DNAs of the rice pathogens in *lanes 2–4* include minichromosome-size DNAs, which hybridize intensely to MGR586. (Hamer et al. 1989a)

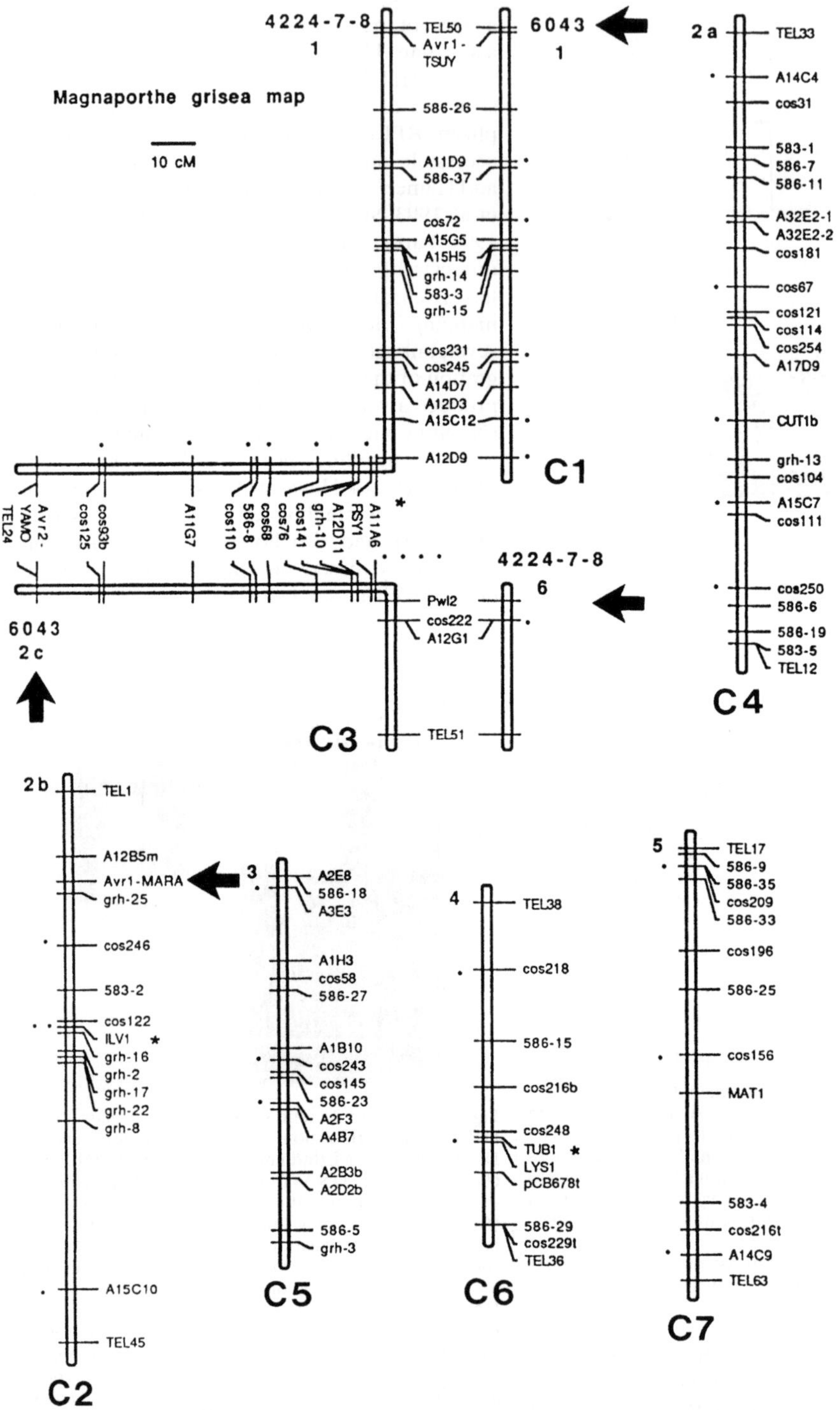
Magnaporthe grisea map
10 cM
4224-7-8
1
6043
1
TEL50
Avr1-
TSUY
586-26
A11D9
586-37
cos72
A15G5
A15H5
grh-14
583-3
grh-15
cos231
cos245
A14D7
A12D3
A15C12
A12D9
C1
2a
TEL33
A14C4
cos31
583-1
586-7
586-11
A32E2-1
A32E2-2
cos181
cos67
cos121
cos114
cos254
A17D9
CUT1b
grh-13
cos104
A15C7
cos111
cos250
586-6
586-19
583-5
TEL12
C4
Avr2-
YAMO
TEL24
cos93b
cos125
A11G7
cos110
586-8
cos68
cos76
cos141
grh-10
A12D11
RSY1
A11A6
4224-7-8
6
Pwl2
cos222
A12G1
TEL51
6043
2c
C3
2b
TEL1
A12B5m
Avr1-MARA
grh-25
cos246
583-2
cos122
ILV1
grh-16
grh-2
grh-17
grh-22
grh-8
A15C10
TEL45
C2
3
A2E8
586-18
A3E3
A1H3
cos58
586-27
A1B10
cos243
cos145
586-23
A2F3
A4B7
A2B3b
A2D2b
586-5
grh-3
C5
4
TEL38
cos218
586-15
cos216b
cos248
TUB1
LYS1
pCB678t
586-29
cos229t
TEL36
C6
5
TEL17
586-9
586-35
cos209
586-33
cos196
586-25
cos156
MAT1
583-4
cos216t
A14C9
TEL63
C7

◄

Fig. 4. Genetic map of the rice blast fungus. Segregation of over 200 markers (not all shown), including cloned genes, random cosmid clones, repetitive DNA elements (labeled *586* or *583*, representing homology to pCB586 or pCB583), a telomeric repeat sequence, and host specificity genes (*arrows*), was scored for 66 progeny of a cross between strains 4224-7-8 and 6043 (Sweigard et al. 1993). Linkage group assignment of markers denoted by (·) was verified by hybridization to blots of chromosome-size DNAs separated on CHEF gels. The parental strains differed by a reciprocal translocation involving chromosomes 1 and 6 in 4224-7-8 and 1 and 2c in 6043. The merged map will have the genome arrangement of strain 6043, and chromosomes will be named as designated by C# in the figure (N. Nitta and S. Leong, pers. comm.)

A versatile transformation system has been developed for *M. grisea* (Parsons et al. 1987; Daboussi et al. 1989; Leung et al. 1990; Kimura et al. 1994). Fungal genes with potential or proven roles in pathogenicity have been cloned (Sweigard et al. 1992a; Talbot et al. 1993b; Keon et al. 1994; Lundqvist et al. 1994; Vidal-Cros et al. 1994), and production of null mutations through gene disruption techniques is now well documented (Sweigard et al. 1992b; Talbot et al. 1993b).

II. *M. grisea* Populations

A. Host Species Specificity Within *M. grisea*

Much effort has been directed toward determining host range for individual strains of the pathogen (Ou 1985; Urashima et al. 1993). Difficulties in performing environmentally sensitive pathogenicity assays, as well as genetic differences between strains tested, probably explain the sometimes contradictory results on host specificity. Molecular analyses are now defining host species-specific subpopulations of *M. grisea* (Table 1; reviewed by Shull and Hamer 1994). Analysis of ribosomal (r) DNA polymorphisms (RFLP and ITS sequences) identified at least six major pathogen haplotypes, each with a restricted set of host species specificities (Biju-Duval 1994). The pathogens of rice, *Eleusine* spp. and wheat, corresponding to rDNA groups 1, 2, and 3, appear closely related, with ITS nucleotide divergence between 0.3 and 1%. Genotypes from rDNA groups 4, 5, and 6 are highly divergent from the previous groups and from each other, with divergences in their ITS sequences in the 6 to 8% range. Analyses of polymorphisms in mitochondrial (mt) DNA sequences independently identified host-specific haplotypes (A.R. Kubelik, J.A. Sweigard, F.G. Chumley, and B. Valent, unpubl. data). Rice pathogens, wheat pathogens, and *Eleusine* pathogens have closely related mtDNA types, Ia, Ib, Ic, and Ie, with minor differences in the sizes of one or two restriction fragments (Table 1). However, the mitochondrial

Table 1. *Magnaporthe grisea* RFLP haplotypes. (After Shull and Hamer 1994)

Composite haplotype	RFLP haplotype methods			Plant host	Geographic location
	rDNA	mtDNA	Single-copy DNA		
a	1	Ia	A, B, C, D, E	*Oryza sativa*	Africa, Asia, South America
b	2	Ib	F	*Triticum aestivum*	South America
c	2	Ie	G, H	*Eleusine* spp.	Africa, Asia
d	3	Ic	I, F	*Eleusine* spp.	Africa, Asia, South America
e	4	IIa	J	*Digitaria* spp.	Africa, North and South America
f	4	IIb	J	*Digitaria* spp.	Asia
g	5	IV	K	*Pennisetum* spp.	Africa, Asia
h	6	III	Weak hybridization	*Pennisetum* spp.	Africa, Asia, North and South America

For each haplotype, at least seven isolates were examined by each of the three RFLP methods. The rDNA haplotypes were generated using four restriction enzymes: *Eco*R I, *Ssp* I, *Sac* II, and *Pst* I. A total of 13–17 fragments were analyzed depending on the isolate (M.H. Lebrun, unpubl. data). The mtDNA haplotypes were generated using seven restriction enzymes, each having a 4-bp recognition sequence, and approximately 50 fragments were analyzed depending on the isolate (A.R. Kubelik, J.A. Sweigard, F.G. Chumley, and B. Valent, unpubl. data). The single-copy DNA haplotypes were generated using four restriction enzymes and three single-copy gene probes, *ILV1*, *CUT1* and *NIA1* (Biju-Duval 1994). Fragments with homology to the *CUT1* gene were highly polymorphic compared to the other two genes, and accounted for the variability within haplotype populations. Note that strains within these composite haplotypes infect additional host species not listed here.

genomes of strains with mtDNA haplotypes I, II, III, and IV lack common restriction fragments and contain sequences with low levels of homology, indicating that these strains have significantly different mitochondrial genomes. Polymorphisms in single-copy sequences seem less useful in defining haplotypes than the rDNA and mtDNA sequences. Interestingly, one subset of *Pennisetum* pathogens shows weak hybridization to the single-copy DNA sequences. At the current level of analysis, rice pathogens belong to a single population, *Eleusine* and *Digitaria* pathogens are each grouped into two closely related populations, and *Pennisetum* pathogens belong to two distantly related populations.

Complementary tools for studying pathogen populations were applied to understanding the origin of wheat blast in Brazil. Blast disease caused by *M. grisea* was first seen on wheat in seven northern counties of the state of Paraná in 1985 (Urashima et al. 1993), and the disease now occurs in all Brazilian wheat-growing areas. Because wheat blast occurred in an area where rice blast was prevalent, it seemed possible that a rice pathogen had mutated to parasitize wheat. However, Urashima et al. (1993) concluded from studies of host specificity and sexual fertility that wheat pathogens were more similar to *Eleusine* pathogens than to pathogens of rice or some other grasses. Fingerprinting studies with MGR586, a repetitive DNA sequence that is amplified in the genomes of rice pathogens relative to other host-specific forms (Fig. 3), demonstrated that wheat pathogens were clearly derived from a population distinct from rice pathogens (Valent and Chumley 1994). Wheat pathogens were included in an *Eleusine* haplotype by studies of rDNA RFLPs, although they fell into a distinct haplotype by studies with mtDNA and single-copy probes (Table 1). Thus, close relatives of strains infecting *Eleusine* spp. appeared to be responsible for wheat blast in Brazil.

B. Clonal Lineages in the Rice Blast Pathogen Population

Low levels of sexual fertility in rice pathogens from the field and geographic isolation of strains of opposite mating type (Kato and Yamaguchi 1982; Notteghem and Silué 1992; Yaegashi and Yamada 1986) suggested that field populations infecting rice were predominantly asexual. Levy et al. (1991) clearly demonstrated the asexual nature of a rice pathogen population by molecular analyses with the middle repetitive DNA sequence subcloned as pCB586 (Hamer et al. 1989a). In rice pathogens, MGR586 fingerprints have 50 or more resolvable *Eco*R I fragments ranging from 0.7 to 20 kb in length (Fig. 5), and these fragments are dispersed among all chromosomes (Figs. 3 and 4; Romao and Hamer 1992). Levy et al. (1991) demonstrated that the pathogen population in the southern United States was clustered into eight distinct groups (lineages) of individuals, with each lineage apparently derived through clonal propagation from a common ancestor. Such lineages, with fingerprints as illustrated in Fig. 5, have now been identified worldwide (Levy et al. 1993; Xia et al. 1993; Zeigler et al. 1994, 1995). This simplified view of the rice pathogen population as sets of discrete lineages is more informative than the previous view of pathotype structure, hundreds of races identified by virulence tests on differential rice cultivars. For example, in Eastern Colombia, seven of the nine lineages identified in a new rice-growing region on the Eastern Plains were unique, and were not represented among the six lineages collected at the Santa Rosa blast resistance breeding nursery approximately 200 km to the west (Levy et al. 1993; Correa-Victoria et al. 1994; Zeigler et al. 1994). The origin of these new clonal populations in an area not previously under rice cultivation is unknown.

Lineage structure within a population would be useful to plant breeders if lineage information predicted the race structure of fungal isolates, and the potential for variation in these isolates. In the original study with fungi from the southern US (Levy et al. 1991), six lineages corresponded to single pathotypes (all isolates within the lineage infected the same spectrum of differential cultivars), and two corresponded to two pathotypes each. In two cases, distinct lineages (with 40% or less fingerprint similarity) contained strains of the same pathotype, suggesting convergent evolution. In subsequent studies (Levy et al. 1993; Xia et al. 1993; Zeigler et al. 1994), the relationship between lineage and pathotype became more complex, although lineages were consistently composed of isolates that infected a limited subset of rice cultivars. Worldwide, some lineages include single or small numbers of pathotypes, and others include multiple pathotypes. Within lineages containing

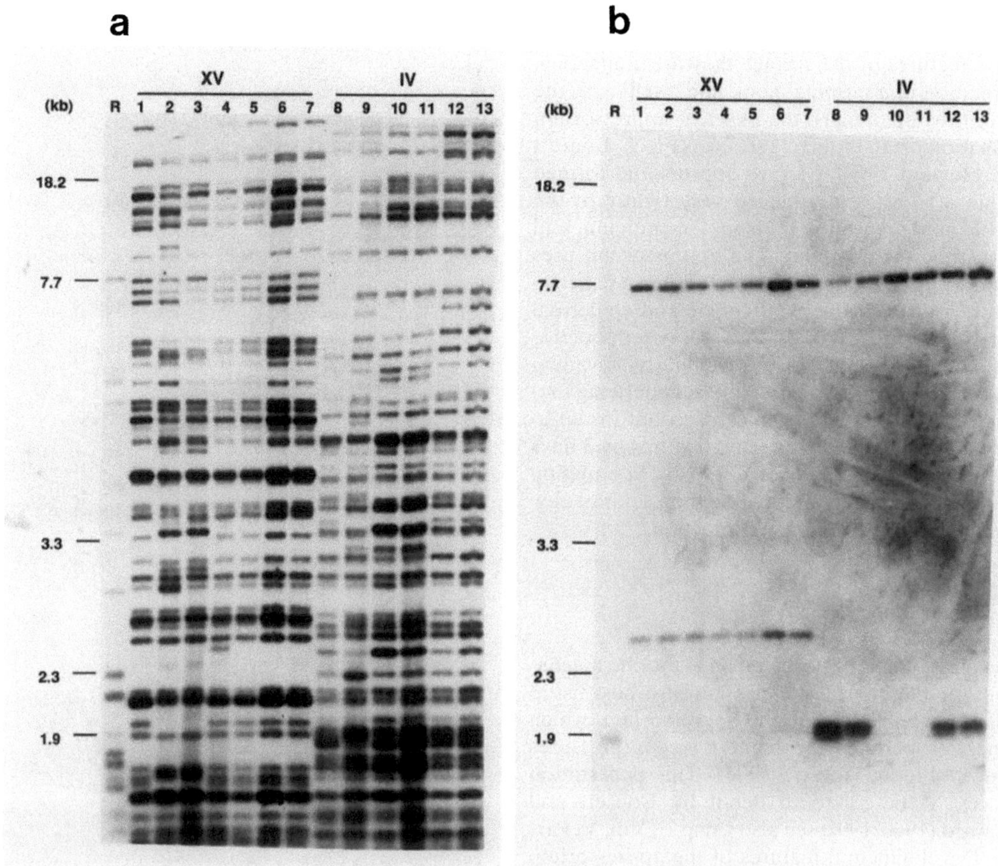

Fig. 5. MGR586 fingerprints corresponding to two clonal lineages from China, and associated AVR gene RFLPs. **a** Isolates within lineages XV and IV show 86 and 88% similarity, respectively, in *Eco*R I restriction fragments. Approximately 40% band similarity occurs between the two lineages. **b** Blots were stripped and probed with the *AVR2-YAMO* gene. The control strain (*lane R*) has the telomeric *AVR2-YAMO* gene on the 1.8 kb *Eco*R I fragment. Isolates in *lanes 1–7* were collected in Zhejiang Province from 1987, 1991, and 1992, and isolates in *lanes 8–13* were collected in Guizhou and Zhejiang Provinces in 1987, 1988, and 1992. (M. Levy, A.K.M. Shahjahan, B. Valent, unpubl. data)

multiple pathotypes, the individual pathotypes appeared related to each other, differing by one or a few specificities toward particular differential cultivars. Current data therefore suggest that individuals within lineages may have a limited potential for mutation to new virulence forms. Lineage exclusion breeding strategies based on pyramiding of lineage-specific resistance genes, effective against all strains of a particular clonal population, may facilitate breeding durable resistance in rice (Levy et al. 1993; Zeigler et al. 1994).

III. Molecular Mechanisms of Pathogenesis

M. grisea executes a series of developmental and metabolic pathways between the time spores land on the waxy leaf surface (Howard 1994; Chaps. 1–3, Vol. V, Part A) and the time sporulating lesions appear (Fig. 2a). Apical cells of conidia contain a compartment of glue, spore tip mucilage (Hamer et al. 1988), which is released upon hydration and functions in attachment to the host surface. Spores germinate, producing specialized germ tubes that sense features of the surface they are contacting. Formation of a terminal hook and swelling at the germ tube apex mark the onset of appressorium formation on an inductive surface (Fig. 6; Bourett and Howard 1990). Mature appressoria, formed within 4 to 8h, are attached very tightly to the cuticle, and are walled off from the conidium and germ tube by a septum. The appressorium uses osmotically generated pressure to force a penetration peg through the plant cuticle and epidermal cell wall (Chap. 3, Vol. V, Part A). After penetration, the fungus grows intracellularly, invading adjacent epidermal cells as well as underlying mesophyll cells. The pathogen can account for up to 10% of the biomass of infected leaf tips by 3 days after inoculation (Talbot et al. 1993b). Sporulating lesions soon appear, and release thousands of new conidia per day to reinitiate infection.

A. Penetration

The blast fungus has evolved an elaborate mechanism for mechanical penetration into host tissue using a specialized cell, the appressorium, which generates perhaps the highest turgor pressures known (Fig. 6; Howard 1994). This penetration process is described in detail by Bourett and Howard (1990, 1992; see also Chap. 3, Vol. V, Part A). Developmental features of the appressorium include the appressorial pore that lies adjacent to the substratum, the pore wall overlay, and the penetration peg with a localized concentration of actin. Extracellular mucilages and adhesives appear to play key roles in the penetration process (Bourett and Howard 1990; Xiao et al. 1994a).

Blocking biosynthesis of the gray pigment DHN-melanin, either genetically or with chemicals, results in appressoria that fail to function in penetration (Chumley and Valent 1990; Lundqvist et al. 1994; Vidal-Cros et al. 1994). Melanin, deposited in a dense layer between the plasma membrane and the appressorial cell wall, acts as a permeability barrier for small solute molecules inside the appressorium (Howard and Ferrari 1989). This is critical for buildup of the tremendous appressorial turgor pressure, estimated at 80 bar, that the fungus uses for physical penetration of nonbiodegradable artificial surfaces (Howard et al. 1991). The solute molecule that mediates

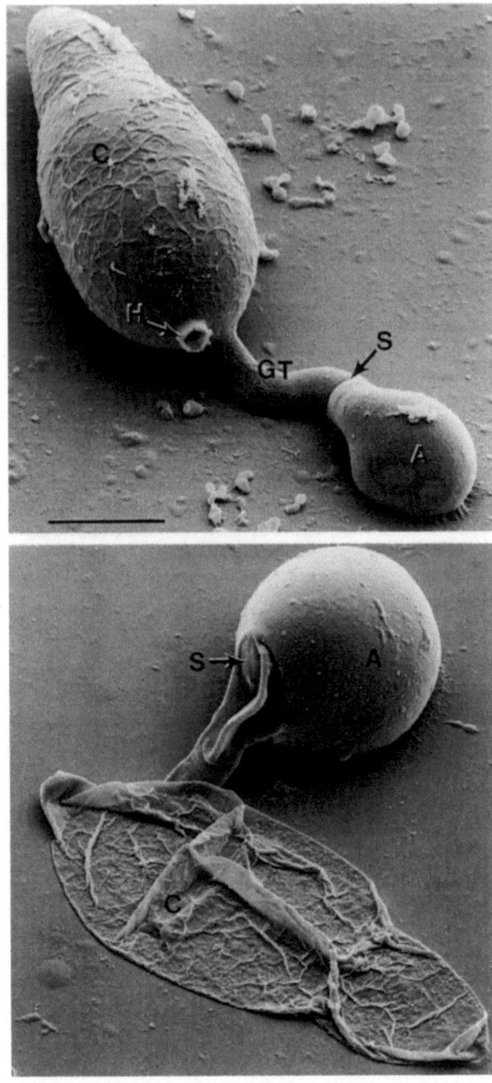

Fig. 6. Cryoscanning electron micrographs of developing (*upper*) and mature (*lower*) infection structures of *Magnaporthe grisea* on a Mylar substrate. An appressorium (*A*) forms at the tip of the germ tube (*GT*) and is isolated by a septum (*S*). After 24h of incubation (*lower panel*), the conidium and germ tube collapse readily in the absence of extracellular free water, whereas the appressorium appears insensitive due to the very high osmotic pressure within this cell. Conidium (*C*); hilum (*H*). Both figures, same magnification; *bar* 5 μm. (Unpublished micrographs courtesy of Richard J. Howard)

pressure buildup may be derived from the breakdown of glycogen, because cytoplasmic glycogen rosettes disappear during the period of pressure buildup before penetration (Bourett and Howard 1990).

Although physical pressure clearly plays a critical role in penetration of the rice surface, it remains probable that cell-surface-modifying enzymes also function in penetration of the host (Howard et al. 1991; Chap. 3, Vol. V, Part A). Sweigard et al. (1992a,b) cloned the *M. grisea* cutinase gene, *CUT1*, and produced fungal strains containing null mutations by gene disruption techniques. *CUT1* is the only *M. grisea* gene that is homologous to the well-studied *Colletotrichum* and *Fusarium* cutinase genes, and it appears to encode the only esterase induced by the presence of cutin. However, loss of function of the *CUT1* gene has no apparent effect on pathogenicity. A small amount of residual cutinase activity in these null mutants suggests that *M. grisea* produces an additional cutinase, encoded by a gene that lacks homology to *CUT1*. This hypothetical second cutinase, as well as enzymes that degrade plant cell walls (Hirayama et al. 1976; Wu et al. 1995), may play a role in penetration into plant tissue.

Various artificial surfaces, including cellophane, Teflon, Mylar and polyvinylchloride, are inductive for appressorial development, and the appressoria that form are capable of penetrating most of these materials (Howard and Ferrari 1989; Bourett and Howard 1990; Howard et al. 1991). There have been several reports on surface features that induce appressorium formation. Lee and Dean (1993a, 1994) suggest that hydrophobicity is the critical factor, but others find no correlation with hydrophobicity (Howard 1994; Xiao et al. 1994b). Xiao et al. (1994b) suggest that hardness is the key inducing factor. No correlation between surface topology and inductiveness has been seen. Conflicting results on inductiveness of hydrophilic glass surfaces (Hamer et al. 1989b; Howard et al. 1991; Lee and Dean 1993a; Jellito et al. 1994; Xiao et al. 1994b) may be due to strain differences, or to the sensitivity of these systems to contamination with various substances that can stimulate appressorium formation (e.g., Uchiyama and Okuyama 1990). Further clarification of the critical substrate features is needed.

Screening mutagenized cultures for appressorium-defective mutants identified a class of mutants, named Smo$^-$ for abnormal *s*pore *mo*rphology (Hamer et al. 1989b). Smo$^-$ mutants exhibit dramatic pleiomorphic alterations in the shapes of conidia (Fig. 7) and appressoria. Mutations at the *SMO1* locus permit germinating conidia to differentiate appressoria on surfaces that are not normally inducive to infection structure formation. Pathogenicity toward rice appears to be reduced in these mutants with regard to the number and size of lesions formed. Other mutants with defects in conidiation show pleiotropic effects on appressorium formation as well (Shi and Leung 1995).

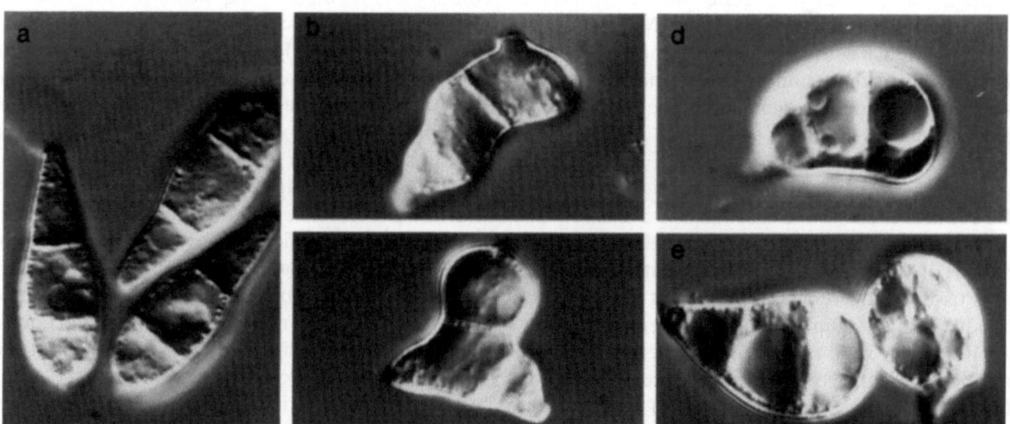

Fig. 7a–e. Differential interference contrast light micrographs of conidia from wild-type and Smo$^-$ strains. **a** Wild-type conidia. **b, c** Conidia from a strain that contains the *smo1-1* mutation. **d, e** Conidia from a strain that contains the less severe *smo1-3* mutation (×1600). (Hamer et al. 1989b)

Talbot et al. (1993b) identified *MPG1*, a gene that appears to play a role in elaboration of infection structures, by cloning fungal cDNAs that are differentially expressed during infection. Mpg1⁻ mutants cause reduced symptoms, producing only a small fraction of the lesions produced by the wild-type pathogen. Germ tubes of these null mutants elongate with multiple hooking and swelling events on normally inductive surfaces. It appears that the mutants undergo the primary differentiation events associated with infection structure formation, but fail to develop mature appressoria. The cloned *MPG1* gene encodes a small, secreted, cysteine-rich protein with characteristics of a fungal hydrophobin (Chaps. 6, 7, Vol. I). The mutants also exhibit an easily wettable phenotype, indicating a change in the hydrophobicity of the fungal surface. *MPG1* is also expressed during conidiation.

Lee and Dean (1993a) presented evidence that cAMP plays a role in mediation of infection structure formation. Exogeneously added cAMP, and select cAMP analogs, stimulate appressorium formation on noninductive surfaces. They hypothesize that the environmental signals from an inductive surface mediate elevated levels of cAMP, which, in turn, trigger signal transduction pathways leading to infection structure formation. Mitchell and Dean (1995) recently cloned the *CPKA* gene encoding the catalytic subunit of a cAMP-dependent protein kinase, the principal target for cAMP-modulated processes in eukaryotes. Mutants produced by targeted gene disruption of the *CPKA* gene fail to produce appressoria, even in response to cAMP. Independently, Sweigard et al. (1994) cloned the same kinase gene using a random insertional mutagenesis strategy to identify genes that are required for pathogenicity.

B. Proliferation

Strains of the pathogen differ in ability to colonize host tissue, as evidenced by different-sized lesions produced in different strain-host interactions (Fig. 1). Relatively little is known about critical factors required for growth of the pathogen inside the host tissue. Heath et al. (1990, 1992) reported differentiation of infection hyphae inside host epidermal cells. The narrow penetration peg enlarges to form a primary hypha that subsequently differentiates into an enlarged bulbous secondary hypha that proliferates throughout the host tissue. The bulbous form of the fungus inside plant tissue differs dramatically from the slender filamentous growth on agar media, and is likely to be adapted for surviving in the hostile environment of the host tissue.

Biochemical studies have suggested complex molecular interactions that might play a role in determining the extent of infection. Components solubilized from hyphal walls, and polymeric chitin were shown to elicit production of the diterpenoid rice phytoalexins (Iwakuma et al. 1990; Ren and West 1992). A heat-labile activity, most likely an endoxylanase, from culture filtrates of the fungus solubilized fragments from rice cell walls that inhibit protein synthesis in rice cells (Bucheli et al. 1990). A variety of phytotoxic fungal metabolites have been suggested to play a role in rice blast pathogenesis, as summarized by Lebrun et al. (1984). Tenuazonic acid (TA, a cyclized acetoacetyl derivative of isoleucine) stands out among the possible phytotoxins because it has reproducibly been associated with the fungus in culture (Lebrun et al. 1990), and, although it has rather low specific activity, has been detected at toxic concentrations in a variety of diseased tissues. TA is believed to play a role in determining the extent of host colonization.

C. Sporulation

Conidiogenesis in *M. grisea* is holoblastic. A conidium is formed by expansion of the conidiophore apex followed by septation to delimit the forming conidium (Howard 1994; Shi and Leung 1995). The conidiophore apex then grows to the side to produce the next conidium, resulting in multiple conidia borne sympodially on mature conidiophores. The previously described mutations that affect appressorium formation also exert major effects on conidiogenesis. Mpg1⁻ mutants produce significantly fewer conidia than do wild-type strains (Talbot et al. 1993b). The Smo⁻ mutants described by Hamer et al. (1989b) produce normal numbers of aberrantly shaped conidia (Fig. 7). Shapes of appressoria and asci, but not ascospores, are affected by mutation at the *SMO* locus, suggesting a defect in depolarization of hyphal tip growth to allow development of differentiated cell types.

Conidiation-defective mutants, Con1⁻ through Con7⁻, were isolated following chemical mutagenesis or insertional mutagenesis via REMI transformation (Shi and Leung 1994, 1995). Two mutants, Con5⁻ and Con6⁻, fail to produce any conidia, and five others show reduced sporulation as well as developmental defects. For example, Con1⁻ mutants form occasional aerial hyphae that produce single elongated terminal conidia ($4 \times 50\,\mu m$). This suggests a defect in depolarization of hyphal tip growth, as with Smo⁻ mutants. Con1⁻ mutants fail to form appressoria, and they are totally nonpathogenic. The Con1⁻ phenotype is epistatic to the Smo⁻ phenotype in double mutants (Shi and Leung 1994).

IV. Genetic Control of Host Specificity

Little is understood about the control of specificity at the host species level. Minor host specificity genes with individually small effects on lesion size on rice (Fig. 1) have been detected in crosses between a rice pathogen and pathogens of weeping lovegrass (Valent et al. 1991). Yaegashi (1978) first identified major genes that have an all-or-nothing effect on ability to infect weeping lovegrass or finger millet. Genetic analyses identified two genes, *PWL1* from a finger millet pathogen and *PWL2* from a rice pathogen, that each exert a major effect on the ability of *M. grisea* strains to infect weeping lovegrass (Valent and Chumley 1991; Sweigard et al. 1995). Cloning and characterization of these genes is described below.

The rice blast system is a classical gene-for-gene system in which AVR genes in the pathogen show a functional correspondence with particular blast resistance genes in the host (Silué et al. 1992a; see also Chaps. 1, 15, 16, this Vol.). AVR genes may encode pathogen molecules that are directly recognized by the corresponding host resistance gene, or they may encode enzymes involved in production of recognized ligands (Chaps. 4–6, Vol. V, Part A; Chap. 1, this Vol.). Recognition triggers host defense responses and stops infection. AVR genes with major effects on the ability to infect particular rice cultivars appear abundant among rice pathogens, and in *M. grisea* pathogens of grasses other than rice (Yaegashi and Asaga 1981; Leung et al. 1988; Valent et al. 1991; Ellingboe 1992; Silué et al. 1992a,b; Smith and Leong 1994; Valent and Chumley 1994). Although many studies have identified single genes controlling rice cultivar specificity, some studies showed complex segregation, with two or more pathogen genes controlling particular host specificities (Silué et al. 1992b; Lau et al. 1993).

A. The *PWL* Host Species Specificity Gene Family

PWL2, the specificity gene from the rice pathogen Guy11 that controlled ability to infect weeping lovegrass, appeared to be genetically unstable in laboratory studies. The frequent appearance of pathogenic mutants derived from nonpathogenic parental strains suggested that *PWL2* functioned to prevent infection of weeping lovegrass, as would be expected for a classical AVR gene. The *PWL2* gene was cloned by chromosome walking from a linked physical marker, A12G1 (Fig. 4; Sweigard et al. 1995), based on the prediction that the functional gene would transform a pathogen of weeping lovegrass into a noninfecting strain. Chromosome walking experiments were complicated by abundant highly repetitive DNA sequences, as well as low copy repeats present in four to six unlinked locations in the pathogen genome. Further analysis, including the study of a *pwl2⁻* point mutation, confirmed that the *PWL2* gene encoded a glycine-rich, hydrophilic polypeptide containing 145 amino acids (Sweigard et al. 1995). A putative signal peptide suggested that the protein might be secreted. However, no clues to a biochemical function of this protein were obtained.

The distribution of DNA sequences with homology to the *PWL2* gene among *M. grisea* pathogens of diverse grass species is shown in Fig. 8. Most pathogens of rice and of *Digitaria* spp. contain sequences with strong homology to *PWL2*. *Eleusine* pathogens have only weak homology. Thus, distantly related *Digitaria* pathogens contain AVR genes in common with rice pathogens, while the more closely related *Eleusine* pathogens have divergent sequences (Table 1).

The *Eleusine* pathogen reported to contain the *PWL1* gene had bands with weak homology to *PWL2* (Fig. 8, Lane 13), and this homology led to the cloning of the fully functional *PWL1* gene (Kang et al. 1995). A second inactive gene, *PWL3*, from the same *Eleusine* pathogen, and an inactive gene, *PWL4*, from a weeping lovegrass pathogen, were also cloned. The cloned *PWL3* gene encodes an inactive protein, and the *PWL4* gene encodes an active protein, but appears not to be expressed.

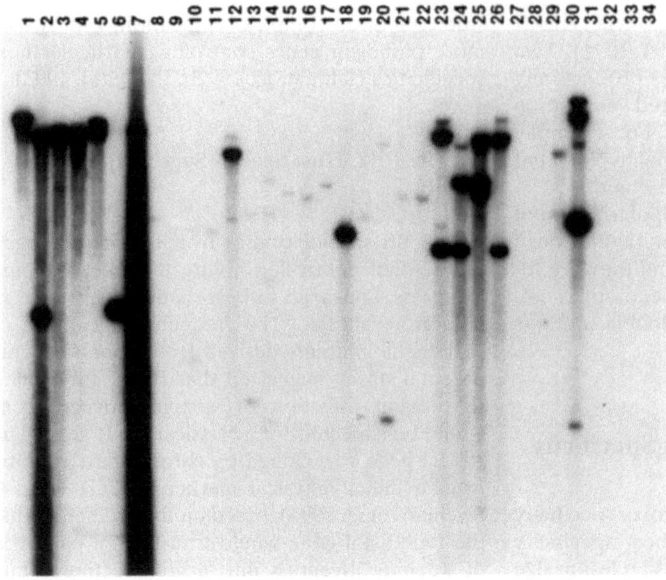

Fig. 8. Distribution of sequences with homology to *PWL2*. Southern analysis in which *Eco*R I digested genomic DNAs from 34 field isolates were probed with the *PWL2* gene. Strains with strong homology include pathogens of: rice (*lanes 1–4*); barley (*lane 5*); *Leersia* (*lane 6*); wheat (*lane 7*; badly degraded here); maize (*lane 12*); *Paspalum* (*lane 18*); *Digitaria* (*lanes 23–26*); and *Pennisetum* (*lane 30*). Strains with weak homology are pathogens of: weeping lovegrass (*lanes 8, 11*); *Eleusine* (*lanes 9–10, 13*); *Panicum* (*lanes 14, 19–20*); *Setaria* (*lanes 15–16, 22*); *Brachiaria* (*lane 17*); *Pennisetum* (*Lane 21, 27–28*); *Digitaria* (*lane 29*); *Leersia* (*lane 31*); ginger (*lane 32*); and *Cyperus* (*lane 33–34*). (Valent and Chumley 1994)

The *PWL1*, *PWL2*, and *PWL3*/*PWL4* genes map to different locations, and the amino acid sequences of the predicted PWL1, PWL3, and PWL4 proteins have 74, 49 and 55% identity, respectively, to the PWL2 protein. *PWL* genes appear to be members of a dynamic rapidly evolving gene family (Kang et al. 1995).

B. The *AVR2-YAMO* Avirulence Gene Family

The AVR gene, *AVR2-YAMO*, which prevents infection of the rice cultivar Yashiro-mochi (Yamada et al. 1976), cosegregated with a cluster of physical markers at the end of a linkage group (Fig. 4). Like *PWL2*, this gene was genetically unstable. That is, spontaneous virulent mutants frequently appeared from avirulent strains in laboratory studies (Valent and Chumley 1994). Spontaneous mutants that had become virulent on rice cultivar Yashiro-mochi showed structural changes in telomeric restriction fragments that mapped with *AVR2-YAMO*, suggesting that the gene resided within 1 to 2kb of the tip of the chromosome (M.J. Orbach, L. Farrall, F.G. Chumley, and B. Valent, unpubl. results). This hypothesis was tested by cloning a telomeric 6.5kb *Bgl* II DNA fragment that corresponded to this chromosome end, and the cloned fragment indeed contained a gene that transformed pathogens of rice cultivar Yashiro-mochi into nonpathogens.

As predicted, the *AVR2-YAMO* gene resides totally within a telomeric 1.5-kb DNA fragment. The gene encodes a protein containing 223 amino acids. Comparison of the amino acid sequence to protein databases revealed the existence of a sequence motif characteristic of neutral zinc metallopeptidases (B. Valent and L. Farrall, unpubl. results). In vitro mutagenesis of residues in this motif destroyed the avirulence-inducing activity of the gene. Although this is consistent with the hypothesis that the *AVR2-YAMO* gene encodes a protease, confirmation of this hypothesis awaits biochemical demonstration of protease activity.

Sequences with strong homology to *AVR2-YAMO* are present in rice pathogens, *Digitaria* pathogens, and *Pennisetum* pathogens. *AVR2-*

YAMO homologues cloned from *Digitaria* pathogens and *Pennisetum* pathogens are functional in preventing rice pathogens from infecting cultivar Yashiro-mochi (S. Kang and B. Valent, unpubl. results). These results confirm genetic predictions that pathogens of grasses other than rice contain AVR genes that confer specificity toward particular rice cultivars (Yaegashi and Asaga 1981; Valent et al. 1991).

Studies of AVR RFLPs are now complementing virulence studies within MGR586 clonal lineages. Initial results suggest that the sizes of restriction fragments homologous to the *AVR2-YAMO* gene are often conserved among strains within a lineage (Fig. 5b). Virulence studies suggest, however, that ability to infect Yashiro-mochi can not be predicted from particular *AVR2-YAMO* RFLPs (M. Levy, A.K.M. Shahjahan, B. Valent, unpubl. results).

V. New Insights into Genetic Instability in *M. grisea*

High mutability is not a general feature of *M. grisea* genes. Mutants defective for either of the melanin biosynthetic genes *ALB1* or *RSY1* (Chumley and Valent 1990), as well as mutants resistant to 6-methylpurine, the sulfonylurea chlorimuron ethyl (Du Pont Agricultural Products; DPX-F6025), or benomyl (Valent and Chumley 1991) occur at normal frequencies for a haploid filamentous fungus. Some host specificity genes appear genetically stable (Valent et al. 1991). However, some genetic loci, in addition to the host specificity genes *PWL2* and *AVR2-YAMO*, are unstable in at least some strains of the fungus. For example, some fungal strains show frequent spontaneous appearance of mutants with altered spore morphology due to mutations in the *SMO1* gene (Hamer et al. 1989b). Many rice pathogens collected from fields around the world show genetic instability for expression of the *BUF1* melanin biosynthetic gene (Chumley and Valent 1990). In the most extreme cases, about 0.25% of the conidia produced by rice pathogens during subculture in the laboratory carry *buf*$^-$ mutations. Many rice pathogens cross with other normally pigmented strains to produce 5–25% *buf*$^-$ progeny (Leung et al. 1988; Chumley and Valent 1990; Valent et al. 1991), indicating that instability also occurs during meiosis.

Potential mechanisms of genetic variation in *M. grisea* include B-like chromosomes, the parasexual cycle, and/or polymorphisms in double-stranded RNAs (Valent and Chumley 1991). In cases studied so far, genetic instability in the fungus appears to be associated with frequent deletion and other mutations occurring in unstable regions of the genome. In other cases, transposable elements account for variability, at least in laboratory studies.

A. Instability Associated with Chromosomal Location

Two of the cloned unstable genetic loci, *PWL2* and *BUF1*, undergo frequent deletions. All spontaneous Pwl2$^-$ mutants examined have deleted the *PWL2* gene and at least 30kb of flanking DNA sequences (Sweigard et al. 1995). Two or three distinct classes of deletions were found among mutants analyzed from different parental strains. In addition, all spontaneous Buf$^-$ mutants examined contained deletions of the *BUF1* gene and adjacent genomic sequences (Valent and Chumley 1994). Some Buf$^-$ mutants had lost homology to all 40kb of genomic sequences included in the hybridization probe. For a particular fungal strain, deletion events appeared to be identical in independently derived mutants that arose during vegetative growth or in the sexual cycle. The large deletions occurring at both *PWL2* and *BUF1* are likely to be due to homologous recombination of flanking repetitive DNA sequences, as is well documented in other organisms (Petes and Hill 1988).

Unlike the relatively uniform deletion events occurring at unstable *BUF1* and *PWL2* loci, a wide variety of genetic events result in inactivation of the *AVR2-YAMO* gene in its telomeric location. For example, analysis of independent mutants derived from a single strain detected point mutations, an insertion (see below), and deletions of various sizes (Fig. 9). This type of instability is characteristic of telomeric regions in other organisms (Horowitz et al. 1984). Interestingly, some strains from lineage IV from China have the same size *AVR2-YAMO* *Eco*R I fragment (approximately 1.8kb) as the Chinese rice pathogen that was the source of this AVR gene (Fig. 5b). It is possible that strains in this MGR lineage have *AVR2-YAMO* in a telomeric location, with consequent genetic instabilities similar to those studied.

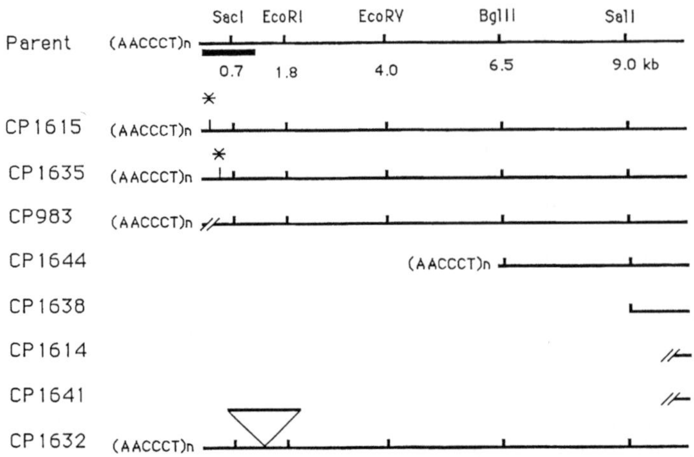

Fig. 9. Diagrammatic representation of molecular events associated with loss of function of *AVR2-YAMO* in eight spontaneous mutants. The telomeric clone contained 20 copies of the conserved telomeric repeat sequence (Blackburn 1990). The *black bar* indicates the location of the gene. Two mutants contained single base-pair changes (*) in the coding sequence: one produced a termination codon and one resulted in a substitution of glutamine for glutamic acid. CP983 had a deletion of approximately 100 bp at the 3′ end of the gene, and CP1632 contained an insertion, approximately 1.8 kb, of a Pot2-like transposable element. Other mutants had deletions ranging in size from 6.5 to 9 kb to larger deletions of undetermined size

B. Transposable Elements

In laboratory studies, transposable elements provide a proven source of genetic variation in *M. grisea.* Four families of transposable elements have been identified, as well as several distinct uncharacterized repetitive DNA sequences. These include a putative retroelement, named MGR583, which is similar to the long interspersed nuclear elements (LINEs) first described in mammals (Hamer et al. 1989a; Valent and Chumley 1991), and two retrotransposons, which contain long terminal repeats and are structurally similar to retroviruses. One retrotransposon, named Fosbury or MAGGY, was discovered independently in three laboratories (Lebrun et al. 1994; Leong et al. 1994; Shull 1995), and the other, named grasshopper, was discovered by Dobinson et al. (1993). Both of these elements are members of the Gypsy family of retrotransposons. Recently, the Pot2 family of inverted repeat transposons, which are related to the Fot1 elements from the phytopathogenic fungus *Fusarium oxysporum*, have been described (Kachroo et al. 1994).

There is evidence for active transposition of two *M. grisea* elements, Fosbury and Pot2. Shull (1995) identified a novel MGR586 RFLP present only in one out of the eight members of a complete tetrad. This RFLP represented an example of a mitotic transposition event in which a Fosbury element had inserted adjacent to an MGR586 sequence. Independently, Lebrun et al. (1994) obtained spontaneous mutations in *M. grisea* strains constructed to utilize nitrate via an *Aspergillus nidulans* nitrate reductase gene (*niaD*) introduced through transformation. Forty spontaneous Nia$^-$ mutants were obtained by selection for chlorate resistance. One mutant had lost function of the *niaD* gene due to an insertion of a 5.6-kb Fosbury element 850 bp upstream from the transcription start site. In the third example, the insertion that inactivated the *AVR2-YAMO* gene (Fig. 9) was a member of the Pot2 family of elements (Kachroo et al. 1994; S. Kang and B. Valent, unpubl. results).

The four described elements have different distributions in the *M. grisea* host-limited populations in Table 1. Like the MGR586 sequences, sequences homologous to the MGR583 LINE-like element have been amplified specifically in rice pathogens, which contain approximately 100 homologous restriction fragments compared to one to a few fragments in pathogens of other grasses (Hamer et al. 1989a). The two Gypsy-like elements are restricted to subpopulations corresponding to a single composite haplotype. The

Fosbury element is restricted to rice pathogens (Shull 1995) and the grasshopper element is restricted to the subpopulation of *Eleusine* pathogens designated composite haplotype c in Table 1 (Dobinson et al. 1993). Pot2 elements occur in approximately equal copy number, approximately 100 per haploid genome, in both rice and nonrice pathogens (Kachroo et al. 1994). Thus, the different repetitive DNA elements described so for have been differentially acquired or selectively amplified in host-specific subpopulations.

VI. Conclusions

Genes that play critical roles in pathogenesis encode proteins that are potential targets for the design of novel, environmentally safe fungicides. Insertional mutagenesis techniques are identifying genes required for pathogenicity or for sporulation (Sweigard et al. 1994; Shi and Leung 1995). Differential cDNA cloning techniques are identifying genes expressed during appressorium formation (Lee and Dean 1993b), or genes expressed during infectious growth of the pathogen in the host plant (Talbot et al. 1993b). Mutational analysis using gene disruption techniques has the potential to identify functions for cloned genes. The complex disease cycle is now accessible to molecular analysis.

Cloning the *PWL* host species specificity genes demonstrates that, at least in some cases, host species specificity is controlled by the same types of interactions that control cultivar specificity. AVR genes, directly or indirectly, result in signals that are recognized by hosts containing corresponding resistance genes, and this recognition triggers effective defense responses. Clearly, AVR genes did not evolve to prevent the pathogen from infecting various genotypes of its host. Instead, it is likely that plants have gained the ability to recognize pathogen molecules with various functions. A major breakthrough will come from discovering what roles individual AVR genes, such as *PWL2* and *AVR2-YAMO*, play for the pathogen. AVR genes with functions that are required for pathogenicity might correspond to durable resistance genes, genes that remain effective in the field because the pathogen must retain the corresponding AVR gene. Population studies that identify AVR genes retained in all strains within particular clonal lineages, in contrast to AVR genes present in some strains and absent in others, might give clues to those with functions critical to pathogen fitness in the field. Thus, studies of AVR genes may also give insight on mechanisms and molecules critical for pathogenesis.

References

Barr ME (1977) *Magnaporthe*, *Telimenella* and *Hyponectria* (Physosporellaceae). Mycologia 69:952–966

Biju-Duval MP (1994) Etude de la structure génétique des populations du champignon pathogène des graminées, *Magnaporthe grisea*. PhD Thesis, University of Paris, Orsay

Blackburn EH (1990) Telomeres: structure and synthesis. J Biol Chem 265:5919–5921

Bonman JM, Vergel de Dios TI, Bandong JM, Lee EJ (1987) Pathogenic variability of monoconidial isolates of *Pyricularia oryzae* in Korea and in the Philippines. Plant Dis 71:127–130

Borromeo ES, Nelson RJ, Bonman JM, Leung H (1993) Genetic differentiation among isolates of *Pyricularia* infecting rice and weed hosts. Phytopathology 83:393–399

Bourett TM, Howard RJ (1990) In vitro development of penetration structures in the rice blast fungus *Magnaporthe grisea*. Can J Bot 68:329–342

Bourett TM, Howard RJ (1992) Actin in penetration pegs of the fungal rice blast pathogen *Magnaporthe grisea*. Protoplasma 168:20–26

Bucheli P, Doares SH, Albersheim P, Darvill A (1990) Partial purification and characterization of heat-labile molecules secreted by the rice blast pathogen that solubilize plant cell wall fragments that kill plant cells. Physiol Mol Plant Pathol 36:159–173

Chumley FG, Valent B (1990) Genetic analysis of melanin-deficient, nonpathogenic mutants of *Magnaporthe grisea*. Mol Plant-Microbe Interact 3:135–143

Correa-Victoria FJ, Zeigler RS, Levy M (1994) Virulence characteristics of genetic families of *Pyricularia grisea* in Colombia. In: Zeigler RS, Leong S, Teng PS (eds) Rice blast disease. CAB International, Wallingford, pp 211–229

Daboussi MJ, Djeballi A, Gerlinger C, Blaiseau PL, Bouvier I, Cassan M, Lebrun MH, Parisot D, Brygoo Y (1989) Transformation of seven species of filamentous fungi using the nitrate reductase gene of *Aspergillus nidulans*. Curr Genet 15:453–456

Dobinson KF, Harris RE, Hamer JE (1993) Grasshopper, a long terminal repeat (LTR) retroelement in the phytopathogenic fungus *Magnaporthe grisea*. Mol Plant-Microbe Interact 6:114–126

Ellingboe AH (1992) Segregation of avirulence/virulence on three rice cultivars in 16 crosses of *Magnaporthe grisea*. Phytopathology 82:597–601

Farman ML, Leong SA (1995) Genetic and physical mapping of telomeres in the rice blast fungus, *Magnaporthe grisea*. Genetics 140:479–492

Hamer JE, Howard RJ, Chumley FG, Valent B (1988) A mechanism for surface attachment in spores of a plant pathogenic fungus. Science 239:288–290

Hamer JE, Farrall L, Orbach MJ, Valent B, Chumley FG (1989a) Host species-specific conservation of a family of repeated DNA sequences in the genome of a fungal plant pathogen. Proc Natl Acad Sci USA 86:9981–9985

Hamer JE, Valent B, Chumley FG (1989b) Mutations at the *SMO* genetic locus affect the shape of diverse cell types in the rice blast fungus. Genetics 122:351–361

Hayashi N, Naito H (1993) Genome mapping of *Magnaporthe grisea*. In: Zeigler RS, Leong S, Teng PS (eds) Rice blast disease. CAB International, Wallingford, p 590

Heath MC, Valent B, Howard RJ, Chumley FG (1990) Interactions of two strains of *Magnaporthe grisea* with rice, goosegrass and weeping lovegrass. Can J Bot 68:1627–1637

Heath MC, Howard RJ, Valent B, Chumley FG (1992) Ultrastructural interactions of one strain of *Magnaporthe grisea* with goosegrass and weeping lovegrass. Can J Bot 70:779–787

Hebert TT (1971) The perfect stage of *Pyricularia grisea*. Phytopathology 61:83–87

Hirayama T, Sudo T, Nagayama H, Matsuda K, Tamari K (1976) Number and interrelation of components of carboxymethyl cellulase enzyme from *Pyricularia oryzae*. Agric Biol Chem 40:2137–2142

Horowitz H, Thorburn P, Haber JE (1984) Rearrangements of highly polymorphic regions near telomeres of *Saccharomyces cerevisiae*. Mol Cell Biol 4:2509–2517

Howard RJ (1994) Cell biology of pathogenesis. In: Zeigler RS, Leong S, Teng PS (eds) Rice blast disease. CAB International, Wallingford, pp 3–22

Howard RJ, Ferrari MA (1989) Role of melanin in appressorium function. Exp Mycol 13:403–418

Howard RJ, Ferrari MA, Roach DH, Money NP (1991) Penetration of hard substrates by a fungus employing enormous turgor pressures. Proc Natl Acad Sci USA 88:11281–11284

Iwakuma T, Ataka Y, Matsuyama N, Wakimoto S (1990) Phytoalexin production elicited by *Pyricularia oryzae* infection and its hyphal wall-component treatment in rice leaves. Ann Phytopathol Soc Jpn 56:665–670

Jellito TC, Page HA, Read ND (1994) Role of external signals in regulating the pre-penetration phase of infection by the rice blast fungus *Magnaporthe grisea*. Planta 194:471–477

Kachroo P, Leong SA, Chattoo BB (1994) Pot2, an inverted repeat transposon from the rice blast fungus *Magnaporthe grisea*. Mol Gen Genet 245:339–348

Kang S, Chumley FG, Valent B (1994) Isolation of the mating-type genes of the phytopathogenic fungus *Magnaporthe grisea* using genomic subtraction. Genetics 138:289–296

Kang S, Sweigard JA, Valent B (1995) The *PWL* host specificity gene family in the blast fungus, *Magnaporthe grisea*. Mol Plant-Microbe Interact 8:949–959

Kato H, Yamaguchi T (1982) The perfect state of *Pyricularia oryzae* Cav. in culture. Ann Phytopathol Soc Jpn 48:607–612

Kato H, Mayama S, Sekine R, Kanazawa E, Izutani Y, Urashima AS, Kunoh H (1994) Microconidium formation in *Magnaporthe grisea*. Ann Phytopathol Soc Jpn 60:175–185

Keon JPR, James CS, Court S, Baden-Daintree C, Bailey AM, Burden RS, Bard M, Hargreaves JA (1994) Isolation of the *ERG2* gene, encoding sterol Δ^8-Δ^7 isomerase, from the rice blast fungus *Magnaporthe grisea* and its expression in the maize smut pathogen *Ustilago maydis*. Curr Genet 25:531–537

Kimura M, Kamakura T, Tao QZ, Kaneko I, Yamaguchi I (1994) Cloning of the blasticidin S deaminase gene (BSD) from *Aspergillus terrus* and its use as a selectable marker for *Schizosaccharomyces pombe* and *Pyricularia oryzae*. Mol Gen Genet 242:121–129

Kiyosawa S (1976) Pathogenic variations of *Pyricularia oryzae* and their use in genetic and breeding studies. SABRAO J 8:53–67

Latterell FM (1975) Phenotypic stability of pathogenic races of *Pyricularia oryzae*, and implications for breeding blast-resistant rice varieties. In: Proc Seminar on Horizontal resistance to blast disease of rice. Centro Internacional de Agricultura Tropical, Cali, Columbia, pp 199–234

Latterell FM, Rossi AE (1986) Longevity and pathogenic stability of *Pyricularia oryzae*. Phytopathology 76:231–235

Lau GW, Chao CT, Ellingboe AH (1993) Interactions of genes controlling avirulence/virulence of *Magnaporthe grisea* on rice cultivar Katy. Phytopathology 83:375–382

Lebrun MH, Orcival J, Duchartre C (1984) Resistance of rice to tenuazonic acid, a toxin from *Pyricularia oryzae*. Rev Cytol Biol Vég Bot 7:249–259

Lebrun MH, Dutfoy F, Gaudemer F, Kunesch G, Gaudemer A (1990) Detection and quantification of the fungal phytotoxin tenuazonic acid produced by *Pyricularia oryzae*. Phytochemistry 29:3777–3783

Lebrun MH, Chumley FG, Valent B (1994) Molecular analysis of spontaneous mutations in *Magnaporthe grisea*. In: Zeigler RS, Leong S, Teng PS (eds) Rice blast disease. CAB International, Wallingford, p 593

Lee YH, Dean RA (1993a) cAMP regulates infection structure formation in the plant pathogenic fungus *Magnaporthe grisea*. Plant Cell 5:693–700

Lee YH, Dean RA (1993b) Stage-specific gene expression during appressorium formation of *Magnaporthe grisea*. Exp Mycol 17:215–222

Lee YH, Dean RA (1994) Hydrophobicity of contact surface induces appressorium formation in *Magnaporthe grisea*. FEMS Microbiol Lett 115:71–76

Leong SA, Farman M, Smith J, Budde A, Tosa Y, Nitta N (1994) Molecular genetic approach to the study of cultivar specificity in the rice blast fungus. In: Zeigler RS, Leong S, Teng PS (eds) Rice blast disease. CAB International, Wallingford, pp 87–110

Leung H, Williams PH (1987) Nuclear division and chromosome behavior during meiosis and ascosporogenesis in *Pyricularia oryzae*. Can J Bot 65:112–123

Leung H, Borromeo ES, Bernardo MA, Notteghem JL (1988) Genetic analysis of virulence in the rice blast fungus *Magnaporthe grisea*. Phytopathology 78:1227–1233

Leung H, Lehtinen U, Karjalainen R, Skinner D, Tooley P, Leong S, Ellingboe A (1990) Transformation of the rice blast fungus *Magnaporthe grisea* to hygromycin B resistance. Cur Genet 17:409–411

Levy M, Romao J, Marchetti MA, Hamer JE (1991) DNA fingerprinting with a dispersed repeated sequence resolves pathotype diversity in the rice blast fungus. Plant Cell 3:95–102

Levy M, Correa-Victoria FJ, Zeigler RS, Xu S, Hamer JE (1993) Genetic diversity of the rice blast fungus in a disease nursery in Colombia. Phytopathology 83:1427–1433

Lundqvist T, Rice J, Hodge CN, Basarab GS, Pierce J, Lindqvist Y (1994) Crystal structure of scytalone dehydratase – a disease determinant of the rice pathogen *Magnaporthe grisea*. Structure 2:937–944

Mitchell TK, Dean RA (1995) The cAMP-dependent protein kinase catalytic subunit is required for appressorium formation and pathogenesis by the rice blast pathogen *Magnaporthe grisea*. Plant Cell 7:1869–1878

Notteghem JL, Silué D (1992) Distribution of mating type alleles in *Magnaporthe grisea* populations pathogenic on rice. Phytopathology 82:421–424

Notteghem JL, Tharreau D, Silué D, Roumen E (1994) Present knowledge of rice resistance genetics and strategies for *Magnaporthe grisea* pathogenicity and avirulence gene analysis. In: Zeigler RS, Leong S, Teng PS (eds) Rice blast disease. CAB International, Wallingford, pp 155–165

Orbach MJ, Chumley FG, Valent B (1996) An electrophoretic karyotype for the rice blast fungus, *Magnaporthe grisea*. Mol Plant-Microbe Interact 9:261–271

Ou SH (1980) Pathogen variability and host resistance in rice blast disease. Annu Rev Phytopathol 18:167–187

Ou SH (1985) Rice diseases, 2nd edn. Commonwealth Mycological Inst, Kew, 380 pp

Parsons KA, Chumley FG, Valent B (1987) Genetic transformation of the fungal pathogen responsible for rice blast disease. Proc Natl Acad Sci USA 84:4161–4165

Petes TD, Hill CW (1988) Recombination between repeated genes in microorganisms. Annu Rev Genet 22:147–168

Ren YY, West CA (1992) Elicitation of diterpene biosynthesis in rice (*Oryza sativa* L.) by chitin. Plant Physiol 99:1169–1178

Romao J, Hamer JE (1992) Genetic organization of a repeated DNA sequence family in the rice blast fungus. Proc Natl Acad Sci USA 89:5316–5320

Rossman AY, Howard RJ, Valent B (1990) *Pyricularia grisea*, the correct name for the rice blast disease fungus. Mycologia 82:509–512

Shi Z, Leung H (1994) Genetic analysis and rapid mapping of a sporulation mutation in *Magnaporthe grisea*. Mol Plant-Microbe Interact 7:113–120

Shi Z, Leung H (1995) Genetic analysis of sporulation in *Magnaporthe grisea* by chemical and insertional mutagenesis. Mol Plant-Microbe Interact 8:949–959

Shull V (1995) Intralineage variation among rice pathogens of *Magnaporthe grisea* may be caused by the transposition of Fosbury, an LTR-containing retrotransposon. PhD Thesis, Purdue University, West Lafayette

Shull V, Hamer JE (1994) Genomic structure and variability in *Pyricularia grisea*. In: Zeigler RS, Leong S, Teng PS (eds) Rice blast disease. CAB International, Wallingford, pp 65–86

Silué D, Notteghem JL, Tharreau D (1992a) Evidence for a gene-for-gene relationship in the *Oryza sativa-Magnaporthe grisea* pathosystem. Phytopathology 82: 577–580

Silué D, Tharreau D, Notteghem JL (1992b) Identification of *Magnaporthe grisea* avirulence genes to seven rice cultivars. Phytopathology 82:1462–1467

Skinner DZ, Budde AD, Farman ML, Smith JR, Leung H, Leong SA (1993) Genome organization of *Magnaporthe grisea*: genetic map, electrophoretic karyotype, and occurrence of repeated DNAs. Theor Appl Genet 87:545–557

Smith JR, Leong SA (1994) Mapping of a *Magnaporthe grisea* locus affecting rice (*Oryza sativa*) cultivar specificity. Theor Appl Genet 88:901–908

Sweigard JA, Chumley FG, Valent B (1992a) Cloning and analysis of *CUT1*, a cutinase gene from *Magnaporthe grisea*. Mol Gen Genet 232:174–182

Sweigard JA, Chumley FG, Valent B (1992b) Disruption of a *Magnaporthe grisea* cutinase gene. Mol Gen Genet 232:183–190

Sweigard JA, Valent B, Orbach MJ, Walter AM, Rafalski A, Chumley FG (1993) Genetic map of the rice blast fungus *Magnaporthe grisea*. In: O'Brien SJ (ed) Genetic maps. Cold Spring Harbor Laboratory Press, Plainview, pp 3.112–3.117

Sweigard JA, Walter AM, Valent B (1994) *Magnaporthe grisea* pathogenicity genes cloned by insertional inactivation mutagenesis. In: Zeigler RS, Leong S, Teng PS (eds) Rice blast disease. CAB International, Wallingford, p 590

Sweigard JA, Carroll AM, Kang S, Farrall L, Chumley FG, Valent B (1995) Identification, cloning and characterization of *PWL2*, a gene from the rice blast fungus controlling host-species specificity. Plant Cell 7:1221–1233

Talbot NJ, Salch Y, Ma M, Hamer JE (1993a) Karyotype variation within clonal lineages of the rice blast fungus *Magnaporthe grisea*. Appl Environ Microbiol 59:585–593

Talbot NJ, Ebbole DJ, Hamer JE (1993b) Identification and characterization of *MPG1*, a gene involved in pathogenicity from the rice blast fungus *Magnaporthe grisea*. Plant Cell 5:1575–1590

Uchiyama T, Okuyama K (1990) Participation of *Oryza sativa* leaf wax in appressorium formation by *Pyricularia oryzae*. Phytochemistry 29:91–92

Urashima AS, Igarashi S, Kato H (1993) Host range, mating type, and fertility of *Pyricularia grisea* from wheat in Brazil. Plant Dis 77:1211–1261

Valent B, Chumley FG (1991) Molecular genetic analysis of the rice blast fungus, *Magnaporthe grisea*. Annu Rev Phytopathol 29:443–467

Valent B, Chumley FG (1994) Avirulence genes and mechanisms of genetic instability in the rice blast fungus. In: Zeigler RS, Leong S, Teng PS (eds) Rice blast disease. CAB International, Wallingford, pp 111–134

Valent B, Farrall L, Chumley FG (1991) *Magnaporthe grisea* genes for pathogenicity and virulence identified through a series of backcrosses. Genetics 127:87–101

Vidal-Cros A, Viviani F, Labesse G, Boccara M, Gaudry M (1994) Polyhydroxynaphthalene reductase involved in melanin biosynthesis in *Magnaporthe grisea*. Eur J Biochem 219:985–992

Wu S-C, Kauffmann S, Darvill AG, Albersheim P (1995) Purification, cloning and characterization of two xylanases from *Magnaporthe grisea*, the rice blast fungus. Mol Plant-Microbe Interact 8:506–514

Xia JQ, Correll JC, Lee FN, Marchetti MA, Rhoads DD (1993) DNA fingerprinting to examine microgeographic variation in the *Magnaporthe grisea* population in two rice fields in Arkansas. Phytopathology 83:1029–1035

Xiao JZ, Ohshimi A, Kamakura T, Ishiyama T, Yamaguchi I (1994a) Extracellular glycoprotein(s) associated with cellular differentiation in *Magnaporthe grisea*. Mol Plant-Microbe Interact 7:639–644

Xiao JZ, Watanabe T, Kamakura T, Ohshimi A, Yamaguchi I (1994b) Studies on the cellular differentiation of

Magnaporthe grisea. Physicochemical aspects of substratum surfaces in relation to appressorium formation. Physiol Mol Plant Pathol 44:227–236

Yaegashi H (1978) Inheritance of pathogenicity in crosses of *Pyricularia* isolates from weeping lovegrass and finger millet. Ann Phytopathol Soc Jpn 44:626–632

Yaegashi H, Asaga K (1981) Further studies on the inheritance of pathogenicity in crosses of *Pyricularia oryzae* with *Pyricularia* sp. from finger millet. Ann Phytopathol Soc Jpn 47:677–679

Yaegashi H, Udagawa S (1978) The taxonomical identity of the perfect state of *Pyricularia grisea* and its allies. Can J Bot 56:180–183

Yaegashi H, Yamada M (1986) Pathogenic race and mating type of *Pyricularia oryzae* from Soviet Union, China, Nepal, Thailand, Indonesia and Colombia. Ann Phytopathol Soc Jpn 52:225–234

Yamada M, Kiyosawa S, Yamaguchi T, Hirano T, Kobayashi T, Kushibuchi K, Watanabe S (1976) Proposal of a new method for differentiating races of *Pyricularia oryzae* Cavara in Japan. Ann Phytopathol Soc Jpn 42:216–219

Zeigler RS, Leong S, Teng PS (eds) (1994) Rice blast disease. CAB International, Wallingford, 626 pp

Zeigler RS, Cuoc LX, Scott RP, Bernardo MA, Chen D, Valent B, Nelson RJ (1995) The relationship between lineage and virulence in *Pyricularia grisea* in the Philippines. Phytopathology 85:443–451

4 *Erysiphe graminis*

H. GIESE[1], S. HIPPE-SANWALD[2], S. SOMERVILLE[3], and J. WELLER[4]

CONTENTS

I. Introduction 55
II. Genetic Basis of Host-Pathogen Interactions .. 55
A. The Pathogen *Erysiphe graminis* 56
1. Life Cycle 56
2. Genome Structure 57
3. Molecular Mapping 57
4. Concept of Formae Specialis 58
5. Avirulence Genes 58
B. Cereals 58
1. Resistance Genes 58
a. Natural and Induced Sources of Resistance 59
b. Mutations to Susceptibility 59
2. Molecular Mapping 59
3. Induced Resistance and Susceptibility 59
C. Summary 60
III. Infection Sequence 60
A. Germination 60
B. Penetration 62
1. Primary Germ Tube 62
2. Appressorium 62
3. Papilla 62
C. Haustorium 64
D. Hypersensitive Necrosis Response 68
E. Summary 68
IV. Host Responses to Infection 69
A. Physiological Aspects 69
B. Biochemical and Molecular Aspects 70
1. Compatible Interaction 70
2. Incompatible Interaction 70
C. Summary 71
V. Conclusions and Future Prospects 72
References 72
Abbreviations 78

I. Introduction

The obligately parasitic, heterothallic ascomycete, *Erysiphe graminis* DC [= *Blumeria graminis* (DC.) E.O. Speer], causes powdery mildew on grasses. Powdery mildew diseases have been of concern to humankind probably since cereals were first cultivated more than 10000 years ago. Today, they are economically significant diseases in major cereal-growing areas of the world (Spencer 1978) and, for this reason, have been studied intensively. Readers are referred to the book edited by Spencer (1978) and to excellent recent reviews by Aist and Bushnell (1991), Bryngelsson and Collinge (1992), Jørgensen (1988, 1992), and Scott (1992) for additional information.

In this chapter, we have summarized a large body of literature relating to the growth and development of *E. graminis* as pathogens of cereal hosts. Most information has been collected on the *E. g. hordei* interaction with barley, and, consequently, this system dominates our discussion. Based on genetic studies, single host resistance and pathogen avirulence genes are known to exhibit dramatic effects on the interaction of *E. graminis* with its host. Thus, many experiments have been directed at answering the question: what is the major cytological, biochemical, or physiological effect of a specific resistance gene and its corresponding avirulence gene? More recently, molecular biological techniques have been employed to answer the same question.

[1] Division of Plant Biology, Risø National Laboratory, DK-4000 Roskilde, Denmark
[2] Institute of Phytopathology, University of Kiel, 24098 Kiel, Germany
[3] Carnegie Institution of Washington, 290 Panama St., Stanford, California 94305, USA
[4] PE Zoogen Lab., 1756 Picasso Drive, Davis, California 95616, USA

II. Genetic Basis of Host-Pathogen Interactions

Based on genetic analyses of host resistance and pathogen avirulence, interactions between *E. graminis* f. sp. and their cereal hosts have been shown to conform to the gene-for-gene hypothesis proposed by Flor (1956). This hypothesis predicts that for each resistance gene in the host there is a corresponding avirulence gene in the pathogen. Pathotypes within each formae specialis of *E. graminis* are identified by their ability to infect

The Mycota V Part B
Plant Relationships
Carroll/Tudzynski (Eds.)
© Springer-Verlag Berlin Heidelberg 1997

plants with different genetically characterized resistance genes.

The gene-for-gene model is the basis for many experiments on the interaction between *E. graminis* and its host plants. Currently, research is directed at the molecular characterization of *E. graminis* avirulence genes and resistance genes from cereals in order to understand the biochemical mechanisms underlying the interaction. These studies are aided by the use of molecular markers such as restriction fragment length polymorphic markers (RFLP) and randomly amplified polymorphic DNA markers (RAPD) that enable map-based cloning of genes. Isolation of corresponding resistance and avirulence genes will facilitate characterization of the interaction at the biochemical level and resolve evolutionary and population genetic questions about the powdery mildew disease.

A. The Pathogen *Erysiphe graminis*

1. Life Cycle

The asexual cycle of *E. graminis* is characterized by germination of haploid conidia on the leaf surface of graminaceous plants followed by haustoria formation and hyphal growth (Fig. 1; see Sect. III for additional detail). Conidiophores are formed from which conidia are dispersed by wind (Fig. 1C). The sexual cycle involves anastomosis between hyphae of different mating types, which leads to a very short dikaryon stage. Bronson (1981) provided evidence that two alleles at one locus determine the mating type in *E. g. tritici*, and experiments by Hiura (1978) indicate that pairs of alleles control mating type both within and between formae specialis. The dikaryon stage is followed by formation of cleistothecia and devel-

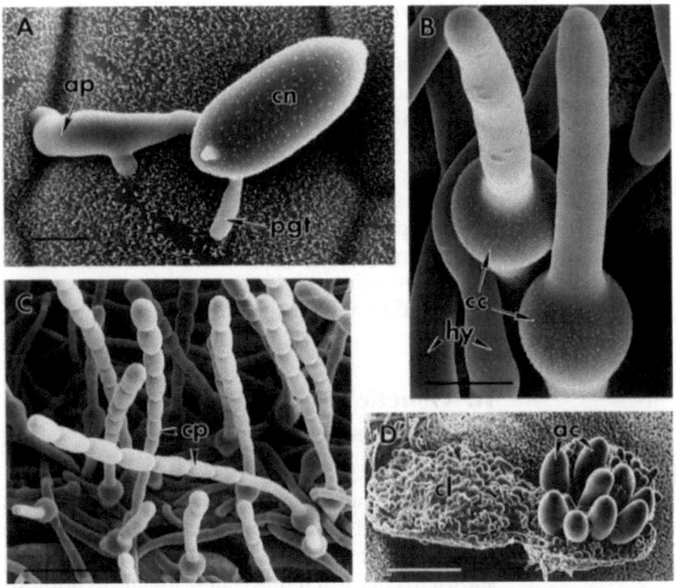

Fig. 1A–D. Scanning electron micrographs illustrating developmental stages of *E. g. hordei* on barley. Methods for cryogenic scanning electron microscopy are given in Klecan et al. (1990). **A** A conidium (*cn*) at 2 days post-inoculation with a primary germ tube (*pgt*) and an appressorial germ tube terminating in a hook-shaped appressorium (*ap*). *Bar* 10 µm. **B** Conidial mother cells (*cc*), which give rise to conidiophores, originate from hyphae (*hy*) growing over the surface of the leaf. *Bar* 10 µm. **C** Hyphae and conidiophores (*cp*) observed at 6 days post-inoculation. *Bar* 50 µm. **D** An open cleistothecium (*cl*) with asci (*ac*). *Bar* 100 µm

opment of asci, each with eight haploid ascospores (Fig. 1D; Spencer 1978). The sexual stage occurs in the autumn and the cleistothecia may produce viable ascospores for many years. Both the asexual conidia and the sexual ascospores are infectious. However, because the asexual phase predominates in agricultural ecosystems, most studies have focused on this phase of the *E. graminis* life cycle.

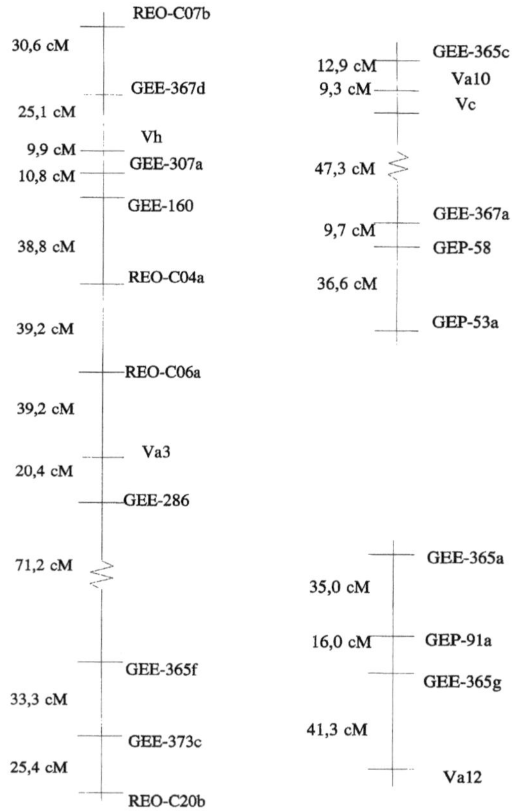

Fig. 2. Genetic map of *E. g. hordei* showing the distribution of five virulence gene loci (*V*) in relation to RFLP (*GEE* and *GEP*) and RAPD (*REO*) markers. The map distances are given in cM (centiMorgans)

2. Genome Structure

By light microscopy, *E. g. hordei* has been shown to have two large and five or six smaller chromosomes (Borbye et al. 1992). Pulsed-field-gel electrophoresis of chromosomes from *E. g. hordei* conidia indicate the presence of some chromosomes larger than 5 Mbp and several smaller chromosomes (Borbye et al. 1992). This is supported by evidence for at least one large linkage group containing nine avirulence gene loci (Jensen et al. 1995) and the identification of two large RFLP and RAPD linkage groups (Fig. 2).

The genome of *E. g. hordei* contains a significant amount of repetitive DNA that is organized as short interspersed nuclear elements (SINE) (Rasmussen et al. 1993; Wei et al. 1996). One element is dispersed throughout the genome with an average frequency of one copy per 19 kbp and represents at least 5% of the genome (Rasmussen et al. 1993), while a second SINE-like element, Eg-R1, occurs in only 50 copies per genome (Wei et al. 1996). In mammalian genomes, SINE elements move by retrotransposition, and it is possible that the *E. g. hordei* SINE elements are capable of transposition also. Thus, the genome of *E. g. hordei* may evolve, and avirulence genes may disperse throughout the genome, by transpositional events and cross-overs between homologous SINE elements. At least one SINE element is also present in the *E. g. tritici* genome.

3. Molecular Mapping

The relative ease with which crosses can be performed and the haploid nature of hyphae and conidia make *E. graminis* amenable to genetic analysis. Data on the segregation of avirulence genes in wheat and barley powdery mildews have been compiled (Jørgensen 1988); and linkage analyses with RFLP (Christiansen and Giese 1990; Brown and Simpson 1994) and RAPD (H. Giese, unpubl.) markers have been carried out in *E. g. hordei*. Avirulence loci are dispersed in the genome (Fig. 2), and there is no evidence for the presence of multigenic or highly polymorphic loci, suggesting that each avirulence gene product interacts with barley in a distinct manner.

In *E. g. hordei*, RFLP markers generally represent multicopy DNA sequences (O'Dell et al. 1989; Christiansen and Giese 1990), and single RFLPs identify up to eight loci. Analysis of yeast artificial chromosome (YAC)-containing DNA of *E. g. hordei* has confirmed the dispersed location of individual cross-hybridizing RFLP fragments (Borbye and Giese 1994). Closely linked molecular markers will provide starting points for map-based cloning of avirulence genes. In addition, molecular markers and good genetic maps will be invaluable tools in population genetic studies of *E. graminis* (Wolfe et al. 1992; McDermott and

McDonald 1993) and allow identification of pathotypes without dependency on test cultivars.

4. Concept of Formae Specialis

Seven *E. graminis* formae speciales have been defined on the basis of their adaptation to particular host species: four on cereals (i.e., *hordei*, *tritici*, *avenae*, and *secalis*) and three on other grasses (i.e., *agropyri*, *bromi*, and *poae*). The presence of homologous repetitive DNA sequences in *E. g. avenae*, *E. g. tritici*, and *E. g. hordei* (O'Dell et al. 1989) indicates a close relationship between these formae speciales, as does the presence of a SINE-like element in both *E. g. tritici* and *E. g. hordei* (Rasmussen et al. 1993). Tosa (1989) crossed *E. g. tritici*, the wheat pathogen, to *E. g. agropyri*, a wheatgrass pathogen. Four avirulence genes segregated in these crosses, with each avirulence gene acting in accordance to the gene-for-gene hypothesis. More recent work from the same group demonstrated that an isolate of *E. g. secalis* contained avirulence genes corresponding to six wheat powdery mildew resistances, *Pm1–Pm4a* (Matsumura and Tosa 1995). Certain pathotypes of *E. g. hordei* infect *Hordeum murinum* as well as cultivated barley (Wahl et al. 1978). Thus, host species specificity may be an inadequate criterion for taxon definition in *E. graminis* (Heath 1991; Andrivon and de Vallavieille-Pope 1992; Tosa 1992).

Both genetic data and the ability of different formae specialis to infect a common host support the hypothesis that species specificity is derived originally from avirulence genes (Tosa 1992). Species specificity caused by factors other than single avirulence genes may develop after an initial separation of *E. graminis* pathotypes. SINE-like elements, as described for *E. g. hordei*, probably contribute to genome reorganization and could promote the development of formae specialis (Rasmussen et al. 1993; Wei et al. 1996).

5. Avirulence Genes

Avirulence genes are operationally defined as genes that limit growth of the pathogen on specific host genotypes within a given host species. This definition may need to be revised in light of the results described above (Wahl et al. 1978; Tosa 1989; Matsumura and Tosa 1995). It is generally believed that an avirulence gene produces an elicitor that interacts with a receptor in the plant. Recognition results in incompatibility, while inactivation of an avirulence allele by mutation gives rise to a virulence allele. The isolation of an avirulence gene from *Cladosporium fulvum*, a fungal pathogen on tomato, has confirmed this hypothesis (Marmeisse et al. 1993; Chap. 1, this Vol.).

The obligate lifestyle of *E. graminis* has limited the identification of genes affecting host-pathogen interactions to those genes that have conditional or nonlethal phenotypes when mutated. Avirulence genes are the only such genes that have been characterized to date. More than 17 avirulence loci have been identified and mapped in *E. g. hordei* (Brown and Simpson 1994; Jensen et al. 1995; Fig. 2). These are likely to be the first genes influencing host-pathogen interactions to be cloned from *E. g. hordei*. Genes encoding essential functions will have to be isolated by alternate methods, such as subtractive hybridization or random large-scale sequencing of cDNA clones (Newman et al. 1994; Justesen et al. 1996).

From an evolutionary point of view, avirulence gene loci are thought to be disadvantageous to the fungus, unless the genes have additional functions unrelated to plant recognition. Studies on the contribution of avirulence genes to fitness have been inconclusive, as *E. graminis* pathotypes carry many avirulence genes, and not all have been well characterized. Not unexpectedly, segregation of avirulence/virulence in offspring from a cross between two pathotypes of *E. g. hordei* indicated the presence of more avirulence genes than predicted on the basis of the parent pathotypes (Jensen et al. 1995). Mutational analysis has not resolved the issue of why avirulence genes are maintained in populations. Partially virulent mutants were recovered from a mutagenized avirulent race (Sherwood et al. 1991). This might indicate a vital function for avirulence genes, as no completely virulent mutants were recovered.

B. Cereals

1. Resistance Genes

Conceptually, several biochemical processes and physiological functions could affect the outcome of host-pathogen interactions in the powdery mildew disease. However, research has focused on disease resistance genes employed in cereal breeding programs. These genes share the property of conferring significant levels of resistance to a limited number of races of *E. graminis* within a specific formae specialis. Most powdery mildew resistance genes that have been studied in detail conform to the gene-for-gene hypothesis (Jørgensen 1988). Genes with minor effects on the interaction or that affect the interaction in a nonspecific manner have not been studied extensively, with the exception of the *ml-o* locus of barley.

a. Natural and Induced Sources of Resistance

A comprehensive knowledge of genes conferring resistance to both local and foreign races of *E. graminis* has provided the foundation for breeding efforts. To this end, the resistance genes present in modern cultivars, as well as those in landraces and wild barley species, are being catalogued (Brown and Jørgensen 1991). New resistance specificities continue to be discovered (Fischbeck and Jahoor 1991).

Thirteen loci conferring resistance to powdery mildew, represented by at least 90 alleles, have been mapped onto the barley genome (Jørgensen 1992; von Wettstein-Knowles 1992). The best-characterized, the *Ml-a* locus on barley chromosome 5 (= 1H), is a complex locus for which more than 30 alleles have been named (Jahoor and Fischbeck 1993). This locus may consist of a multigene family rather than a single, highly polymorphic gene (Giese 1981; Mahadevappa et al. 1994). Several additional powdery mildew resistance loci cluster to the same arm of chromosome 5, including *Ml-n*, *Ml-nn*, *Ml-k*, *Ml-ra*, and *Ml-at*. Four resistance loci are distributed along chromosome 4 (= 4H) [i.e., *Ml-g*, *ml-o*, *N182*, and *Ml(CP)*]. The remaining powdery mildew resistance loci are dispersed in the genome or unmapped.

The genes for race-specific resistance to powdery mildew are generally dominant, each having a unique, highly specific and positive function. Mutation, which generally leads to a loss of gene function, has not led to any new resistances, with the exception of the *ml-o* resistance gene (Jørgensen 1983, 1991). A natural source of *ml-o* resistance was described subsequently in Ethiopian landraces of barley (Negassa 1985). Resistance encoded by *ml-o* is unique among powdery mildew resistance genes in barley. Plants carrying *ml-o* alleles are resistant to all races of *E. g. hordei* and exhibit the same resistance reaction with only minor quantitative differences. Necrotic leaf spotting is associated with the *ml-o* allele and the severity of this pleiotropic effect varies with the particular allele (Kjær et al. 1990).

b. Mutations to Susceptibility

A popular mechanistic model of the gene-for-gene hypothesis is that a resistance gene product interacts with a pathogen avirulence gene product (Aist and Bushnell 1991). Modifications of either the host or the pathogen that prevent or reduce this interaction are predicted to result in increased compatibility between the organisms. To test this hypothesis, a resistant barley cultivar with the *Ml-a12* resistance gene was mutagenized and screened for susceptible mutants (Torp and Jørgensen 1986). The mutants exhibited partial but not full susceptibility to one race of *E. g. hordei*. These observations support the hypothesis that the *Ml-a* gene product plays an active role in resistance. In addition, new genes, *Nar1* and *Nar2*, were identified (Freialdenhoven et al. 1994). These two genes are required for the normal expression of *Ml-a12*-mediated resistance, and thus are thought to define components of a signal transduction pathway leading to the activation of defenses, including the hypersensitive necrosis response (Torp and Jørgensen 1986; Freialdenhoven et al. 1994).

2. Molecular Mapping

High-density maps incorporating molecular markers are being generated for the cereals. Such maps are potentially useful both in the context of marker-directed plant breeding and in the map-based cloning of resistance genes. It is noteworthy that in barley RFLP markers 0.7 cM from *Ml-a* (Schüller et al. 1992) and without 2.4 cM of *ml-o* (Hinze et al. 1991) have been identified.

Based on the synteny among Triticeae genomes, extrapolation among genomes may provide approximate gene locations and gene orders. As an example, 60% of rice single-copy sequences hybridized to single-copy sequences in wheat and barley (Moore et al. 1993). Thus, resistance genes identified in one grass species may be utilized to recover homologous genes from the remaining grass species. Minor genes contributing to durable resistance are unlikely to be the same among the cereals, as they do not map to the same chromosomes in barley and wheat (Moore et al. 1993).

3. Induced Resistance and Susceptibility

Systemically induced (= systemically acquired) resistance has been induced in barley with avirulent or virulent isolates of *E. g. hordei*

(Hwang and Heitefuss 1982), culture filtrates from *Bacillus subtilis* (Stenzel et al. 1985) or 2,6-dichloroisonicotinic acid (Kogel et al. 1994). Induced resistance was measured as a reduction in colony number and size, and paralleled resistance mechanisms observed in genetically resistant hosts. At the molecular level, mRNA for pathogenesis-related protein 1b (= PR1b), a peroxidase, and a chitinase accumulated in barley following chemical treatments that induced systemic resistance (Muradov et al. 1993; Kogel et al. 1994).

In addition, local induced resistance, which operates over a distance of one or a few cells, has been reported. At high inoculation densities, the proportion of fungal conidia that successfully establish a haustorium is much lower than at low inoculation densities (Thordal-Christensen and Smedegaard-Petersen 1988a). Similarly, axenically grown seedlings support higher rates of successful penetration and haustorium formation by the powdery mildew pathogen than nonaxenic plants (Sahashi et al. 1989). These results suggest that either *E. g. hordei* or microorganisms elicit local induced resistance to *E. g. hordei* (Kunoh et al. 1988a; Kobayashi et al. 1995). In some experiments, the timing of induced resistance was correlated with the formation of primary germ tubes by *E. g. hordei* (Shiraishi et al. 1989), and local induced resistance appears to be associated with a papilla-based resistance mechanism (Sahashi and Shishiyama 1986). A papilla-regulating extract, which was shown to be potassium phosphate, potentiates susceptible barley to respond with pronounced papillae formation following infection by *E. g. hordei* (Yokoyama et al. 1991; Inoue et al. 1994). This compound, in conjunction with Ca^{2+}, may play a role in the establishment of local induced resistance (Inoue et al. 1994).

A parallel phenomenon of local induced susceptibility has been reported in which prior inoculation with a virulent race of powdery mildew will render a cell susceptible to subsequent infection by races that are normally avirulent or nonpathogenic on barley (Ouchi et al. 1976; Kunoh et al. 1985a). At short time intervals, induced susceptibility is expressed only in cells directly infected by a virulent propagule. By 24h after inoculation, induced susceptibility spreads to adjacent epidermal cells. To explain induced susceptibility, it has been suggested both that *E. g. hordei* releases a suppressor that counteracts the local induction of resistance (Komura et al. 1990) and that *E. g. hordei* depletes host nutrients, thereby limiting defensive responses (Yamaoka et al. 1994).

C. Summary

A large number of single genes governing avirulence in *E. graminis* or resistance in cereals have been described. In no case is the biochemical basis for avirulence or resistance known. This remains one of the most important limitations to understanding and ultimately manipulating disease resistance. Under current practices, introduction of resistance genes in the host plants results in strong selection for virulent pathotypes, and current agricultural practices render a resistance gene effective for only about 5 years. With a deeper understanding of the nature of disease development and of the basis for allelic variation at avirulence and resistance loci, better strategies for genetically engineering cereals for durable disease resistance can be developed.

III. Infection Sequence

The cytology of the infection stages of the Erysiphales in general, and of the host-parasite interaction of *E. graminis* on cereals in particular, has been investigated extensively (Johnson et al. 1979; Kunoh 1982; Carver 1988; Carver et al. 1995; Nicholson and Kunoh 1995). In contrast, progress in understanding the structure and biochemical composition of the host-parasite interface, as well as mechanisms of resistance, has been limited by the obligate lifestyle of these pathogens. As the infection takes place only in the epidermis, superior preservation of ultrastructural details following low temperature methods of sample preparation is possible. These methods, in conjunction with immunocytochemical detection techniques, provide new tools to analyze molecular events associated with the development of fungal infection and resistance mechanisms (Hippe-Sanwald 1993, 1995; see also Chap. 3, Vol. V, Part A). In this section, the infection sequences of *E. g. hordei* on barley in relation to distinctive host responses governed by the *Ml-a* and *ml-o* loci are emphasized.

A. Germination

E. graminis is an obligate pathogen, and two complementary hypotheses to account for this lifestyle have been proposed. *E. graminis* may be dependent on an essential nutrient from the host plant or the host may provide a developmental cue to the parasite (e.g., a signal leading to the formation of haustoria). As *E. graminis* will form appressoria on cellulose films (Kobayashi et al. 1991), no unique host features or signals are required in the first stages of development.

The asexual cycle begins when a conidium is released from a conidiophore and lands on a suitable plant host surface (Fig. 3). The earliest event described is the contact-stimulated release of a protein-containing fluid from conidia (Kunoh et al. 1988b; Nicholson and Kunoh 1995). This fluid contains esterases, including a cutinase that dissolves amorphous materials from the leaf surface, as visualized by cryogenic scanning electron microscopy (Nicholson et al. 1988; Kunoh et al. 1990; Pascholati et al. 1992). The cutinase activity may facilitate attachment to the leaf surface or aid rapid penetration of the epidermis (Pascholati et al. 1992; Nicholson and Kunoh 1995). Alternatively, cutin monomers may serve as a carbon source for the conidium or they may act as signal molecules triggering either defensive responses in the host or developmental responses in the pathogen (Kunoh et al. 1990). In later experiments, these authors proposed that the liquid exudate makes the leaf surface more hydrophilic and hence more suitable for appressorium formation (Nicholson et al. 1993).

Conidia of *E. graminis* show optimal germination under conditions of high relative humidity. However, free water will cause the conidia to lose their viability. Germination does not occur while conidia are attached to the condiophore. However, contact with a solid surface is not required for conidial germination, suggesting that germination depends upon removal of a maternal suppressor (Aist and Bushnell 1991; Carver et al. 1995). Upon germination, *E. graminis* conidia form a short primary germ tube within 1–2 h (Fig. 1A, pgt). While conidia may produce more than one short germ tube, only the primary germ tube is capable of anchoring the conidium to plant surfaces (Kunoh 1982; Carver 1988). The primary germ tube is capable of taking up Ca^{2+}, silicon, and water from the host cell, which may be important to survival of the conidium (Kunoh 1982; Carver and Bushnell 1983; Aist and Gold 1987). A larger appressorial germ tube and appressorium are formed within 5–8 h (Fig. 1A). Prior to maturation of the appressorium, no differences in pathogen development on the near-isogenic resistant and susceptible barley lines are observed (Aist and Bushnell 1991; Hippe-Sanwald et al. 1992).

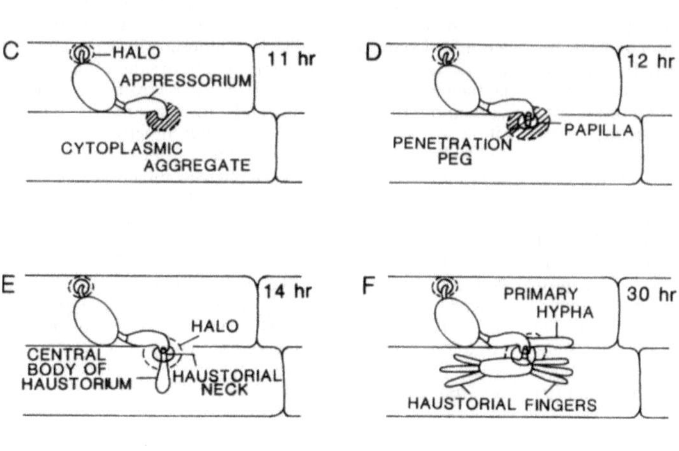

Fig. 3A–F. Line drawing of the infection cycle for compatible and incompatible interactions, giving the timing of major events in the infection sequence. (Aist and Bushnell 1991)

B. Penetration

1. Primary Germ Tube

Barley responds to the initial penetration attempt by the primary germ tube with nonspecific defensive responses, such as aggregation of cytoplasm and formation of an autofluorescent cell wall halo and a papilla, at the penetration site (Figs. 5, 7). The early induction of plant defensive responses during the attempted penetration by the primary germ tube occurs in both compatible and incompatible interactions (Thordal-Christensen and Smedegaard-Petersen 1988a,b; Boyd et al. 1994; Carver et al. 1995). Esterases, which are secreted by the conidium prior to germination, may play a role in eliciting these responses from the host (Kunoh et al. 1990; Nicholson and Kunoh 1995).

2. Appressorium

A penetration peg, which forms from the undersurface of the appressorium, is the primary mechanism by which the *E. graminis* gains entry to the host epidermal cell. In the host cell below the penetration peg, a cytoplasmic aggregate usually forms (Kunoh 1982; Kunoh et al. 1985b; Carver et al. 1995). These aggregates contain polyphenols, lipid, protein including hydrolytic enzymes, and callose but not suberin, cellulose, lignin, or pectin (Takahashi et al. 1985; Russo and Bushnell 1989). Cytoplasmic aggregates are thought to support the formation of halos and papillae (Kunoh et al. 1985b; Aist and Bushnell 1991). Kobayashi et al. (1993) suggest that the fungal propagule releases a signal or elicitor prior to penetration, which induces physiological changes in barley cells, causing formation of cytoplasmic strands and cytoplasmic aggregates.

Halos, locally modified regions of the host cell wall, form at penetration sites. They may provide a seal against excessive water loss around the penetration peg (Aist 1983). These structures have been implicated in plant resistance to powdery mildew (Aist and Bushnell 1991). Opaline silica in association with cuticular lipids, as well as hydrolytic enzymes and phenolic compounds, are reported to occur in halos and may impede growth of the penetration peg as it traverses the host cell wall (Kunoh et al. 1985b; Takahashi et al. 1985).

In barley with *ml-o* resistance alleles, halos form earlier and attain a larger size than halos that form in the susceptible *Ml-o* line after infection (Skou et al. 1984). However, halos are observed in susceptible and resistant plants, independent of whether penetration is successful. Thus, the reduction of water loss may be the only benefit the plant cell derives from halos (Aist et al. 1988).

3. Papilla

Barley forms cell wall appositions or wound plugs, called papillae, at encounter sites with *E. g. hordei* primary germ tubes and appressorial penetration pegs (Figs. 4, 5; Aist 1983; Takahashi et al. 1985; Kunoh et al. 1986; Smart et al. 1986). The presence of phenolics, which account for autofluorescence (Carver et al. 1991), callose, silicon compounds,

Fig. 4a–f. Transmission electron micrographs illustrating papillae in interactions between *E. g. hordei* and either a resistant barley line or a near isogenic susceptible line. (Hippe-Sanwald et al. 1992). **a** Oblique section of a susceptible barley coleoptile epidermal cell infected with *E. g. hordei*. One plugged penetration site (*ph*) encased in a papilla (*p*) is correlated with a failed penetration attempt by *E. g. hordei*. In the host cell wall (*hocw*), an unplugged penetration hole (*ph*) is located close to a second papilla (*p*) and a young haustorium (*h*). Host cytoplasm (*hocy*) in the vicinity of the infection sites is visible. *Bar* 5 µm. **b** An enlarged view of **a**. An unplugged infection peg (*ph*) located in the middle of a papilla (*p*) and surrounded by host cytoplasm (*hocy*) is pictured. The extrahaustorial matrix (*ema*) is closely attached to the haustorium (*h*). Along most of the haustorium, no space is visible in between extrahaustorial membrane (*em*) and haustorial cell wall (*hcw*). *Bar* 1 µm. **c** Infection site in epidermal tissue of a resistant barley line. An oblique section of the host epidermal cell wall (*hocw*) with a penetration hole (*ph*) and a papilla (*p*) is shown. Local darkening of the host cell wall (see *brackets*) and of the material forming the papilla is observed. *Bar* 2 µm. **d** An oversized papilla (*p*) observed in resistant coleoptile tissue. The papilla is composed of aggregates of diverse electron-dense material. On both sides of the papilla, the periclinal plant cell wall (*hocw*) is covered by a layer of material resembling the papilla in structure and composition. An aggregation of dense material between the host cell wall and plasma membrane (*pl*) can be observed in an adjacent cell. *Bar* 2 µm. **e** A plugged infection peg (*ph*) found within a papilla (*p*) in resistant tissue. The infection peg is filled with an electron-dense material of homogeneous density. The papilla consists of small, electron-dense, granular bodies. Lipid-like droplets (*l*) are embedded in a network of fibrillar material (*f*). *Bar* 0.5 µm. **f** A plugged penetration hole (*ph*) formed by *E. g. hordei* in resistant epidermal tissue. The papilla (*p*) is notable for its large size and distinct, layered structure. The host cytoplasm (*hocy*) appears highly degenerated in this interaction. *Bar* 2 µm

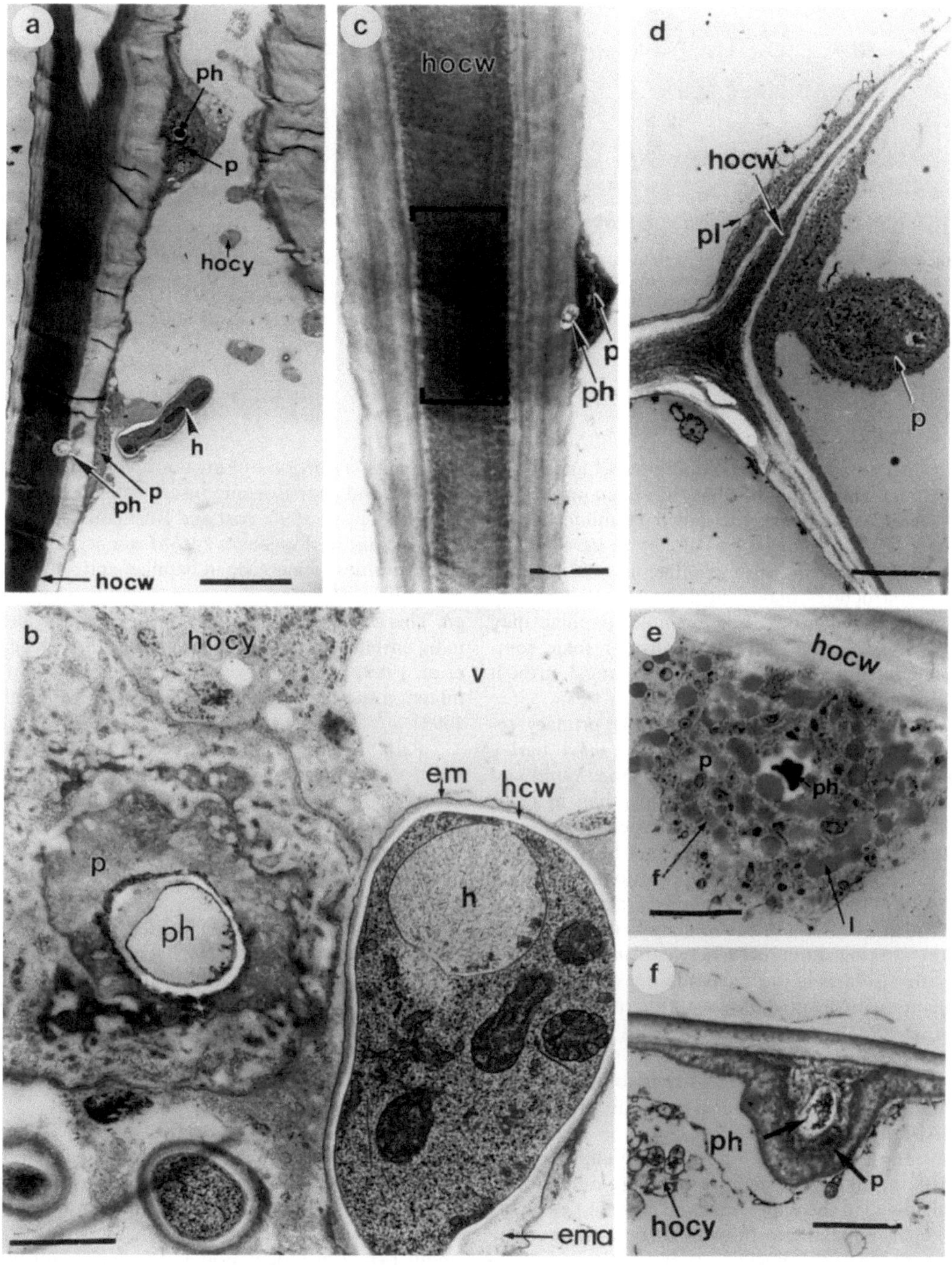
a
ph
p
hocy
h
ph
p
hocw
c
hocw
p
ph
d
hocw
pl
p
b
hocy
v
em
hcw
p
ph
h
ema
e
hocw
p
ph
f
l
f
ph
p
hocy

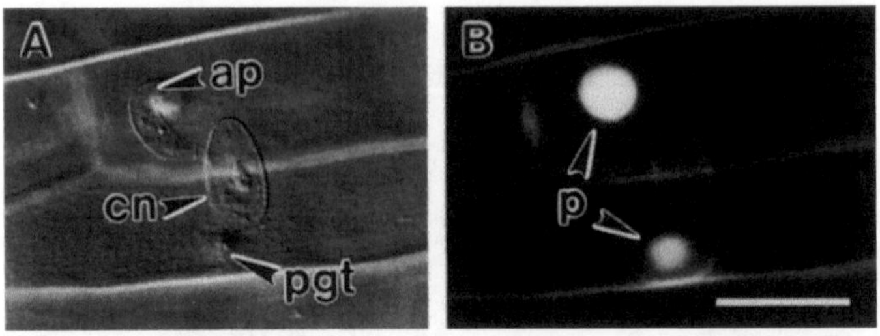

Fig. 5A,B. Light micrographs of papillae formed in an incompatible interaction between resistant barley and *E. g. hordei*. *Bar* 10 µm. **A** DICT image showing a conidium (*cn*) with a primary germ tube (*pgt*) and an appressorium (*ap*). **B** Fluorescence image from the same site. Note the callose-containing papillae (*p*) associated with attempted penetrations by the primary germ tube and the penetration peg that develops from the appressorium. Methods for aniline-blue staining of callose are given in Eschrich and Currier (1964)

Ca^{2+}, potentially toxic compounds, and basic staining materials in papillae, has been implicated in resistance of barley to powdery mildew (Aist 1983; Aist and Israel 1986; Wei et al. 1994). However, the exact role of papillae is unclear, and these structures may contribute to resistance by several mechanisms. For example, papillae may restrict penetration mechanically, or toxic compounds in papillae may inhibit fungal growth (Koga et al. 1980; Aist et al. 1988).

Papillae are thought to be the primary resistance mechanism operating in *ml-o* barley genotypes. Papillae formation is more rapid and extensive in resistant *ml-o* barley than in near-isogenic susceptible *Ml-o* genotypes (Skou et al. 1984). In addition, *ml-o* barley lines have a greater tendency to form cell wall appositions and develop these structures spontaneously in axenic plants (Wolter et al. 1993). Microanalysis did not reveal any striking differences between *ml-o* and *Ml-o* barley isolines in the accumulation of various elements, with the possible exception of Mg^{2+} (Zeyen et al. 1992). However, papillae in barley carrying the *ml-o* resistance allele contain a light-absorbing, phenylpropanoid-rich component that may be involved in the *ml-o* resistance mechanism (Aist et al. 1988).

Barley carrying *Ml-g* resistance alleles permit little or no development of powdery mildew conidia beyond the appressorium stage. In these genotypes, papillae are thought to be an important, but not the sole, mechanism preventing establishment of haustoria (Görg et al. 1993). In contrast, in *Ml-a1*-mediated resistance, papillae formation is a nonspecific defensive response, which is effective against a portion of fungal propagules in both resistant and near-isogenic susceptible lines (Figs. 4, 5; Koga et al. 1990; Aist and Bushnell 1991). In the resistant barley leaves (*Ml-a1* genotype), silicon continues to increase in papillae and adjacent cell wall areas after callose deposition is completed. This process is thought to occur passively and to be unrelated to the direct action of *Ml-a1* (Koga et al. 1988). The significance of ions localized at infection sites is unclear (Kunoh 1990; Zeyen et al. 1992).

Papillae formation has been examined by electron microscopy in resistant barley, with *Ml-a1* resistance alleles, and in a near-isogenic susceptible line. Plugged as well as unplugged penetration sites were observed (Fig. 4). In general, the papillae were characterized by a layered structure (Fig. 4b,f). The papillae were often irregular in shape and of varying size (Fig. 4d–f). Local darkening of the host coleoptile cell wall and of papillae was observed frequently in the resistant line (Fig. 4c,d; Zeyen and Bushnell 1979; Hippe-Sanwald et al. 1992).

C. Haustorium

Haustorium formation and subsequent development of hyphae and conidiophores occur only on living epidermal cells and have not been induced to form in vitro. The haustorium begins to form at about 14 h after inoculation and matures over the next 2–3 days to form a characteristic, lobed structure with numerous finger-like projections. Haustoria of *E. graminis* are essential organs for attachment and nutrient supply (Fig. 6; Manners and Gay 1983; Manners 1989), and their establishment is a prerequisite for formation of hyphae on the surface of the epidermis (Fig. 1C; Masri and Ellingboe 1966). The initiation of haustorium formation is correlated with the hypersensitive necrosis response in *Ml-a*-mediated resistance (Figs. 6, 7; Koga et al. 1990; Hippe-Sanwald et al. 1992; Görg et al. 1993).

With low temperature sample preparation for electron microscopy, the haustoria that develop in susceptible barley appear well defined and smoothly contoured (Figs. 4b, 6a–c; Hippe 1985; Hippe-Sanwald et al. 1992). One striking feature of young haustoria is the proliferation of vesicular bodies, primarily in the periphery of the haustorial central body and in the tips of haustorial fingers (Fig. 6b,c). The haustorial cell wall consists of chitin, as demonstrated by its sensitivity to chitinase treatment (Toyoda et al. 1991). Monoclonal antibodies raised to isolated pea/*E. pisi* haustorial complexes, identified a 62-kDa glycoprotein located in the haustorial plasma membrane and a second glycoprotein located in haustorial and hyphal cell walls (Mackie et al. 1991). This general approach, when coupled with cloning of the corresponding cDNAs, promises to uncover many of the components of the haustorium and the pea/*E. pisi* interface (Green et al. 1995).

The extrahaustorial membrane forms the interface between the pathogen and the host (Figs. 4b, 6d). This specialized membrane is joined to the host plasma membrane at the neck band region. Numerous investigations have shown that extrahaustorial membrane differs from the host plasma membrane structurally and biochemically (Manners and Gay 1983; Green et al. 1995). The extrahaustorial membrane of intact pea/*E. pisi* haustorial complexes excludes uranyl-phosphate precipitates and horseradish peroxidase, suggesting that this membrane is semipermeable in nature (Gay and Manners 1987). Histochemical techniques have demonstrated that the extrahaustorial membrane does not contain ATPase activity, suggesting that transport across this membrane is passive (Manners and Gay 1983). No specific transporters or permeases have been localized to the extrahaustorial membrane.

The extrahaustorial matrix separates the extrahaustorial membrane from the haustorial body (Figs. 4b, 6a,d). It is assumed that the extrahaustorial matrix contributes to recognition and signaling between host and pathogen, in addition to its role in nutrient transfer (Manners and Gay 1983). The matrix area is thought to include gel-like glucans (Gay and Manners 1987). In addition, antibodies were used to localize threonine hydroxyproline-rich glycoprotein (THRGP)-like molecules to the extrahaustorial matrix of both rust and powdery mildew haustorial complexes (Fig. 8; Hippe-Sanwald 1993; Hippe-Sanwald et al. 1994). This result suggests that the extracellular matrix may represent a modified cell wall barrier between host and pathogen. No differences between incompatible or compatible interactions were found in the spatial pattern of THRGP-like epitopes. Thus, no role for THRGPs in resistance to powdery mildew was indicated (Hippe-Sanwald et al. 1994).

With fluorescent potentiometric indicator dyes, the physiological status of haustorial mitochondria can be monitored in living *E. g. hordei* (Bushnell et al. 1987; Mendgen and Nass 1988). D-glucose stimulated respiration and thereby increased the membrane potential more effectively than sucrose or other sugars (Mendgen and Nass 1988). Because *E. g. hordei* hyphae have α-glucosidase activity capable of metabolizing sucrose (Donaldson and Jørgensen 1988) and because sucrose is a common transport metabolite in plants, this sugar has been considered the major metabolite transferred to *E. graminis*. Although the results of Mendgen and Nass (1988) do not contradict this hypothesis, they do suggest that glucose can be effectively utilized by *E. g. hordei*.

Little else is known about the metabolism of *E. graminis* aside from information arising from the studies of fungicides. Hollomon and colleagues have ascertained that *E. g. hordei* does not have enzymes for de novo purine synthesis, although it does possess enzymes for the purine salvage pathway (Hollomon 1984). *E. g. hordei* will readily take up adenosine and convert it to inosine via the enzyme adenosine deaminase (Butters et al. 1985). This enzyme is particularly sensitive to inhibition by the fungicide, ethirimol. Ethirimol resistance is associated with reduced fitness under field conditions, suggesting that ethirimol target site(s) serve a vital function in *E. g. hordei* (Hollomon 1981). *E. graminis* is also sensitive to fungicides of the sterol biosynthesis inhibitor class, suggesting that sterol biosynthesis is similar to that found in other fungi. Haustorium formation is particularly sensitive to sterol biosynthesis inhibitors, while ethirimol inhibits appressorium formation (Hollomon 1982).

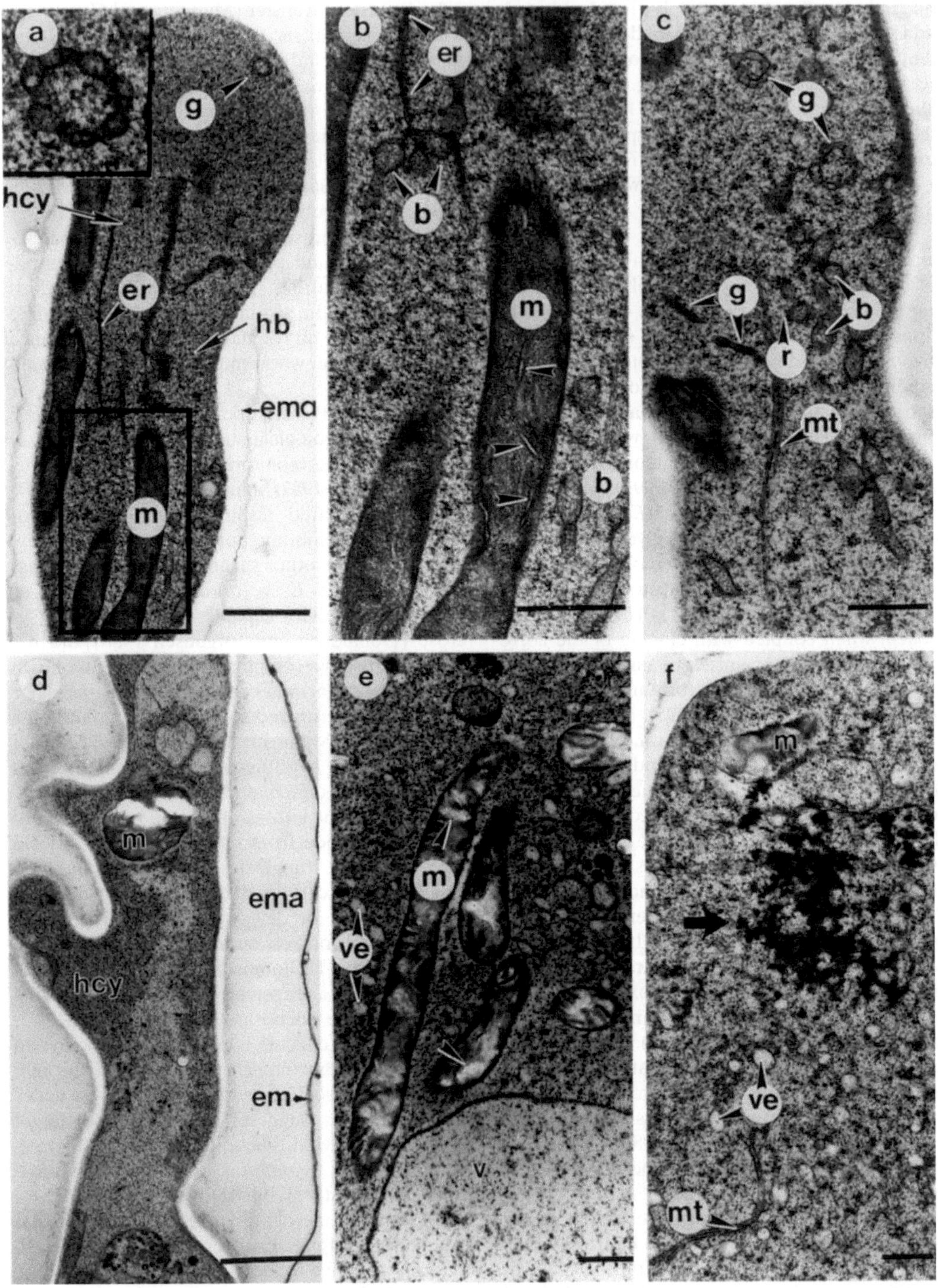
a
g
hcy
er
hb
ema
m
b
er
b
m
b
c
g
g
b
r
mt
d
m
ema
hcy
em
e
m
ve
v
f
m
ve
mt

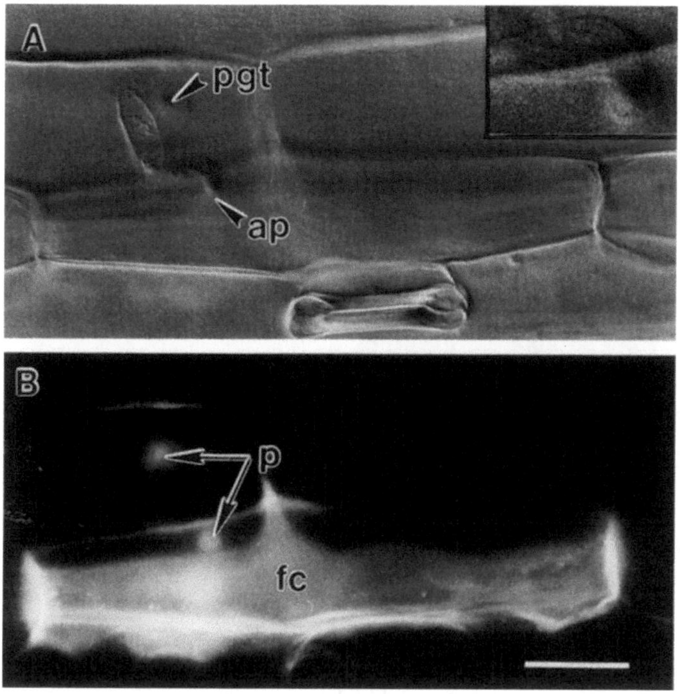

Fig. 7A,B. Light micrographs of an autofluorescent cell from an incompatible interaction between barley and *E. g. hordei*. *Bar* 10 μm. **A** DICT image of a conidium with a primary germ tube (*pgt*) and an appressorium (*ap*). *Inset* displays the appressorium at higher magnification. **B** Fluorescent image of the same site illustrating an autofluorescent cell (*fc*) and papillae (*p*)

Fig. 6a–f. Composite of transmission electron micrographs illustrating haustoria formed in interactions between *E. g. hordei* and a susceptible line (**a–c**) or a near-isogenic resistant barley line (**d–f**). (Hippe-Sanwald et al. 1992). **a** An overview on the fine structure of a haustorial body (*hb*) surrounded by the extrahaustorial matrix (*ema*). Within a homogeneously structured cytoplasm (*hcy*), composed of ribosomes and a fibrillar matrix, the following organelles are visible: mitochondria (*m*), Golgi-like bodies (*g*), and endoplasmic reticulum (*er*). *Inset* displays an annular Golgi-like body found lying at the periphery of the haustorium at higher magnification. *Bar* 1 mm. **b** A section of **a** under higher magnification. Mitochondria (*m*) showing cristae (*arrowheads*) in a dense matrix are illustrated. Some enlarged areas of the endoplasmic reticulum (*er*) showing pronounced blebbing (*b*) of the membrane system are also presented. *Bar* 0.5 μm. **c** An example of the fine structure of a young haustorium of *E. g. hordei* in a susceptible epidermal cell. Microtubule sections (*mt*), rod-like as well as annular Golgi-like structures (*g*), and an aggregation of vesicles (*b*) are visible at the periphery of the haustorium. Ribosomes (*r*) are attached to the surface of these vesicle-like structures. *Bar* 0.5 μm. **d** An overview of the cytoplasm (*hcy*) of a haustorium observed in resistant tissue. The center of the haustorial body looks cleared, which is indicative of advanced lysis. A loss of cellular integrity and compartments is evident. Remnants of a mitochondrion (*m*) are seen. The extrahaustorial membrane (*em*) is separated from the haustorial body and the extrahaustorial matrix (*ema*) looks completely degraded. *Bar* 1 μm. **e** A damaged haustorium of *E. g. hordei* from an incompatible interaction displayed at higher magnification. The mitochondria (*m*) are degenerated; the mitochondrial matrix appears cleared and fragments of smooth, lysed cristae are seen (*arrowheads*). The haustorial cytoplasm surrounding the mitochondria contains many empty-looking vesicles (*ve*) and a large vacuole (*v*). *Bar* 0.5 μm. **f** A typical haustorium of *E. g. hordei* in an advanced stage of degeneration in resistant barley tissue. An accumulation of electron-dense material (*large arrow*) can be observed. Lysed mitochondria (*m*), microtubules (*mt*) and a large number of small vesicles (*ve*) are found in this region. *Bar* 0.5 μm

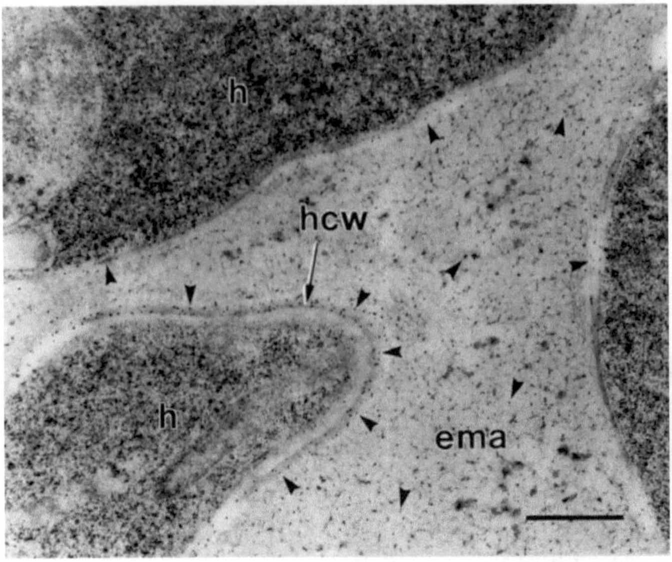

Fig. 8. Details of the extrahaustorial matrix (*ema*) of powdery mildew haustorial bodies (*h*) in susceptible barley epidermal tissue. Note the homogeneous distribution of THRGP-like epitopes over the extrahaustorial matrix (*ema*) and outer parts of the haustorial cell wall (*hcw*) as indicated by the *arrowheads*. Methods for the immunogold localization of THRGP are given in Hippe-Sanwald et al. 1994. *Bar* 1 μm

D. Hypersensitive Necrosis Response

The hypersensitive necrosis response is defined as all observable events associated with death of plant cells in response to pathogen attack (Wright and Heale 1988). Hypersensitive necrosis occurs only when a haustorium is present. Usually, only the epidermal cell directly under attack dies, but adjacent epidermal and mesophyll cells may also die. The hypersensitive necrosis is strongly correlated with *Ml-a-* but not *ml-o*-mediated resistance to powdery mildew in barley (Koga et al. 1990; Zeyen et al. 1995). In *Ml-g*-resistant barley, hypersensitive necrosis also occurs, but its role in resistance is unclear (Görg et al. 1993).

Cytological events described in cells of resistant *Ml-a* barley undergoing hypersensitive necrosis include a halt in cytoplasmic streaming, the accumulation of electron-dense material, disruption of the host nucleus and breakdown of the tonoplast (Bushnell 1981; Koga et al. 1988; Aist and Bushnell 1991). Increases in autofluorescence associated with the hypersensitive necrosis response probably derive from the release of phenols from the vacuole (Fig. 7) (Aist and Israel 1986; Koga et al. 1988; Carver et al. 1991). This observation is supported by inhibitor studies, which suggest that products of phenylalanine ammonia lyase and cinnamyl alcohol dehydrogenase are required for *Ml-a1*-mediated resistance but not for *ml-o5*-based resistance (Zeyen et al. 1995). In some epidermal cells, the ultrastructural changes are so drastic that they give the appearance of a hypersensitive cell death of the host cell, whereas in other cells the changes are very slight (Hippe-Sanwald et al. 1992).

The few haustoria found in *Ml-a1*-resistant barley tissue show clear evidence of degeneration (Fig. 6d–f; Hippe-Sanwald et al. 1992). The absence of distinct endoplasmic reticulum and Golgi-like cisternae, the formation of vacuoles, and the occurrence of a distended extrahaustorial matrix are characteristic of intermediate stages of haustorial degeneration (Fig. 6d,f). This process of degeneration is not observed in haustoria developing in susceptible barley (Hippe-Sanwald et al. 1992). The lysis of mitochondria was the most striking early feature of haustorial degeneration observed in the incompatible interaction between powdery mildew and *Ml-a1* resistant barley (Fig. 6e,f; Hippe-Sanwald et al. 1992).

E. Summary

Cytological studies suggest that different barley resistance loci activate distinct defensive mechanisms in response to infection by *E. g. hordei.*

Resistance mediated by *ml-o* is an example of a nonspecific resistance mechanism based on papilla formation. *Ml-g* confers race-specific resistance at the appressorial germ tube stage, in part, by papilla formation. In contrast, resistance mediated by the *Ml-a* locus is correlated with the hypersensitive necrosis response and a halt of fungal growth during haustorium formation. It is noteworthy that although more than 13 distinct powdery mildew resistance loci have been described in barley, the resistance responses of only three of these loci have been characterized. Analysis of additional resistance loci may suggest alternative steps in the infection sequence that can be interrupted, and thereby broaden our understanding of the interactions between *E. graminis* and its hosts.

From analyses of compatible interactions between *E. graminis* and its cereal hosts, critical steps in the infection sequence may be identified. For example, given the parallel function of the extrahaustorial membrane to that of vesicular-arbuscular mycorrhizae (Bonfante-Fasolo and Perotto 1992) or the peribacteroid membrane in legume-*Rhizobium* interactions (Verma et al. 1992), it may prove informative to ask whether features of these two interfaces also occur at the powdery mildew interface. These components may serve as targets of future attempts to control powdery mildew diseases and provide an alternate disease control strategy to the use of race-specific resistance genes.

IV. Host Responses to Infection

To illustrate the physiological consequences of *E. graminis* infection, near-isogenic resistant and susceptible barley lines were inoculated with the identical race of powdery mildew (Smedegaard-Petersen and Stølen 1981). Seed yield was reduced to 73% in the susceptible isoline and to 95% in the resistant isoline compared to uninoculated plants. This experiment demonstrates that there is an energy cost associated with resistance, although this cost is clearly less significant than losses associated with disease. The second conclusion from this experiment is that *E. g. hordei* does not devastate the barley plants, which distinguishes infections by obligate fungal pathogens from other more destructive pathogens. Experiments outlined below describe physiological consequences of disease on susceptible plants, and biochemical and molecular changes in resistant cereals. Fewer studies detailing physiological changes in incompatible interactions and biochemical and molecular changes during compatible interactions have been published.

A. Physiological Aspects

Changes in basic physiological processes, like respiration and photosynthesis, have been measured in powdery mildew-infected plants to assess the basis for yield reductions in diseased plants. Increases in dark respiration rates have been measured in both resistant and susceptible barley. A transient burst of respiration followed inoculation of the resistant plants and a sustained increase in oxygen uptake began at 48 h after inoculation in susceptible barley (Millerd and Scott 1956; Smedegaard-Petersen and Stølen 1981). Increases in respiratory activity have been attributed to increases in alternative oxidase-based respiration (CN-insensitive respiration) (McAinsh et al. 1989, 1991) or to increases in both alternative oxidase and cytochrome oxidase-based respiratory pathways (Farrar and Rayns 1987). *E. graminis* does not contribute significantly to measured levels of respiration in infected leaves. The transient burst of respiration between 1 and 3 days after inoculation in resistant barley may support increased demands for energy and C-skeletons by the defensive responses. In contrast, in susceptible barley, the sustained increase in respiration most likely contributes both energy and C-skeletons to pathogen growth.

In powdery mildew-infected plants, photosynthesis rates decrease as the powdery mildew disease develops, with early decreases occurring prior to visible chlorotic symptoms (Edwards 1970). There is a progressive loss of photosystem components and chlorophyll during the course of infection. No one component of the photosynthetic apparatus is preferentially lost or becomes rate-limiting (Holloway et al. 1992). As noted below, decreases in photosynthesis are preceded by reduced expression of genes encoding components of the photosynthetic apparatus (Higgins et al. 1985). Increases in photosynthesis in distal, uninfected leaves of infected plants have been measured at 6–9 days after infection (Walters and Ayres 1983; Williams and Ayres 1981). Such increases most likely reflect increased sink strength by infected leaves and may partially offset reduced photosynthesis in infected leaves.

The physiological changes outlined above are thought to occur in part because the pathogen provides a strong sink that directs metabolites to itself and away from its host. The limited number of biochemical studies focused on photosynthetic processes tend to support this view (Scholes et al. 1994; Wright et al. 1995a,b). For example, changes in starch and sucrose in host plants and changes in the metabolic regulator, fructose-2,6-bisphosphate, are consistent with decreased synthesis of sucrose in photosynthetically compromised susceptible leaves (Klecan and Buchanan 1988).

Mineral nutrition is altered in powdery mildew-infected plants. Barley root growth is more dramatically reduced than shoot growth in mildew-infected plants, and this change in root mass will influence the uptake of water and minerals from the soil (Walters and Ayres 1981). Increases in phosphate uptake and decreases in nitrogen uptake relative to control uninoculated plants have been reported in powdery mildew-infected plants. Increases in phosphate uptake are driven in part by the accumulation of polyphosphates by *E. graminis* (Zulu et al. 1991). Decreased nitrate uptake by roots is accompanied by increased evolution of ammonium (Sadler and Scott 1974) and reductions in total N accumulation in roots and shoots (Murray and Ayres 1986). Reduced N accumulation is reflected in reduced grain N content in seed produced by diseased plants (Smedegaard-Petersen and Stølen 1981).

Deleterious changes like those outlined above are exacerbated in cereals subject to environmental stresses, such as low phosphate (Zulu et al. 1991), drought (Ayres 1982), or low temperature (McAinsh et al. 1991).

B. Biochemical and Molecular Aspects

1. Compatible Interaction

One rapid set of changes observed in susceptible barley prior to the appearance of deleterious symptoms is a specific and preferential decrease in the translation of host genes (Manners and Scott 1985) and reduced transcript accumulation for the large and small subunits of Rubisco and the 32-kDa protein (Higgins et al. 1985). Although *E. g. hordei* only attacks epidermal cells, mesophyll cells respond to infection. The authors speculate that the suppression of host gene expression in susceptible genotypes permits *E. graminis* to evade host defensive responses (Manners and Scott 1985). In addition, preferential increases in enzymes of the ascorbate-glutathione cycle were observed in susceptible infected barley, leading the authors to propose that suppression of oxidative damage is correlated with susceptibility (El-Zahaby et al. 1995). In both resistant and susceptible genotypes at 24 h after inoculation, a change in the attachment of the host plasma membrane to the cell wall occurs (Lee-Stadelmann et al. 1984). These authors demonstrated reduced permeability of epidermal cells of infected host tissues, which was correlated with the occurrence of altered plasmolysis (Lee-Stadelmann et al. 1991). Thus, the susceptible host exhibits a number of metabolic and physiological changes during powdery mildew disease development, implying that it is not a passive participant in the interaction.

2. Incompatible Interaction

A variety of approaches have been taken to determine molecular components of host resistance to powdery mildew. Differential screens have been used to identify cDNA clones for genes that are expressed at higher levels in inoculated resistant than in uninoculated cereals. In a second approach, researchers have characterized the expression of genes shown to be induced in response to pathogens in other diseases. From these experiments, a number of genes of unknown function (Davidson et al. 1987, 1988), as well as peroxidases (Kerby and Somerville 1989; Schweizer et al. 1989; Rebmann et al. 1991; Thordal-Christensen et al. 1992; Boyd et al. 1994), β-glucanase (Jutidamrongphan et al. 1991), PR1b (Muradov et al. 1993), PR-R (Boyd et al. 1994), a thaumatin-like PR protein (Bryngelsson et al. 1994), glutathione-S-transferase (Dudler et al. 1991), HSP90 (Walter-Larsen et al. 1993), chitinase, phenylalanine ammonia lyase (Boyd et al. 1994; Clark et al. 1994), a germin-like oxalate oxidase (Zhang et al. 1995), and a 14-3-3 protein kinase regulator-like protein (Brandt et al. 1992) have been identified (see Chap. 6, Vol. V, Part A).

For several of these genes, increases in mRNA accumulation or of enzymatic activity were similar in compatible and incompatible interactions between barley and *E. g. hordei* at early time points (<24 h after inoculation), However, at later time points when differences in the growth of avirulent and virulent pathogen isolates can be detected, increased levels of peroxidase, chitinase, PR-R, β-glucanase, and several anonymous genes were observed in incompatible interactions (Davidson et al. 1987; Clark et al. 1993; Boyd et al. 1994; Båga et al. 1995). Davidson et al. (1988) monitored the expression of six genes in resistant genotypes with either the *Ml-a*, *Ml-k*, or *Ml-p* resistance alleles and the corresponding near-isogenic susceptible lines and found similar increases in expression. These results and the timing of changes in expression suggest that these genes are components of nonspecific defense responses and not directly related to the action of a given resistance locus. In support of this statement, several of these genes were induced in a number of grasses following infection by a variety of fungal pathogens, but not by viral or bacterial pathogens (Scott et al. 1990; Jutidamrongphan et al. 1991).

The mRNAs for several of these genes, including peroxidase, chitinase, phenylalanine ammonia lyase, HSP90, and the 14-3-3-like protein, accumulated in two phases. One peak in accumulation occurred at about 5–6 h after inoculation and the second at about 12–15 h after inoculation. The two peaks of mRNA accumulation coincided with two attempted penetration events by *E. graminis* initiated by the primary germ tube and the appressorium (Brandt et al. 1992; Thordal-Christensen et al. 1992; Clark et al. 1993, 1994; Walter-Larsen et al. 1993; Boyd et al. 1994). Phenylalanine ammonia lyase and peroxidases may contribute to the deposition of autofluorescent phenolic compounds in walls and papillae adjacent to penetration sites (Aist and Bushnell 1991; Scott-Craig et al. 1995; Zeyen et al. 1995). Specific roles for the other genes in the papilla response have not been evaluated.

These data suggest that barley responds to inoculation with avirulent isolates of powdery mildew by increased expression of a number of defense-related genes, such as chitinases, glucanases, PR proteins, and peroxidases (Chap. 6, Vol. V, Part A). In addition, novel defense-related genes have been identified, including the HSP90 and 14-3-3 genes. In some interactions, the expression of defense-related genes was correlated with incompatible interactions, and in other examples, the expression of these genes was indistinguishable between incompatible and compatible interactions.

The role of antimicrobial compounds in resistance to powdery mildew has been explored. Resistant barley accumulates increased amounts of thionins, small toxic polypeptides, in response to attack by powdery mildew (Bohlmann et al. 1988; Chap. 6, Vol. V, Part A). It has been postulated that higher thionin levels in walls and papillae at penetration sites may contribute to resistance to powdery mildew (Ebrahim-Nesbat et al. 1993). Some barley genotypes constitutively produce gramine, an indole alkaloid. This compound inhibits germination by *E. g. hordei*, but no clear demonstration of its role in resistance has been made (Wippick and Wink 1985). Finally, one group has reported that barley produces a phytoalexin capable of inhibiting *E. g. hordei* germination (Oku et al. 1975). The production of this phytoalexin was correlated with the expression of resistance at early time points (<24 h after inoculation).

Jasmonic acid and related compounds are believed to act as signaling molecules in plants responding to wounding or pathogen attack (Chap. 5, Vol. V, Part A) Schweizer et al. (1993) demonstrated that application of jasmonic acid inhibited powdery mildew infections of barley. However, this protective effect was associated with the direct inhibition of appressorium formation. Liu and Bushnell (1986) noted that appressorium development in vitro was similarly inhibited by kinetin. The direct effect of these plant regulatory compounds on *E. g. hordei* development suggests a linkage between pathogen development and physiological status of the host.

C. Summary

Infection and disease development exert deleterious effects on host plants, which are attributed to the diversion of nutrients to the pathogen and away from the host. Many physiological measurements are consistent with this view. However, the pathogen may exert more specific deleterious effects on the host. The results from Scott's group suggest that the pathogen acts to specifically block gene expression of at least some of the genes required for photosynthesis (Higgins et al. 1985). Such results indicate a level of complexity in host-pathogen interactions that has not been fully explored.

Biochemical and physiological changes associated with the expression of resistance have focused on *Ml-a-* and *ml-o*-mediated resistance.

Resistance mechanisms governed by other genes are less well characterized. In addition to race-specific resistance mechanisms, a number of nonspecific responses to inoculation have been monitored. Chromosome walking methods may lead to the recovery of avirulence genes, which are thought to encode race-specific elicitors. With these genes, questions about when and how avirulence gene products activate resistance can be addressed. In addition to race-specific elicitors, it is clear that several nonspecific elicitors are generated during the course of infection, and it will be important to establish the relative contribution of each of these signals and how they act in conjunction with race-specific signals.

V. Conclusions and Future Prospects

Because of the economic importance of powdery mildew, these diseases have been studied extensively during most of this century, providing a broad base of genetic and cytological information. The recent cloning of several plant disease resistance genes and the apparent conservation of sequence motifs in resistance genes from a wide range of plant species (for review see Staskawicz et al. 1995) offers the promise that cereal powdery mildew resistance genes will be cloned shortly. From the sequences of these genes, questions about the biochemical function of resistance genes can be answered. What role does the barley *Ml-a* or *ml-o* gene product play in resistance? Is the *Ml-a* gene product a plasma membrane receptor, as it is often pictured in mechanistic models? Additional questions about the evolution of resistance genes can be addressed. For example, what is the basis for the highly polymorphic nature of the *Ml-a* resistance gene, and what impact does it have on the functioning of the *Ml-a* gene product? How similar are the 13 or more powdery mildew resistance genes in barley? Cytological and biochemical analyses of *Ml-a* and *ml-o* barley lines would suggest that these two resistance genes are likely to operate via distinctly different mechanisms. Sequence information should resolve this issue.

In addition, large-scale sequencing of expressed sequence tags of important pathogens, like *E. graminis*, will make available information about growth and development of this organism that is currently technically inaccessible due to its obligate lifestyle. At present, as many as 30% of randomly sequenced cDNA clones can be assigned a tentative function based on sequence similarity to known genes in the public databases (Newman et al. 1994). With time, this percentage value will increase. Such information may resolve the long-standing question of the basis for the obligate dependence of *E. graminis* on living host tissues.

As methods for detailed microscopic analyses that combine biochemical and molecular probes with the ability to monitor the events in the infection court improve, our ability to analyze host-pathogen interactions in this system will increase. Methods such as immunoelectron microscopy or confocal microscopy of living specimens stained with fluorescent indicator dyes will provide novel opportunities to study this highly specialized and coadapted host-pathogen system.

Acknowledgments. We would like to thank Drs. Aist and Bushnell and Plenum Publishing for permission to use Fig. 3. Financial support of the Danish Agricultural and Veterinary Research Council and the Royal Veterinary and Agricultural University to H. Giese; of the Dr. Christian Dräger Stiftung (Lübeck, Germany) to S. Hippe-Sanwald; and of the National Science Foundation (MCB91-19107) and the US Department of Energy (DE-FG02-91-ER20021) to S.C. Somerville is gratefully acknowledged. This is manuscript #1235 of the Carnegie Institution of Washington, Department of Plant Biology.

References

Aist JR (1983) Structural responses as resistance mechanisms. In: Bailey JA, Deverall BJ (eds) The dynamics of host defense. Academic Press, New York, pp 33–70

Aist JR, Bushnell WR (1991) Invasion of plants by powdery mildew fungi, and cellular mechanisms of resistance. In: Cole GT, Hoch HC (eds) The fungal spore and disease initiation in plants and animals. Plenum Press, New York, pp 321–345

Aist JR, Gold RE (1987) Prevention of fungal ingress. In: Nishimura S, Vance CP, Doke N (eds) Molecular determinants of plant diseases. Springer, Berlin Heidelberg New York, pp 47–58

Aist JR, Israel HW (1986) Autofluorescent and ultraviolet-absorbing components in cell walls and papillae of barley coleoptiles and their relationship to disease resistance. Can J Bot 64:266–272

Aist JR, Gold RE, Bayles CJ, Morrison GH, Chandra S, Israel HW (1988) Evidence that molecular components of papillae may be involved in *ml-o* resistance to powdery mildew. Physiol Mol Plant Pathol 33:17–32

Andrivon D, de Vallavieille-Pope C (1992) Infection attempts of cultivated barley (*Hordeum vulgare*) with isolates of *Erysiphe graminis* collected from *Hordeum murinum* in southwestern Europe. Mycol Res 96:1029–1032

Ayres PG (1982) Water stress modifies the influence of powdery mildew on root growth and assimilate import in barley. Physiol Plant Pathol 21:283–293

Båga M, Chibbar RN, Kartha KK (1995) Molecular cloning and expression analysis of peroxidase genes from wheat. Plant Mol Biol 29:647–662

Bohlmann H, Clausen S, Behnke S, Giese H, Hiller C, Reimann-Philipp U, Schrader G, Varkholt V, Apel K (1988) Leaf-specific thionins of barley – a novel class of cell wall proteins toxic to plant-pathogenic fungi and possibly involved in the defence mechanism of plants. EMBO J 7:1559–1565

Bonfante-Fasolo P, Perotto S (1992) Plant and endomycorrhizal fungi: the cellular and molecular basis of their interaction. In: Verma DPS (ed) Molecular signals in plant-microbe communications. CRC Press, Boca Raton, pp 445–470

Borbye L, Giese H (1994) Genome manipulation in recalcitrant species: construction and characterization of a yeast artificial chromosome (YAC) library from *Erysiphe graminis* f. sp. *hordei*, an obligate fungal pathogen of barley. Gene 144:107–111

Borbye L, Linde-Laursen I, Christiansen SK, Giese H (1992) The chromosome complement of *Erysiphe graminis* f. sp. *hordei* analyzed by light microscopy and field inversion gel electrophoresis. Mycol Res 96:97–102

Boyd LA, Smith PH, Green RM, Brown JKM (1994) The relationship between the expression of defense-related genes and mildew development in barley. Mol Plant-Microbe Interact 7:401–410

Brandt J, Thordal-Christensen H, Vad K, Gregersen PL, Collinge DB (1992) A pathogen-induced gene of barley encodes a protein showing high similarity to a protein kinase regulator. Plant J 2:815–820

Bronson CR (1981) The influence of unnecessary virulence genes on the reproductive fitness of *Erysiphe graminis* f. sp. *tritici*. PhD Thesis, Michigan State University, East Lansing

Brown JKM, Simpson CG (1994) Genetic analysis of DNA fingerprints and virulences in *Erysiphe graminis* f. sp. *hordei*. Curr Genet 26:172–178

Brown JKM, Jørgensen J Helms (1991) A catalogueof mildew resistance genes in European barley varieties. In: Jørgensen J H (ed) Integrated control of cereal mildews: virulence patterns and their change. Risø National Laboratory, Roskilde, Denmark, pp 263–285

Bryngelsson T, Collinge DB (1992) Biochemical and molecular analysis of the response of barley to infection by powdery mildew. In: Shewry PR (ed) Barley: genetics, biochemistry, molecular biology and biotechnology. CAB International, Wallingford, pp 459–480

Bryngelsson T, Sommer-Knudsen J, Gregersen PL, Collinge DB, Ek B, Thordal-Christensen H (1994) Purification, characterization, and molecular cloning of basic PR-1-type pathogenesis-related proteins from barley. Mol Plant-Microbe Interact 7:267–275

Bushnell WR (1981) Incompatibility conditioned by the *Ml-a* gene in powdery mildew of barley: the halt in cytoplasmic streaming. Phytopathology 71:1062–1066

Bushnell WR, Mendgen K, Liu Z (1987) Accumulation of potentiometric and other dyes in haustoria of *Erysiphe graminis* in living host cells. Physiol Mol Plant Pathol 31:237–250

Butters JA, Burrell MM, Hollomon DW (1985) Purine metabolism in barley powdery mildew and its host. Physiol Plant Pathol 27:65–74

Carver TLW (1988) Pathogenesis and host-parasite interaction in cereal powdery mildew. In: Hess W, Singh R, Singh U, Weber D (eds) Experimental and conceptual plant pathology, vol 2. Gordon and Beach, New York, pp 351–381

Carver TLW, Bushnell WR (1983) The probable role of primary germ tubes in water uptake before infection by *Erysiphe graminis*. Physiol Plant Pathol 23:229–240

Carver TLW, Robbins MP, Zeyen RJ (1991) Effects of two PAL inhibitors on the susceptibility and localized autofluorescent host cell responses of oat leaves attacked by *Erysiphe graminis* DC. Physiol Mol Plant Pathol 39:269–287

Carver TLW, Ingerson-Morris SM, Thomas BJ, Zeyen RJ (1995) Early interactions during powdery mildew infection. Can J Bot 73 Suppl 1:S632–S639

Christiansen SK, Giese H (1990) Genetic analysis of the obligate parasitic barley powdery mildew fungus based on RFLP and virulence loci. Theor Appl Genet 79:705–712

Clark TA, Zeyen RJ, Smith AG, Bushnell WR, Szabo LJ, Vance CP (1993) Host response gene transcript accumulation in relation to visible cytological events during *Erysiphe graminis* attack in isogenic barley lines differing at the *Ml-a* locus. Physiol Mol Plant Pathol 43:283–298

Clark TA, Zeyen RJ, Smith AG, Carver TLW, Vance CP (1994) Phenylalanine ammonia lyase mRNA accumulation, enzyme activity and cytoplasmic responses in barley isolines, differing at *Ml-a* and *Ml-o* loci, attacked by *Erysiphe graminis* f. sp. *hordei*. Physiol Mol Plant Pathol 44:171–185

Davidson AD, Manners JM, Simpson RS, Scott KJ (1987) cDNA cloning of mRNAs induced in resistant barley during infection by *Erysiphe graminis* f. sp. *hordei*. Plant Mol Biol 8:77–85

Davidson AD, Manners JM, Simpson RS, Scott KJ (1988) Altered host gene expression in near-isogenic barley conditioned by different genes for resistance during infection by *Erysiphe graminis* f. sp. *hordei*. Physiol Mol Plant Pathol 32:127–139

Donaldson IA, Jørgensen JH (1988) Barley powdery mildew "invertase" is an α-glucosidase. Carlsberg Res Commun 53:421–430

Dudler R, Hertig C, Rebmann G, Bull J, Mauch F (1991) A pathogen-induced wheat gene encodes a protein homologous to glutathione-S-transferases. Mol Plant-Microbe Interact 4:14–18

Ebrahim-Nesbat F, Bohl C, Heitefuss R, Apel K (1993) Thionin in cell walls and papillae of barley in compatible and incompatible interactions with *Erysiphe graminis* f. sp. *hordei*. Physiol Mol Plant Pathol 43:343–352

Edwards HH (1970) Biphasic inhibition of photosynthesis in powdery mildewed barley. Plant Physiol 45:594–597

El-Zahaby HM, Gullner G, Király Z (1995) Effects of powdery mildew infection of barley on the ascorbate-glutathione cycle and other antioxidants in different host-pathogen interactions. Phytopathology 85:1225–1230

Eschrich W, Currier HB (1964) Identification of callose by its diachrome and fluorochrome reactions. Stain Technol 39:303–307

Farrar JF, Rayns FW (1987) Respiration of leaves of barley infected with powdery mildew: increased engagement of the alternative oxidase. New Phytol 107:119–125

Fischbeck G, Jahoor A (1991) The transfer of genes for mildew resistance from *Hordeum spontaneum*. In: Jørgensen JH (ed) Integrated control of cereal mildews: virulence patterns and their change. Risø National Laboratory, Roskilde, pp 247–255

Flor HH (1956) The complementary genic system in flax and flax rust. Adv Genet 8:29–54

Freialdenhoven A, Scherag B, Hollricher K, Collinge DB, Thordal-Christensen H, Schulze-Lefert P (1994) *Nar1* and *Nar2*, two loci required for *Mla12*-specified race-specific resistance to powdery mildew in barley. Plant Cell 6:983–994

Gay JL, Manners JM (1987) Permeability of the haustorium-host interface in powdery mildews. Physiol Mol Plant Pathol 30:389–399

Giese H (1981) Powdery mildew resistance genes in the *Ml-a* and *Ml-k* regions on barley chromosome 5. Hereditas 95:51–62

Görg R, Hollricher K, Schulze-Lefert P (1993) Functional analysis and RFLP-mediated mapping of the *Ml-g* resistance locus in barley. Plant J 3:857–866

Green JR, Pain NA, Cannell ME, Jones GL, Leckie CP, McCready S, Mendgen K, Mitchell AJ, Callow JA (1995) Analysis of differentiation and development of the specialized infection structures formed by biotrophic fungal plant pathogens using monoclonal antibodies. Can J Bot 73 Suppl 1:S408–S417

Heath MC (1991) The role of gene-for-gene interactions in the determination of host species specificity. Phytopathology 81:127–130

Higgins CM, Manners JM, Scott KJ (1985) Decrease in three messenger RNA species coding for chloroplast proteins in leaves of barley infected with *Erysiphe graminis* f. sp. *hordei*. Plant Physiol 78:891–894

Hinze K, Thompson RD, Ritter E, Salamini F, Schulze-Lefert P (1991) Restriction fragment length polymorphism-mediated targeting of the *ml-o* resistance locus in barley (*Hordeum vulgare*). Proc Natl Acad Sci USA 88:3691–3695

Hippe S (1985) Ultrastructure of haustoria of *Erysiphe graminis* f. sp. *hordei* preserved by freeze-substitution. Protoplasma 129:52–61

Hippe-Sanwald S (1993) The impact of freeze substitution in biological electron microscopy. Microsc Res Tech 24:400–422

Hippe-Sanwald S (1995) Low temperature techniques as a tool in plant pathology. Scanning Microsc 9:881–899

Hippe-Sanwald S, Hermanns M, Somerville SC (1992) Ultrastructural comparison of incompatible and compatible interactions in the barley powdery mildew disease. Protoplasma 168:27–40

Hippe-Sanwald S, Marticke KH, Kieliszewski MJ, Somerville SC (1994) Immunogold localization of THRGP-like epitopes in the haustorial interface of obligate, biotrophic fungi on monocots. Protoplasma 178:138–155

Hiura U (1978) Genetic basis of formae speciales in *Erysiphe graminis* DC. In: Spencer DM (ed) The powdery mildews. Academic Press, New York, pp 101–128

Hollomon DW (1981) Genetic control of ethirimol resistance in a natural population of *Erysiphe graminis* f. sp. *hordei*. Phytopathology 71:536–540

Hollomon DW (1982) The effects of tridemorph on barley powdery mildew: Its mode of action and cross sensitivity relationships. Phytopathol Z 105:279–287

Hollomon DW (1984) Antifungal activity of substituted 2-aminopyrimidines. In: Trinci APJ, Ryley JF (eds) Mode of action of antifungal agents. Cambridge University Press, New York, pp 185–206

Holloway PJ, Maclean DJ, Scott KJ (1992) Electron transport in thylakoids isolated from barley leaves infected by the powdery mildew fungus (*Erysiphe graminis* DC. ex Merat f. sp. *hordei* Marchal). New Phytol 120:145–151

Hwang BK, Heitefuss R (1982) Induced resistance of spring barley to *Erysiphe graminis* f. sp. *hordei*. Phytopathol Z 103:41–47

Inoue S, Macko V, Aist JR (1994) Identification of the active component in the papilla-regulating extract from barley leaves. Physiol Mol Plant Pathol 44:441–453

Jahoor A, Fischbeck G (1993) Identification of new genes for mildew resistance of barley at the *Ml-a* locus in lines derived from *Hordeum spontaneum*. Plant Breed 110:116–122

Jensen J, Jensen HP, Jørgensen JH (1995) Linkage studies of barley powdery mildew virulence loci. Hereditas 122:197–209

Johnson LEB, Bushnell WR, Zeyen RJ (1979) Binary pathways for analysis of primary infection and host response in populations of powdery mildew fungi. Can J Bot 57:497–511

Jørgensen JH (1983) Experience and conclusions from the work at Risø on induced mutations for powdery mildew resistance in barley. In: Induced mutations for disease resistance in crop plants II. International Atomic Energy Agency, Vienna, pp 73–87

Jørgensen JH (1988) *Erysiphe graminis*, powdery mildew of cereals and grasses. Adv Plant Pathol 6:137–157

Jørgensen JH (1991) Experiences from mutation studies on cereal disease resistance. In: Plant mutation breeding for crop improvement, vol 2. International Atomic Energy Agency, Vienna, pp 81–91

Jørgensen JH (1992) Sources and genetics of resistance to fungal pathogens. In: Shewry PR (ed) Barley: genetics, biochemistry, molecular biology and biotechnology. CAB International, Wallingford, pp 441–458

Justesen A, Somerville SC, Christensen SK, Giese H (1996) Isolation and characterization of two novel genes expressed in germinating conidia of the obligate biotroph *Erysiphe graminis* f. sp. *hordei*. Gene 170:131–135

Jutidamrongphan W, Andersen JB, Mackinnon G, Manners JM, Simpson RS, Scott KJ (1991) Induction of β-1,3-glucanase in barley in response to infection by fungal pathogens. Mol Plant-Microbe Interact 4:234–238

Kerby K, Somerville S (1989) Enhancement of specific intercellular peroxidases following inoculation of barley with *Erysiphe graminis* f. sp. *hordei*. Physiol Mol Plant Pathol 35:323–337

Kjær B, Jensen HP, Jensen J, Jørgensen J Helms (1990) Associations between three *ml-o* powdery mildew resistance genes and agronomic traits in barley. Euphytica 46:185–193

Klecan AL, Buchanan BB (1988) Powdery mildew infection alters fructose-2,6-bisphosphate content and sucrose to starch ratio in leaves of wheat plants actively forming grain. Physiol Mol Plant Pathol 32: 221–227

Klecan AL, Hippe S, Somerville SC (1990) Reduced growth of *Erysiphe graminis* f. sp. *hordei* induced by *Tilletiopsis pallescens*. Phytopathology 80:325–331

Kobayashi I, Tanaka C, Yamaoka N, Kunoh H (1991) Morphogenesis of *Erysiphe graminis* conidia on artificial membranes. Trans Mycol Soc Jpn 32:187–198

Kobayashi Y, Kobayashi I, Kunoh H (1993) Recognition of a pathogen and a nonpathogen by barley coleoptile cells. II. Alteration of cytoplasmic strands in coleoptile cells caused by the pathogen, *Erysiphe graminis*, and the nonpathogen, *E. pisi*, prior to their penetration. Physiol Mol Plant Pathol 43:243–254

Kobayashi I, Watanabe H, Kunoh H (1995) Induced accessibility and enhanced inaccessibility at the cellular level in barley coleoptiles. XIV. Evidence for elicitor(s) and suppressor(s) of host inaccessibility from *Erysiphe graminis*. Physiol Mol Plant Pathol 46:445–456

Koga H, Mayama S, Shishiyama J (1980) Correlation between the deposition of fluorescent compounds in papillae and resistance in barley against *Erysiphe graminis hordei*. Can J Bot 58:536–541

Koga H, Zeyen RJ, Bushnell WR, Ahlstrand GG (1988) Hypersensitive cell death, autofluorescence, and insoluble silicon accumulation in barley leaf epidermal cells under attack by *Erysiphe graminis* f. sp. *hordei*. Physiol Mol Plant Pathol 32:395–409

Koga H, Bushnell WR, Zeyen RJ (1990) Specificity of cell type and timing of events associated with papilla formation and the hypersensitive reaction in leaves of *Hordeum vulgare* attacked by *Erysiphe graminis* f. sp. *hordei*. Can J Bot 68:2344–2352

Kogel KH, Beckhove U, Dreschers J, Münch S, Rommé Y (1994) Acquired resistance in barley. The resistance mechanism induced by 2,6-dichloroisonicotinic acid is a phenocopy of a genetically based mechanism governing race-specific powdery mildew resistance. Plant Physiol 106:1269–1277

Komura T, Kobayashi I, Yamaoka N, Kunoh H (1990) Induced accessibility and enhanced inaccessibility at the cellular level in barley coleoptiles. VIII. Cytological evidence for suppressor(s) of host inaccessibility from *Erysiphe graminis*. Physiol Mol Plant Pathol 37:409–416

Kunoh H (1982) Primary germ tubes of *Erysiphe graminis* conidia. In: Asada Y, Bushnell WR, Ouchi S, Vance CP (eds) Plant infection: the physiological and biochemical basis. Japan Scientific Societies Press, Tokyo, pp 45–59

Kunoh H (1990) Ultrastructure and mobilization of ions near infection sites. Annu Rev Phytopathol 28:93–111

Kunoh H, Hayashimoto A, Harui M, Ishizaki H (1985a) Induced susceptibility and enhanced resistance at the cellular level in barley coleoptiles. I. The significance of timing of fungal invasion. Physiol Mol Plant Pathol 27:43–54

Kunoh H, Kuno K, Ishizaki H (1985b) Cytological studies of the early stages of powdery mildew in barley *Hordeum vulgare* and wheat. XI. Autofluorescence and haloes at penetration sites of appressoria of *Erysiphe graminis hordei* and *Erysiphe pisi* on barley coleoptiles. Can J Bot 63:1535–1539

Kunoh H, Aist JR, Israel HW (1986) Elemental composition of barley coleoptile papillae in relation to their ability to prevent penetration by *Erysiphe graminis*. Physiol Mol Plant Pathol 29:69–78

Kunoh H, Katsuragawa N, Yamaoka N, Hayashimoto A (1988a) Induced accessibility and enhanced inaccessibility at the cellular level in barley coleoptiles. III. Timing and localization of enhanced inaccessibility in a single coleoptile cell and its transfer to an adjacent cell. Physiol Mol Plant Pathol 33:81–93

Kunoh H, Yamaoka N, Yoshioka H, Nicholson RL (1988b) Preparation of the infection court by *Erysiphe graminis* I. Contact-mediated changes in morphology of the conidium surface. Exp Mycol 12:325–335

Kunoh H, Nicholson RL, Yosioka H, Yamaoka N, Kobayashi I (1990) Preparation of the infection court by *Erysiphe graminis*: Degradation of the host cuticle. Physiol Mol Plant Pathol 36:397–407

Lee-Stadelmann OY, Bushnell WR, Stadelmann EJ (1984) Changes in plasmolysis form in epidermal cells of *Hordeum vulgare* infected by *Erysiphe graminis*: evidence for increased membrane-wall adhesion. Can J Bot 62: 1714–1723

Lee-Stadelmann OY, Bushnell WR, Curran CM, Stadelmann EJ (1991) Permeability to nonelectrolytes as an indicator of change in membrane lipid structure in *Hordeum vulgare* cells infected by *Erysiphe graminis* f. sp. *hordei*. Physiol Mol Plant Pathol 38:163–178

Liu Z, Bushnell WR (1986) Effects of cytokinins on fungus development and host response in powdery mildew of barley. Physiol Mol Plant Pathol 29:41–52

Mackie AJ, Roberts AM, Callow JA, Green JR (1991) Molecular differentiation in pea powdery-mildew haustoria. Identification of a 62-kDa N-linked glycoprotein unique to the haustorial plasma membrane. Planta 183:399–408

Mahadevappa M, DeScenzo RA, Wise RP (1994) Recombination of alleles conferring specific resistance to powdery mildew at the *Ml-a* locus in barley. Genome 37: 460–468

Manners JM (1989) The host-haustorium interface in powdery mildews. Aust J Plant Physiol 16:45–52

Manners JM, Gay JL (1983) The host-parasite interface and nutrient transfer in biotrophic parasitism. In: Callow JA (ed) Biochemical plant pathology. Wiley, Chichester, pp 163–195

Manners JM, Scott KJ (1985) Reduced translatable messenger RNA activities in leaves of barley infected with *Erysiphe graminis* f. sp. *hordei*. Physiol Plant Pathol 26:297–308

Marmeisse R, van den Ackerveken GFJM, Goosen T, de Wit PJGM, van den Broek HWJ (1993) Disruption of the avirulence gene *avr9* in two races of the tomato pathogen *Cladosporium fulvum* causes virulence on tomato genotypes with the complementary resistance gene *Cf9*. Mol Plant-Microbe Interact 6:412–417

Masri SS, Ellingboe AH (1966) Primary infection of wheat and barley by *Erysiphe graminis*. Phytopathology 56: 389–395

Matsumura K, Tosa Y (1995) The rye mildew fungus carries avirulence genes corresponding to wheat genes for resistance to races of the wheat mildew fungus. Phytopathology 85:753–756

McAinsh MR, Ayres PG, Hetherington AM (1989) The respiration of protoplasts from leaves of barley infected by powdery mildew (*Erysiphe graminis* f. sp. *hordei*). Plant Sci 64:221–230

McAinsh MR, Ayres PG, Hetherington AM (1991) The effects of infection by powdery mildew (*Erysiphe graminis* f. sp. *hordei*) and low temperature on the respiratory activity of winter barley. Physiol Mol Plant Pathol 39:13–23

McDermott JM, McDonald BA (1993) Gene flow in plant pathosystems. Annu Rev Phytopathol 31:353–373

Mendgen K, Nass P (1988) The activity of powdery-mildew haustoria after feeding the host cells with different sugars, as measured with a potentiometric cyanine dye. Planta 174:283–288

Millerd A, Scott K (1956) Host pathogen relations in powdery mildew of barley. II. Changes in respiratory pattern. Aust J Biol Sci 9:37–44

Moore G, Gale MD, Kurata N, Flavell RB (1993) Molecular analysis of small grain cereal genomes. Bio/Technology 11:584–589

Muradov A, Petrasovits L, Davidson A, Scott KJ (1993) A cDNA clone for a pathogenesis-related protein 1 from barley. Plant Mol Biol 23:439–442

Murray AJS, Ayres PG (1986) Uptake and translocation of nitrogen by mildewed barley seedlings. New Phytol 104:355–365

Negassa M (1985) Geographic distribution and genotypic diversity of resistance to powdery mildew of barley in Ethiopia. Hereditas 102:113–121

Newman T, de Bruijn F, Green P, Keegstra K, Kende H, McIntosh L, Ohlrogge J, Raikhel N, Somerville S, Thomashow M, Retzel E, Somerville C (1994) Genes galore: a summary of the methods for accessing the results from large-scale partial sequencing of anonymous *Arabidopsis* cDNA clones. Plant Physiol 106: 1241–1255

Nicholson RL, Kunoh H (1995) Early interactions, adhesion and establishment of the infection court by *Erysiphe graminis*. Can J Bot 73 (Suppl 1):S609–S615

Nicholson RL, Yoshioka H, Yamaoka N, Kunoh H (1988) Preparation of the infection court by *Erysiphe graminis*. II. Release of esterase enzyme from conidia in response to a contact stimulus. Exp Mycol 12:336–349

Nicholson RL, Kunoh H, Shiraishi T, Yamada T (1993) Initiation of the infection process by *Erysiphe graminis*: Conversion of the conidial surface from hydrophobicity to hydrophilicity and influence of the conidial exudate on the hydrophobicity of the barley leaf surface. Physiol Mol Plant Pathol 43:307–318

O'Dell M, Wolfe MS, Flavell RB, Simpson CG, Summers RW (1989) Molecular variation in populations of *Erysiphe graminis* on barley, oats and rye. Plant Pathol 38:340–351

Oku H, Ouchi S, Shiraishi T, Komoto Y, Oki K (1975) Phytoalexin activity in barley powdery mildew. Ann Phytopathol Soc Jpn 41:185–191

Ouchi S, Oku H, Hibino C (1976) Localization of induced resistance and susceptibility in barley leaves inoculated with the powdery mildew fungus. Phytopathology 66: 901–905

Pascholati SF, Yoshioka H, Kunoh H, Nicholson RL (1992) Preparation of the infection court by *Erysiphe graminis* f. sp. *hordei*: cutinase is a component of the conidial exudate. Physiol Mol Plant Pathol 41:53–59

Rasmussen M, Rossen L, Giese H (1993) SINE-like properties of a highly repetitive element in the genome of the obligate parasitic fungus *Erysiphe graminis* f. sp. *hordei*. Mol Gen Genet 239:298–303

Rebmann G, Hertig C, Bull J, Mauch F, Dudler R (1991) Cloning and sequencing of complementary DNAs encoding a pathogen-induced putative peroxidase of wheat (*Triticum aestivum* L.). Plant Mol Biol 16:329–332

Russo VM, Bushnell WR (1989) Responses of barley cells to puncture by microneedles and to attempted penetration by *Erysiphe graminis* f. sp. *hordei*. Can J Bot 67:2912–2921

Sadler R, Scott KJ (1974) Nitrogen assimilation and metabolism in barley leaves infected with the powdery mildew fungus. Physiol Plant Pathol 4:235–247

Sahashi N, Shishiyama J (1986) Increased papilla formation, a major factor of induced resistance in the barley – *Erysiphe graminis* f. sp. *hordei* system. Can J Bot 64:2178–2181

Sahashi N, Tsuji H, Shishiyama J (1989) Barley plants grown under germ-free conditions have increased susceptibility to two powdery mildew fungi. Physiol Mol Plant Pathol 34:163–170

Scholes JD, Lee PJ, Horton P, Lewis DH (1994) Invertase: understanding changes in the photosynthetic and carbohydrate metabolism of barley leaves infected with powdery mildew. New Phytol 126:213–222

Schüller C, Backes G, Fischbeck G, Jahoor A (1992) RFLP markers to identify the alleles on the *Ml-a* locus conferring powdery mildew resistance in barley. Theor Appl Genet 84:330–338

Schweizer P, Hunziker W, Mösinger E (1989) cDNA cloning, in vitro transcription and partial sequence analysis of mRNAs from winter wheat (*Triticum aestivum* L.) with induced resistance to *Erysiphe graminis* f. sp. *tritici*. Plant Mol Biol 12:643–654

Schweizer P, Gees R, Mösinger E (1993) Effect of jasmonic acid on the interaction of barley (*Hordeum vulgare* L.) with the powdery mildew *Erysiphe graminis* f. sp. *hordei*. Plant Physiol 102:503–511

Scott KJ (1992) The molecular analysis of barley resistance to powdery mildew. In: Shewry PR (ed) Barley: genetics, biochemistry, molecular biology and biotechnology. CAB International, Wallingford, pp 481–498

Scott KJ, Davidson AD, Jutidamrongphan W, Mackinnon G, Manners JM (1990) The activation of genes of wheat and barley by fungal phytopathogens. Aust J Plant Physiol 17:229–238

Scott-Craig JS, Kerby KB, Stein BD, Somerville SC (1995) Expression of an extracellular peroxidase that is induced in barley (*Hordeum vulgare*) by the powdery mildew pathogen (*Erysiphe graminis* f. sp. *hordei*). Physiol Mol Plant Pathol 47:407–418

Sherwood JE, Slutsky B, Somerville SC (1991) Induced morphological and virulence variants of the obligate barley pathogen *Erysiphe graminis* f. sp. *hordei*. Phytopathology 81:1350–1357

Shiraishi R, Yamaoka N, Kunoh H (1989) Association between increased phenylalanine ammonia-lyase activity and cinnamic acid synthesis and the induction of temporary inaccessibility caused by *Erysiphe graminis* primary germ tube penetration of the barley leaf. Physiol Mol Plant Pathol 34:75–83

Skou JP, Jørgensen JH, Lilholt U (1984) Comparative studies on callose formation in powdery mildew compatible and incompatible barley. Phytopathol Z 109: 147–168

Smart MG, Aist JR, Israel HW (1986) Structure and function of wall appositions. 1. General histochemistry of papillae in barley coleoptiles attacked by *Erysiphe graminis* f. sp. *hordei*. Can J Bot 64:793–801

Smedegaard-Petersen V, Stølen O (1981) Effect of energy-requiring defense reactions on yield and grain quality in a powdery mildew-resistant barley cultivar. Phytopathology 71:396–399

Spencer DM (1978) The powdery mildews. Academic Press, London

Staskawicz BJ, Ausubel FM, Baker BJ, Ellis JG, Jones JDG (1995) Molecular genetics of plant disease resistance. Science 268:661–667

Stenzel K, Steiner U, Schönbeck F (1985) Effect of induced resistance on the efficiency of powdery mildew haustoria in wheat and barley. Physiol Plant Pathol 27:357–367

Takahashi K, Aist JR, Israel HW (1985) Distribution of hydrolytic enzymes at barley powdery mildew encounter sites: Implications for resistance associated with papilla formation in a compatible system. Physiol Plant Pathol 27:167–184

Thordal-Christensen H, Smedegaard-Petersen V (1988a) Comparison of resistance-inducing abilities of virulent and avirulent races of *Erysiphe graminis* f. sp. *hordei* and a race of *Erysiphe graminis* f. sp. *tritici* on barley. Plant Pathol 37:20–27

Thordal-Christensen H, Smedegaard-Petersen V (1988b) Correlation between induced resistance and host fluorescence in barley inoculated with *Erysiphe graminis*. J Phytopathol 123:34–46

Thordal-Christensen H, Brandt J, Cho BH, Rasmussen SK, Gregersen PI, Smedegaard-Petersen V, Collinge DB (1992) cDNA cloning and characterization of two barley peroxidase transcripts induced differentially by the powdery mildew fungus *Erysiphe graminis*. Physiol Mol Plant Pathol 40:395–409

Torp J, Jørgensen JH (1986) Modification of barley powdery mildew resistance gene *Ml-a12* by induced mutation. Can J Genet Cytol 28:725–731

Tosa Y (1989) Evidence on wheat for gene-for-gene relationship between formae speciales of *Erysiphe graminis* and genera of gramineous plants. Genome 32:918–924

Tosa Y (1992) A model for the evolution of formae specialis and races. Phytopathology 82:728–730

Toyoda H, Matsuda Y, Yamaga T, Ikeda S, Morita M, Tamai T, Ouchi S (1991) Suppression of the powdery mildew pathogen by chitinase microinjected into barley coleoptile epidermal cells. Plant Cell Rep 10:217–220

Verma DPS, Miao GH, Lee NG, Cheon CI, Hong Z (1992) Root nodule organogenesis and formation of the peribacteroid membrane compartment. In: Nester EW, Verma DPS (eds) Advances in molecular genetics of plant-microbe interactions. Kluwer, Dordrecht, pp 343–352

von Wettstein-Knowles P (1992) Cloned and mapped genes: Current status. In: Shewry PR (ed) Barley: genetics, biochemistry, molecular biology and biotechnology. CAB International, Wallingford, pp 73–98

Wahl I, Eshed N, Segal A, Sobel Z (1978) Significance of wild relatives of small grains and other wild grasses in cereal powdery mildew. In: Spencer DM (ed) The powdery mildews. Academic Press, New York, pp 83–100

Walter-Larsen H, Brandt J, Collinge DB, Thordal-Christensen H (1993) A pathogen-induced gene of barley encodes a HSP90 homologue showing striking similarity to vertebrate forms resident in the endoplasmic reticulum. Plant Mol Biol 21:1097–1108

Walters DR, Ayres PG (1981) Phosphate uptake and translocation by roots of powdery mildew infected barley. Physiol Plant Pathol 18:195–205

Walters DR, Ayres PG (1983) Changes in nitrogen utilization and enzyme activities associated with CO_2 exchanges in healthy leaves of powdery mildew-infected barley. Physiol Plant Pathol 23:447–459

Wei YD, De Neergaard E, Thordal-Christensen H, Collinge DB, Smedegaard-Petersen V (1994) Accumulation of a putative guanidine compound in relation to other early defense reactions in epidermal cells of barley and wheat exhibiting resistance to *Erysiphe graminis* f. sp. *hordei*. Physiol Mol Plant Pathol 45:469–484

Wei YD, Collinge DB, Smedegaard-Petersen V, Thordal-Christensen H (1996) Characterization of the transcript of a new class of retroposon-type repetitive element cloned from the powdery mildew fungus, *Erysiphe graminis*. Mol Gen Genet 250:477–482

Williams GM, Ayres PG (1981) Effects of powdery mildew and water stress on CO_2 exchange in uninfected leaves of barley. Plant Physiol 68:527–530

Wippick C, Wink M (1985) Biological properties of alkaloids. Influence of quinolizidine alkaloids and gramine on the germination and development of powdery mildew, *Erysiphe graminis* f. sp. *hordei*. Experientia 41: 1477–1479

Wolfe MS, Brändle U, Koller B, Limpert E, McDermott JM, Küller K, Schaffner D (1992) Barley mildew in Europe: population biology and host resistance. Euphytica 63:125–139

Wolter M, Hollricher K, Salamini F, Schulze-Lefert P (1993) The *ml-o* resistance alleles to powdery mildew infection in barley trigger a developmentally controlled defense mimic phenotype. Mol Gen Genet 239:122–128

Wright AJ, Heale JB (1988) Host responses to fungal penetration in *Erysiphe graminis* f. sp. *hordei* infections in barley. Plant Pathol 37:131–140

Wright DP, Baldwin BC, Shephard MC, Scholes JD (1995a) Source-sink relationships in wheat leaves infected with powdery mildew. I. Alterations in carbohydrate metabolism. Physiol Mol Plant Pathol 47:237–253

Wright DP, Baldwin BC, Shephard MC, Scholes JD (1995b) Source-sink relationships in wheat leaves infected with powdery mildew. II. Changes in the regulation of the Calvin cycle. Physiol Mol Plant Pathol 47: 255–267

Yamaoka N, Toyoda K, Kobayashi I, Kunoh H (1994) Induced accessibility and enhanced inaccessibility at the cellular level in barley coleoptiles. XIII. Significance of haustorium formation by the pathogen *Erysiphe graminis* for induced accessibility to the non-pathogen *E. pisi* as assessed by nutritional manipulations. Physiol Mol Plant Pathol 44:217–225

Yokoyama K, Aist JR, Bayles CJ (1991) A papilla-regulating extract that induces resistance to barley powdery mildew. Physiol Mol Plant Pathol 39:71–78

Zeyen RJ, Bushnell WR (1979) Papilla response of barley epidermal cells caused by *Erysiphe graminis*: rate and method of deposition determined by microcinematography and transmission electron microscopy. Can J Bot 57:898–913

Zeyen RJ, Ahlstrand GG, Carver TLW (1992) X-ray microanalysis of frozen-hydrated, freeze-dried, and critical point dried leaf specimens: determination of soluble and insoluble chemical elements at *Erysiphe graminis* epidermal cell papilla sites in barley isolines containing *Ml-o* and *ml-o* alleles. Can J Bot 71:284–296

Zeyen RJ, Bushnell WR, Carver TLW, Robbins MP, Clark TA, Boyles DA, Vance CP (1995) Inhibiting phenylalanine ammonia lyase and cinnamyl-alcohol dehydrogenase suppresses *Mla1* (HR) but not *mlo5* (non-HR) barley powdery mildew resistances. Physiol Mol Plant Pathol 47:119–140

Zhang Z, Collinge DB, Thordal-Christensen H (1995) Germin-like oxalate oxidase, a H_2O_2-producing enzyme, accumulates in barley attacked by the powdery mildew fungus. Plant J 8:139–145

Zulu JN, Farrar JF, Whitbread R (1991) Effects of phosphate supply on the phosphorus status, dry mass and photosynthesis of wheat infected with powdery mildew. New Phytol 118:453–461

Abbreviations

ac	ascus
ap	appressorium
b	membrane blebbing
cc	conidial mother cell
cl	cleistothecium
cM	centiMorgans
cn	conidium
cp	conidiophore
em	extrahaustorial membrane
ema	extrahaustorial matrix
er	endoplasmic reticulum
f	fibrillar material
fc	fluorescent cell
g	Golgi body
h	haustorium
hb	haustorial body
hcw	haustorial cell wall
hcy	haustorial cytoplasm
hocw	host cell wall
hocy	host cytoplasm
hy	hypha
kbp	kilo base pairs
kDa	kilo Daltons
l	lipid-like body
m	plant or fungal mitochondrion
Mbp	mega base pairs (10^6 base pairs)
mt	microtubule
p	papilla
pgt	primary germ tube
ph	penetration hole
pl	plasma membrane
PR	pathogenesis related
r	ribosome
RAPD	randomly amplified polymorphic DNA
RFLP	restriction fragment length polymorphism
SINE	short interspersed nuclear element
THRGP	threonine-hydroxyproline-rich glycoprotein
v	plant or fungal vacuole
ve	vesicle
YAC	yeast artificial chromosome

5 The Uredinales

K. Mendgen

CONTENTS

I. Introduction ... 79
II. Recent Studies on the Life Cycle of the Rust Fungi ... 80
III. Long-Distance Transport of Rust Spores and the Economic Importance of Rust Diseases ... 81
IV. The Infection Process of Basidiospores, Aeciospores, and Urediospores ... 83
V. Molecular Studies on Rust Fungi ... 84
VI. Cytology and Physiology of the Host-Parasite System ... 85
A. Cytology of the Monokaryotic Infection ... 85
B. Cytology of the Dikaryotic Infection ... 86
C. A New Physiological Unit, the Fungus in the Host Plant ... 87
D. Physiology and Molecular Biology of the Aegricorpus ... 87
VII. Methods of Rust Control ... 88
VIII. Biological Control of Weeds with Rusts ... 88
IX. Conclusions ... 89
References ... 89

I. Introduction

Rust-infected wheat leaf fragments have been found in a storage jar from the Late Bronze Age filled with wheat grains (Kislev 1982). This report, together with numerous others of plant diseases in the ancient literature, obviously indicating rust fungus infections, emphasizes the importance of these pathogens for human food production (Lehmann et al. 1937; Chester 1946; Schafer et al. 1984). In his book on agriculture, Duhamel du Monceau (1779) mentions stem rust "as an extreme nuisance: the most beautiful grains are reduced to nothing as soon as the stem is completely covered by rust".

The first comprehensive description of rust fungi, containing some 120 species, was published by Unger (1833). He found different rust fungi or rust spores on most plant families and observed that the extent of rust infection very often correlated with humid and foggy climatic conditions. He studied cross sections through infected leaves, noted the degeneration of the chlorophyll in the cells of the diseased area, and described a matrix between the leaf cells from which the different spore forms arise. Along with the scientific community at his time, he believed in spontaneous generation of life. The matrix was assumed to be formed by diseased plant sap which, under the influence of the plant, produces the spores. The failure of his infection experiments with spores from *Puccinia asarina* and a rust from *Salix* supported his belief.

De Bary (1853) was the first to notice the importance of the germ pore on spores for the production of germ tubes of rusts and smuts. He discussed the importance of tip growth for the ability of a fungus to grow through the stomatal opening. Excellent pictures of teliospore development reflect his skills for correct observation and the improvement of the microscope at that time. A few years later, De Bary (1865) elucidated the life cycle of *P. graminis*, the stem rust of wheat, coined the term teleutospore as the final spore form in the life cycle, and defined autoecious and heteroecious rust fungi.

In the book *Die Getreideroste*, Eriksson and Henning (1896) described the specialization of the rust fungi on cereals. Stakman and Piemeisal (1917) found races of wheat stem rust. This result was the basis for an effective breeding program for resistance in cereals and other plants. In 1927, Craigie discovered heterothallism of wheat stem rust and elucidated the function of the pycnia (spermogonia) as sexual organs. This was the final step in the explanation of the rust life cycle.

The gene-for-gene concept by Flor (1956), resulted from experiments with flax rust, *Melampsora lini*, and explained the interaction of genes for avirulence in the rust fungus and genes for resistance in plants. Today, rust fungi represent an

Fakultät für Biologie, Lehrstuhl Phytopathologie, Universität Konstanz, Universitäts str. 10, D-78464 Konstanz, Germany

The Mycota V Part B
Plant Relationships
Carroll/Tudzynski (Eds.)
© Springer-Verlag Berlin Heidelberg 1997

important subject for basic research, since these fungi are unique in the way they recognize the stomatal opening of their host leaves (Hoch and Staples 1987, 1991, Chaps. 2, 3, Vol. V, Part A) or produce highly differentiated haustoria within host cells.

This chapter on Uredinales cannot cover the complete literature published even for a single year on rust fungi. It mainly presents key references from the past few years and cites review articles and books for access to the more specialized literature on most aspects of rust fungi. A general introduction into the life of rust fungi can be found in the books by Littlefield (1981) and Alexopoulos et al. (1996). A treatise on most questions considering cereal rusts has been edited by Bushnell (Bushnell and Roelfs 1984) and Roelfs (Roelfs and Bushnell 1985). More general morphological, taxonomic and phylogenetic information on rusts has been published by Gäumann (1959), Cummins and Hiratsuka (1983), Preece and Hick (1990) and Swann and Taylor (1995).

II. Recent Studies on the Life Cycle of the Rusts

Rust fungi exhibiting all spore forms are called macrocyclic. The following numbers are assigned to the various fruiting structures (sori) and spores in the life cycle: O – pycnia and pycniospores (= spermogonia and spermatia), I – aecia and aeciospores (= aecidia and aecidiospores), II – uredia and urediospores (= uredinia and uredospores or urediniospores), III – telia and teliospores (= teleutosori and teleutospores), IV – basidiospores. The existence of different spore names is due to historical reasons, when the origin and the tasks of the different spore forms was not yet known. The introduction of new names is sometimes difficult to understand (Savile 1968).

Rusts restricted to a single host are autoecious, such as *Uromyces viciae-fabae* (Fig. 1) on *Vicia faba*, rusts infecting an alternate host during the life cycle are heteroecious, such as *Puccinia graminis* (Pers.) on barberry and cereals (Arthur 1962). This rust changes nuclear condition with the host plant. The macrocyclic rusts produce repetitive infections in an asexual cycle by means of urediospores.

A life cycle which deletes the uredial stage is termed demicyclic. Sometimes, pycnia with pycniospores are also missing. A microcyclic rust fungus deletes both aecia and uredia (uredinia), leaving the telial and pycnial stage. The latter may be missing also. In the abbreviated life cycle, all microcyclic forms are autoecious. Thus, the general scheme as shown in Fig. 1 has many variations. A more extensive description can be found in the reviews by Petersen (1974) and Cummins and Hiratsuka (1983).

Fertilization generally occurs in the spermogonial stage. However, exchange or complementation of nuclei can also be carried out by dikaryotic urediospores (Groth 1988).

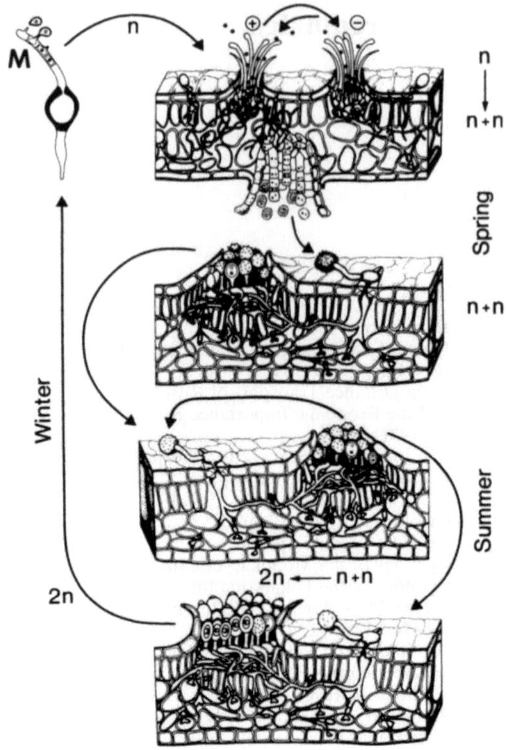

Fig. 1. Life cycle of *Uromyces viciae-fabae*. After overwintering on residual plant material, teliospores germinate, meiosis (M) occurs, and the metabasidium produces haploid (n) basidiospores of two mating types (+,−); basidiospores are ejected and land on the host leaf, differentiate infection structures, and pycnia are produced on the upper side of the leaf; after spermatization (pycniospores may be transported by insects) dikaryotization occurs in aecial primordia (n → n + n), and the aecium differentiates on the lower side of the leaf; aeciospores (n + n) infect and urediospores are produced; *Summer*: urediospores infect and produce urediospores repetitively; in autumn, uredia differentiate into telia; the nuclei fuse (n + n → 2n) during sporogenesis and teliospores develop. (F. Meinhard, A. Klink, U. Labitzke, and M. Hahn contributed to the diagram)

Teliospores or teleutospores represent the final spore form of the rusts and, as a rule, provide the basis for rust nomenclature (Mendgen 1984). For a recent example see the comparison of anamorphs and teleomorphs for the delimitation of *Phakopsora*, *Cerotelium*, and *Physopella* (Ono et al. 1992). Sometimes this nomenclatural rule is difficult to apply. For example, *P. pampeana* (Speg), the rust of cultivated *Capsicum* spp., produces *Aecidium*-like and *Puccinia*-like sori. Both spore types from these sori produce basidia containing four basidiospores. Basidiospores of both types of spores produce pycnia, followed, after spermatization, by aecidoid and, a few weeks later, by telioid sori. Thus, this rust is both an *Endophyllum* and a *Puccinia* species. The unicellular *Endophyllum* spores are responsible for the rapid dissemination and dispersion in the field.

The Gicellular *Puccinia*-like spores exhibit dormancy and thus can survive hot and rainy periods in summer. This is an interesting alternative for a tropical rust to survive unfavorable conditions (Hennen et al. 1984; Figueiredo et al. 1987). The survival strategy is different from the well-known example of *P. graminis* (Pers.), which is long-cycled and heteroecious. *P. graminis* is native in the north temperature zone, where heteroecism is common. This fungus shows an adaptation to seasonal climatic fluctuations, especially weather conditions during spring and winter (Anikster and Wahl 1979).

As mentioned above, *Endophyllum* aeciospores are generally produced in typical aecia. The nuclei fuse in these aeciospores and, upon germination, a metabasidium is produced where meiosis takes place and basidiospores develop. However, some races of *Endophyllum euphorbiae sylvaticae* remain haploid during their whole life cycle (Gäumann 1959).

From the examples given above, it can be assumed that the genes responsible for spore morphology and the genes that govern nuclear behavior can act independently. Other factors also act independently. Urediospores and aeciospores of *U. vignae* differ in genesis and morphology, but their behavior, physiology, and development after germination are very similar (Stark-Urnau and Mendgen 1993).

Plants and their rusts are excellent examples of coevolution. The rusts on ferns and conifers are believed to be descendants of ancient rusts. Nearly all rusts on ferns are both heteroecious and macrocyclic. It can be speculated that the extraordinarily complex heteroecious macrocyclic life cycle evolved only once. Reductions could have easily occurred at any stage of evolution (Laundon 1973). Wahl et al. (1984) discussed host-parasite coevolution processes for the cereal rust fungi. Since genetic recombination occurs in teliospores, its influence on diversity and distribution of virulences can be studied in the derived aecial populations and is typical for various regions (Groth and Roelfs 1982; Kolmer 1992; McCain et al. 1992). In centers for coevolution, genes responsible for plant defenses and genes for fungal virulence have accumulated. Redistribution of the main host or the alternate host by human influence has given rise to subsequent independent evolution of the gametophytic or the sporophytic generation (Anikster and Wahl 1979; Anikster 1984).

This diversification has complicated the species concept in rusts. Obviously, morphological and physiological aspects need to be evaluated if a species or a forma specialis is to be defined (Anikster 1984; Savile 1984). A modern concept is the virulence formula to indicate race specificity and the interaction with the host plant according to the avirulence/virulence pattern of a rust (Roelfs et al. 1992).

III. Long-Distance Transport of Rust Spores and the Economic Importance of Rust Disease

Long-distance spread of rust spores is an important factor for the damage that rust diseases induce (Nagarajan and Singh 1990). In the USA, spore movements occur with prevailing winds from the southwest to the northwest. Such winds may transport many urediospores of *P. graminis* Pers. f. sp. *tritici*, since a single wheat leaf infected with this fungus can produce more than a million urediospores (Zadoks and Schein 1979; Burdon 1987). This long-distance transport occurs annually (800 km) across the North American Great Plains (Roelfs et al. 1992) and nearly annually (2000 km) from Australia to New Zealand (Luig 1985). Losses can be significant (McGrath and Pennypackers 1991).

Peanut rust (*P. arachidis* Speg) was confined to tropical America and China until 1969 and suddenly appeared in southern Asia and Oceania during the early 1970s, on the Ivory Coast in 1976, and in Senegal in 1980 (Savary et al. 1988). About 80% of the world production of peanuts comes from developing countries, most of it from the semiarid tropics. Here, yields are low, around 800–900 kg ha^{-1}, compared with approximately 2500 kg ha^{-1} in the developed world. This is because the farmers cannot afford the cost of crop protection chemicals. Rust accounts for about 50% of the losses, sometimes for much more (Subrahmanyam et al. 1985).

Puccinia melanicephala, the sugar cane rust, has been present in scattered locations in Africa and Asia and was of minor significance for many years. During July 1978, this rust first appeared in the Dominican Republic and 1 year later it had spread to almost all sugarcane-growing areas in the Americas. In retrospective analysis of mesocale meteorology during the 1-month period before July 1978, it was found that a Saharan dust cloud traveled through the Atlantic trade winds from the coast of West Africa to the Caribbean area within 5 days in June 1978 (Purdy 1985). Assuming a terminal deposition velocity of 12.2 mm s^{-1}, a urediospore at an altitude of 3000 m could travel for 2.3 days. With a uniform wind speed of 10 m s^{-1}, horizontal transport of 2000 km would occur during 2.3 days. In the Caribbean region, the spores met the highly susceptible cane cultivar B 4362, which had been grown in the region for more than 20 years. Therefore, the rust met no resistance and caused losses of some 40% (Comstock et al. 1992).

The coffee leaf rust (*Hemileia vastatrix*) affecting wild coffee was discovered in 1861 in the Lake Victoria region of Kenya. After appearing in Ceylon 1867, the rust literally had extinguished the coffee trees from the island by 1882. In *Coffea arabica*, a single rust pustule may cause premature abscission of the leaf (Javed 1982). Rapid defoliation may follow and repeated infection by rust will cause death of a tree. The rust spread through Southeast Asia in the following years and finally reached Bahia, Brazil, early in 1970. Two years later, the rust fungus invaded Paraguay and Argentinia. It reached Nicaragua in 1976, Peru in 1979, and was detected in Ecuador and Mexico during 1981. Wind alone was not always responsi-

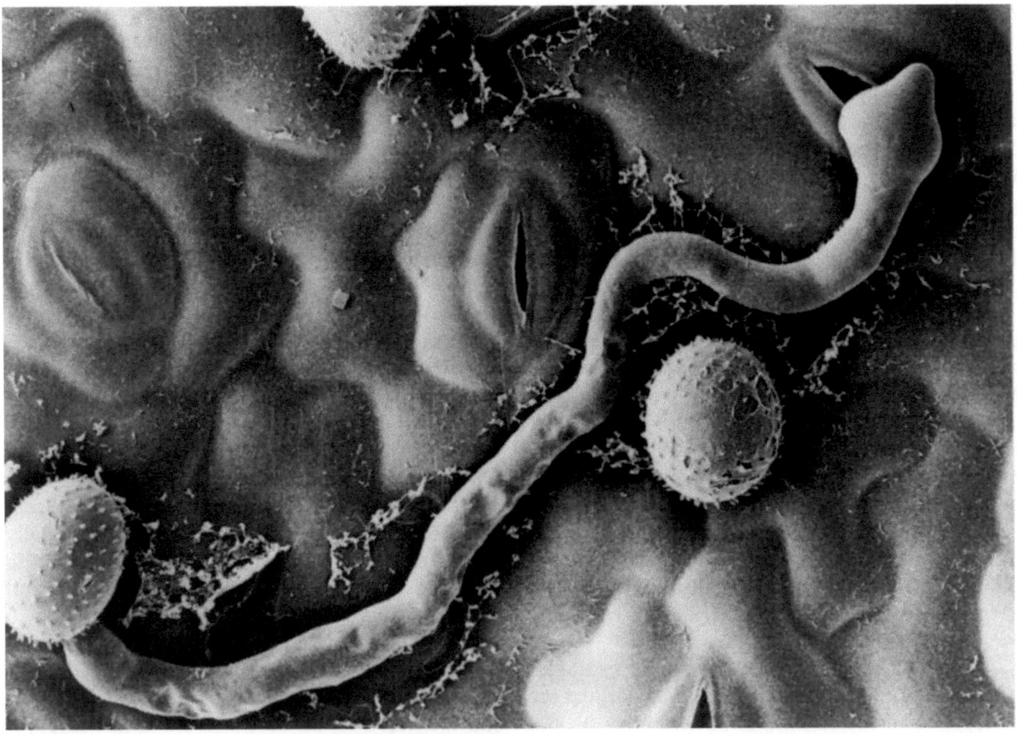

Fig. 2. Spore, germ tube, and appressorium (over the stomatal opening) of *Uromyces viciae-fabae* on the host leaf (*Vicia faba*) as seen with a low-temperature scanning electron microscope at −150 °C. Magnification 900×

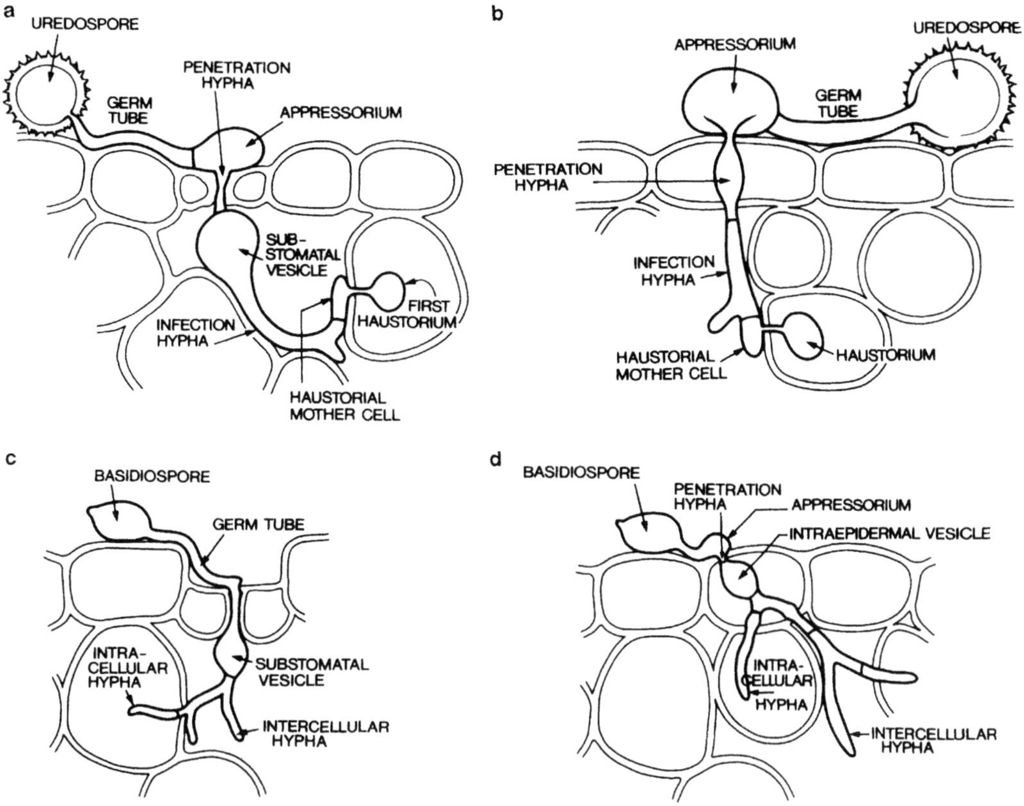

Fig. 3a–d. Diagrammatic representation of direct and stomatal penetration by rusts in the haploid and in the dikaryotic stage (not drawn to scale). **a** *Uromyces viciae-fabae*, dikaryon. ***b*** *Phakopsora pachyrhizi*, dikaryon. **c** *Cronartium ribicola*, monokaryon. **d** *Uromyces appendiculatus*, monokaryon

ble for distribution of this disease. It could not have spread this way from Brasilia to Nicaragua because of prevailing natural geographical barriers. The introduction into Nicaragua constitutes a good example of spread by migrant workers, who carried seedlings and thus seem to have imported the disease (Schrieber and Zentmyr 1984). Altogether, it took coffee rust about 120 years to travel around the world.

IV. The Infection Process of Basidiospores, Aeciospores, and Urediospores

Rust fungi offer the unique possibility to the study of infection structures of monokaryotic and dikaryotic stages of the same fungus on one host plant or, in the case of heteroecious rusts, on two host plants from different plant families. In addition, rust fungi may penetrate the plant cuticle directly or through stomatal openings (Fig. 2; Chap. 2, Vol. V, Part A). Below, infection strategies of urediospores, and basidiospores are compared (Fig. 3).

The spores differ considerably in their morphology. Recent descriptions have been published in books by Littlefield and Heath (1979), Littlefield (1981), and Preece and Hick (1990). The physiology and the effect of germination stimulators or inhibitors are reviewed by French (1992; see also Chap. 2, Vol. V, Part A). Basidiospores are smooth, thin-walled, and do not withstand dry periods. Aeciospores have a thick wall and a typical surface ornamentation with small knob-like structures. Urediospores have a thick, darkly pigmented wall with germ pores and typical spine-like ornamentations.

The development of infection structures from these spores has been reviewed recently (Gold and Mendgen 1991; Mendgen and Deising 1993; Mendgen et al. 1996). Upon landing on a wet leaf, or under humid conditions, a gelatinous matrix forms an adhesion pad below the spores. Cutinase, esterases, and other unspecific factors contribute to the adhesion process of bean rust urediospores. After germination, germ tubes adhere firmly to the plant surface, a prerequisite for the ability of bean rust germ tubes to recognize the stomatal opening (Epstein et al. 1985; Clement et al. 1993). The germling, including the appressorium, is shown in Fig. 2. The recognition process of the dikaryon may be quite precise. Differences in height of the outer ledge of the stoma within a few microns are recognized (Hoch et al. 1987). For penetration, the guard cells can be pushed apart (Fig. 3a). However, the dikaryon of the soybean rust (*Phakopsora* spp.) produces appressoria on any hard surface and penetrates epidermal cells directly (Fig. 3b; Koch and Hoppe 1988). All appressoria derived from dikaryotic spores (urediospores and aecidiospores) appear similar in morphology (Kwon and Hoch 1991; Swann and Mims 1991). Also the development of infection structures derived from urediospores and aecidiospores of the cowpea rust does not differ appreciably (Stark-Urnau and Mendgen 1993). In contrast, most haploid basidiospores barely differentiate typical appressoria and penetrate epidermal cells directly (Fig. 3d; Morin et al. 1992); however *Cronartium* spp. basidiospores produce a long germ tube and penetrate through the stomata (Fig. 3c). Studies on the process of stomatal recognition by this monokaryon, such as have been done with the dikaryon of *Uromyces appendiculatus*, would be interesting.

The penetration hypha of the dikaryon is highly differentiated to perform its tasks. It reaches through the stoma in the case of most rusts, or through the epidermal cell in the case of *Phacopsora* spp., and is delimited by a septum (Koch and Hoppe 1988; Terhune et al. 1993). In the haploid stage, the penetration hypha appears less differentiated. A septum is missing in most cases and a plug is produced which separates the more or less differentiated appressorium from the vesicle and the infection hypha in the host leaf. In all cases, however, the chitin of the fungal wall appears masked by glycoproteins and glucans starting from the differentiation of the penetration hypha (Deising and Siegrist 1995; Deising et al. 1996). Surface carbohydrate patterns on hyphae of infection structures are typical either for the monokaryon or for the dikaryon, and this is independent of whether or not a change of the host plant is involved (Freytag and Mendgen 1991).

The physiology of enzyme secretion by rust infection structures is unique. The dikaryon of *U. viciae-fabae* produces wall-degrading enzymes during differentiation. Starting with the growth of the appressorium, acidic cellulases are secreted in the absence of the substrate. During infection hypha development, pectin methyl esterase is synthesized, also in the absence of the substrate. During haustorium mother cell development, polygalacturonate lyase formation is controlled by both the differentiation stage and the availability of substrate (Deising et al. 1995). Corresponding studies of enzyme production by basidiospore-derived infection structures are not yet available.

V. Molecular Studies on Rust Fungi

Molecular methods have dramatically increased our knowledge of the genome of rust fungi during the past few years. Reassociation kinetics suggest a haploid genome size of 4.7×10^7 bp for *P. sorghi*. Half of it represented a moderately repetitive fraction (Anderson et al. 1992). *P. graminis* f. sp. *tritici* has a haploid genome size of about 6.7×10^7 bp with about 64% unique, 4% foldback, and 30% repetitive DNA sequences. The repetitive sequences have a total complexity of 390 kb and are repeated on an average of 52 times. The average amount of DNA per ungerminated urediospore was 435 fg. The spores were in G 2 (Backlund and Szabo 1993). The mitochondrial genome of *P. graminis* f. sp. *avenae* has a size of 80.2 kb, with an average of 21 copies per urediospore germling (Sock et al. 1994). Based on light microscopy, the haploid chromosome number of *P. graminis* f. sp. tritici was assumed to be n = 6. However, recent ultrastructurally derived reconstructions of sectioned fungal pachytene nuclei provided a definitive karyotype of n = 18 (Boehm et al. 1992). A similar number was found for *Melampsora lini* (Boehm and Bushnell 1992). This technology, applied for the analysis of pachytene centromeres, allowed the discrimination of the phylogenetically related moss parasite,

Eocronartium muscicola, from the Uredinales (Boehm and McLaughlin 1989, 1991).

Biochemical and molecular methods also help to differentiate rust fungi from one another. The rusts that attack cowpea (*Vigna sinensis*) or beans (*Phaseolus vulgaris*) have been considered to be either identical (Boerema and Verhoeven 1979), or different varieties of the same species (Arthur 1962), or separate species (Fromme 1924). After separation of proteins from urediospores by two-dimensional isoelectric focusing polyacrylamide gel electrophoresis, *Uromyces appendiculatus* (the bean rust) and *Uromyces vignae* (the cowpea rust) shared only 183 polypeptides out of 335 polypeptides (Kim et al. 1985). This result was confirmed by DNA hybridization techniques (Braithwaite et al. 1991). In addition, nuclear DNA content was used as a tool in fungal taxonomy. It had already been shown that variants observed after artificial culture of stem rust exhibited differences in genome sizes (Williams and Mendgen 1975). Fluorescence of nuclei in binucleate basidiospores and mononucleate pycniospores as a measure for relative amounts of DNA content among the rusts ranged from 0.8 to 6.8 times that of *P. graminis*. Isolates of *P. recondita* and *P. hordei* from different telial hosts differed considerably in DNA content (Eilam et al. 1992). Also, DNA restriction fragment length polymorphisms and the internal transcribed spacer region in the nuclear ribosomal repeat unit were useful to discriminate rust species and races of *P. graminis* f. sp. *tritici* (Martinez et al. 1991; Anderson and Pryor 1992; Zambino and Szabo 1992, 1993; Gardes and Bruns 1993). Polymorphic DNA analysis revealed relationships among diverse genotypes in Australian and American collections of *U. appendiculatus* (Maclean et al. 1995).

During differentiation of infection structures, the development of each morphological unit is correlated with the appearance or disappearance, increase or decrease in concentration of certain polypeptides (Staples and Macko 1984; Deising and Mendgen 1992). By cascade hybridization several differential-specific genes have been isolated (Bhairi et al. 1989). The infection structure gene (INF 24) from *Uromyces appendiculatus* had no homologues in nonrust fungi (Xuei et al. 1992). INF 56 represented a member of a multigene family preferentially expressed during development of appressoria (Xuei et al. 1993). Genes expressed only during growth of the fungus within the host plant have been characterized in flax after infection with rust (*Melampsora lini*) and in stem rust (*P. graminis tritici*) (Liu et al. 1993; Roberts and Pryor 1995). These results demonstrate that molecular methods open a new door for the characterization of rust gene products possibly influencing the virulence of the fungus. A transformation system may help to elucidate the role of such genes (Bhairi and Staples 1992). The detection of double-stranded RNA viruses has added a new dimension to the possible biocontrol of rust fungi (Zhang et al. 1994).

VI. Cytology and Physiology of the Host-Parasite System

A. Cytology of the Monokaryotic Infection

In the monokaryotic stage, hyphae grow inter- and intracellularly. The fungal cytoplasm is dense. Endoplasmic reticulum-derived, tubular vesicular complexes suggest high protein synthesis rates. Golgi equivalents, mitochondria, filasomes, and microtubules are numerous (Mims et al. 1988; Mims and Richardson 1989; Gold and Mendgen 1991). Hyphae are covered with an amorphous, electron-opaque extracellular matrix which closely contacts host cells. Hyphae enter mesophyll cells to form intracellular hyphae (also called monokaryotic haustoria) without evident modification of the fungal protoplast (Littlefield and Heath 1979; Harder 1984; Hopkin and Reid 1988). Even the host vascular system is invaded, which does not occur in the dikaryotic stage of rusts (Larous and Lösel 1993b). The different strategies in the monokaryon and dikaryon for the acquisition of host assimilates are discussed by Larous and Lösel (1993a). Host cell responses as a result of fungal penetration are quite variable (Heath 1989; Stark-Urnau and Mendgen 1995; Heath and Skalamera 1997).

The mononucleate hyphae finally aggregate and form the pycnium, a structure in which the gametes of the rust fungi are produced (Harder 1984). The sporophores are uninucleate and elongated. Spores are produced in chains (Mims et al. 1976; Harder and Chong 1978; Metzler 1981). Fertilization occurs by means of pycniospores and flexuous hyphae which reach out of the pycnium. Following dikaryotization, the aecium is produced as the fruiting structure that produces aeciospores and reinitiates the dikaryotic phase. The dikaryotization process is poorly understood (Harder 1984). It is known, however, that the dikaryotic primary aeciosporophores arise from multinucleate fusion cells (Rijkenberg and Truter 1974).

B. Cytology of the Dikaryotic Infection

After the development of the infection hypha and during the differentiation of the primary haustorium, intercellular hyphal growth starts with a branching of the infection hypha. Growth and branching of hyphae continue together with production of haustorial mother cells and haustoria until a certain leaf area – the amount of hyphae depends very much on the different rusts – is colonized. Composition and structure of the cytoplasm are comparable to the monokaryon (Littlefield and Heath 1979; Harder 1984; Welter et al. 1988; Gold and Mendgen 1991). The hyphal tip has, similar to the situation in other fungi, a Spitzenkörper (Hoch and Staples 1983). Nuclei and nuclear division have been studied in detail (Harder 1984; Swann and Mims 1991; Boehm and Bushnell 1992; Boehm et al. 1992).

Haustorial differentiation starts with the development of a haustorial mother cell at the hyphal tip. Basically, the cytoplasmic contents do not differ from those of intercellular hyphae. However, mitochondria undergo a change in conformation and distribution. They round up and migrate to the periphery of the protoplast. Nuclei are more compact and the ER increasingly forms tubular vesicular complexes (Harder 1984; Knauf et al. 1989; Harder and Chong 1991). Walls become thicker and include additional layers. The septum is composed of more layers than normal hyphal septa. A transfer apparatus is associated with this septum and the whole structure seems to support the subsequent penetration of the host cells.

Roughly in the middle of the contact area with the plant cell wall, a penetration hypha develops. The penetration hypha elongates to form the haustorial neck. The neck wall differs from that in the area of host cell wall penetration and from the wall of the subsequently formed haustorial body. The neck wall is characterized by a neck ring, a structure with high mineral content (iron, phosphorus, and silicon) and a zone with peroxidase activity. Evidence suggests that the neck ring has a function similar to that of the Casparian strip in root endodermal cells in controlling apoplastic flow. It would force substances from the infected plant cell symplastically into the haustorium (Heath 1976). The haustorial body develops at the end of the neck, increases in size during aging, and becomes a multilobed, rigid structure (Hahn and Mendgen 1992). The haustorium has a layered wall, is surrounded by the extrahaustorial matrix and is embedded within the cytoplasm of the host plant (Fig. 4). This matrix contains various substances, including α- and β-linked sugars and polysaccharides (Harder and Chong 1991). Although the detection of β-linked sugars in this area indicates a contribution from the fungus, the bulk of material seems to come from the host cell: the extrahaustorial membrane smoothly surrounds the haustorium of both monokaryotic and dikaryotic infections. Many coated extensions of this membrane reach into the host cytoplasm and suggest endo- and exocytotic events. Although continuous with the host plasmalemma, the extrahaustorial membrane is highly modified in several

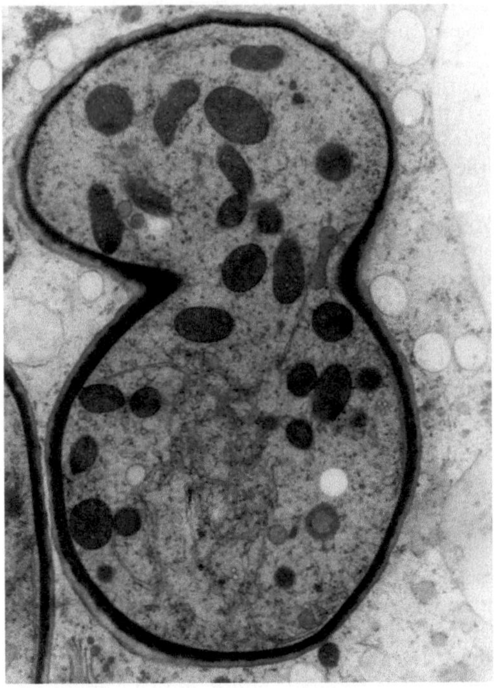

Fig. 4. Haustorium of *Uromyces viciae-fabae* (dikaryon) after high-pressure freezing and freeze substitution. The haustorium is smoothly embedded within the cytoplasm of the host plant. Magnification 12250×. (U. Bachem and K. Mendgen, unpubl.)

respects. Freeze-etch studies have revealed a loss of intramembrane particles, increased thickness, loss of ATPase activity (Woods and Gay 1987), and a reduced reactivity with stains for polysaccharides. The meaning of these modifications remains to be explained (Harder and Chong 1991).

Some days after infection, fungal hyphae aggregate in an intercellular space underneath the epidermis. These cells form the base of the fruiting structure. Elongated, sporogenous cells give rise to stalk cells and urediospores. Accessory cells with no obvious function mingle with the urediospores. The succession of urediospores from a sporogenous cell and its development is described in detail by Harder (1984). Later in the season, such sporogenous cells also produce teliospores (Mendgen 1984).

C. A New Physiological Unit, the Fungus in the Host Plant

The pathogen and the host, both together, produce a symbiotic structure, the aegricorpus (Loegering 1984). It is a result of coevolutionary processes (Heath 1991). A study of the pathogen or the host alone cannot explain the physiology of the diseased plant. With biochemical and physiological methods, it is difficult to discriminate the contributions of each partner. Molecular methods have improved the situation by analyzing the activity of single genes during infection (Liu et al. 1993). They will help us find and analyze plant disease resistance genes (Bennetzen and Hulbert 1992). Genetic studies of the interaction by Flor (1971) resulted in the gene-for-gene concept. Recent advances of this theory are summarized by Knott (1989). It tells us about the outcome of the interaction of two clones, the host and the parasite, in which classical diploid inheritance occurs. Some problems with this model still exist (Loegering 1984), although it is generally accepted (Chaps. 1, 15, 16, this Vol.).

D. Physiology and Molecular Biology of the Aegricorpus

Rust fungi are biotrophic parasites. The infected host cells undergo many changes, but continue to live during fungal development (Heath and Skalamera 1996). Only in incompatible combinations is hypersensitive cell death frequently observed. This event seems to interfere with the growth of haustorial pathogens requiring living plant cells. Programmed cell death is reminiscent of apoptosis in animals (Ryerson and Heath 1996). The recent characterization of the flax rust resistance gene L 6 is a first step in the elucidation of a signal receptor in the plant membrane that seems to be involved in the recognition process between plant and pathogen (Ellis et al. 1995; Lawrence et al. 1995). This result will help to explain the steps from recognition of the fungus to activation of defense and the final outcome, the hypersensitive reaction.

It is generally assumed that the new physiologic status of the host benefits the fungus. Most obvious are green islands, which develop in the colonized tissue when the surrounding tissue becomes chlorotic (Dekhuijzen 1976; Bushnell 1984; Durbin 1984). Even if the central area of the pustule becomes necrotic, a green ring often will remain. In the infected cells, organelles (e.g., the nucleus) change size and movements dramatically (Bushnell 1984; Xu and Mendgen 1991), and very often migrate towards the site of fungal penetration. Changes in gene expression of the host plant during the early stage of infection and the increase of the intermediary metabolism of the plant reflect these changes (Chakravorty and Scott 1982). Cytokinins (Durbin 1984), auxins (Pegg 1976) and ethylene (Daly 1976; Paradies et al. 1979) are produced in high concentrations. General defense mechanisms (e.g., lignification) of the plant are induced (Moerschbacher et al. 1988; Fink et al. 1991). Autofluorescence of cell walls and papillae are observed (Ryerson and Heath 1992). Phytoalexins may contribute to the defense of the rust fungus (Mayama et al. 1991). On the other hand, photosynthesis and chlorophyll content decrease and a considerable drop in activity of xanthine oxidase is observed (Berghaus and Reisener 1985; Montalbini 1992). This is followed by a loss of epidermal cell turgor and a reduction in the efficiency of the roots to replace water losses (Paul and Ayres 1985; Berryman et al. 1991). The interaction between host and parasite might be regulated by race specific or unspecific elicitors and suppressors of plant defense reactions induced by the fungus (Beissmann and Kogel 1991; Kogel and Beissmann 1991; Beissmann et al. 1992; Heath 1995; Yamamoto 1995).

With the successful culture of rust fungi on relatively simple media containing mainly sugars, minerals, and amino acids (Maclean 1982; Williams 1984), obligate parasitism has lost many of its mysteries. It seems that the increased metabolic activity of the host mainly provides a generous supply of simple substrates such as sugars and amino acids (Durbin 1984). A well-balanced supply of such substrates induced the production of infective urediospores and teliospores, but no haustoria on artificial media (Kuck and Reisener 1985). In addition, rust fungi can be grown on epidermal monolayers (Mendgen and Dressler 1983) and in liquid culture (Fasters et al. 1993).

VII. Methods of Rust Control

The first known effective means to control rusts was mandated by a law in Rouen, France, from 1660, requiring the destruction of barberries near grain fields. The Connecticut barberry law of 1726, and subsequent barberry laws in other states and countries, are proofs of farmers' powers of observation. This method of control is still valuable today (Roelfs 1982), but only for *Puccinia graminis*.

Currently, chemical control and breeding for resistance go together. Buchenauer (1982) evaluated the use of the more established fungicides such as dithiocarbamates, and compared them with carboxylic acid anilides, morpholine derivates, Indar, and inhibitors of ergosterol biosynthesis. Azole fungicides in general and the recently introduced triazole fungicides exhibit high biological activity at low application rates (Kuck and Scheinpflug 1986; Saur et al. 1991). Rowell (1985) described the different in vitro tests for fungicides, such as spore germination and mycelial growth tests, or in vivo tests, such as seedling assays and field evaluation. In Western Europe, intensive cropping systems, high yield and high grain prices permit economic use of chemical control of rusts. However, genetic resistance is important under any circumstance. Most rust-resistant cereal cultivars have remained so for 5 years or more, which is about the life span of an active breeding program (Roelfs et al. 1992). Others have remained resistant for many years. A discussion of this problem can be found in several reviews (Röbbelen and Sharp 1978; Dyck and Kerber 1985; Parlevliet 1985; Knott 1989). Recent progress in the establishment of durable resistance was explained by Steffensen (1992) and McIntosh (1992). An overview on genes for resistance against wheat rusts is available (McIntosh et al. 1995). As an alternative to the pure line cultivars, intraspecific mixtures of different genotypes or interfield diversity together with a regional deployment of resistance genes was discussed (Mundt and Browning 1985). The evolution of a superrace capable of overcoming many Sr genes for resistance against rust may destroy such hopes (Gates and Loegering 1991). Also, cultural methods such as early sowing to avoid stem rust or the induction of resistance to rust by avirulent races (Van Asch et al. 1992) should not be forgotten. The detection of chemically induced resistance in wheat opens new dimensions to rust control (Görlach et al. 1996).

Biological control of rusts with hyperparasitic fungi has been tried often (Kranz 1981), but is still in its infancy. Of 24 species of fungi in 12 genera screened for their hyperparasitic activity towards *Puccinia coronata* on oat seedlings, 7 *Verticillium* spp. and *Acremonium implicatum* were able to colonize uredial sori (Leinhos and Buchenauer 1992). In addition, *Erwinia* spp., *Bacillus* spp., and *Trichoderma* spp. have been recommended for rust control (Govindasamy and Balasubramanian 1989; Kempf and Wolf 1989; Rytter et al. 1989). However, because of the high humidity requirements of hyperparasitic fungi, such methods are not effective in temperate or arid conditions (Grabski and Mendgen 1986). In the tropics, better results may be obtained (Saksiriat and Hoppe 1990).

VIII. Biological Control of Weeds with Rusts

Biological control of weeds has a long and sometimes very successful history. Examples with rusts are the control of skeleton weed (*Chondrilla juncea*) by *P. chondrillina* in Australia and California (Burdon 1987; Supkoff et al. 1988) and blackberry (*Rubus* spp.) by *Phragmidium violaceum* (Templeton et al. 1979). Other examples are cited by Watson (1991) and Evans and Elliston (1990). According to Charudattan (1991) some 107 fungal taxa, including many rust fungi, are presently under investigation for the biological control of weeds. Currently, the control of diffuse and spotted knapweeds (*Centaurea diffusa* and *C. maculosa*) with *Puccinia jaceae* appears promising, since it forms heavy stem infections early in the season. This rust has a restricted and stable host range and the danger of damaging safflower (*Carthamus tinctorius*) is restricted to a short period during the seedling stage of the plant and thus appears to be of minor importance (Mortensen et al. 1991).

Cyperus esculentus was only recently introduced into The Netherlands. Its growth capacity and its insensitivity to many herbicides caused considerable losses. Integration of biological control with *P. canaliculata* and subsequent treatment with herbicides resulted in a rapid decline of the weed in the USA (Callaway et al. 1987). However, since this rust also infects *Cy. fuscus*, it was not introduced for use in Europe (Scheepens and Hoogerbrugge 1991).

IX. Conclusions

Rust fungi are a group of parasites highly adopted to their host plants. They may be useful as mycoherbicides but, in most cases, are deleterious to many crops. The control of rusts by either genetical resistance or the use of chemicals and, to a smaller extent, also by cultural methods is still a major problem. Cultural methods, such as the interruption of the life cycle or cultivation to remove volunteer plants from an earlier crop are effective, but by far not sufficient means of control. Several fungicides are effective, safe, and easy to use. However, they may represent a problem for our environment. Some pathogens have become resistant to fungicides and other methods are preferable. Therefore, the use of resistant cultivars is of major importance and adds no extra cost to farmers. However, resistant cultivars are developed and released and, as soon as production increases rapidly, new, virulent races appear. The breeding cycle then starts over again. Therefore, research on the interaction of rusts with their host plants should be increased. Physiological and cytological studies have helped us to understand this interaction much better. Recent progress in the molecular biology of the host-parasite interaction will give us a new handle to modify host genes in a more efficient way. The goal should be to explain the action of avirulence genes of the rusts and, consequently, the action of resistance genes in the host. Thus, breeding for resistance may become more efficient.

Acknowledgments. Dr. Matthias Hahn, Konstanz, Dr. Stefan Wirsel, Konstanz, and Prof. Dr. Roland Rohringer, Göttingen, read the manuscript.

References

Alexopoulos CJ, Mims CW, Blackwell M (1996) Introductory mycology, 4th edn. Wiley, New York, 868 pp
Anderson PA, Pryor AJ (1992) DNA restriction fragment length polymorphisms in the wheat stem rust fungus, *Puccinia graminis tritici.* Theor Appl Gen 83:715–719
Anderson PA, Tyler BM, Pryor A (1992) Genome complexity of the maize rust fungus, *Puccinia sorghi.* Exp Mycol 16:302–307
Anikster Y (1984) The formae speciales. In: Bushnell WR, Roelfs AP (eds) The cereal rusts, vol I. Academic Press, Orlando, pp 115–130
Anikster Y, Wahl I (1979) Coevolution of the rust fungi on Gramineae and Liliaceae and their hosts. Annu Rev Phytopathol 17:367–403
Arthur JC (1962) Manual of the rusts in United States and Canada. Hafner, New York
Backlund JE, Szabo LJ (1993) Physical characteristics of the genome of the phytopathogenic fungus *Puccinia graminis.* Curr Gen 24:89–93
Beissmann B, Kogel KH (1991) Identification and characterization of suppressors. In: Linsken HF, Jackson JF (eds) Plant toxin analysis, vol 13. Springer, Berlin Heidelberg New York, pp 259–275
Beissmann B, Engels W, Kogel K, Marticke KH, Reisener HJ (1992) Elicitor-active glycoproteins in apoplastic fluids of stem-rust-infected wheat leaves. Physiol Mol Plant Pathol 40:79–89
Bennetzen JL, Hulbert SH (1992) Organization, instability and evolution of plant disease resistance genes. Plant Mol Biol 20:575–578
Berghaus R, Reisener HJ (1985) Changes in photosynthesis of wheat plants infected with wheat stem rust (*Puccinia graminis* f. sp. *tritici*). Phytopathol Z 112:165–172
Berryman CA, Eamus D, Farrar JF (1991) Variations in epidermal cell turgor of rust-infected barley seedlings. New Phytol 119:535–540
Bhairi SM, Staples RC (1992) Transient expression of the β-glucoronidase gene introduced into *Uromyces appendiculatus* uredospores by particle bombardment. Phytopathology 82:986–989
Bhairi SM, Staples RC, Freve P, Yoder OC (1989) Characterization of an infection structure-specific gene from the rust fungus *Uromyces appendiculatus.* Gene 81:237–243
Boehm EWA, Bushnell WR (1992) An ultrastructural pachytene karyotype for *Melampsora lini.* Phytopathology 82:1212–1218
Boehm EWA, McLaughlin DJ (1989) Phylogeny and ultrastructure in *Eocronartium muscicola*: meiosis and basidial development. Mycologia 81:98–114
Boehm EWA, McLaughlin DJ (1991) An ultrastructural karyotype for the fungus *Eocronartium muscicola* using epifluorescence preselection of pachytene nuclei. Can J Bot 69:1309–1320
Boehm EWA, Wenstrom JC, McLaughlin DJ, Szabo LJ, Roelfs AP, Bushnell WR (1992) An ultrastructural

pachytene karyotype for *Puccincia graminis* f. sp. *tritici.* Can J Bot 70:401–413

Boerema GH, Verhoeven AA (1979) Check-list for scientific names of common parasitic fungi. Ser 2c. Fungi on field crops: pulse (legumes) and forage crops (herbage legumes). Neth J Plant Pathol 85:151–185

Braithwaite KS, Manners JM, Maclean DJ, Irwin JAG (1991) A molecular approach to distinguish the bean rust and siratro rust pathogens, DNA. Aust J Bot 39: 527–534

Buchenauer H (1982) Chemical and biological control of cereal rust. In: Scott KJ, Chakravorty AK (eds) The rust fungi. Academic Press, London, pp 247–279

Burdon JJ (1987) Diseases and plant population biology. Cambridge University Press, Cambridge

Bushnell WR (1984) Structural and physiological alterations in susceptible host tissue. In: Bushnell WR, Roelfs AP (eds) The cereal rusts, vol I. Academic Press, Orlando, pp 477–507

Bushnell WR, Roelfs AP (eds) (1984) The cereal rusts. Academic Press, Orlando

Callaway MB, Phatak SC, Wells HD (1987) Interactions of *Puccinia canaliculata* (Schw.) Lagerh. with herbicides on tuber production and growth of *Cyperus esculentus* L. Trop Pest Manage 33:22–26

Chakravorty AK, Scott KJ (1982) Biochemistry of host rust interactions. In: Scott KJ, Chakravorty AK (eds) The rust fungi. Academic Press, London, pp 179–222

Charudattan R (1991) The mycoherbicide approach with plant pathogens. In: TeBeest DO (ed) Microbial control of weeds. Chapman and Hall, New York, pp 24–57

Chester KS (1946) The nature and prevention of the cereal rusts as exemplified in the leaf rust of wheat. Chronica Botanica Company, Walthan, Massachusetts

Clement JA, Martin SG, Porter R, Butt TM, Beckett A (1993) Germination and the role of extracellular matrix in adhesion of urediniospores of *Uromyces viciae-fabae* to synthetic surfaces. Mycol Res 97:585–593

Comstock JC, Shine JM Jr, Raid RN (1992) Effect of rust on sugarcane growth and biomass. Plant Dis 76:175–177

Craigie JH (1927) Experiments on sex in rust fungi. Nature 120:116–117

Cummins GB, Hiratsuka Y (eds) (1983) Illustrated genera of rust fungi. The American Phytopathological Society, St Paul

Daly JM (1976) The carbon balance of diseased plants: changes in respiration, photosynthesis and translocation. In: Heitefuss R, Williams PH (eds) Physiological plant pathology. Springer, Berlin Heidelberg New York, pp 450–479

DeBary A (1853) Untersuchungen über Brandpilze und die durch sie verursachten Krankheiten der Pflanzen. GWF Müller, Berlin

DeBary A (1865) Neue Untersuchungen über die Uredineen, insbesondere die Entwicklung der *Puccinia graminis* und den Zusammenhang derselben mit *Aecidium berberidis*. Monatsber Königl Preuss Akad Wiss Berlin 1865:15–50

Deising H, Mendgen K (1992) Differentiation specific enzyme production and wall modification during rust infection. In: Tudzynski P, Stahl U (eds) Molecular biology of filamentous fungi. Verlag Chemie, Weinheim, pp 27–44

Deising H, Siegrist J (1995) Chitin deacetylase activity of the rust *Uromyces viciae-fabae* is controlled by fungal morphogenesis. FEMS Microbiol Lett 127:207–212

Deising H, Frittrang AK, Kunz S, Mendgen K (1995) Regulation of pectin methylesterase and polygalacturonate lyase activity during differentiation of infection structures in *Uromyces viciae-fabae*. Microbiology 141:561–571

Deising H, Heiler S, Rauscher M, Xu H, Mendgen K (1996) Cellular aspects of rust infection structure differentiation: Spore adhesion and fungal morphogenesis. In: Nicole M, Gianinazzi-Pearson V (eds) Histology, ultrastructure and molecular cytology of plant-microorganism interactions. Kluwer, Dordrecht pp 135–156

Dekhuijzen HM (1976) Endogenous cytokinins in healthy and diseased plants. In: Heitefuss R, Williams PH (eds) Physiological plant pathology. Springer, Berlin Heidelberg New York, pp 526–559

Duhamel du Monceau M (1779) Éléments d'agriculture, tome 1. Veuve Desaint, Paris

Durbin RD (1984) Effects of rust on plant development in relation to the translocation of inorganic and organic solutes. In: Bushnell WR, Roelfs AP (eds) The cereal rusts, vol I. Academic Press, Orlando, pp 509–528

Dyck PL, Kerber ER (1985) Resistance of the race-specific type. In: Roelfs AP, Bushnell WR (eds) The cereal rusts, vol II. Academic Press, Orlando, pp 469–500

Eilam T, Bushnell WR, Anikster Y, McLaughlin DJ (1992) Nuclear DNA content of basidiospores of selected rust fungi as estimated from fluorescence of propidium iodide-stained nuclei. Phytopathology 82:705–712

Ellis JG, Lawrence GJ, Finnegan EJ, Anderson PA (1995) Contrasting complexity of two rust resistance loci in flax. Proc Natl Acad Sci USA 92:4185–4188

Epstein L, Laccetti L, Staples RC, Hoch HC, Hoose WA (1985) Extracellular proteins associated with induction of differentiation in bean rust uredospore germlings. Phytopathology 75:1073–1976

Eriksson J, Henning E (1896) Die Getreideroste. Ihre Geschichte und Natur sowie Massregeln gegen dieselben. Norstedt and Soner, Stockholm

Evans HC, Elliston CA (1990) Classical biological control of weeds with microorganisms: past, present, prospects. Aspects Appl Biol 24:39–49

Fasters MK, Daniels U, Moerschbacher BM (1993) A simple and reliable method for growing the wheat stem rust fungus, *Puccinia graminis* f. sp. *tritici*, in liquid culture. Physiol Mol Plant Pathol 42:259–265

Figueiredo MB, Pimental CPV, Russomanno OMR, Coutinho LN (1987) Biologia da espécie biteliomórphica *Puccinia pampeana* Speg. *Endophyllum pampeanum* (Speg.) Lindq., ferrugem da pimentae do pimentao (*Capsicum* spp.). Arq Inst Biol Sao Paulo 54:1–10

Fink W, Haug M, Deising H, Mendgen K (1991) Early defence responses of cowpea (*Vigna sinensis* L.) induced by non-pathogenic rust fungi. Planta 185:246–254

Flor HH (1956) The complementary gene systems in flax and flax rust. Adv Genet 8:29–54

Flor HH (1971) Current status of the gene-for-gene concept. Annu Rev Phytopathol 9:275–296

French RC (1992) Volatile chemical germination stimulators of rust and other fungal spores. Mycologia 84:277–288

Freytag S, Mendgen K (1991) Carbohydrates on the surface of urediniospore- and basidiospore-derived infection structures of heteroecious and autoecious rust fungi. New Phytol 119:527–534

Fromme FD (1924) The rust of cowpeas. Phytopathology 14:67–79

Gardes M, Bruns TD (1993) ITS primers with enhanced specificity for basidiomycetes – application to the identification of mycorrhizae and rusts. Mol Ecol 2:113–118

Gates JE, Loegering WQ (1991) Mutation to wider virulence in *Puccinia graminis* f. sp. *tritici*: evidence for the existence of loci which allow the fungus to overcome several host stem rust resistance genes simultaneously. Appl Environ Microbiol 57:2332–2336

Gäumann E (1959) Die Rostpilze Mitteleuropas. Bücheler, Bern

Gold RE, Mendgen K (1991) Rust basidiospore germlings and disease initiation. In: Cole GT, Hoch HC (eds) The fungal spore and disease initiation in plants and animals. Plenum, New York, pp 67–99

Görlach J, Volrath S, Knauf-Beiter G, Hengy G, Beckhove U, Kogel K, Oostendorp M, Staub T, Ward E, Kessmann H, Ryals J (1996) Benzthiadiazole, a novel class of inducers of systemic acquired resistance, activates gene expression and disease resistance in wheat. Plant Cell 8:629–643

Govindasamy V, Balasubramanian R (1989) Biological control of groundnut rust *Puccinia arachidis* by *Trichoderma harzianum*. Z Pflanzenkr Pflanzenschutz 96:337–345

Grabski GC, Mendgen K (1986) Die Parasitierung des Bohnenrostes *Uromyces appendiculatus* var. *appendiculatus* durch den Hyperparasiten *Verticillium lecanii*: Untersuchungen zur Wirterkennung, zur Penetration und Abbau der Rostpilzsporen. J Phytopathol 115:116–123

Groth JV (1988) *Uromyces appendiculatus*, rust of *Phaseolus* beans. Adv Plant Pathol 6:389–410

Groth JV, Roelfs AP (1982) Effect of sexual and asexual reproduction on race abundance in cereal rust populations. Phytopathology 72:1503–1507

Hahn M, Mendgen K (1992) Isolation by Con A binding of haustoria from different rust fungi and comparison of their surface qualities. Protoplasma 170:95–103

Harder DE (1984) Developmental ultrastructure of hyphae and spores. In: Bushnell WE, Roelfs AP (eds) The cereal rusts, vol I. Academic Press, Orlando, pp 333–373

Harder DE, Chong J (1978) Ultrastructure of spermatium ontogeny in *Puccinia coronata avenae*. Can J Bot 50: 395–403

Harder DE, Chong J (1991) Rust haustoria. In: Mendgen K, Lesemann D-E (eds) Electron microscopy of plant pathogens. Springer, Berlin Heidelberg New York, pp 235–250

Heath MC (1976) Ultrastructural and functional similarity of the haustorial neckband of rust fungi and the Casparian strip of vascular plants. Can J Bot 54:2484–2489

Heath M (1989) A comparison of growth and plant responses in cowpea and bean cultivars inoculated with urediospores and basidiospores of the cowpea rust fungus. Physiol Mol Plant Pathol 34:415–426

Heath MC (1991) Evolution of resistance to fungal parasitism in natural ecosystems. New Phytol 119:331–343

Heath MC (1995) Signal exchange between higher plants and rust fungi. Can J Bot 73:S616–S623

Heath MC, Skalamera D (1997) Cellular interactions between plants and biotrophic fungal parasites. In: Callow JA (ed) Advances in botanical research. Academic Press, London pp 196–225

Hennen JF, Figueiredo MB, Pimental CP, Russomanno OMR (1984) The life cycle and taxonomy of *Puccinia pampeana* Speg. and *Endophyllum pampeanum* (SPEG.) Lindq. on *Capsicum* spp. and other Solanaceae. Rep Tottori Mycol Inst Jpn 22:209–220

Hoch HC, Staples RC (1983) Ultrastructural organization of the nondifferentiated uredospore germling of *Uromyces phaseoli* variety *typica*. Mycologia 75:795–824

Hoch HC, Staples RC (1987) Structural and chemical changes among the rust fungi during appressorium development. Annu Rev Phytopathol 25:231–247

Hoch HC, Staples RC (1991) Signaling for infection structure formation in fungi. In: Cole GT, Hoch TC (eds) The fungal spore and disease initiation in plants and animals. Plenum Press, New York, pp 25–46

Hoch HC, Staples RC, Whitehead B, Comeau J, Wolf ED (1987) Signaling for growth orientation and cell differentiation by surface topography in *Uromyces*. Science 235:1659–1662

Hopkin AA, Reid J (1988) Cytological studies of the M-haustorium of *Endocronartium harknessii*: morphology and ontogeny. Can J Bot 66:974–988

Javed ZUR (1982) Field trials with new and recommended fungicides for leaf rust control during 1981. Kenya Coffee 47:199–205

Kempf H-J, Wolf G (1989) *Erwinia herbicola* as a biocontrol agent of *Fusarium culmorum* and *Puccinia recondita* f. sp. *tritici* on wheat. Phytopathology 79:990–994

Kim WK, Heath MC, Rohringer R (1985) Comparative analysis of proteins of *Uromyces phaseoli* var. *typica*, *U. phaseoli* var. *vignae*, and *U. viciae-fabae*: polypeptide mapping by two-dimensional electrophoresis. Can J Bot 63:2144–2149

Kislev ME (1982) Stem rust of wheat 3300 years old found in Israel. Science 216:993–994

Knauf G, Welter K, Müller M, Mendgen K (1989) The haustorial host-parasite interface in rust-infected bean leaves after high-pressure freezing. Physiol Mol Plant Pathol 34:519–530

Knott DR (1989) The wheat rusts – breeding for resistance. Springer, Berlin Heidelberg New York

Koch E, Hoppe H-H (1988) Development of infection structures by the direct-penetrating soybean rust fungus (*Phakopsora pachyrhizi* Syd.) on artificial membranes. J Phytopathol 122:232–244

Kogel KH, Beissmann B (1991) Isolation and characterization of elicitors. In: Linskens HF, Jackson JF (eds) Plant toxin analysis. Springer, Berlin Heidelberg New York, pp 239–257

Kolmer JA (1992) Effect of sexual recombination in two populations of the wheat leaf rust fungus *Puccinia recondita*. Can J Bot 70:359–363

Kranz J (1981) Hyperparasitism of biotrophic fungi. In: Blakeman JP (ed) Microbial ecology of the phylloplane. Academic Press, London, pp 327–352

Kuck KH, Scheinpflug H (1986) Biology of sterol-biosynthesis inhibiting fungicides. In: Haug G, Hoffmann H (eds) Chemistry of plant protection, vol 1. Springer, Berlin Heidelberg New York, pp 65–96

Kuck KH, Reisener HJ (1985) In vitro sporulation of race 32 of *Puccinia graminis* Pers. f. sp. *tritici* Erikss. & Henn. Physiol Mol Plant Pathol 27:259–268

Kwon YH, Hoch HC (1991) Temporal and spatial dynamics of appressorium formation in *Uromyces appendiculatus*. Exp Mycol 15:116–131

Larous L, Lösel DM (1993a) Strategies of pathogenicity in monokaryotic and dikaryotic phases of rust fungi, with special reference to vascular infection. Mycol Res 97: 415–420

Larous L, Lösel DM (1993b) Vascular infection by *Puccinia menthae* and other rust fungi. Mycol Rcs 97:409–414

Laundon GF (1973) Uredinales. In: Ainsworth GC, Sparrow FK, Sussman AS (eds) The fungi, an advanced treatise, vol IVB. Academic Press, London, pp 247–279

Lawrence GJ, Finnegan EJ, Ayliffe MA, Ellis JG (1995) The L6 gene for flax rust resistance is related to the *Arabidopsis* bacterial resistance gene RPS2 and the tobacco viral resistance gene N. Plant Cell 7:1195–1206

Lehmann E, Kummer H, Dannenmann H (1937) Der Schwarzrost, seine Geschichte, seine Biologic und seine Bekämpfung in Verbindung mit der Berberitzenfrage. Lehmann, München

Leinhos GME, Buchenauer H (1992) Hyperparasitism of selected fungi on rust fungi of cereal. Z Pflanzenkr Pflanzenschutz 99:482–498

Littlefield LJ (1981) Biology of the plant rusts: an introduction. Iowa State University Press, Ames

Littlefield LJ, Heath MC (1979) Ultrastructure of rust fungi. Academic Press, New York

Liu Z, Szabo LJ, Bushnell WR (1993) Molecular cloning and analysis of abundant and stage-specific mRNAs from *Puccinia graminis*. Mol Plant-Microb Interact 6:84–91

Loegering WQ (1984) Genetics of the pathogen-host association. In: Bushnell WR, Roelfs AP (eds) The cereal rusts, vol I. Academic Press, Orlando, pp 165–192

Luig NH (1985) Epidemiology in Australia and New Zealand. In: Roelfs AP, Bushnell WR (eds) The cereal rusts, vol II. Academic Press, Orlando, pp 301–328

Maclean DJ (1982) Axenic culture and metabolism of rust fungi. In: Scott KJ, Chakravorty AK (eds) The rust fungi. Academic Press, London, pp 37–120

Maclean DJ, Braithwaite KS, Irwin JAG, Manners JM, Groth JV (1995) Random amplified polymorphic DNA reveals relationships among diverse genotypes in Australian and American collections of *Uromyces appendiculatus*. Phytopathology 85:757–765

Martinez JP, Groth JV, Young ND (1991) Development of a restriction fragment length polymorphism (RFLP) map for *Uromyces appendiculatus*. Phytopathology 81:1208

Mayama S, Bordin APA, Morikawa T, Tani T (1991) Role of phytoalexins in host defense reactions. In: Patil SS, Ouchi S, Mills D, Vance C (eds) Molecular strategies of pathogens and host plants. Springer, Berlin Heidelberg New York, pp 203–213

McCain JW, Groth JV, Roelfs AP (1992) Are asexual populations of the bean rust fungus less diverse than sexual ones? Phytopathology 82:1139

McGrath MT, Pennypackers SP (1991) Reduction in the rate and duration of grain growth in wheat due to stem rust and leaf rust. Phytopathology 81:778–787

McIntosh RA (1992) Pre-emptive breeding to control wheat rusts. Euphytica 63:103–113

McIntosh RA, Wellings CR, Park RF (1995) Wheat rusts. An atlas of resistance genes. Kluwer, Dordrecht

Mendgen K (1984) Development and physiology of teliospores. In: Bushnell WR, Roelfs AP (eds) The cereal rusts, vol I. Academic Press, Orlando, pp 375–398

Mendgen K, Deising H (1993) Infection structures of fungal plant pathogens – a cytological and physiological evaluation, Tansley Review. New Phytol 124:193–213

Mendgen K, Dressler E (1983) Culturing *Puccinia coronata* on a cell monolayer of the *Avena sativa* coleoptile. Phytopathol Z 108:226–234

Mendgen K, Hahn M, Deising H (1996) Mechanisms and morphogenesis of penetration by plant-pathogenic fungi. Annu Rev Phytopathol 34:367–386

Metzler B (1981) Pyknidialstruktur und Pyknosporgenese bei *Gymnosporangium fuscum* DC. Z Mykol 47:271–280

Mims CW, Richardson EA (1989) Ultrastructure of appressorium development by basidiospore germlings of the rust fungus *Gymnosporangium juniperi-virginianae*. Protoplasma 148:111–119

Mims CW, Seabury F, Thurston EL (1976) An ultrastructural study of spermatium formation in the rust fungus *Gymnosporangium juniperi-virginianae*. Am J Bot 63: 997–1002

Mims CW, Roberson RW, Richardson EA (1988) Ultrastructure of freeze-substituted and chemically fixed basidiospores of *Gymnosporangium juniperi-virginianae*. Mycologia 80:356–364

Moerschbacher BM, Noll U, Flott BE, Reisener HJ (1988) Lignin biosynthetic enzymes in stem rust infected, resistant and susceptible near-isogenic wheat lines. Physiol Mol Plant Pathol 33:33–46

Montalbini P (1992) Changes in xanthine oxidase activity in bean leaves induced by *Uromyces phaseoli* infection. J Phytopathol 134:63–74

Morin L, Brown JF, Auld BA (1992) Teliospore germination, basidiospore formation and the infection process of *Puccinia xanthii* on *Xanthium occidentale*. Mycol Res 96:661–669

Mortensen K, Harris P, Kim WK (1991) Host ranges of *Puccinia jaceae*, *P. centaureae*, *P. acroptili*, and *P. carthami*, and the potential value of *P. jaceae* as a biological control agent for diffuse knapweed (*Centaurea diffusa*) in North America. Can J Plant Pathol 13:71–80

Mundt CC, Browning JA (1985) Genetic diversity and cereal rust management. In: Roelfs AP, Bushnell WR (eds) The cereal rusts, vol II. Academic Press, Orlando, pp 527–560

Nagarajan S, Singh DV (1990) Long-distance dispersion of rust pathogens. Annu Rev Phytopathol 28:139–153

Ono Y, Buriticá P, Hennen JF (1992) Delimitation of *Phakopsora*, *Physopella* and *Cerotelium* and their species on Leguminosae. Mycol Res 96:825–850

Paradies I, Hümme B, Hoppe HH, Heitefuss R, Elstner EF (1979) Induction of ethylene formation in bean (*Phaseolus vulgaris*) hypocotyl segments by preparations isolated from germ tube cell walls of *Uromyces phaseoli*. Planta 146:193–197

Parlevliet JE (1985) Resistance of the nonrace-specific type. In: Roelfs AP, Bushnell WR (eds) The cereal rusts, vol II. Academic Press, Orlando, pp 501–525

Paul ND, Ayres PG (1985) Water relations and growth of rust-infected groundsel (*Puccinia lagenophorae* Cooke; *Senecio vulgaris* L.) during and after exposure to freezing soil temperatures. Physiol Plant Pathol 27:185–196

Pegg GF (1976) Endogenous auxins in healthy and diseased plants. In: Heitefuss R, Williams PH (eds) Physiological plant pathology. Springer, Berlin Heidelberg New York, pp 560–616

Petersen RH (1974) The rust fungus life cycle. Bot Rev 40:453–513

Preece T, Hick AJ (1990) An introductory scanning electron microscope atlas of rust fungi. Farrand Press, London

Purdy LH (1985) Sugarcane rusts. In: Roelfs AP, Bushnell WR (eds) The cereal rusts, vol II. Academic Press, Orlando, pp 237–256

Rijkenberg FHJ, Truter SJ (1974) The ultrastructure of sporogenesis in the pycnial stage of *Puccinia sorghi*. Mycologia 66:319–326

Röbbelen G, Sharp EL (1978) Mode of inheritance, interaction and application of genes conditioning resistance to yellow rust. Fortschr Pflanzenzücht 9:1–218

Roberts K, Pryor A (1995) Isolation of a flax (*Linum usitatissimum*) gene induced during susceptible infection by flax rust (*Melampsora lini*). Plant J 8:1–8

Roelfs AP (1982) Effects of barberry eradication on stem rust in the United States. Plant Dis 66:177–181

Roelfs AP, Bushnell WR (eds) (1985) The cereal rusts. Academic Press, Orlando

Roelfs AP, Singh RP, Saari EE (1992) Rust diseases of wheat: concepts and methods of disease management. CIMMYT, Mexico

Rowell JB (1985) Evaluation of chemicals for rust control. In: Roelfs AP, Bushnell WR (eds) The cereal rusts, vol II. Academic Press, Orlando, pp 561–589

Ryerson DE, Heath MC (1992) Fungal elicitation of wall modifications in leaves of *Phaseolus vulgaris* L. cv. Pinto II. Effects of fungal wall components. Physiol Mol Plant Pathol 40:283–298

Ryerson DE, Heath MC (1996) Cleavage of nuclear DNA into oligonucleosomal fragments during cell death induced by fungal infection or by abiotic treatments. Plant Cell 8:393–402

Rytter JL, Lukezic FL, Craig R, Moorman GW (1989) Biological control of geranium rust by *Bacillus subtilis*. Phytopathology 79:367–370

Saksiriat W, Hoppe HH (1990) *Verticillium psalliotae*, an effective mycoparasite of the soybean rust fungus *Phakopsora pachyrhizi* Syd. J Plant Dis Protect 97:622–633

Saur R, Gold R, Ammermann E (1991) BAS 480 F, ein neues, breit wirksames Fungizid zur Bekämpfung von Getreidekrankheiten. Med Fac Landbouww Rijksuniv Gent 56:479–489

Savary S, Bosc J-P, Noirot M, Zadoks JC (1988) Peanut rust in West Africa: a new component in a multiple pathosystem. Plant Dis 72:1001–1009

Savile DBO (1968) The case against "Uredium". Mycologia 60:459–464

Savile DBO (1984) Taxonomy of the cereal rust fungi. In: Bushnell WR, Roelfs AP (eds) The cereal rusts, vol I. Academic Press, Orlando, pp 79–112

Schafer JF, Roelfs AP, Bushnell WR (1984) Contribution of early scientists to knowledge of cereal rusts. In: Bushnell WR, Roelfs AP (eds) The cereal rusts, vol I. Academic Press, Orlando, pp 3–38

Scheepens PC, Hoogerbrugge A (1991) Host specificity of *Puccinia canaliculata*, a potential biocontrol agent for *Cyperus esculentus*. Neth J Plant Pathol 97:245–250

Schrieber E, Zentmyr GA (1984) Distribution and spread of coffee rust in Latin America. In: Fulton RH (ed) Coffee rust in the Americas. The American Phytopathological Society, St Paul, pp 1–14

Sock J, Rohringer R, Kolmer JA (1994) Mitochondrial DNA of *Puccinia graminis* f. sp. *avenae*: molecular cloning, restriction map, and copy number. Phytopathology 84:49–55

Stakman EC, Piemeisal FJ (1917) Biological forms of *Puccinia graminis* on cereals and grasses. J Agric Res 10:429–495

Staples RC, Macko V (1984) Germination of urediospores and differentiation of infection structures. In: Bushnell WR, Roelfs AP (eds) The cereal rusts, vol I. Academic Press, Orlando, pp 255–289

Stark-Urnau M, Mendgen K (1993) Differentiation of aecidiospore- and uredospore-derived infection structures on copwea leaves and on artificial surfaces by *Uromyces vignae*. Can J Bot 71:1236–1242

Stark-Urnau M, Mendgen K (1995) Sequential deposition of plant glycoproteins and carbohydrates into the host-parasite interface of *Uromyces vignae* and *Vigna sinensis*. Evidence for endocytosis and secretion. Protoplasma 186:1–11

Steffenson BJ (1992) Analysis of durable resistance to stem rust in barley. Euphytica 63:153–167

Subrahmanyam P, Reddy LJ, Gibbons RW, McDonald D (1985) Peanut rust: a major threat to peanut production in the semiarid tropics. Plant Dis 69:813–819

Supkoff DM, Joley DB, Marois JJ (1988) Effect of introduced biological control organisms on the density of *Chondrilla juncea* in California. J Appl Ecol 25:1089–1095

Swann EC, Mims CW (1991) Ultrastructure of freeze-substituted appressoria produced by aeciospore germlings of the rust fungus *Arthuriomyces peckianus*. Can J Bot 69:1655–1665

Swann E, Taylor J (1995) Phylogenetic perspectives on basidiomycete systematics: evidence from the 18S rRNA gene. Can J Bot 73:S862–S868

Templeton GE, Te-Beest DO, Smith RJ Jr (1979) Biological weed control by mycoherbicides. Annu Rev Phytopathol 17:301–310

Terhune BT, Bojko RJ, Hoch HC (1993) Deformation of stomatal guard cell lips and microfabricated artificial topographies during appressorium formation by *Uromyces*. Exp Mycol 17:70–78

Unger F (1833) Die Exantheme der Pflanzen und einige mit diesen verwandte Krankheiten der Gewächse, pathogenetisch und nosographisch dargestellt. Carl Gerold, Wien

Van Asch MAJ, Rijkenberg FHJ, Coutinho TA (1992) Resistance induced in wheat by an avirulent race of *Puccinia recondita* f. sp. *tritici*. Plant Dis 4:412–415

Wahl I, Anikster Y, Manisterski J, Segal A (1984) Evolution at the center of origin. In: Bushnell WR, Roelfs AP (eds) The cereal rusts, vol I. Academic Press, Orlando, pp 39–77

Watson AK (1991) The classical approach with plant pathogens. In: TeBeest DO (ed) Microbial control of weeds. Chapman and Hall, New York, pp 3–23

Welter K, Müller M, Mendgen K (1988) The hyphae of *Uromyces appendiculatus* within the leaf tissue after high pressure freezing and freeze substitution. Protoplasma 147:91–99

Williams PG (1984) Obligate parasitism and axenic culture. In: Bushnell WR, Roelfs AP (eds) The cereal rusts, vol I. Academic Press, Orlando, pp 399–430

Williams PG, Mendgen K (1975) Cytofluorometry of DNA in uredospores of *Puccinia graminis* f. sp. *tritici*. Trans Br Mycol Soc 64:23–28

Woods AM, Gay JH (1987) The interface between haustoria of *Puccinia poarum* (monokaryon) and *Tussilago farfara*. Physiol Mol Plant Pathol 30:167–185

Xu H, Mendgen K (1991) Early events in living epidermal cells of cowpea and broad bean during infection with basidiospores of the cowpea rust fungus. Can J Bot 69:2279–2285

Xuei X, Bhairi S, Staples RC, Yoder OC (1992) Differentiation-specific genes of rust fungi have limited distribution among fungi. Exp Mycol 16:320–323
Xuei XL, Bhairi S, Staples RC, Yoder OC (1993) INF 56 represents a family of differentiation-specific genes from *Uromyces appendiculatus*. Curr Gen 24:84–88
Yamamoto H (1995) Pathogenesis and host-parasite specificity in rusts. In: Kohmoto K (ed) Pathogenesis and host specificity in plant diseases. Pergamon, Oxford, pp 203–215
Zadoks J, Schein R (1979) Epidemiology and plant disease management. Oxford University Press, New York
Zambino PJ, Szabo LJ (1992) Relatedness of rusts of cereals and grasses from rDNA sequence analysis. Phytopathology 82:1142
Zambino PJ, Szabo LJ (1993) Phylogenetic relationships of selected cereal and grass rusts based on rDNA sequence analysis. Mycologia 85:401–414
Zhang R, Dickinson MJ, Pryor A (1994) Double-stranded RNAs in the rust fungi. Annu Rev Phytopathol 32:115–133

6 Genetics and Molecular Biology of the Fungal Partner in the Ectomycorrhizal Symbiosis *Hebeloma cylindrosporum* × *Pinus pinaster*

J.-C. DEBAUD, R. MARMEISSE, and G. GAY

CONTENTS

I. Introduction 95
A. Roles of Ectomycorrhizal Symbiosis in Ecosystems 95
B. Nutritional and Hormonal Relationships Between the Two Partners 96
C. Towards a Genetic and Molecular Analysis of the Ectomycorrhizal Symbiosis; Characteristics of Model Species 96
II. *Hebeloma cylindrosporum*, a Model Ectomycorrhizal Basidiomycete 97
A. *Hebelomas* as Mycorrhizal Fungi 97
B. Taxonomic Position, Habitat, and Geographic Distribution of *H. cylindrosporum* 97
C. Life Cycle 98
1. Spore Germination and Monokaryotic Mycelia 98
2. Dikaryotic Mycelia and the Breeding System 99
3. In Vitro Fruiting 101
D. Physiological Characteristics 101
III. Quantitative Genetic Variation of Fungal Traits 103
A. Methods for Studying Quantitative Traits 103
B. Variation in Enzyme Activities and Metabolite Production 103
1. Interspecific Variability in the Genus *Hebeloma* 103
2. Intraspecific Variability in *H. cylindrosporum* 104
a) Variability Between Unrelated Strains 104
b) Variability Within the Progeny of a Single Strain 105
C. Intraspecific Variation in Ectomycorrhizal Ability and Infectivity 106
IV. Fungal Mutants Altered in Symbiotic-Related Functions 106
A. Protoplasts and Mutagenesis 107
B. Fungicide-Resistant Mutants 107
C. Auxin-Overproducer Mutants 108
1. Selection and Biochemical Characterization 108
2. Effect of IAA Overproduction in Mycorrhizal Ratio and Morphogenesis 108
D. Auxotrophic mutants 109
E. Nitrate Nonutilizers 110
V. Genetic Transformation of *H. cylindrosporum* 110
VI. Conclusions 111
References 112

Université Claude Bernard Lyon I, Laboratoire d'Ecologie Microbienne du Sol (UMR CNRS 5557), Bât. 405, 43 Boulevard du 11 Novembre 1918, 69622 Villeurbanne Cedex, France

I. Introduction

A. Roles of Ectomycorrhizal Symbiosis in Ecosystems

Mycorrhizal symbiosis exists in all terrestrial ecosystems and plays important functions in both natural and human-altered environments. Angiosperm species are mostly endomycorrhizal, whereas the ectomycorrhizal symbiosis involves only about 3% of all phanerogams but a large number of tree species. Harley and Harley (1987a,b) mentioned that mycorrhizas occur in 75–80% of the vascular plants of the British flora. According to Read (1992), ectomycorrhizal trees are the natural dominants of the boreal and temperate forests and are important also in tropical rainforests. The fungi that form ectomycorrhizas belong to the Ascomycotina (several hundred species) and to the Basidiomycotina (around 5000 known species). Few ectomycorrhizal fungi are Zygomycotina or Imperfect Fungi. Altogether, ectomycorrhizal species represent approximately 10% of the 69000 known fungal species (Hawksworth 1991). Mycorrhizal fungi form a diverse group of organisms but have common characteristics: they all form mycorrhizas on the root systems of their host plants; they frequently improve the mineral nutrition of the plant partner (especially phosphorus and nitrogen nutrition); and they can also increase the water uptake and protect their host plants against root pathogens.

One of the present challenges in mycorrhizal research is to establish a relation between the functions of each mycorrhizal type and its contribution to the dynamics of the ecosystem in which it predominates. As emphasized by Carroll (1992),

The Mycota V Part B
Plant Relationships
Carroll/Tudzynski (Eds.)
© Springer-Verlag Berlin Heidelberg 1997

"fungal mutualists seldom do just one thing. One species of mycorrhizal fungus may harvest water and nitrogen, provide protection from root pathogens and alleviate competition all in a single tree, although not necessarily simultaneously". A great majority of mycorrhizal fungi can form mycorrhizas with different host species, and most plants can have different fungal partners on their root systems simultaneously. Consequently, the functioning of such associations under natural conditions is, in most cases, the result of a "multipartner" symbiosis in which the different fungal clones or species may play different and complementary roles. It has also been demonstrated that different host plants can be interconnected by a common mycelium, leading to development of plant communities in which there is evidence for hyphal transport of nutrients between plants and subsequently a competition mediated by mycorrhizas (see Miller and Allen 1992; Read 1992; Trappe and Luoma 1992; Chaps. 7, 11, this Vol.). It must also be noticed that, in natural ecosystems, these complex interactions can be affected by other microorganisms present in the rhizosphere (Garbaye and Bowen 1989; Garbaye 1994): "mycorrhization helper bacteria" can stimulate fungal growth as well as mycorrhiza formation via volatile substances (Garbaye and Duponnois 1992).

B. Nutritional and Hormonal Relationships Between the Two Partners

The mycorrhizal association between the roots of an autotrophic plant and the hyphae of a heterotrophic fungus is a mutualistic symbiosis characterized by exchanges resulting in beneficial effects for each of the two partners (see Smith and Smith 1990 for an overview). The main benefit for the fungus is the receipt of carbohydrates from the host plant: 10–15% of the total photosynthates produced by a plant can be transferred to the fungal partners present on its root system. The fungal associate may also receive other organic compounds such as vitamins or growth substances (Harley and Smith 1983), some of which may be required for the differentiation of the fruiting bodies.

One of the main effects of the symbiosis on the host plant is the improvement of its mineral nutrition and growth. This results from a better exploitation of soil resources by the fungus: acid or enzymatic solubilization of nutrients present in the substrate, uptake, assimilation, storage, and finally transfer of these nutrients to the host. Numerous physiological and biochemical studies on the effects of the ectomycorrhizal fungi on the nutrition of their host plants, mostly using pure fungal cultures, mycorrhizal plants, or excised mycorrhizas, have been carried out under controlled conditions (Harley and Smith 1983; Gianinazzi-Pearson and Smith 1993). Improvement of phosphate nutrition is generally considered as the most important (positive) effect of the symbiosis (see Bolan 1991; Harley and Smith 1983; Hetrik 1989; Gianinazzi-Pearson and Gianinazzi 1989). Ectomycorrhizal fungi also improve nitrogen nutrition of their host plants. All of these fungi can use ammonium whereas there is considerable variation in their ability to use nitrate as sole nitrogen source. They can also use organic nitrogen such as amino acids, and at least some of them can hydrolyze proteins and chitin (Read et al. 1989; Read 1992). Nitrogen metabolism in free-living ectomycorrhizal fungi and in ectomycorrhizas has been reviewed by Martin and Botton (1993), with particular emphasis on the N-assimilating enzymes and their changes in distribution and activity during the symbiosis.

Hormonal relationship between the two partners has received little attention. Some ectomycorrhizal fungi produce cytokinins and gibberellins in pure culture, whereas others produce ethylene. It seems that most – if not all – ectomycorrhizal fungi are able to produce auxin (see Harley and Smith 1983; Gay et al. 1989). Fungal hormones may affect root growth and branching, allowing the production of short roots potentially able to be transformed into mycorrhizas. Fungal auxin may also play a morphogenetic role during ectomycorrhiza formation, by modifying the interfaces in the Hartig net or by increasing cortical cell radial growth (Gea et al. 1994).

C. Towards a Genetic and Molecular Analysis of the Ectomycorrhizal Symbiosis; Characteristics of Model Species

Different hypotheses have been put forth to explain the role of fungal physiological properties (e.g., enzyme activities) in the ectomycorrhizal symbiosis, but several key questions will have to be addressed for a better understanding of this plant-fungus relationship. Two research approaches may contribute to the clarification of the roles of fungal functions in the establishment and functioning of ectomycorrhizas. The first one consists in modifying fungal pathways supposed to play a role during the symbiosis, either by mutation or by DNA-mediated transformation, and then comparing the effect of the modification on the functioning of the symbiosis (Chap. 12, Vol. V, Part A). The second approach seeks to estimate intraspecific variation of a given fungal property at the populational level and to correlate this variation with variations in effects on the host plant during the symbiotic association.

Until recently, almost nothing was known about genome organization, gene regulation, and expression in ectomycorrhizal fungi except by analogy to well-studied saprotrophic Ascomycotina and Basidiomycotina. The main problem for the genetic study of ectomycorrhizal fungi has been the difficulty in obtaining fruiting bodies in the laboratory.

Future progress in understanding some aspects of the plant-fungus relationship will depend on our ability to concentrate on well-characterized model species which can easily be manipulated in the laboratory. Most ectomycorrhizal fungi are difficult to study in the laboratory; some cannot be cultivated; basidiospore germination is frequently difficult or impossible to induce; and fruiting cannot be controlled. Thus, most of the physiological studies have been performed with many different wild strains which are not well characterized genetically.

A few ectomycorrhizal Basidiomycetes (e.g., *Laccaria bicolor*, *Laccaria laccata*, and *Hebeloma cylindrosporum*) have proved to be suitable model species for carrying out genetic investigations of the plant-fungus relationships. Different prerequisites have been fulfilled with these fungi. They can be cultivated in vitro with or without the host plant under controlled conditions. Their complete life cycle can be obtained under controlled conditions (under sterile conditions in the case of *H. cylindrosporum*). Spores can be germinated in vitro, allowing the isolation of homokaryotic mycelia. Their breeding systems have been determined, allowing crosses to be performed. Uninucleate protoplasts can be produced and regenerated from monokaryotic mycelia, thus allowing the production of mutants. Genetic transformation systems have also been used to modify biochemical pathways. Finally, molecular methods have been developed to characterize strains of these fungi for population studies and for following the fates of selected and/or modified strains in the field (Gardes et al. 1991; Henrion et al. 1992; Marmeisse et al. 1992b).

In this chapter we present the main characteristics of *H. cylindrosporum* that make this fungus a suitable model organism for investigating the role of fungal properties in the mycorrhizal symbiosis; we further provide preliminary results which demonstrate the usefulness of this approach.

II. *Hebeloma cylindrosporum*: a Model Ectomycorrhizal Basidiomycete

A. *Hebelomas* as Mycorrhizal Fungi

Hebeloma (Basidomycotina, Agaricales) is an important genus in the family Cortinariaceae. Many, but not all *Hebeloma* species are ectomycorrhizal fungi. In the field, these fungi are found in association with gymnosperms (*Abies*, *Picea*, *Pinus* spp., etc.) and/or angiosperms (*Betula*, *Fagus*, *Fraxinus*, *Populus*, *Quercus*, *Salix* spp., etc.). Most of the species which have been studied do not show a strict host-plant specificity. Thus, *Hebeloma* species can be induced to form mycorrhizas with tree species which are not their natural host plants in the field. For example, the European alpine species *Hebeloma alpinum* and *Hebeloma marginatulum* associated with the rosaceous alpine *Dryas octopetala*, as well as with other alpine shrubs in their native habitat (*Salix retusa* and *S. reticulata*), form mycorrhizas in vitro with the American pine *Pinus virginiana* (Hacskaylo and Bruchet 1972). Similarly, *H. cylindrosporum*, associated with *Pinus pinaster* along the temperate Atlantic coast of France, has been successfully induced to form mycorrhiza with American gymnosperm species like *Pinus banksiana* or *Larix laricina* (Mac Afee and Fortin 1988) and also with a wide range of other species from different dicot families: the North-American alpine *Dryas integrifolia* (Melville et al. 1987, 1988), *Betula* species (Giltrap 1982), and *Castanea sativa* (Valjalo 1979).

B. Taxonomic Position, Habitat, and Geographic Distribution of *H. cylindrosporum*

H. cylindrosporum was first described by Romagnesi (1965) and has been placed in the subgenus *Hebeloma*. This species, which produces cylindrical basidiospores, can easily be distinguished from closely related *Hebeloma* species (Bruchet 1980). Although a systematic survey has not been carried out, this species seems to be distributed only along the European Atlantic coast. All the strains isolated in our laboratory were collected in the forest of Les Landes, east of Bordeaux (44°50 N 1°10 W), where the basidiocarps appear in late October–November. The species has also been described in similar habitats in Portugal (J. Guinberteau, pers. comm.) and in The Netherlands (Jansen 1982). The fungus is a pioneer species which forms ectomycorrhizas with several species of *Pinus*, including *P. pinaster* and *P. sylvestris*. Considering the density of fruiting bodies observed in the field, *H. cylindrosporum*, together with *Rhizopogon luteolus* and *Thelephora terrestris*, appears to be one of the dominant fungal species associated with pine trees colonizing the unstable environment of sand dunes along the coast. Basidiocarps of *H. cylindrosporum* are very seldom found in older stands of *P. pinaster* characterized by a developing humus layer. This species could therefore be classified as an early-stage mycorrhizal fungus as opposed to late-stage fungi like many *Boletus*, *Suillus*, or *Russula* species, which colonize older *P. pinaster* forest stands. Although *H. cylindrosporum* is not widely used in forestry practice to inoculate trees produced in commercial nurseries, it has been shown to significantly improve the growth of different tree species (Le Tacon and Valdenaire 1980; Le Tacon et al. 1985).

Fig. 1. In vitro fruiting of *Hebeloma cylindrosporum*. Mature basidioscarp in axenic culture with 3-month-old *Pinus pinaster* plants. Fruiting is obtained under controlled conditions on synthetic medium in 1-l Erlenmeyer flasks. (Debaud and Gay 1987)

C. Life Cycle

H. cylindrosporum is so far the only ectomycorrhizal species for which the different stages of the life cycle including fruit body production (Fig. 1) have been characterized and obtained in axenic culture under laboratory conditions. The mycelia of many symbiotic species are difficult to grow on semisynthetic or synthetic media and their basidiospores do not germinate on such substrates. For most of the well-studied species like *Laccaria* species, *Pisolithus arhizus*, *Suillus collinitus*, *T. terrestris*, or *Tuber maculatum*, fruit body production is occasional and has been reported to occur only under greenhouse conditions, after symbiotic association with a host plant grown in containers or pot cultures (see Godbout and Fortin 1980). The mating systems of many of these species is still unclear and can be complicated for some of them (e.g., *L. laccata*) by the occurrence of somatic incompatibility systems which prevent crosses between strains collected in different populations to be performed (Fries and Mueller 1984; Kropp and Fortin 1988; Doudrick and Anderson 1989).

In the case of *H. cylindrosporum*, the control and the characterization of the different stages of the life cycle in vitro have allowed a whole range of genetic analyses to be performed, e.g., the selection and characterization of mutant strains, the analysis of fungal quantitative genetic traits, and the study of recombinant-DNA.

1. Spore Germination and Monokaryotic Mycelia

Basidiospores produced by the mature fruit bodies contain two identical haploid nuclei resulting from a post-meiotic mitotic division (Fig. 2). The percentage germination of the basidiospores (Fig. 3) in pure culture is low (2–4%) and decreases rapidly after spore discharge. The homokaryotic mycelia produced are monokaryotic: all the nuclei are genetically identical, and there is only one nucleus per cell except for the apical cell, which

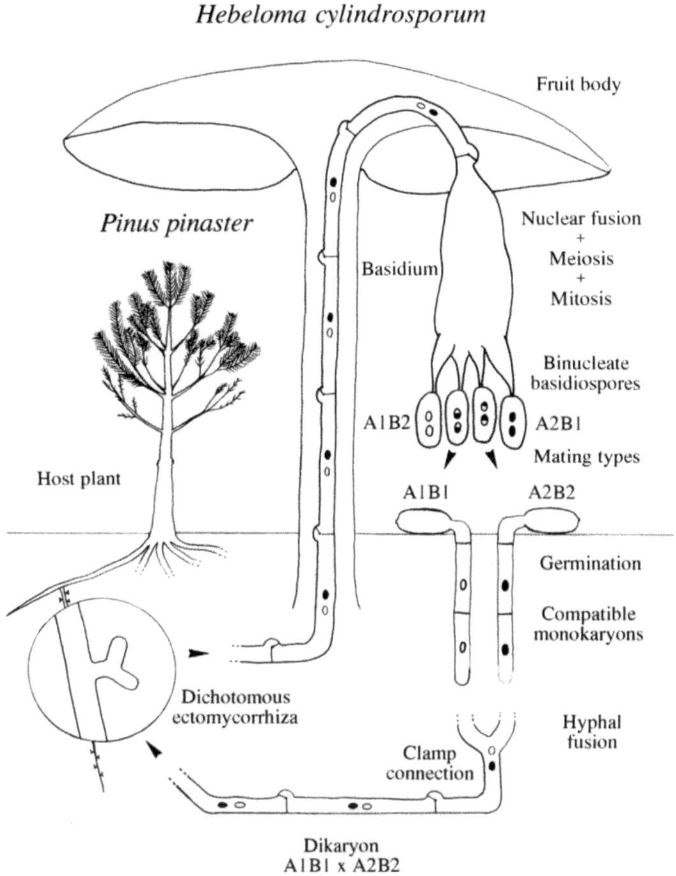

Fig. 2. Life cycle of *Hebeloma cylindrosporum*. This ectomycorrhizal basidiomycete shows a typical tetrapolar mating system. Spore germination gives rise to monokaryotic mycelia belonging to four mating types. Crossing between two compatible monokaryons results in dikaryon formation. The dikaryotic mycelium forms ectomycorrhizae on *Pinus pinaster* root systems, which allows the formation of sporulating basidiocarps

contains from three to ten. These monokaryotic mycelia are sterile; they do not produce asexual spores and have never been isolated as such in the field. Only dikaryotic mycelia which develop after mating between two compatible monokaryons form ectomycorrhizas in the field (Figs. 4, 5).

2. Dikaryotic Mycelia and the Breeding System

The dikaryotic mycelia of *H. cylindrosporum* have two nuclei per cell and are easily identified by the presence of a clamp connection on each hyphal septum (Fig. 6). These mycelia are similar to the dikaryons of the well-studied saprophytic basidiomycetes *Coprinus cinereus* and *Schizophyllum commune*, and we can assume that the events leading to the formation of the dikaryons (Casselton 1978) are similar in the three species. Thus, it is presumed that in *H. cylindrosporum* the two nuclei present in each cell of the dikaryons are genetically different and originate from the two parental monokaryons. As in *C. cinereus* and *S. commune*, the compatibility between monokaryons of *H. cylindrosporum* is controlled by two genetically unlinked mating type genes (A and B). Pairings between sib-monokaryons of the progeny of a single dikaryotic strain demonstrated the segregation of the alternate alleles of the A and B genes present in the two nuclei of the dikaryon (Debaud et al. 1986). The sib-monokaryons fall into four mating types, each characterized by a unique combination of the A and B alleles. Only crosses between monokaryons having different alleles at both the A and B loci (e.g., between A1B1 and A2B2 in Fig. 2) result in dikaryon formation. Crosses between monokarytic progeny of

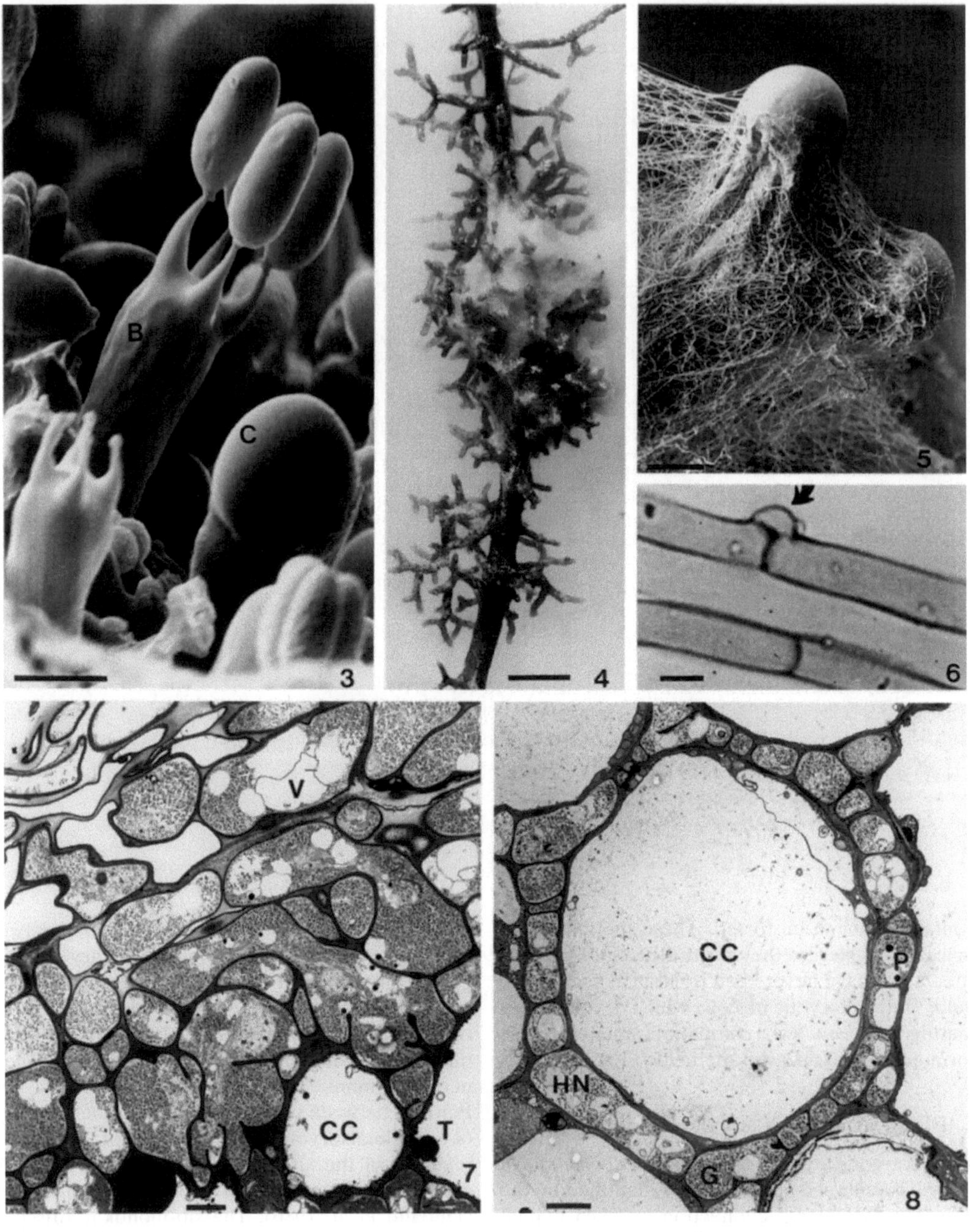

different wild dikaryotic strains have shown that there are more than two different alleles at the A and B mating type genes (Debaud et al. 1986). All crosses between monokaryons from four different strains and the monokaryons from the previously studied HC1 strain were compatible, indicating that the four strains had no mating type gene allele in common with the HC1 strain. With a fifth strain, some of the crosses were incompatible and it was concluded that this strain shared the same A2 allele with the HC1 strain. Multiallelism at the mating type loci has been interpreted as favoring outbreeding versus inbreeding and could be of great importance in the colonization strategy of this species.

The bifactorial multiallelic mating system which characterizes *H. cylindrosporum* has also been reported to operate in other ectomycorrhizal basidiomycete species, e.g., *L. laccata* and *L. bicolor* (Fries and Mueller 1984) and several *Suillus* species (Fries and Neumann 1990; Fries and Sun 1992). However, other ectomycorrhizal species have adopted other mating strategies such as secondary homothallism in *Laccaria altaica* (Fries and Mueller 1984).

3. In Vitro Fruiting

In the field, differentiation of sporulating basidiocarps of *H. cylindrosporum* has been observed only in association with its habitual host plant *P. pinaster*, although fruiting has also been obtained under nursery conditions in association with other gymnosperms. On semisynthetic or synthetic agar media in the absence of the host plant, dikaryotic mycelia of *H. cylindrosporum* can occasionally differentiate compact mycelial clusters or small (1–2mm high) primordia after an extended culture period (6 months). Under these culture conditions, very rare sterile immature fruit bodies have also been observed (Debaud and Gay 1987). A constant and continuous supply of carbohydrates and of other metabolites by the host plant could be a key factor influencing fruit body differentiation in mycorrhizal fungi.

In association with *P. pinaster* seedlings, grown on a sterile vermiculite-based substrate in a 1-l Erlenmeyer flask, fruit bodies of *H. cylindrosporum* start appearing 1.5 months after the inoculation of a 1-month-old seedling. The fruiting period can last for more than 4 months and an average of 3.5 mature basidiocarps are produced per flask over a 1-year culture period. The fruit bodies usually appear in the substrate near the mycorrhizas, their final size (stipe from 5 to 10cm in height, pileus 1 to 2cm in diameter) is smaller than the average size of fruit bodies collected in the field. Within 1 to 2 weeks, a basidiocarp produces approximately between 10^7 to 10^8 basidiospores.

The ability to fruit seems to be a stable attribute of the dikaryotic strains and does not appear to decline after several years (up to 14) of preservation in a culture collection stored at 4°C. Of 11 wild dikaryotic strains of *H. cylindrosporum* tested, 10 were successful in differentiating fruit bodies. Basidiocarps were obtained not only from wild-collected dikaryons; many different dikaryons resulting from crosses performed in the laboratory between wild-type, mutated, or DNA-transformed monokaryons were also successfully induced to fruit. Spores collected from these different dikaryotic strains were used to study the genetic basis of quantitative fungal traits, the segregation of mutant phenotypes, and the meiotic stability of DNA-transformed strains.

D. Physiological Characteristics

Mycorrhizal associations between *H. cylindrosporum* and *P. pinaster* can be induced in Erlenmeyer flasks using a sterile vermiculite substrate watered with a synthetic medium under controlled environmental (light and temperature) conditions. Four-week-old *P. pinaster* seedlings

Fig. 3. SEM view of the hymenium *Hebeloma cylindrosporum* showing cystidia (*C*) and basidia (*B*) with typical cylindrical basidiospores. *Bar* 5 μm

Fig. 4. Dichotomously branched ectomycorrhizae on the root system of *Pinus pinaster* inoculated by *Hebeloma cylindrosporum*. *Bar* 2 mm

Fig. 5. SEM of the apical region of a *Pinus pinaster* × *Hebeloma cylindrosporum* ectomycorrhiza showing the fungal sheath and abundant extramatrical hyphae. *Bar* 100 μm

Fig. 6. *Hebeloma cylindrosporum* hyphae. The monokaryotic mycelium has simple septa whereas the dikaryotic mycelium shows clamp connections (*arrow*). *Bar* 5 μm

Fig. 7. TEM of the fungal sheath of a *Dryas octopetala* × *Hebeloma alpinum* ectomycorrhiza. The external hyphae are vacuolated (*V*) whereas internal hyphae contain dense cytoplasm and abundant glycogen (*G*) and polyphosphate granules (*P*). External cortical cells (*CC*) contain deposits of tannins (*T*) *Bar* 2 μm. (Debaud et al. 1981b)

Fig. 8. TEM of the Hartig net of a *Dryas octopetala* × *Hebeloma marginatulum* ectomycorrhiza showing a cortical cell (*CC*) totally surrounded by a uniseriate layer of hyphae forming the Hartig net (*HN*). Hyphal cells contain glycogen (*G*) and polyphosphate granules (*P*). *Bar* 2 μm. (Debaud et al. 1981b)

are inoculated with several fungal inocula cut out of the margin of an actively growing colony on an agar medium. After an initial incubation period (2 to 3 weeks), during which the mycelium colonizes the substrate, the mycorrhizas start to differentiate. Synthetic mycorrhizas can also be obtained, under axenic conditions, in petri dishes containing an agar medium covered with a filter paper sheet. Different clones of *P. pinaster*, selected and micropropagated in vitro, are now being used as host plants to obtain synthetic mycorrhizas.

Ectomycorrhizas are characterized by a thin fungal sheath (Fig. 7) and intercellular hyphae between cortical cells forming the Hartig net (Fig. 8). The ectomycorrhizas produced on *P. pinaster* by genetically characterized monokaryotic and dikaryotic mycelia of *H. cylindrosporum* belonging to the progeny of the HC1 strain have been compared (Debaud et al. 1988). All four mating types of this species formed normal dichotomously branching ectomycorrhizas with *P. pinaster.* These were similar at the ultrastructural level of organization to those formed with the same host plant by the wild parental dikaryons. Monokaryotic mycorrhizas were also functional, as demonstrated by an ultrastructural localization of acid phosphatase activity similar to that in dikaryotic mycorrhizas. However, as emphasized with *Laccaria* spp. by Kropp et al. (1987), Kropp and Fortin (1988), Wong et al. (1989), and Wong and Fortin (1990), dikaryons are generally more infective than monokaryons.

Different enzymes involved in fungal mineral nutrition and thought to play a role in the mycorrhizal symbiosis have been studied in *H. cylindrosporum.* Levels of phosphatase activity of *H. cylindrosporum* are comparable to those of other ectomycorrhizal species such as *Laccaria*, *Suillus*, or *Pisolithus* species. *H. cylindrosporum* can use different forms of phosphate substrates, including sodium phytate and sodium polyphosphate, the phytase activity representing ca. 2% of the PNPPase (para-nitrophenyl phosphatase) activity (Mousain et al. 1988) and the polyphosphatase activity ca. 52% (Deransart et al. 1990). These authors demonstrated a positive correlation between the secreted PNPPase and polyphosphatase activities so that this latter could be inferred from the simple measurement of the PNPPase activity. The purification of a phosphatase secreted by a dikaryotic strain of *H. cylindrosporum* allowed the preparation of polyclonal antibodies (Deransart et al. 1990) which can be used to study the regulation of the fungal phosphatase activities during symbiotic association. The characterization of the different storage forms of phosphorus present in the mycelium obtained under conditions of pure culture shows that *H. cylindrosporum* stores little polyphosphate (4 to 9% of total P), and that most of the phosphorus in phosphate-starved mycelia is stored in the form of orthophosphate (Rolin et al. 1984). These results confirm those reported by Martin et al. (1983) using ^{31}P nuclear magnetic resonance to study polyphosphate metabolism in intact ectomycorrhizal fungi.

Other enzymes active in ectomycorrhizal fungi are those involved in nitrogen nutrition (Botton and Chalot 1995), i.e., glutamate dehydrogenase (GDH) and nitrate reductase. Ammonium assimilation in filamentous fungi is predominantly channeled through the GDH/glutamine synthetase pathway, although the glutamine synthetase/glutamate synthetase pathway may also operate. Evidence from enzyme assays in mycelial cell-free extracts, electrophoresis patterns, and immunological assays consistently showed that the *Hebeloma* NADP-GDH was highly active in spruce ectomycorrhizas (Dell et al. 1989), where the GDH/glutamine synthase pathway is the main assimilating pathway for ammonium. The GDH activity in mycorrhizas was largely associated with the extramatrical hyphae and the mantle, whereas a lowest activity was found in the Hartig net. In contrast, the fungal GDH activity was strongly repressed when *Fagus sylvatica* was the host plant (Dell et al. 1989).

H. cylindrosporum mycelia grow on synthetic media containing nitrate as sole nitrogen source; nitrate and nitrite reductases of this fungus have been studied by Plassard et al. (1984a,b). The affinity of the *Hebeloma* nitrate reductase for NO_3^- is comparable to the value found for the higher plant nitrate reductases, suggesting that ectomycorrhizal fungi and host plants may have similar nitrate-reducing capabilities. The regulation of the nitrate reductase in *H. cylindrosporum* has been studied by Scheromm et al. (1990a,b), who showed that this regulation differs from that of the nitrate reductase of the saprophytic ascomycetes *Neurospora crassa* and *Aspergillus nidulans* (Tomsett 1989). *H. cylindrosporum* mycelia have a nitrate reductase activity in the absence of nitrate and even in the presence of NH_4^+, suggesting that the expression of its nitrate reductase gene is constitutive. The nitrate uptake rate of *P. pinaster* plants was increased about twofold when associated with *H. cylindrosporum* and this rate was higher than the sum of the rates measured on the two individual symbiotic partners grown separately (Plassard et al. 1994).

III. Quantitative Genetic Variation of Fungal Traits

A. Methods for Studying Quantitative Traits

The physiology of mycorrhizal fungi has been studied in depth for many years (Harley and Smith 1983), but little is known about the genetic structure of fungal populations. Very few studies have been performed on their genetic variability, especially at the intraspecific level, and on the inheritance of quantitative traits in mycorrhizal fungi.

The variation observed between individuals in a natural population can result from genetic differences, as well as from environmental effects and genotype/environment interactions. Additive effects between different genes acting on the same trait and dominant effects between different alleles at a single locus can be distinguished. Gene interactions are also known in which the phenotypic expression of a pair of alleles is affected by another pair of alleles at a different locus. As a result, the genotypic variance (VG) of a character in a population can be divided into two main items, the additive (parental) component (VP) and the interactive one (VI) such that VG = VP + VI. The nonadditive component takes into account nonallelic interactions (epistasis) and allelic interactions (dominance). The ratio between the additive variance and the genotypic variance (VP/VG) is generally used as a measure of the degree of inheritance or heritability of quantitative traits. In selection practice, high heritability values indicate that crosses between selected parents will produce offspring with, on average, phenotypic values similar to or better than their parents.

Environmental effects include variations which do not result from genetic differences. In natural populations they include factors such as growth in different places, under different nutritional conditions or, in the case of mycorrhizal fungi, with different host plants. Parameters called environmental variance and genotype/environment interactions have been introduced to measure interactions in which the relative performances of different genotypes are affected by environmental conditions. As a consequence, the total phenotypic variance (VT) observed in noncontrolled environmental conditions can be partitioned into three main items: genotypic (VG), environmental (VE), and interactive (VGE) such that: VT = VG + VE + VGE.

Simchen and Jinks (1964) have developed biometrical methods to quantify the genetic variation (VG) and its components (VP and VI) in dikaryotic basidiomycetes. A modification of these methods was introduced by Williams et al. (1976) to estimate the environmental component (VE) of the phenotypic variation and the genotype/environment interaction (VGE).

Work on quantitative genetics of ectomycorrhizal fungi has been carried out using various physiological traits such as enzyme activities and metabolite production as well as infectivity towards host roots and ability to form ectomycorrhizas. Variability in such traits has been quantified among different species, among different unrelated strains of the same species isolated within a very limited geographic area, and, most rigorously, among monokaryotic and dikaryotic progenies isolated in the laboratory from fruiting bodies produced by a single dikaryon.

B. Variation in Enzyme Activities and Metabolite Production

The fungal properties studied in the case of *H. cylindrosprum* were enzyme activities thought to play a role in either the establishment or the functioning of the symbiotic association. They were indole acetic acid (IAA) synthesizing activity, glutamate dehydrogenase (GDH), nitrate reductase, and acid phosphatase activities of the mycelia.

1. Interspecific Variability in the Genus *Hebeloma*

Interspecific variability of IAA synthesizing activity has been studied by Gay and Debaud (1987) using 12 *Hebeloma* species having different ecophysiological characteristics (Fig. 9). Indol acetic acid production by these mycelia ranged from 50.7 to 354.4 nmol IAA synthesized h^{-1} mg^{-1} protein. This variability is of the same order of magnitude as intergeneric or interspecific variabilities reported for various physiological properties studied in other ectomycorrhizal fungi (Harley and Smith 1983; Wong and Fortin 1990). No correlation could be detected between the IAA-synthesizing activity of the mycelia and their taxonomic position within the genus *Hebeloma*, their geographic origin, or the host plant under which they had been collected.

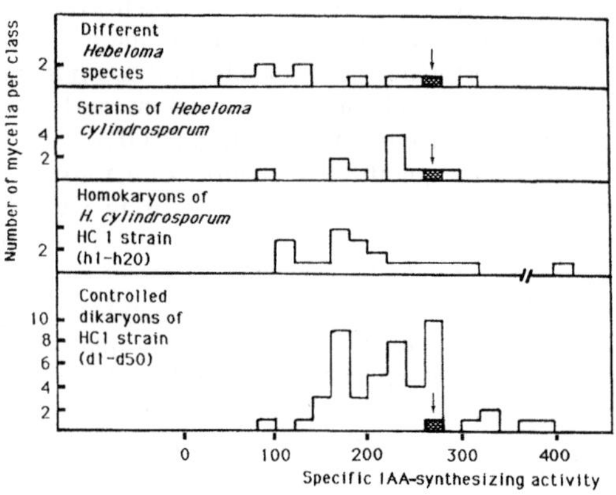

Fig. 9. Frequency diagram showing specific IAA-synthesizing activity of different *Hebeloma* spp. wild strains, different wild strains of *Hebeloma cylindrosporum*, and different homokaryotic and controlled dikaryotic mycelia belonging to the progeny of the HC1 laboratory fruiting strain of *Hebeloma cylindrosporum*. Specific IAA-synthesizing activity of the mycelia was expressed as nmol IAA synthesized $h^{-1} mg^{-1}$ fungal protein. *Arrows* indicate the HC1 original dikaryon of *Hebeloma cylindrosporum*. (Redrawn from Gay and Debaud 1987)

2. Intraspecific Variability in *H. cylindrosporum*

Studies with genetically different wild mycelia of the same species have also demonstrated a wide intraspecific variability in ectomycorrhizal fungi (Lundeberg 1970; Calleja et al. 1980; Cline et al. 1987; Ho and Trappe 1987; Zhu et al. 1988; Debaud et al. 1989; Ho 1989; Lamhamedi et al. 1990; Sen 1990). In particular, intraspecific variability in infectivity of *L. bicolor* has been reported by Kropp et al. (1987), Kropp and Fortin (1988), and Wong et al. (1989).

In *H. cylindrosporum*, intraspecific variability has been studied at two levels: between wild dikaryons isolated from fruit bodies collected in the field and within mono- and dikaryotic progenies isolated after laboratory fruiting of one of these dikaryons.

a) Variability Between Unrelated Strains

The interstrain variability in *H. cylindrosporum* has been studied using 11 wild dikaryotic strains isolated under *P. pinaster* in a very limited geographic area in Les Landes. Variability of IAA-synthesizing activity of these strains was of the same range of magnitude as that recorded at the interspecific level (Fig. 9) The variation in GDH, nitrate reductase, or acid phosphatase activities of these dikaryons was comparable with that recorded for IAA production (Wagner et al. 1988, 1989; Meysselle et al. 1991). The variation ranged from one to five times (Table 1), and no correlation could be established between the geographic ori-

Table 1. Range of intraspecific variation of different enzyme activities of the ectomycorrhizal fungus *Hebeloma cylindrosporum*

	Interstrain variation			Intrastrain variation				Synthesized dikaryons showing an activity higher than that of the parental dikaryon (%)
	11 Wild dikaryotic mycelia		HC1 Parental dikaryon	20 Monokaryotic mycelia		50 Dikaryotic mycelia		
	Range of variation	Average		Range of variation	Average	Range of variation	Average	
GDH activity (nkat. /mg prot.)	1.5–11.6	7.78	9.14	0.19–9.0	2.85	4– 19.8	8.5	36
NR activity (nmol NO2 / h/mg prot.)	201–700	345	397	51–510	211	72–689	344	28
IAA synthesizing activity (n mol/h/mg prot.)	85–321	222	279	108–561	220	94–437	241	30

Abbreviations: GDH, glutamate dehydrogenase; NR, nitrate reductase.

gins of the different strains and the level of their enzyme activites.

b) Variability Within the Progeny of a Single Strain

To estimate the magnitude and the components of the genetic variation within these populations, studies were performed under controlled environments to keep the environmental variance to a minimum; thus, VE accounts for uncontrolled minor variations and is considered as the "error" in variance analyses. Where control of the environmental conditions is difficult (e.g., in the nursery or in the field), environmental effects and genotype/environment interactions must be taken into account. To obtain progeny, five monokaryons of each of the four mating types were randomly chosen among 216 sib-monokaryons from single-spore germinations. Each monokaryon was paired with the five monokaryons of the compatible mating type, thus generating five half-sib dikaryons. As a result, two groups of 25 synthesized dikaryons were produced.

Gay and Debaud (1987) emphasized the magnitude of intraspecific genetic variation by demonstrating that the magnitude of variation in IAA production within mono- or dikaryotic progenies of a single dikaryon of *H. cylindrosporum* was of the same order as that recorded at the interspecific level, between different *Hebeloma* species (Fig. 9). A similar wide range of variation within a progeny of a single dikaryon was seen for activities of GDH (Wagner et al. 1988), nitrate reductase (Wagner et al. 1989), and acid phosphatase (Meysselle et al. 1991) with a range of variation from 1 to 15 times. This was of the same order of magnitude as the variation recorded by Kropp (1990) between monokaryotic mycelia of *L. bicolor* and by Ducamp and Olivier (1989) in the case of *Suillus granulatus*.

For all these activites of the 50 synthesized dikaryons, the additive and interactive components of the variation have been estimated. Each of these components accounted for ca. 50% of the genetic variation in the cases of the IAA synthesizing activity or of the GDH activity (Table 2). By contrast, more than 85% of the genetic variation of the nitrate reductase or of the acid phosphatase activites resulted from interactions between parental genotypes. This high level of interaction explains the absence of correlation between the activites of the synthesized dikaryons of *H. cylindrosporum* and those of their component monokaryons.

At least 30% of the synthesized dikaryons had enzyme activites higher than those of the original parental dikaryon HC1 (Table 1). Crosses between sib-monokaryons can therefore generate dikaryons with increased physiological capacities. However, the presence of a significant nonadditive component (Table 2) will prevent the design of a rapid selection scheme based only on the study of the performances of the parental monokaryons.

In an attempt to determine if variations in fungal GDH, nitrate reductase, or phosphatase activites could be correlated to variations in growth rate of the mycelia, the intraspecific variability of fungal growth was compared with the variation of these enzyme activites (Meysselle et al. 1986; Wagner et al. 1986; Gay et al. 1993). Less variation for growth between mycelia was detected compared to the variations in enzyme activities previously reported. According to Gay et al. (1993), this could indicate that, for processes like growth, which are likely to be controlled by many more genes than the activity of an enzyme, the presence of numerous genetic interactions tends to generate mostly average phenotypes. The variation in growth rates of dikaryons also involves additive and interactive effects, most of the genetic variation resulting from interactive effects.

Gay et al. (1993) showed that no correlation could be detected between GDH activity of the dikaryons and their growth rate on ammonium-containing media. The same applied for nitrate reductase activity compared to growth rates on nitrate, and also for acid phosphatase activity and growth rates on media with different concentrations of phosphate. This indicates that these enzymes, although considered key enzymes involved in the mineral nutrition of the fungus, are not

Table 2. Estimates of the components of variation of glutamate dehydrogenase within synthesized dikaryons of *Hebeloma cylindrosporum*. (Wagner et al. 1988)

Source	VT	VE	VG	
			VI	VP
h1–h5 × h6–h10 (A1B2 × A2B1)	6.92 100%	0.98 14.16%	2.96 42.77%	2.98 43.06%
h11–h15 × h16–h20 (A1B1 × A2B2)	19.01 100%	1.15 6.05%	7.61 40.03%	10.25 53.92%

the only limiting factors for fungal growth. These results question the value of studies which try to reduce a complex physiological activity like fungal growth or improvement of a plant mineral nutrition to the level of simple and single enzyme activites like GDH, nitrate reductase, or acid phosphatases. As suggested by Meysselle et al. (1991) in the case of phosphate assimilation, a mycorrhizal fungus which has a high ability to assimilate soil nutrients and a poor growth would not use the assimilated nutrients for its own growth and could presumably have a high ability to provide soil nutrients to its host plant.

Another complex question concerns the phenotypic expression of genetic variation under field conditions characterized by high environmental variation, i.e., what is the relative importance of the genetic and environmental variations as components of the phenotypic variation? In an attempt to estimate these components, Gay et al. (1993) studied the intraspecific variation of mycelial growth of 50 synthesized dikaryons cultivated under different nutritional conditions frequently used to grow mycorrhizal fungi in pure culture or in symbiotic association. The results showed that nutritional conditions have a tremendous effect on the phenotypic expression of fungal genomes. The highest total variation occurred when the mycelia were grown on a minimal medium containing nitrate as sole nitrogen source. The variation was ten times lower when the mycelia were grown on an organic nutrient-rich medium and two times lower on the widely used Modified Melin-Norkrans medium. The components of the phenotypic variation for growth of the 50 dikaryons cultivated under four different nutrient conditions were: VG = 26.39, VE = 3483.6, and VGE = 22.97. This high value of the environmental component indicates that, under the range of culture conditions frequently used to grow mycorrhizal fungi in pure culture or in symbiotic association, the phenotypic expression of fungal genes is much more affected by nutrient conditions than by the genotype of the mycelia per se.

These results obviously demonstrate the importance of genetic as well as environmental factors in field experiments. They also demonstrate that further studies are necessary for a better understanding of the influence of environmental conditions, including host-plant variability, on the expression of fungal biodiversity during the symbiosis.

C. Intraspecific Variation in Ectomycorrhizal Ability and Infectivity

There is a need to correlate laboratory measurements of fungal activites to field performances of selected mycelia. To this purpose, the mycorrhizal activity of 25 out of the 50 synthesized dikaryons of *H. cylindrosporum* previously used to study intraspecific variation in enzyme activites was estimated under nursery conditions (R. Duponnois et al., unpubl.). With *Pseudotsuga menziesii* the results obtained after a 1-year culture period demonstrated a wide range of variation of infectivity and efficiency within the 25 dikaryons studied. This variation was of the same order of magnitude as that recorded for enzyme activites. It was comparable to that recorded by Kropp et al. (1987), Kropp and Fortin (1988), and Wong et al. (1989) for different populations of *L. bicolor*. However, no correlation could be established between the variation of the mycorrhizal ratio or efficiency under nursery conditions and the variations recorded in the levels of the different enzyme activities of these mycelia.

As was previously emphasized for growth rates, mycorrhiza formation in the field is the end-result of numerous complex interactions which are likely to be extremely sensitive to environmental conditions. For this reason, it appears difficult to predict the exact action of individual fungal genes on the establishment or functioning of the mycorrhizal symbiosis under field conditions. From the results obtained with *H. cylindrosporum*, it is obvious that many different physiological criteria must be estimated in order to be in a position to predict the mycorrhizal performances of a strain and to carry out a selection program.

IV. Fungal Mutants Altered in Symbiotic-Related Functions

The characterization of either plants or fungal strains genetically altered in symbiotic functions such as infection mechanisms and mycorrhizal differentiation, or nutrient uptake, assimilation, and translocation would constitute a significant step towards a better understanding of the symbiotic association.

Most woody plants involved in the ectomycorrhizal associations grow slowly and have long generation times; further, they have rather large and poorly characterized diploid or polyploid genomes, and screening of mutants appears difficult. Use of potentially interesting mutant plants would require vegetative propagation, which can be a difficult, tedious, and time-consuming task for most tree species. For these reasons, very few tree mutants have been characterized, with the notable exception of poplar; however, none of these mutants are related to the symbiotic processes. Such is not the case with endomycorrhizal herbaceous plants, for which nonmycorrhizal mutants have been described. Some pea (*Pisum sativum*) mutants originally selected for their inability to differentiate nodules with *Rhizobium* also fail to be infected by arbuscular mycorrhizal fungi (Duc et al. 1989; Peterson and Bradbury 1995).

Fungi are better candidates for the induction and characterization of mutations, although some problems must be overcome with ectomycorrhizal species. Asexual spores, preferred for use in mutagenesis, have not been reported in any of these species. A suitable alternative is the use of protoplasts. Some mutations can also be selected after submitting hyphal fragments or thalli to a mutagenic treatment.

If the fungal phenotype selected for is the inability to infect the host plant, the screening procedures would involve testing thousands of different mycelia on individual plants. Instead, the different mutants which have been selected in *H. cylindrosporum* are mutants for which selection techniques exist and could easily be adapted to the particular biology of mycorrhizal species. The different mutants fall into two classes:

- mutants unable to use or to produce a specific compound such as a nutrient source or a metabolite; such strains might allow the study of exchanges between the two partners,
- mutants altered in the metabolism of a compound or an enzyme thought to play a function in the differentiation of the mycorrhizas.

Mutations in *H. cylindrosporum* can be induced in monokaryotic strains and the resulting phenotype can be directly tested on the plant. Crosses between mutant strains and compatible monokaryons give heterozygous dikaryons which can be used to test the dominance of the mutation. Fruiting and spore production after meiosis can be used to select recombinant monokaryotic mycelia in order to study the segregation of the selected phenotype. This can also be used to "remove" unwanted secondary mutations coselected after mutagenesis and to transfer the selected mutation into a different genomic background.

A. Protoplasts and Mutagenesis

The production of viable protoplasts has been reported for a limited number of ectomycorrhizal species. This is the case for *L. bicolor* (Kropp and Fortin 1986), *H. cylindrosporum*, *H. edurum*, *H. sinapizans*, *Suillus bellinii* (Hébraud and Fèvre 1988), *H. circinans*, and *L. laccata* (Barret et al. 1989). For some other species, like *Cenococcum geophilum* and *Pisolithus tinctorius* (Barrett et al. 1989), the protoplasts produced could not be regenerated.

Different factors influencing protoplast production by various strains of *H. cylindrosporum* have been worked out and optimized (Hébraud and Fèvre 1988). The large sizes of the uninucleate (monokaryons) or binucleate (dikaryons) fungal cells probably explain the high percentage (up to 40%) of enucleate protoplasts. With the monokaryotic strain h1 of *H. cylindrosporum*, yields of about 1–4 $\times$ 10^7 protoplasts/ml of lytic solution are obtained. On a nutrient-rich medium supplemented with 0.6M sucrose, 2% of these protoplasts regenerate to give new thalli. UV light irradiation has so far been the only mutagenic treatment used to induce mutations in *H. cylindrosporum*.

B. Fungicide-Resistant Mutants

The first mutants selected in *H. cylindrosporum* were fungicide-resistant mutants (Hébraud and Fèvre 1988). In a case study, these authors could obtain up to one fungicide-resistant mutant amongst 10^6 protoplasts surviving the mutagenic treatment. Mutants resistant to benomyl or to compounds of the anilide group of fungicides (benodanil, carboxine, and oxycarboxine) were recovered. In ascomycete species, the target site of benomyl is microtubule assembly, and many mutants resistant to this molecule were shown to have mutations in the β-tubulin gene. In several of these species (*N. crassa* and *A. nidulans*), the cloning of the β-tubulin mutant alleles has led to the development of a gene transfer system characterized by the positive selection of the transformants on benomyl-containing media. Wild-type mycelia of *H. cylindrosporum* are inhibited by rather high concentrations of benomyl (about $10\,\mu g\,ml^{-1}$) and it is not known if the target site for this fungicide is also microtubule assembly and if among the selected mutants some have mutations in their β-tubuline gene. If such were the case, cloning of these alleles would constitute an opportunity for the development of a homologous transformation system with the positive selection of the transformed mycelia. Fungicide resistance can also be used as marker to follow the fate of selected strains introduced in a soil where different mycorrhizal species and strains of *H. cylindrosporum* are present.

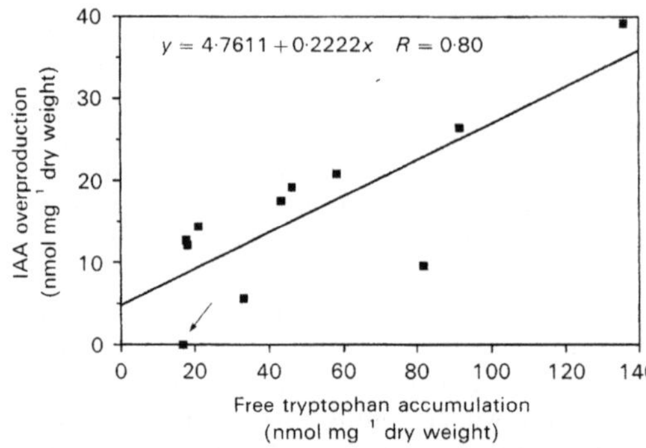

Fig. 10. Correlation between IAA overproduction and free tryptophan accumulation in mycelia of IAA-overproducing mutants of *Hebeloma cylindrosporum*. *Arrow* indicates the wild-type monokaryon. (Redrawn from Durand et al. 1992)

C. Auxin-Overproducer Mutants

1. Selection and Biochemical Characterization

IAA-overproducer mutants of *H. cylindrosporum* were isolated and characterized by Durand et al. (1992) in order to clarify the role of fungal IAA in ectomycorrhiza morphogenesis and functioning. The selection procedure used by these authors was based on previous results (Rouillon et al. 1986; Gay et al. 1989) showing that ectomycorrhizal fungi produce high levels of IAA only when tryptophan was supplied exogenously. Because of the lack of a direct selection method, the IAA-overproducer mutants were selected by using a two-step procedure: fluoroindole-resistant (FIR) mutants altered in the regulation of the tryptophan biosynthetic pathway were first obtained from two compatible monokaryotic mycelia. These mutants are tryptophan overproducers: they accumulate free tryptophan in their cells and presumably convert it into IAA. In a second step, IAA overproducers were identified amongst FIR mutants by screening for IAA production in the absence of any precursor like tryptophan in the culture medium. Therefore, the selected mutants have an improved ability to metabolize endogenous tryptophan into IAA. A significant correlation was detected between IAA overproduction by the mutants and their ability to accumulate intracellular free tryptophan (Fig. 10). By studying the activity of anthranilate synthetase and tryptophan synthase (the key enzymes of the tryptophan branch of the shikimate pathway), it was evident that different mutations could lead to tryptophan accumulation and subsequently to IAA overproduction. Anthranilate synthetase and tryptophan synthase activities of the wild-type monokaryon are feedback-inhibited by tryptophan. Selection for resistance to fluoroindole gave mutants lacking feed-back inhibition of anthranilate synthetase and/or tryptophan synthase. All the studied IAA-overproducer mutants were able to establish mycorrhizal association with *P. pinaster*.

2. Effect of IAA Overproduction on Mycorrhizal Ratio and Morphogenesis

The ectomycorrhizal activity of the IAA-overproducer mutants and of their mono- and dikaryotic progeny has been studied (Gay et al. 1994). These mutants formed significantly more mycorrhizas than did the corresponding wild-type mycelia. The in vitro fruiting of a dikaryon heterozygous for a FIR mutation allowed the isolation of recombinant IAA-overproducer sib-monokaryons. These monokaryons also show a higher mycorrhizal activity than the wild-type monokaryons, indicating that the level of IAA produced plays an important role in mycorrhiza formation. This was confirmed by the high number of mycorrhizas formed by dikaryons homozygous for a FIR mutation.

Moreover, IAA overproduction appeared to have an important effect on the morphogenesis of the Hartig net. The main modifications concerned the width of the Hartig net. In mycorrhizas formed by wild-type monokaryons, the Hartig net is char-

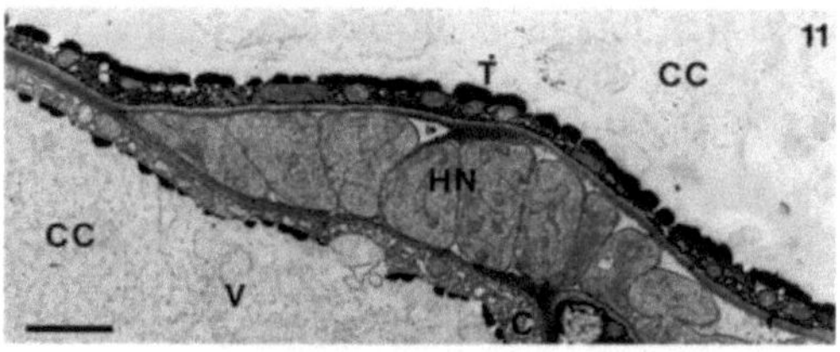

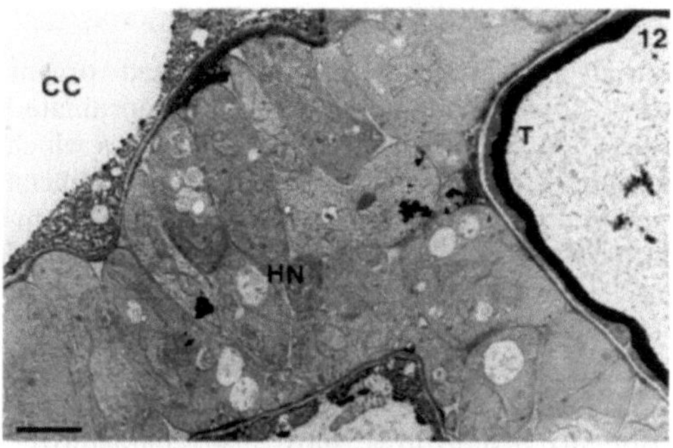

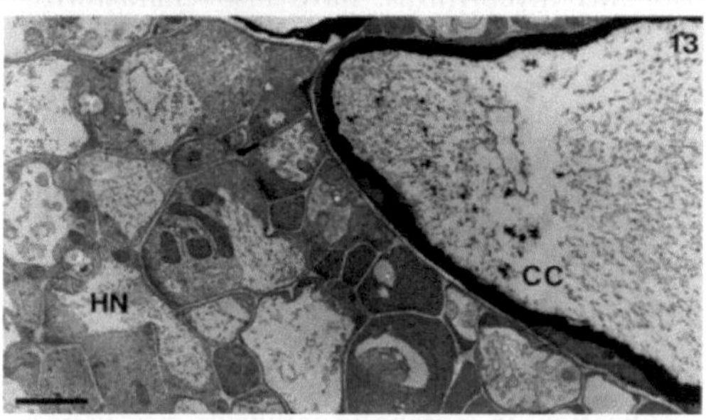

Fig. 11. Typical uniseriate Hartig net in a ectomycorrhiza formed by a wild-type monokaryon of *Hebeloma cylindrosporum* associated with *Pinus pinaster*. *CC* Cortical cell; *V* vacuole; *T* tannin deposit; *C* cortical cell cytoplasm; *HN* Hartig net. *Bar* 2 μm. (L. Gea and B. Vian, unpubl.)

Fig. 12. Early stage of development of a multilayered Hartig net produced by an IAA-overproducer monokaryotic mutant of *Hebeloma cylindrosporum*. *CC* Cortical cell; *T* tannin deposit; *HN* Hartig net. *Bar* 2 μm. (Gea and Vian, unpubl.)

Fig. 13. Detail of a mature hypertrophic Hartig net in a 2-week-old mycorrhiza formed by an IAA-overproducer mutant of *Hebeloma cylindrosporum* associated with *Pinus pinaster*. Root cortical cells are separated by two to seven layers of hyphae. *CC* Cortical cell; *HN* Hartig net. *Bar* 2 μm. (Gea and Vian, unpubl.)

acterized by a unique layer of hyphae between the cortical cells (Fig. 11) and generally restricted to the outermost layer of cortical cells. The mycorrhizas formed by the IAA-overproducer mutants presented substantial separation of root cells by a pluriseriate Hartig net (Fig. 12), which frequently reached the endodermis (Gea et al. 1994). The Hartig net was more developed in the outer cortex (Fig. 13), where its average width between two cortical cells ranged from two to seven hyphae, depending on the monokaryon. Surprisingly, electron microscopy even revealed limited penetration of living host cortical cells in mycorrhizas formed by an IAA-overproducer monokaryon (Gea et al. 1994). These intracellular hyphae were surrounded by invaginated host plasmalemma and by interfacial material. According to Gea et al. (1994), these observations suggest that fungal IAA may facilitate Hartig net development and intracellular fungal penetration by affecting host cell wall metabolism and structure. These results demonstrate that the use of mutants offers a suitable approach to identify critical interactions between the two partners at the origin of mycorrhiza morphogenesis.

D. Auxotrophic Mutants

The nutritional interactions between the fungus and its host plant imply the bidirectional transfer of organic and inorganic molecules between the root cortical cells and the fungal Hartig net. Several studies (Martin and Botton 1993) have demonstrated that free-living and symbiotic mycelia can convert ammonium and nitrate into organic nitrogenous molecules like amino acids, which accumulate in the fungal cells. These amino acids could be transferred to the plant either directly or after further transformation. Auxotrophic mutants blocked in the synthesis of key fungal metabolites could be used to elucidate the nature and spectrum of molecules exchanged between the two partners of the symbiosis.

Several auxotrophic mycelia have been selected in *H. cylindrosporum* (R. Marmeisse et al. unpubl.). UV-irradiated protoplasts were regenerated in a nutrient-rich medium and the resulting mycelia transferred onto a minimal medium with ammonium and glucose as the only nitrogen and carbon sources. The nutritional requirements of the mycelia which failed to grow on this medium were characterized using a standard auxanographic test. Among 3000 mycelia analyzed, 17 auxotrophs were recovered, and the specific requirements of 8 of them could be identified. The remaining 9 had multiple requirements. None of the mutants selected corresponds to the main amino acids which accumulate at significant levels in the mycelia of mycorrhizal species (glutamate, glutamine, and arginine). Most of the nutritional requirements identified correspond to molecules with rather complex biosynthetic pathways (e.g., adenine, methionine, or histidine). As expected for auxotrophs, all the selected mutations were recessive, and heterozygous dikaryons could grow on minimal medium. The selection of auxotrophs of a mycorrhizal species is feasible, although the development of enrichment or of positive selection techniques are needed to characterize mutants blocked in the biosynthesis of amino acids involved in mycorrhiza nitrogen nutrition.

E. Nitrate Nonutilizers

Another class of fungal mutants potentially useful to study the nutritional interactions between the two partners are mutants unable to use a particular nutrient source found in the soil such as phosphates, proteins, ammonium, or nitrate ions. Studies on nitrate assimilation by *H. cylindrosporum* have shown that strains of this species are characterized by the constitutive synthesis of the nitrate reductase enzyme even in the absence of nitrate and in the presence of ammonium in the culture media (Plassard et al. 1984a). Some mycorrhizal fungi showing better growth on nitrate than on ammonium do not accumulate nitrate in their cells and could simultaneously assimilate ammonium and nitrate when grown on a medium containing both these nitrogen sources.

Mutants of *H. cylindrosporum* unable to use nitrate as the only nitrogen source and lacking nitrate reductase activity have been selected (Marmeisse et al. in preparation). Selection of these mutants was performed using the positive screening protocol used to select nitrate reductase-deficient mutants in ascomycete fungi like *A. nidulans* or *N. crassa* (Cove 1976; Tomsett and Garrett 1980). In a first step, mutants resistant to a lethal concentration of chlorate are selected. In a second step, strains which fail to grow on nitrate-containing medium are selected among the chlorate-resistant mutants. This protocol allowed us to select several nitrate nonutilizers in different monokaryons of *H. cylindrosporum*, all of them lacking nitrate reductase activity. These mutants could be used to study the contribution of the fungus to the use of nitrate by mycorrhizal plants by comparing nitrate assimilation of uninoculated plants with that of plants inoculated either with a wild-type strain or with an isogenic mutant blocked in nitrate assimilation.

V. Genetic Transformation of *H. cylindrosporum*

Ectomycorrhizas are highly organized organs whose differentiation necessitates the coordinated expression of both plant and fungal genes which may not be normally expressed in nonmycorrhizal roots or in free-living mycelia. This view is supported by studies from Hilbert and Martin (1988) and Simoneau et al. (1993), who identified a class of polypeptides (ectomycorhizins) which accumulate specifically in ectomycorrhizas. More recently, Tagu et al. (1993), working on *Eucalyptus/Pisolithus* mycorrhizas, have also characterized and cloned several cDNA sequences which correspond to plant and fungal genes whose transcription is induced or repressed in the mycorrhizas (Chap. 12, Vol. V, Part A).

The use of molecular techniques is not limited to the study of differentiation mechanisms but could also be extended to the study of the nutritional interactions between the fungus and the plant. It should be possible in a near future to produce transgenic fungal strains with specific modifications in metabolic pathways, e.g., in pathways leading to the synthesis of amino acids, in order to identify the molecules transferred from the fungus to the plant. For such reasons, DNA-mediated transformation systems have been worked out for ectomycorrhizal fungi (Lemke et al. 1995; Marmeisse et al. 1995). Transformation systems for filamentous fungi were developed in the 1980s and have significantly contributed to the molecular dissection of many different metabolic as well as differentiation processes. Such transformation systems have found numerous applications, and their use is not limited to the introduction into the fungal chromosomes of extra copies of a cloned gene in order to increase the production of the corresponding gene product. Among the other applications of a gene-transfer system of some interest for mycorrhizal species can be mentioned: gene cloning by complementation of available mutants; gene disruption following transformation with a modified cloned gene; and study of gene regulation after fusion of their promoters to reporter genes coding for enzymes whose activites are easy to detect and quantify.

Genetic transformation of ectomycorrhizal species has only been reported for three species: *L. laccata* (Barrett et al. 1990), *H. cylindrosporum* (Marmeisse et al. 1992a), and *Paxillus involutus* (Bills et al. 1995). In the absence of suitable auxotrophic mutants, which could be transformed to prototrophy using the corresponding genes cloned in a plasmid vector, the approach used in these three species involved developing a selection system based on plasmids conferring hygromycin resistance. The plasmid pAN7.1 used to transform *H. cylindrosporum* and *L. laccata* was constructed by Punt et al. (1987) and contains the *E. coli* hygromycin phosphotransferase gene, conferring hygromycin B resistance, fused to promoter and terminator sequences from the ascomycete *A. nidulans*. Plasmids were introduced into protoplasts of *H. cylindrosporum* and *L. laccata* using polyethylene glycol and $CaCl_2$, whereas particle bombardment was used to transform intact cells of *P. involutus*. Transformants were selected by adding to the regeneration medium a lethal concentration of the antibiotic. With *H. cylindrosporum*, transformants were consistently selected at a low frequency of one to five per μg of transforming DNA. All the hygromycin-resistant mycelia analyzed had integrated several copies of the transforming plasmid into their genomes. The transformants differed with respect to their level of resistance to hygromycin, but no correlation could be established between the level of resistance and the number of copies of the transforming plasmid integrated into the genomes. Important for future physiological studies, it was shown that the transforming plasmids were stably integrated into the chromosomes and were not lost after extended periods of growth of the mycelia on media without the antibiotic. Furthermore, these transformants retained their ability to form mycorrhizas and to mate with compatible homokaryons, forming dikaryons and mature basidiocarps.

This positive selection system, which has the advantage of working on wild-type strains, was used to introduce other genes in the genome of *H. cylindrosporum* by cotransformation. In these experiments, a plasmid containing the gene to be introduced is mixed to pAN7.1 before transformation. The transformants are first selected on hygromycin plates and their genomic DNA are then screened by hybridization for the presence of the second plasmid. This system was used to introduce two different plasmids which contained genes presenting a potential interest in the study of the differentiation and of the functioning of the mycorrhizas. One contained two genes of the tryptophan pathway and the other the NADP-GDH gene from *Coprinus cinereus*. Genes of the tryptophan pathway are interesting, because this amino acid is the main limiting factor for IAA production by mycorrhizal fungi (Gay et al. 1989; Durand et al. 1992). In these cotransformation experiments, up to 70% of the selected transformants turned out to be co-transformants. These results clearly show that genetic manipulations of a mycorrhizal species are possible, but more work is needed in order to improve the efficiency of the transformation system. Of particular interest would be the construction of plasmids based on sequences isolated from *H. cylindrosporum* to ensure better expression of the antibiotic resistance gene and to study the frequency of homologous integrations of a DNA sequence in this species. This is a prerequisite for performing gene disruption experiments in order to study some aspects of the plant-fungus relationship.

VI. Conclusions

Until recently, genetic variation in the fungal partner and the molecular bases of the ectomycorrhizal symbiosis have been poorly understood. A genetic approach appears necessary for a better understanding of the plant-fungus relationships. The accomplishment of such studies requires appropriate model organisms which can be easily manipulated in vitro. Molecular characterization of the wild types and manipulated strains is also necessary. The resulting knowledge of the mycorrhizal relationship could lead to more commercial applications, by using either more efficient strains selected for improved host plant growth or strains selected for the production of edible fruiting bodies, of e.g. *Tuber*, *Boletus*, or *Cantharellus* species.

As demonstrated with *H. cylindrosporum*, the mycelia belonging to the progeny of a single dikaryotic strain display a wide range of quantitative variation in their physiological properties such as enzyme activities related to nitrogen and phosphorus metabolism or in hormone production. Further studies are necessary to correlate these properties, measured in vitro, with the efficiency and field distribution of the strains.

The use of mutants will facilitate study of the role of fungal biochemical pathways or metabolites in either the establishment or functioning of the ectomycorrhizal symbiosis. Such is the case of auxin-overproducer mutants of *H. cylindrosporum*, which have been used to demonstrate a positive correlation between IAA overproduction, mycorrhizal ratio, and morphogenesis on *P. pinaster*. Further studies with these mutants should lead to a better understanding of the host cell wall modifications following infection and of possible roles of changes in membrane permeability in the modification of the nature and volume of exchanges between the two partners. Understanding of the nutritional aspect of the plant-fungus relationship should also benefit from the use of mutants affected in nitrogen or phosphate metabolism. Auxotrophic mutants could give interesting insights into the exchange of minute amounts of organic compounds such as amino acids, growth factors, or vitamins. A more challenging task would be the production of fungal mutants blocked at different stages of mycorrhiza differentiation, from the initial steps of recognition between the two partners to Hartig net and mantle differentiations.

As recently demonstrated with *H. cylindrosporum* and *L. laccata*, ectomycorrhizal fungi can be genetically transformed by using cloned genes, and cotransformants with several genes of interest in their genomes can be obtained. Such gene manipulations, as well as the use of mutants, could clarify the function of specific fungal properties during the symbiosis and provide information about which properties might be exploited for the construction of more efficient strains. Moreover, the molecular characterization of the fungal partner at the infraspecific level, as performed with *H. cylindrosporum* (Marmeisse et al. 1992b), will help to determine the role of ectomycorrhizal biodiversity in terrestrial ecosystems.

Acknowledgments. The authors wish to thank Dr. Lorna Casselton (Oxford) and Prof. K. Esser (Bochum) for their helpful discussions and encouraging support during stays in their laboratories by J.-C. Debaud and R. Marmeisse (both supported by European Community grants). The genetic and molecular work on *Hebeloma cylindrosporum* could not have been performed without their help. The work of G. Gay on fungal auxin was supported by Eurosilva funds.

The authors wish also to thank Dr. B. Vian and L. Gea for the unpublished TEM illustrations, Colette Raffier for her valuable technical assistance, Dr. R. Huguency and the CMEABG of Lyon 1 University for the SEM illustrations, and P. Audenis for the photographs.

References

Barret V, Lemke PA, Dixon RK (1989) Protoplast formation from selected species of ectomycorrhizal fungi. Appl Microbiol Biotechnol 30:381–387

Barret V, Dixon RK, Lemke PA (1990) Genetic transformation of a mycorrhizal fungus. Appl Microbiol Biotechnol 33:313–316

Bills SN, Richter DL, Podila GK (1995) Genetic transformation of the ectomycorrhizal fungus *Paxillus involutus* by particle bombardment. Mycol Res 99:557–561

Bolan NS (1991) A critical review on the role of mycorrhizal fungi in the uptake of phosphorus by plants. Plant Soil 134:189–207

Botton B, Chalot M (1995) Nitrogen assimilation: enzymology in ectomycorrhizas. In: Varma A, Hock B (eds) Mycorrhiza, structure, function, molecular biology and biotechnology. Springer, Berlin Heidelberg New York, pp 325–363

Bruchet G (1980) *Hebeloma cylindrosporum* Romagnesi. Rev For Fr 32:294–295

Calleja M, Mousain D, Lecouvreur B, D'Auzac J (1980) Influence de la carence phosphatée sur les activités phosphatases acides de trois champignons mycorhiziens: *Hebeloma edurum* Metrod., *Suillus granulatus* (Fr. ex L.) O. Küntze et *Pisolithus tinctorius* (Pers.) Coker et Couch. Physiol Vég 18:489–504

Carroll GC (1992) Fungal mutualism. In: Carroll GC, Wicklow DT (eds) The fungal community. Its organization and role in the ecosystem. Marcel Dekker, New York, pp 327–354

Casselton LA (1978) Dikaryon formation in higher basidiomycetes. In: Smith JE, Berry DR (eds) The filamentous fungi, vol 3. Edward Arnold, London, pp 275–297

Cline ML, France RC, Reid CPP (1987) Intraspecific and interspecific growth variation of ectomycorrhizal fungi at different temperatures. Can J Bot 65:869–875

Cove DJ (1976) Chlorate toxicity in *Aspergillus nidulans*: studies of mutants altered in nitrate assimilation. Mol Gen Genet 146:147–159

Debaud J-C, Gay G (1987) In vitro fruiting under controlled conditions of the ectomycorrhizal fungus *Hebeloma cylindrosporum* associated with *Pinus pinaster*. New Phytol 105:429–435

Debaud J-C, Pepin R, Bruchet G (1981a) Etude des ectomycorhizes de *Dryas octopetala*. Obtention de synthèses mycorhiziennes et de carpophores d'*Hebeloma alpinum* et *H. marginatulum*. Can J Bot 59:1014–1020

Debaud J-C, Pepin R, Bruchet G (1981b) Ultrastructure des ectomycorhizes synthétiques à *Hebeloma alpinum* et *Hebeloma marginatulum* de *Dryas octopetala*. Can J Bot 59:2160–2166

Debaud J-C, Gay G, Bruchet G (1986) Intraspecific variability in an ectomycorrhizal fungus: *Hebeloma cylindrosporum*. 1- Preliminary studies on in vitro fruiting, spore germination and sexual comportment. In: Gianinazzi-Pearson V, Gianinazzi S (eds) Physiological and genetic aspects of mycorrhizae. INRA Publ, Paris, pp 581–588

Debaud J-C, Gay G, Prevost A, Lei J, Dexheimer J (1988) Ectomycorrhizal ability of genetically different homokaryotic and dikaryotic mycelia of *Hebeloma cylindrosporum*. New Phytol 108:323–328

Debaud J-C, Gay G, Bruchet G (1989) Genetics of the fungal partner as a tool to study and improve the mycorrhizal symbiosis. In: Nardon P, Gianinazzi-Pearson V, Grenier AM, Margulis L, Smith DC (eds) Endocytobiology IV. INRA Publ, Paris, pp 107–110

Debaud J-C, Marmeisse R, Gay G (1995) Intraspecific genetic variation in ectomycorrhizal fungi. In: Hock B, Varma AK (eds) Mycorrhiza: function, molecular biology and biotechnology. Springer, Berlin Heidelberg New York, pp 79–113

Dell B, Botton B, Martin F, Le Tacon F (1989) Glutamate dehydrogenases in ectomycorrhizas of spruce (*Picea excelsa* L.) and beech (*Fagus sylvatica* L.). New Phytol 11:683–692

Deransart C, Chaumat E, Cleyet-Marel J-C, Mousain D, Labarere J (1990) Purification assay of phosphatases secreted by *Hebeloma cylindrosporum* and preparation of polyclonal antibodies. Symbiosis 9:185–194

Doudrick RL, Anderson NA (1989) Incompatibility factors and mating competence of two *Laccaria* spp. (Agaricales) associated with black spruce in northern Minnesota. Phytopathology 79:694–700

Duc G, Trouvelot A, Gianinazzi-Pearson V, Gianinazzi S (1989) First report on non-mycorrhizal plant mutants (myc^-) obtained in pea (*Pisum sativum* L.) and faba bean (*Vicia faba* L.) Plant Sci 60:215–222

Ducamp M, Olivier JM (1989) Comparaison des activités phosphatases acides d'hétérocaryons et d'homocaryons de *Suillus granulatus*. Agronomie 9:295–304

Durand N, Debaud J-C, Casselton LA, Gay G (1992) Isolation and preliminary characterization of 5-fluoroindole-resistant and IAA-overproducer mutants of the ectomycorrhizal fungus *Hebeloma cylindrosporum* Romagnesi. New Phytol 121:545–553

Fries N, Mueller GM (1984) Incompatibility systems, cultural features and species circumscriptions in the ectomycorrhizal genus *Laccaria* (Agaricales). Mycologia 76:633–642

Fries N, Neumann W (1990) Sexual incompatibility in *Suillus luteus* and *S. granulatus*. Mycol Res 94:64–70

Fries N, Sun YP (1992) The mating system of *Suillus bovinus*. Mycol Res 96:237–238

Garbaye J (1994) Transley review no 76. Helper bacteria: a new dimension to the mycorrhizal symbiosis. New Phytol 128:197–210

Garbaye J, Bowen JD (1989) Stimulation of ectomycorrhizal infection of *Pinus radiata* by some microorganisms associated with the mantles of ectomycorrhizas. New Phytol 112:383–388

Garbaye J, Duponnois R (1992) Specificity and function of Mycorrhization Helper Bacteria (MHB) associated with the *Pseudotsuga menziesii-Laccaria laccata* symbiosis. Symbiosis 14:335–344

Gardes M, White TJ, Fortin JA, Bruns TD, Taylor JW (1991) Identification of indigenous and introduced symbiotic fungi in ectomycorrhizae by amplification of nuclear and mitochondrial ribosomal DNA. Can J Bot 69:180–190

Gay G, Debaud J-C (1987) Genetic study on indole-3-acetic acid production by ectomycorrhizal *Hebeloma* species: inter- and intraspecific variability in homo- and dikaryotic mycelia. Appl Microbiol Biotechnol 26:141–146

Gay G, Rouillon R, Bernillon J, Favre-Bonvin J (1989) IAA biosynthesis by the ectomycorrhizal fungus *Hebeloma hiemale* as affetred by different precursors. Can J Bot 67:2235–2239

Gay G, Marmeisse R, Fouillet P, Bouletreau M, Debaud J-C (1993) Genotype/nutrition interactions in the ectomycorrhizal fungus *Hebeloma cylindrosporum* Romagnesi. New Phytol 123:335–343

Gay G, Normand L, Marmeisse R, Debaud J-C (1994) Auxin overproducer mutants of *Hebeloma cylindrosporum* Romagnesi have increased mycorrhizal activity. New Phytol 128:645–657

Gea L, Normand L, Vian B, Gay G (1994) Structural aspects of ectomycorrhizae of *Pinus pinaster* (Ait.)Sol. formed by an IAA-overproducer mutant of the fungus *Hebeloma cylindrosporum*. New Phytol 128:659–670

Gianinazzi-Pearson V, Gianinazzi S (1989) Phosphorus metabolism in mycorrhizas. In: Boddy L, Marchant R, Read DJ (eds) Nitrogen, phosphorus and sulphur utilization by fungi. Cambridge University Press, Cambridge, pp 227–241

Gianinazzi-Pearson V, Smith SE (1993) Physiology of mycorrhizal mycelia. Adv Plant Pathol 9:55–82

Giltrap NJ (1982) *Hebeloma* spp. as mycorrhizal associates of birch. Trans Br Mycol Soc 79:157–160

Godbout C, Fortin JA (1990) Cultural control of basidiome formation in *Laccaria bicolor* with container-grown white pine seedlings. Mycol Res 94:1051–1058

Hacskaylo E, Bruchet G (1972) *Hebelomas* as mycorrhizal fungi. Bull Torrey Bot Club 99:17–20

Harley JL, Harley EL (1987a) A check-list of mycorrhiza in the British flora. New Phytol 105:1–102

Harley JL, Harley EL (1987b) A check-list of mycorrhiza in the British flora – addenda, errata, an index. New Phytol 107:741–749

Harley JL, Smith SE (1983) Mycorrhizal symbiosis. Academic Press, London, 234 pp

Hawksworth DL (1991) The fungal dimension of biodiversity: magnitude, significance and conservation. Mycol Res 95:641–655

Hébraud M, Fèvre M (1988) Protoplast production and regeneration from mycorrhizal fungi and their use for isolation of mutants. Can J Microbiol 34:157–161

Henrion B, Le Tacon F, Martin F (1992) Rapid identification of genetic variation of ectomycorrhizal fungi by amplification of ribosomal RNA genes. New Phytol 122:289–298

Hetrik BAD (1989) Acquisition of phosphorus by VA mycorrhizal fungi and the growth response of their host plants. In: Boddy L, Marchant R, Read DJ (eds) Nitrogen, phosphorus and sulphur utilization by fungi. Cambridge University Press, Cambridge, pp 205–226

Hilbert J-L, Martin F (1988) Regulation of gene expression in ectomycorrhizas. I: Protein changes and the presence of ectomycorrhiza-specific polypeptides in the *Pisolithus-Eucalyptus* symbiosis. New Phytol 110:339–346

Ho I (1989) Acid phosphatase, alkaline phosphatase and nitrate reductase activity of selected ectomycorrhizal fungi. Can J Bot 67:750–753

Ho I, Trappe JM (1987) Enzymes and growth substances of *Rhizopogon* species in relation to mycorrhizal host and infrageneric taxonomy. Mycologia 79:553–558

Jansen AE (1982) *Lactarius hysginus* en *Hebeloma cylindrosporum* in Nederland. Coolia 25:62–67

Kropp BR (1990) Variation in acid phosphatase activity among progeny from controlled crosses in the ectomycorrhizal fungus *Laccaria bicolor*. Can J Bot 68:864–866

Kropp BR, Fortin JA (1986) Formation and regeneration of protoplasts from the ectomycorrhizal basidiomycete *Laccaria bicolor*. Can J Bot 64:1224–1226

Kropp BR, Fortin JA (1988) The incompatibility system and relative ectomycorrhizal performance of monokaryons and reconstituted dikaryons of *Laccaria bicolor*. Can J Bot 66:289–294

Kropp BR, McAfee BJ, Fortin JA (1987) Variable loss of ectomycorrhizal ability in monokaryotic and dikaryotic cultrues of *Laccaria bicolor*. Can J Bot 65:500–504

Lamhamedi MS, Fortin JA, Kope HH, Kropp BR (1990) Genetic variation in ectomycorrhiza formation by *Pisolithus arhizus on Pinus pinaster* and *Pinus banksiana*. New Phytol 115:689–698

Lemke PA, Singh NK, Temann UA (1995) Genetic transformation of ectomycorrhizal fungi. In: Varma A, Hock B (eds) Mycorrhiza, structure, function, molecular biology and biotechnology. Springer, Berlin Heidelberg New York, pp 137–156

Le Tacon F, Valdenaire JM (1980) La mycorhization contrôlée en pépinière. Premiers résultats obtenus à la pépinière du fonds forestier national de Peyrat-le-Château (Haute Vienne) sur épicéa (*Picea excelsa*) et pin Douglas (*Pseudotsuga douglasii*). Rev For Fr 32:281–293

Le Tacon F, Jung G, Mugnier J, Michelot P, Mauperin C (1985) Efficiency in a forest nursery of an ectomycorrhizal fungus inoculum produced in a fermenter and entrapped in polymeric gels. Can J Bot 63:1664–1668

Lundeberg (1970) Utilization of various nitrogen sources, in particular bound soil nitrogen, by mycorrhizal fungi. Stud For Suec 79:1–95

Mac Afee BJ, Fortin JA (1988) Comparative effects of the soil microflora on ectomycorrhizal inoculation of conifer seedlings. New Phytol 108:443–449

Marmeisse R, Gay G, Debaud J-C, Casselton LA (1992a) Genetic transformation of the symbiotic basidiomycete fungus *Hebeloma cylindrosporum*. Curr Genet 22:41–45

Marmeisse R, Debaud J-C, Casselton LA (1992b) DNA probes for species and strain identification in the ectomycorrhizal fungus *Hebeloma*. Mycol Res 96:161–165

Marmeisse R, Gay G, Debaud JC (1995) Genetics of ectomycorrhizal fungi and their transformation. In: Stocchi V, Bonfante P, Nuti M (eds) Biotechnology of ectomycorrhizae, molecular approaches. Plenum Press, New York, pp 99–114

Martin F, Botton B (1993) Nitrogen metabolism of ectomycorrhizal fungi and ectomycorrhiza. Adv Plant Pathol 9:83–102

Martin F, Tagu D (1995) Ectomycorrhizal development: a molecular perspective. In: Varma A, Hock B (eds) Mycorrhiza, structure, function, molecular biology and biotechnology. Springer, Berlin Heidelberg New York, pp 29–58

Martin F, Canet D, Marchal JP, Larher F (1983) Phosphorus 31 nuclear magnetic resonance study of phosphate metabolism in intact ectomycorrhizal fungi. Plant Soil 71:469–476

Melville LH, Massicotte HB, Peterson RL (1987) Ontogeny of early stages of ectomycorrhizae synthesized between *Dryas integrifolia* and *Hebeloma cylindrosporum*. Bot Gaz 148:332–341

Melville LH, Massicotte HB, Ackerley CA, Peterson RL (1988) An ultrastructural study of modifications in *Dryas integrifolia* and *Hebeloma cylindrosporum* during ectomycorrhiza formation. Bot Gaz 149:408–418

Miller SL, Allen EB (1992) Mycorrhizae, nutrient translocation and interaction between plants. In: Allen MF (ed) Mycorrhizal functioning: an integrative plant-fungal process. Chapman and Hall, New York, pp 301–332

Meysselle J-P, Gay G, Debaud J-C (1986) Intraspecific variability in an ectomycorrhizal fungus: *Hebeloma cylindrosporum*. 3- Soluble phosphate utilization by sib-monokaryons and wild dikaryons. In: Gianinazzi-Pearson V, Gianinazzi S (eds) Physiological and genetic aspects of mycorrhizae. INRA Publ, Paris, pp 597–604

Meysselle J-P, Gay G, Debaud J-C (1991) Intraspecific genetic variation of acid phosphatase activity in monokaryotic and dikaryotic populations of the ectomycorrhizal fungus *Hebeloma cylindrosporum*. Can J Bot 69:808–813

Mousain D, Bousquet N, Polard C (1988) Comparaison des activités phosphatases d'homobasidiomycètes ectomycorhiziens en culture in vitro. J Eur Pathol For 18:299–309

Peterson RL, Bradbury SM (1995) Use of plant mutants, intraspecific variants, and non-hosts in studying mycorrhiza formation and function. In: Varma A, Hock B (eds) Mycorrhiza, structure, function, molecular biology and biotechnology. Springer, Berlin Heidelberg New York, pp 157–180

Plassard C, Mousain D, Salsac L (1984a) Mesure in vitro de l'activité nitrate réductase dans les thalles de *Hebeloma cylindrosporum*, champignon basidiomycète. Physiol Vég 22:67–74

Plassard C, Mousain D, Salsac L (1984b) Mesure in vitro et in vivo de l'activité nitrite réductase dans les thalles de *Hebeloma cylindrosporum*, champignon basidiomycète. Physiol Vég 22:147–154

Plassard C, Barry D, Eltrop L, Mousain D (1994) Nitrate uptake in maritime pine (*Pinus pinaster*) and the ectomycorrhizal fungus *Hebeloma cylindrosporum*: effect of ectomycorrhizal symbiosis. Can J Bot 72:189–197

Punt PJ, Oliver RP, Dingemanse MA, Pouwels PH, Van den Hondel CAMJJ (1987) Transformation of *Aspergillus* based on the hygromycin B resistance marker from *Escherichia coli*. Gene 56:117–124

Read DJ (1991) Mycorrhizas in ecosystems. Experientia 47:376–391

Read DJ (1992) The mycorrhizal fungal community with special reference to nutrient mobilization. In: Carroll GC, Wicklow DT (eds) The fungal community. Its organization and role in the ecosystem. Marcel Dekker, New York, pp 631–652

Read DJ (1993) Mycorrhiza in plant communities. Adv Plant Pathol 9:1–26

Read DJ, Leake JR, Langdale AR (1989) The nitrogen nutrition of mycorrhizal fungi and their host plants. In: Boddy L, Marchant R, Read DJ (eds) Nitrogen, phosphorus and sulphur utilization by fungi. Cambridge University Press, Cambridge, pp 181–204

Read DJ, Lewis DH, Fitter AH, Alexander IJ (1992) Mycorrhizas in ecosystems. CAB International, Oxon

Rolin D, Le Tacon F, Larher F (1984) Characterization of the different forms of phosphorus in the mycelium of the ectomycorrhizal fungus *Hebeloma cylindrosporum* cultiveated in pure culture. New Phytol 98:335–343

Romagnesi H (1965) Etudes sur le genre *Hebeloma*. Bull Soc Mycol Fr 81:321–344

Rouillon R, Gay G, Bernillon J, Favre-Bonvin J, Bruchet G (1986) Analysis by HPLC-mass spectromerty of the indole compounds released by the ectomycorrhizal fungus *Hebeloma hiemale* in pure culture. Can J Bot 64:1893–1987

Scheromm P, Plassard C, Salsac L (1990a) Regulation of nitrate reductase in the ectomycorrhizal basidiomycete *Hebeloma cylindrosporum* Romagnesi, cultured on nitrate or ammonium. New Phytol 114:441–447

Scheromm P, Plassard C, Salsac L (1990b) Nitrate assimilation of maritime pine (*Pinus pinaster* Soland *in* Ait.) ectomycorrhizal with *Hebeloma cylindrosporum* Romagnesi. New Phytol 114:93–98

Sen R (1990) Intraspecific variation in two species of *Suillus* from Scots pine (*Pinus sylvestris* L.) forests based on somatic incompatibility and isozyme analyses. New Phytol 114:607–616

Simchen G, Jinks JL (1964) The determination of dikaryotic growth rate in the basidiomycete *Schizophyllum commune*: a biometrical analysis. Heredity 19:629–649

Simoneau P, Viemont JD, Moreau J-C, Strullu DG (1993) Symbiosis-related polypeptides associated with the early stages of ectomycorrhiza organogenesis in birch (*Betula pendula* Roth). New Phytol 124:495–504

Smith SE, Smith FA (1990) Structure and function of the interfaces in biotrophic symbioses as they relate to nutrient transport. New Phytol 114:1–38

Tagu D, Python M, Crétin C, Martin F (1993) Cloning symbiosis-related cDNAs from eucalypt ectomycorrhiza by PCR-assisted differential screening. New Phytol 125:339–343

Tomsett AB (1989) The genetics and biochemistry of nitrate assimilation in ascomycete fungi. In: Boddy L, Marchant R, Read D (eds) Nitrogen, phosphorus and sulphur utilization by fungi. Cambridge University Press, Cambridge, pp 33–57

Tomsett AB, Garrett RH (1980) The isolation and characterization of mutants defective in nitrate assimilation in *Neurospora crassa*. Genetics 95:649–660

Trappe JM, Luoma DL (1992) The ties that bind: fungi in ecosystems. In: Carroll GC, Wicklow DT (eds) The fungal community. Its organization and role in the ecosystem. Marcel Dekker, New York, pp 631–652

Valjalo J (1979) Etude de la nature et du rôle de l'association ectomycorhizienne chez le châtaignier: cas de l' *Hebeloma cylindrosporum* (Romagnesi). In: Delmas J (ed) Mushroom Science X Part I. Centre for Agricultural Publications and Documentation, Wageningen, pp 903–918

Wagner F, Gay G, Debaud J-C (1986) Intraspecific variability in an ectomycorrhizal fungus: *Hebeloma cylindrosporum*. 2.-Ammonium and nitrate utilization by sib-monokaryons and wild dikaryons. In: Gianinazzi-Pearson V, Gianinazzi S (eds) Physiological and genetic aspects of mycorrhizae. INRA Publ, Paris, pp 589–596

Wagner F, Gay G, Debaud J-C (1988) Genetic variability of glutamate dehydrogenase activity in monokaryotic and dikaryotic mycelia of the ectomycorrhizal fungus *Hebeloma cylindrosporum*. Appl Microbiol Biotechnol 66:588–594

Wagner F, Gay G, Debaud J-C (1989) Genetic variation of nitrate reductase activity in mono- and dikaryotic populations of the ectomycorrhizal fungus *Hebeloma cylindrosporum* Romagnesi. New Phytol 113:259–264

Williams S, Verma MM, Jinks JL, Brasier CM (1976) Variation in a natural population of *Schizophyllum commune*. II Variation within the extreme isolates for growth rate. Heredity 37:365–375

Wong KKY, Fortin JA (1990) Root colonization and intraspecific mycobiont variation in ectomycorrhiza. Symbiosis 8:197–231

Wong KKY, Piché Y, Montpetit D, Kropp B (1989) Differences in the colonization of *Pinus banksiana* roots by sib-monokaryotic and dikaryotic strains of ectomycorrhizal *Laccaria bicolor*. Can J Bot 67:1717–1726

Zhu H, Higginbotham KO, Dancik BP (1988) Intraspecific genetic variability of isoenzymes in the ectomycorrhizal fungus *Suillus tomentosus*. Can J Bot 66:588–594

7 Vesicular-Arbuscular Mycorrhizal (VAM) Fungi

R.G. Linderman

CONTENTS

I. Introduction 117
II. VAM Fungal Life Cycle 118
A. VAM-Fungal Spores and Germination 118
B. Germ-Tube and Hyphal Growth and Root Penetration 120
C. Root Phase of VAM Fungi 121
D. Soil Phase of VAM Fungi 121
III. VAM Connection Between Root and Soil 122
A. Rhizodeposition of Carbon to Soil 122
B. Soil Aggregation 122
C. Nutrient and Water Uptake and Transport by VAM-Fungal Hyphae 123
IV. Effects of VAM on Plants 123
A. Physiological and Morphological Effects 123
B. Ecological Effects 124
V. Significance of VA Mycorrhizae in Agriculture and Forestry 125
VI. Summary and Conclusions 125
References 125

I. Introduction

The roots of most plants are symbiotically associated with specialized soil fungi that form mycorrhizae. It is well known that mycorrhizae can benefit the growth and health of plants, but it is not widely known or appreciated just how critical and normal this association is to the well-being of plants, especially in natural and, even more so, in disturbed ecosystems. As D.H. Marx and N.C. Schenck (1983) stated, "on most plants mycorrhizae are as much a natural component of roots as chloroplasts are components of leaves." Most plants have evolved with mycorrhizal fungi and their very existence depends on the mycorrhizal association (Malloch 1987; Simon et al. 1993).

The mycorrhizal association that occurs on most land plants is typified by the presence within the root cortex of distinctive fungal structures known as arbuscules and vesicles; thus the symbiosis is called a vesicular-arbuscular (VA) mycorrhiza. Since the association may not always form vesicles, some prefer the term arbuscular mycorrhizae, but for this discussion, VA mycorrhizae or VAM will be used.

VAM occur on the majority of land plants, including mosses (Parke and Linderman 1980) and ferns (Cooper 1976), many tropical and temperate trees (Kough et al. 1985), and even epiphytes (Janos 1993). Malloch's (1987) analysis of VAM ancestry indicates that they date as far back as the origin of vascular plants. Recently, Simon et al. (1993) amplified and sequenced genes encoding the small-subunit rRNA from spores of 12 VAM fungi representative of the main taxonomic groups. Their analyses supported the monophyletic origin of VAM fungi, and also the morphologically based classification of the Zygomycotina order Glomales into three families (Bentivenga and Morton 1994): Glomaceae, Acaulosporaceae, and Gigasporaceae. Furthermore, they used the small-subunit rRNA sequences as a molecular clock to infer dates of divergence, and dated the origin of VAM fungi at 353–462 million years ago. These findings are consistent with the hypothesis that VAM fungi facilitated the colonization of land by ancient plants. Fossils resembling hyphae, vesicles, and spores of extant VAM fungi in roots also support this idea (Pirozynski and Dalpé 1989).

The evidence strongly supports the contention that mycorrhizae, and especially VA mycorrhizae, coevolved with land plants such that the plant and fungal symbionts became highly dependent on each other. VAM fungi are obligate biotrophs that must complete their life cycles in association with living plant roots. While some aspects of the VA mycorrhizal symbiosis seem simple, it is certainly a highly evolved, finely tuned relationship that has withstood the test of time. Our understanding of the VAM symbiosis, however, is in its infancy, although the literature ex-

Horticultural Crops Research Laboratory, 3420 N.W. Orchard Avenue, Corvallis, Oregon 97330, USA

The Mycota V Part B
Plant Relationships
Carroll/Tudzynski (Eds.)
© Springer-Verlag Berlin Heidelberg 1997

plosion in recent years (Klironomos and Kendrick 1993) is testimony to the widespread acceptance of its importance to plant growth and health as well as ecosystem health. The intent of this chapter is to expose the interested but less informed reader to the highlights of the VAM symbiosis while summarizing current views for more informed mycologists.

II. VAM Fungal Life Cycle

VAM fungi are obligate symbionts whose simple life history consists of somatic and reproductive phases (Figs. 1 and 2). The somatic phase involves growth and differentiation of vegetative hyphae within roots and soil, intraradical arbuscules or hyphal coils and vesicles, and auxiliary cells in the soil. The reproductive phase involves the production of thick-walled chlamydospores either intraradically (i.e., *Glomus intraradices*) or on hyphae within the soil matrix. New root colonization can be initiated from spores in the soil, from root fragments that contain vesicles (Biermann and Linderman 1983) or spores, or from hyphal outgrowth from existing colonized live roots. In many habitats spore production is minimal, but inoculum potential may still be high due to propagules in the vegetative or somatic phases.

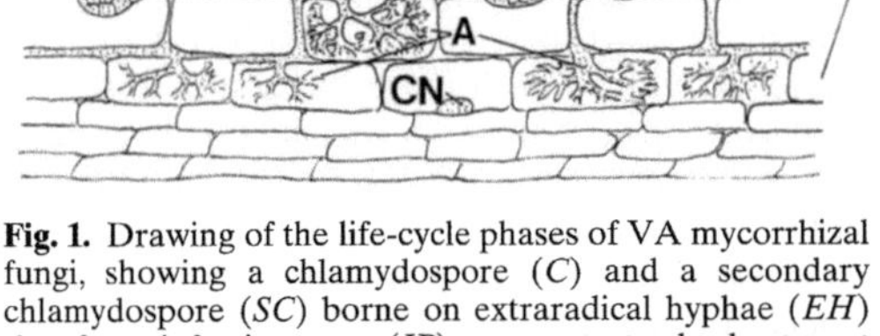

Fig. 1. Drawing of the life-cycle phases of VA mycorrhizal fungi, showing a chlamydospore (*C*) and a secondary chlamydospore (*SC*) borne on extraradical hyphae (*EH*) that form infection pegs (*IP*) to penetrate the host root epidermis (*RE*), or root hairs (*RH*) may be directly infected (*RHI*). Intraradical hyphae (*IH*) grow intercellularly in the root cortex (*RC*) and form vesicles (*V*) or invade cortical cells to form highly branched arbuscules (*A*). Cell nuclei (*CN*) may also be apparent in stained preparations

A. VAM-Fungal Spores and Germination

Spores function both as agents of long-term survival and dispersal (Fig. 2a,b). They range in size from 50 to 500 μm in diameter, and may contain up to several thousand nuclei (Burggraaf and Beringer 1987; Cooke et al. 1987; Viera and Glenn 1990). Spores vary in their structural features – wall layering, hyphal attachment, and size – and mode of production. They are typically produced on extraradical hyphae that extend into the soil, either singly, in clusters, or even in small sporocarps on the soil surface, e.g., *Glomus versiforme*. Chlamydospores are formed within roots colonized by *Glomus intraradices*.

VAM-fungal spores are capable of very long-term survival in soil or root fragments, especially if they are dried gradually. Young (1982) reported the viability of spores of representatives of several VAM fungal genera stored in dry soil for up to 28 years. Their thick walls provide the means to survive degradation from environmental extremes and parasitic or saprophytic microbes. They may be dispersed in soil by wind or water, moved within the soil profile by soil animals (including insects and earthworms), or within the ecosystem by birds or small rodents (Rabatin and Stinner 1991).

Spores of VAM fungi have been produced under aseptic conditions, but only in association with plant roots grown in root cultures (Mugnier and Mosse 1987; Becard and Fortin 1988). J.A. Fortin and his colleagues (pers. comm.) have recently modified cultural conditions to allow production of thousands of spores in petri dishes (Fig. 3).

Spore germination occurs near host roots in response to both host and environmental factors, as well as intrinsic fungal factors. Tommerup (1985) demonstrated that spores of some species of VAM fungi may be strongly affected by dormancy factors, while other species appear to lack such features. Spores germinate in response to root exudates (Graham 1982), but the specific components that trigger germination are not well known. There are many possible compounds included in the root exudate, both diffusible and volatile. For example, Whitfield et al. (1981) isolated 270 distinct organic compounds in the root exudate of *Acacia pulchella* R. Br., including organic acids, alcohols, aldehydes, ketones, esters, phenols, terpenoids, and miscellaneous compounds. Assuming that a similar array of substances could exude from most plant roots, it would be

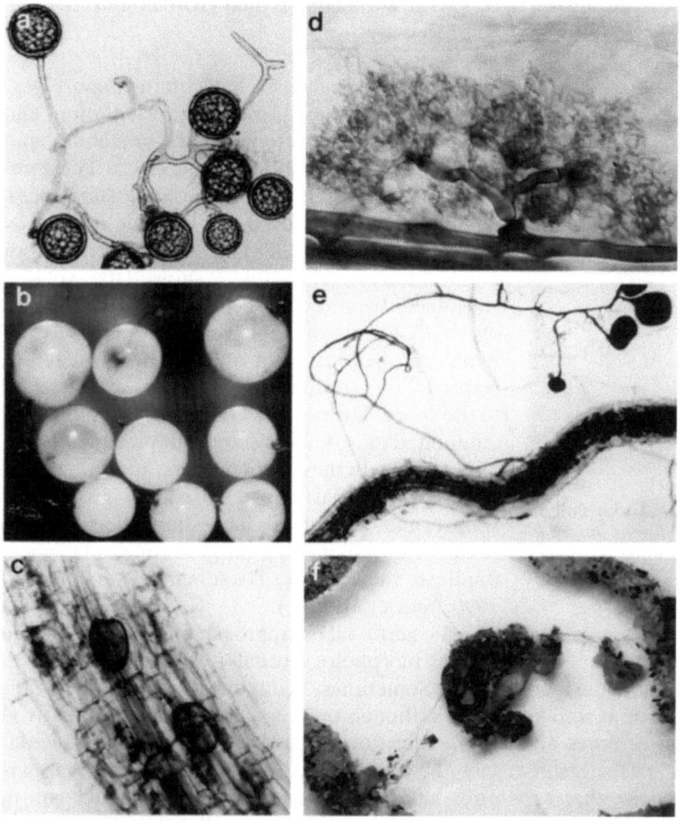

Fig. 2a–f. Photographs of life-cycle stages of VA mycorrhizal fungi: spore phase showing spores of **a** *Glomus fasciculatum* and **b** *Gigaspora margarita* extracted from soil; intraradical phase showing **c** vesicles and **d** arbuscule in root cortical cells; and extraradical phase showing **e** spore-bearing VAM fungal hyphae. (Photo courtesy G.J. Bethlenfalvay) and **f** soil aggregate formed in association with VAM fungal hyphae

difficult to single out specific stimulatory substances. However, with the recent development of methods to study spore germination and growth in vitro (Mugnier and Mosse 1987; Becard and Fortin 1988), it has been shown that volatile compounds may play a key role in fungus-root communication.

Volatiles elicit microbial responses at very low concentrations (Linderman and Gilbert 1975; Koske and Gemma 1992) and apparently are a major means of communication between VAM fungi and other soil microbes and plant roots. Volatile substances have the advantage as communication signals between microbes and roots in soil because they are generally neutral and do not readily bind to soil particles. Such substances cross air spaces in soil and therefore traverse greater distances than compounds dissolved in the soil solution, although volatiles are highly soluble in water and are readily utilized by microbes.

Volatile compounds produced by some soil bacteria and fungi can stimulate VAM fungal spore germination (Tommerup 1985; Azcón-Aguilar et al. 1986; Mayo et al. 1986; Azcón 1987; Mugnier and Mosse 1987; Tylka et al. 1991), and volatiles orient resultant germ tubes toward nearby roots (Koske 1982; St. John et al. 1983; Gemma and Koske 1988). For nonhost plant species, root volatiles may inhibit spore germination (El-Atrach et al. 1989). High levels of CO_2 (5%) were shown to inhibit spore germination (Le Tacon et al. 1983), but recent results indicate that lower levels (2%) are not inhibitory, especially in the presence of flavonoid compounds in root exudates (Becard and Piché 1989; Gianinazzi-Pearson et al. 1990; Tsai and Phillips 1991). Some bacteria closely associated with VAM fungal spores produce metabolites that may stimulate or inhibit spore germination (Mayo et al. 1986). The seemingly complex or inconsistent regulation of spore germi-

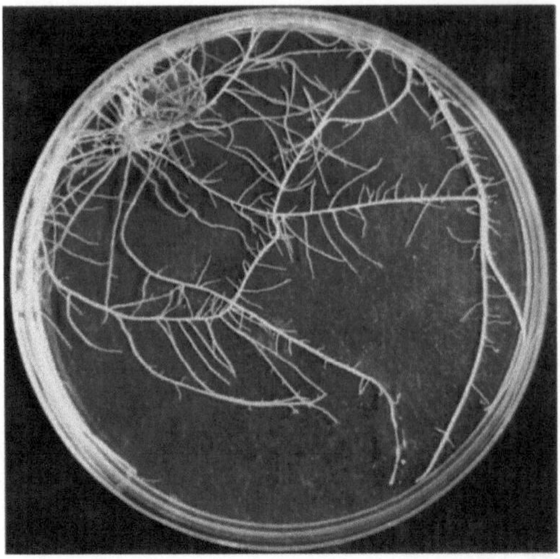

Fig. 3. Carrot (*Daucus carota*) transformed root culture system where roots are inoculated with *Glomus intraradices*. Roots become colonized internally, forming arbuscules and vesicles, and then form extraradical hyphae on which spores are produced. (Photo courtesy J.A. Fortin)

nation for different VAM fungi remains unclear. Koske and Gemma (1992) suggest that small-spored VAM fungi with fewer food reserves are more dependent on other soil microbes for germination stimuli than large-spored species that require no external stimulus for germination.

Germination of VAM fungal spores may also be inhibited in some soils by metabolites from microorganisms (Daniels and Trappe 1980; Tommerup 1985; Wilson et al. 1989). Extracts from suppressive soils inhibited spore germination, but the specific inhibitors were not identified (Tommerup 1985). In some cases, other microorganisms may antagonize VAM fungal spores as hyperparasites (Siqueira et al. 1984; Paulitz and Linderman 1991; Lee and Koske 1994).

VAM-fungal spore germination may also be affected by physical and chemical factors in the spores and soil (Siqueira et al. 1985). These factors include self-inhibiting dormancy factors, pH, temperature, moisture, light and radiation, aeration, and inorganic ions. In addition, carbohydrates, organic acids, and amino acids, tested as possible substrates for growth in vitro, suppress spore germination. Of course, different VAM fungal species or strains may vary considerably in their response to these factors.

B. Germ-Tube and Hyphal Growth and Root Penetration

Root exudates influence VAM-fungal spore germination, germ tube growth and orientation, and hyphal growth rate, frequency of branching, and orientation (Koske and Gemma 1992). These responses may be due to water-soluble substances, volatile substances, or both. Elegant studies by Becard and Fortin (1988) using transformed-root organ cultures in vitro (Fig. 3) showed that volatile substances from roots function in the early stages of VAM formation. The volatiles increased the hyphal growth rate, but only in the presence of water-soluble substances, and only if still attached to the spore. The use of KOH traps to remove CO_2 eliminated the effects of some, but presumably not all, volatiles. In these experiments, addition of low levels of CO_2 partially restored the growth response, and later studies have shown that 2% CO_2 in the presence of flavonoids in root exudates will duplicate the response (Gianinazzi-Pearson et al. 1990; Nair et al. 1991).

As germ tubes approach the host root, the hyphal morphology usually changes and growth slows. Sometimes, highly branched infection fans form, although these apparently are not involved in the infection process per se. Once hyphae contact the root surface, recognition factors apparently take over, although the exact mechanism remains unclear (Harley and Smith 1983). There is little doubt that recognition of compatible symbionts is required before a symbiotic relationship can occur. The communication between the symbionts is refined and involves chemical (soluble and volatile) components of root exudates as well as physical tactile responses that guide the hyphae to regions of the root where penetration is possible. Anderson (1988) suggested that numerous plant and fungal genes are involved in the recognition process at various stages of development of the symbiotic association. She suggested that phenolic compounds in root exudates may induce gene expression in VAM-fungal hyphae prior to contacting root tissue, and that failure of the fungus to respond to those signals would preclude establishment of a symbiotic relationship. Non-VAM host roots may allow initial infection by VAM fungal hyphae, but subsequent in compatibility checks halt the spread of the fungus (Malajczuk et al. 1981; Parra-Garcia et al. 1992). Some believe that such chemical defense against invasion by an incompatible fungus is of host origin, but others suggest that it is the fungus that rejects the nonhost as being incapable of supplying the substrates it needs to establish this obligate relationship (Harley and Smith 1983; Koske and Gemma 1992).

When VAM-fungal hyphae contact susceptible roots, they appear to select specific sites on the root surface, where they produce appressoria and penetration hyphae, usually near the root tip where suberization has not yet occurred. They are guided to these sites by specific signal molecules bound to the host cell wall and by other compounds exuded from the root (Anderson 1992). The chance of successful contact between VAM fungal hyphae and host roots is increased by fungus-stimulated root branching (Gemma and Koske 1988) that creates more potential infection sites. Also, roots apparently are attracted to spores or hyphae due to an unknown volatile organic compound(s) released by the fungus (Gemma and Koske 1988). Thus, at this stage of interaction between the symbionts, the exudation of a chemical signal by one partner triggers the production of response signals by the other in a cyclic exchange analogous to the production of sex hormones by some fungi (Koske and Gemma 1992). The signals leading to appressorium formation are clearly chemical and not thigmotropic (Giovannetti et al. 1993). Appressorium formation does not occur on roots of nonhost plants, probably due to chemical substances produced by the roots (Gianinazzi-Pearson and Gianinazzi 1992). Actual penetration of the root by VAM fungi is by means of a penetration hypha or infection peg from the appressorium that penetrates either by force or by production of cell-wall dissolving enzymes.

C. Root Phase of VAM Fungi

Once inside the root, VAM-fungal hyphae branch and spread within and between cortical cells. As they penetrate cortical cell walls, hyphae invaginate the root cell plasmalemma, where they form either hyphal coils or elaborate and extensively branched structures called arbuscules (Fig. 2d). The arbuscule is the major exchange site between the two symbionts: carbon from the host to the fungus, mineral nutrients from the fungus to the host. Arbuscules increase the exchange surface area manifold without breaching the host plasmalemma, which grows to maintain contact with the entire surface of the highly branched arbuscule. Arbuscules are ephemeral structures, lasting from a few days to 2 weeks before they degenerate (Harley and Smith 1983). While they are active, they occupy most of the cell volume, pushing aside the host nucleus and nucleolus, which are enlarged. During this phase, bidirectional exchange of metabolites between the arbuscule and host cytoplasm takes place; similar exchange occurs with intracellular hyphal coils or intercellular hyphae.

Most VAM fungi also produce vesicles borne on intercellular hyphae in the older regions of infections within the cortex (Fig. 2c), which seem to function as storage structures, since they contain large amounts of lipid. They apparently can, however, function as propagules in root fragments to initiate new infections (Biermann and Linderman 1983).

D. Soil Phase of VAM Fungi

After primary infection, germ tube hyphae may branch outside the root, grow in any direction from the initial penetration site, and initiate more infections. Hyphae branch and grow outside the root, produce spores (Fig. 2e), and colonize the soil, creating a hyphal network resulting in an amazing amount of fungal biomass. Miller and Jastrow (1994) summarized the reports on the length of VAM-fungal hyphae in soils of different ecosystems; their own measurements in a chronosequence of tallgrass prairie reconstruction indicated the hyphal length was as high as $47\,m\,cm^{-3}$.

Single-plant experiments indicate that extraradical hyphae can extend into the soil some 7–9 cm to absorb nutrients (Rhodes and Gerdemann 1975; Camel et al. 1991). In multiple or mixed plant systems, extraradical hyphae constitute an extensive interconnecting, anastomosing network (Tisdall 1991; Read 1992; Miller and Jastrow 1994). In terms of ecosystem stability and sustainability in agriculture, some feel that the VAM-fungal hyphal network in soil constitutes the most important component of the VAM symbiosis (Bethlenfalvay 1992). Clearly, these hyphae are the major conduits connecting plant roots with the physical, chemical, and biological components of soil.

III. VAM Connection Between Root and Soil

A. Rhizodeposition of Carbon to Soil

The flow of energy-rich compounds from plant roots into the soil is a fundamental process in the soil ecosystem (Lynch and Whipps 1990; Tisdall 1991; Bethlenfalvay 1992; Finlay and Söderström 1992). VAM fungi play a major role in this process, due not only to changes in the pattern of root exudation, but also to the presence of exudates from soil hyphae of VAM fungi. The rate of exudation from mycorrhizal roots has been negatively correlated with the phosphorus content of roots and the phospholipid content of root cell membranes. Changes in phosphorus status of plant roots due to VAM formation and increased phosphorus uptake result in a decrease in the amount of root exudation (Ratnayake et al. 1978; Graham et al. 1981; Schwab et al. 1983). These changes, plus VAM-fungal hyphal exudates (Vancura et al. 1989), cause qualitative changes in the rhizosphere microflora. Because these changes appear to be persistent, the term mycorrhizosphere has been used to describe the new microbial composition and equilibrium (Linderman 1988a). Jakobsen and Rosendahl (1990) have documented the flow of ^{14}C-labeled substances from photosynthate in plant leaves into roots, root and soil VAM-fungal hyphae, and the mycorrhizosphere soil. They showed that 43.3% of the photo-assimilated C was partitioned to the root system, and 70% of that was lost as CO_2 and organic exudates. Below-ground root activity and respiration in VAM plants accounted for 27% of the photoassimilated carbon, five times more than that of the non-VAM control plants. Mycorrhizal metabolism consumed 20% of the below-ground carbon. These results demonstrate that VAM fungi dramatically increase the flow of carbon from the plant into the soil matrix that supports microbial activity. It might appear that the carbon flowing through and exuded by extraradical hyphae is "lost" from the point of view of the plant biomass, but it is considered by some researchers to be a major investment in the physical and biological health of the ecosystem (Bethlenfalvay 1992).

The organic compounds in exudates from root tissue and soil hyphae of VAM fungi are the basis for the mycorrhizosphere effect. The amount and composition of those exudates accounts for the selective enrichment for specific microbes from the background soil (Ames et al. 1984; Meyer and Linderman 1986; Secilia and Bagyaraj 1987). Some bacteria and actinomycetes occur only in the mycorrhizosphere soil (as compared to non-VAM rhizosphere soil), and there is specificity of selection depending on the species of VAM fungus (Secilia and Bagyaraj 1987). Whether the specificity is due to the soil hyphae of the VAM fungus or to its alteration of the root exudation pattern has not been determined. It seems plausible, however, that the total of the resultant combination of mycorrhizosphere microbes associated with any plant species in specific soils strongly influences the plant's growth and survival in that system (Linderman 1988a,b). Once equilibrium has been reached, there simply is no way to distinguish between effects of VAM fungi, of their associates, or their combination; the effects are simply the consequences of the VAM symbiosis.

B. Soil Aggregation

Soil hyphae of VAM fungi comprise a tremendous biomass that greatly influences soil structure by means of the formation of water-stable aggregates (Tisdall 1991; Miller and Jastrow 1992). Hyphal exudates containing a mucilage of polysaccharides bind clay particles and form microaggregates. Macroaggregates form by the binding of microaggregates; their physical and nutritional composition enriches for specific microorganisms (Fig. 2f). It is largely unknown what these organisms contribute to the stabilization of the aggregates, to the growth of VAM fungal hyphae (Sutton and Sheppard 1976; Burns and Davies 1986), or to plant growth and health. The fact that VAM fungal hyphae persist even for several months after the host plant dies indicates that they are more important stabilizers than saprophytic fungi (Tisdall 1991). The VAM fungal hyphal length is directly proportional to the stability of macroaggregates in soil, and that length, as indicated above, may be as high as $50\,m\,g^{-1}$ of stable aggregate (Tisdall 1991). Mycorrhizal fungi are unique among nonpathogenic fungi in having direct access to a host plant energy supply that can give them a competitive advantage for hyphal growth in soil.

C. Nutrient and Water Uptake and Transport by VAM-Fungal Hyphae

The VAM-fungal hyphal network is a living extension of the host root that increases the efficiency of nutrient and water uptake over roots alone (Harley and Smith 1983). The most dramatic effect of increased uptake of nutrients is with those elements that normally are relatively immobile in soil such as P, Zn, and Cu. However, the greatly increased absorptive surface area provided by soil hyphae of VAM fungi facilitates the uptake of even relatively mobile nutrients such as N and Ca (Rhodes and Gerdemann 1978; Ames et al. 1983). The potential for water uptake by mycorrhizal roots is similarly increased by means of the absorptive surface area provided by soil hyphae. The influence of extraradical hyphae on plant water relations has been demonstrated in studies by Hardie (1985) and Faber et al. (1991), who showed dramatic changes when these hyphae were severed. In the latter study, where root and hyphal uptake were separated, it was shown that when hyphal links from moist soil to roots were maintained, plants transpired 35% more water than when the hyphae were severed. Under soil drought conditions, the capacity to tap a greater volume of soil for water provides considerable advantage to the mycorrhizal plant.

Plants of the same or different species may share common VAM fungal symbionts. Although numerous studies have demonstrated the interplant transfer of C, N, and P through hyphal connections, it is still unclear to what extent nutrients are shared and what impact that may have on ecosystem stability (Finlay and Söderström 1992). Interplant transfer appears to occur only when there is some physiological imbalance that creates a source-sink relationship (Frey and Schuepp 1993; Martins 1993). The extensive hyphal network also may increase inoculum potential and contribute to early root colonization and establishment of seedlings that can tap almost immediately into water and nutrient supplies (Birch 1986).

IV. Effects of VAM on Plants

A. Physiological and Morphological Effects

The most obvious and well-known effect of VAM on plants is the growth enhancement that occurs under nutrient-deficient conditions. Soil hyphae of VAM fungi are able to bridge the zone of nutrient depletion around roots and acquire nutrient elements such as P, Zn, and Cu which are bound to soil particles and do not readily flow to the roots. Because of the increased capacity for VAM plants to acquire nutrients, striking growth differences can occur between VAM and non-VAM plants (Fig. 4a,b,c). Measurement of the nutritional status of VAM plant tissues clearly demonstrates the effects of improved nutrient uptake. VAM may also regulate the uptake of mineral elements (such as Mn) that are at toxic levels in soil (Fig. 4d).

The establishment of the VAM symbiosis induces a number of other physiological changes that may be related to improved mineral nutrition. Many studies have suggested that most VAM effects can be traced to improved nutrition, usually involving phosphorus, because adding P often gives the same response as inoculating with VAM fungi (Graham 1987). These effects include increased rate of photosynthesis, hydraulic conductivity, and transpiration (Allen et al. 1981); altered membrane permeability (Graham et al. 1981), and elevated levels of phytohormones (Allen et al. 1980, 1982). The effect of increased drought tolerance of VAM compared to non-VAM plants remains controversial. The controversy is not about the relative increase in drought tolerance in VAM plants, but rather about the mechanism involved, i.e., whether the effect is related to improved phosphorus uptake (Nelson and Safir 1982) or some other physiological change such as osmoregulation (Auge et al. 1986; Davies et al. 1992). Studies carefully conducted to assure phosphorus equivalency in VAM and non-VAM plants have demonstrated the benefit of VAM resulting from non-P-related physiological changes, increased uptake of water by extraradical hyphae, or both (Hardie 1985; Faber et al. 1991; Davies et al. 1992).

There is also controversy about possible increased disease resistance of VAM vs. non-VAM plants and about the mechanisms which might be involved (Linderman 1994). Some reports clearly indicate that increased phosphorus nutrition has the same effect as inoculation with VAM fungi, while others suggest a physiological basis for resistance where P was not a variable between VAM and non-VAM plants. Increased levels of phytoalexin-like compounds in VAM plants would support the latter mechanism (Morandi et al. 1984). Changes in other plant constituents that

Fig. 4a–d. Plant growth effects of VA mycorrhizae under nutrient deficiency (P) or excess/toxicity (Mn) conditions. **a** Japanese maple and **b** geranium grown under P-limiting conditions with (*left*) or without (*right*) VA mycorrhizae. **c** Western red cedar seedlings grown in a fumigated nursery bed showing nonmycorrhizal stunted seedlings and normal seedlings with VA mycorrhizae. **d** Soybean plants grown under Mn toxicity conditions with and without VA mycorrhizae. (Photo courtesy G.J. Bethlenfalvay)

suppress fungal pathogens have also been noted, even in the absence of differences in tissue levels of P (see Linderman 1994). Qualitative changes in the mycorrhizosphere microbes that are antagonistic to pathogens appear also to result from VAM-induced changes in the plant host physiology. In addition, VAM-induced plant changes can result in increased lignification or other anatomical changes that block the development of root pathogens (see Linderman 1994).

B. Ecological Effects

VA mycorrhizae have profound effects on the ecology of the mycorrhizosphere due to the induced changes in root exudation patterns as well as to the presence of the VAM soil mycelium. Furthermore, the VAM-induced physiological and morphological changes in plants can alter their competitiveness and ability to survive in the ecosystem (Goodwin 1992).

Ecological effects of VAM on plants frequently can be traced to VAM-induced microbial changes and interactions in the mycorrhizosphere (Linderman 1988a). Mycorrhizosphere microbes appear to influence the extent of extraradical hyphal development, which can affect soil drought tolerance and soil aggregate formation (Sutton and Sheppard 1976; Hardie 1985; Tisdall 1991). Furthermore, VAM can interact synergistically with other microbes in the mycorrhizosphere to enhance both plant growth and survival (Linderman 1992). For example, nitrogen-fixing bacteria (both free-living and nodulating) in combination with VAM are more effective at nitrogen fixation than in the absence of VAM, primarily due to increased numbers and activity of nodules. Furthermore, there is specificity in the interendophyte compatibility between VAM fungal strains and bacterial strains, independent of nutritional effects induced by the VAM (Ianson and Linderman 1993). Whether the interaction occurs in the mycorrhizosphere soil or within host tissues remains in question.

V. Significance of VA Mycorrhizae in Agriculture and Forestry

A strong case can be made for the importance of VA mycorrhizae in undisturbed ecosystems whose nutritional and microbial balance accounts for sustained plant growth and health. This case also offers an explanation for loss of sustainability when the system is disturbed and degraded by agricultural and forestry practices of recent times (Bethlenfalvay and Linderman 1992). Cultural practices involving cultivation, application of fertilizers and biocides, certain crop rotations, and selection of crop genotypes with reduced dependency on VAM have reduced the impact of VAM fungi in agriculture and reforestation. VAM effects have been replaced in most agroecosystems by external inputs and, as a consequence, VAM fungal numbers and diversity have decreased. It has been argued, therefore, that restoration of VAM in crop production systems is critical in the move to reduce inputs of fertilizers and biocides that are deleterious to the environment and to the inhabitants of Earth. VAM fungi link plants with the soil and play a crucial role in facilitating both microbial and plant functions, mediating exchanges between them. It is doubtful that any other group of fungi could be so credited.

VI. Summary and Conclusions

Evidence of vesicular-arbuscular mycorrhizal fungi occurs in fossils of prehistoric plants, supporting the idea that they have existed as long as land plants, and in fact probably facilitated the colonization of land by plants. They have a simple life cycle that includes (1) a spore phase consisting of large, thick-walled spores that can survive in soil for many years, germinate, and initiate new root infections, (2) an intraradical phase typified by hyphae bearing storage bodies called vesicles, and haustoria-like arbuscules where metabolites are exchanged between plant and fungal symbionts, and (3) an extraradical phase involving extensive hyphae that grow into the soil, contribute to a mycorrhizosphere effect, cause soil aggregates to form, and create a biological link between the soil and plants and between plants.

Spore germination is stimulated by root exudates, but is also influenced by soil microbial metabolic products, often volatile compounds. Volatiles also influence the orientation of germ tubes and appear to be involved in recognition between the fungal and host symbionts. Root surface materials are involved in the final determination of a compatible association.

VAM cause few obvious morphological changes in roots, but they can influence root size and architecture, root tissue lignification, and root absorptive capacity by means of extraradical hyphae. They also cause significant physiological effects on host plants, including altered nutrition, chemical constitutents, phytohormone production, membrane permeability, rate of photosynthesis and partitioning of photosynthate, and increased osmoregulation and disease resistance. These plant physiological changes account for the capacity of VAM plants to grow and survive in nondisturbed ecosystems, exposed to environmental stresses, and without artificial inputs. In contrast, disturbed cultivated agroecosystems that have reduced benefits from VAM require extensive inputs of nutrients and biocides to maintain high levels of plant productivity.

The most obvious benefit of VAM is the enhanced plant growth (and even survival) under nutrient-limiting or toxic conditions due to increased uptake of immobile nutrients from soil or regulation of uptake or impact of toxic elements in soil. VAM plants are better able to avoid environmental and physiological stresses, including soil drought stress, root disease effects, and transplant shock. Without these benefits, most plants simply could not survive without artificial assistance. VA mycorrhizae have evolved with most plants, and orchestrate the microbial composition and activities in the mycorrhizosphere. Essentially, they are the link that connects plants with the soil, both physically and microbiologically. They are more than a conduit, however. They are the very fiber of the below-ground ecosystem that keeps it alive and functioning.

References

Allen MF, Moore TS, Christensen M (1980) Phytohormone changes in *Bouteloua gracilis* infected by

vesicular-arbuscular mycorrhizae: I. Cytokinin increases in the host plant. Can J Bot 58:371–374
Allen MF, Smith WK, Moore TS, Christensen M (1981) Comparative water relations and photosynthesis of mycorrhizal and non-mycorrhizal *Bouteloua gracilis*. New Phytol 88:683–693
Allen MF, Moore TS, Christensen M (1982) Phytohormone changes in *Bouteloua gracilis* infected by vesicular-arbuscular mycorrhizae. II. Altered levels of gibberclin-like substances and abscisic acid in the host plant. Can J Bot 60:468–471
Ames RN, Reid CPP, Porter LK, Cambardella C (1983) Hyphal uptake and transport of nitrogen from two ^{15}N-labelled sources by *Glomus mosseae*, a vesicular-arbuscular mycorrhizal fungus. New Phytol 95:381–396
Ames RN, Reid CPP, Ingham ER (1984) Rhizosphere bacterial population responses to root colonization by vesicular-arbuscular mycorrhizal fungus. New Phytol 96:555–563
Anderson AJ (1988) Mycorrhizae – host specificity and recognition. Phytopathology 78:375–378
Anderson AJ (1992) The influence of the plant root on mycorrhizal formation. In: Allen MF (ed) Mycorrhizal functioning: an integrative plant-fungal process. Chapman & Hall, New York, pp 37–64
Auge RM, Schekel KA, Wample RL (1986) Greater leaf conductance of well-watered VA mycorrhizal rose plants is not related to phosphorus nutrition. New Phytol 103:107–116
Azcón R (1987) Germination and hyphal growth of *Glomus mosseae* in vitro: effects of rhizosphere bacteria and cell-free culture media. Soil Biol Biochem 19:417–419
Azcón-Aguilar C, Diaz-Rodriguez R, Barea JM (1986) Effect of soil microorganisms on spore germination and growth of the vesicular-arbuscular mycorrhizal fungus *Glomus mosseae*. Trans Br Mycol Soc 86:337–340
Becard G, Fortin A (1988) Early events of vesicular-arbuscular mycorrhiza formation on RiT-DNA transformed roots. New Phytol 108:211–218
Becard G, Piché Y (1989) Fungal growth stimulation by CO_2 and root exudates in vesicular-arbuscular mycorrhizal symbiosis. Appl Environ Microbiol 55:2320–2325
Bentivenga, SP, Morton JB (1994) Systematics of Glomalean endomycorrhizal fungi: current views and future directions. In: Pfleger FL, Linderman RG (eds) Mycorrhizae and plant health. APS Press, St Paul, pp 283–308
Bethlenfalvay GJ (1992) Mycorrhizae in the agricultural plant-soil system. Symbiosis 14:413–425
Bethlenfalvay GJ, Linderman RG (eds) (1992) Mycorrhizae in sustainable agriculture. ASA Spec Publ No 54, American Society of Agronomy, Crop Science Society of America, Soil Science Society of America, Madison, Wisconsin, Preface pp viii–xiii
Biermann B, Linderman RG (1983) Use of vesicular-arbuscular mycorrhizal roots, intraradical vesicles and extraradical vesicles as inoculum. New Phytol 95:97–105
Birch CPD (1986) Development of VA mycorrhizal infection in seedlings in semi-natural grassland turf. In: Gianinazzi-Pearson V, Gianinazzi S (eds) Proc 1st Eur Symp on Mycorrhizas. INRA, Paris, pp 233–239
Burggraaf AJP, Beringer JE (1987) Nuclear division and VA-mycorrhizal in-vitro culture. In: Sylvia DM, Hung LL, Graham JH (eds) Mycorrhizae in the next decade. Practical applications and research priorities. Proc 7th N A Conf on Mycorrhizae, Gainesville, p 190
Burns RG, Davies JA (1986) Microbiology of soil structure. Biol Agric Hortic 3:95–113
Camel SB, Reyes-Solis MG, Ferrera-Cerrato R, Franson RL, Brown MS, Bethlenfalvay GJ (1991) The growth of mycorrhizal mycelium through bulk soil. Soil Sci Soc Am J 58:389–393
Cooke JC, Gemma JN, Koske RE (1987) Observations of nuclei in vesicular-arbuscular mycorrhizal fungi. Mycologia 79:331–333
Cooper KM (1976) A field survey of mycorrhizas in New Zealand ferns. NZ J Bot 14:169–181
Daniels BA, Trappe JM (1980) Factors affecting spore germination of the vesicular-arbuscular mycorrhizal fungus, *Glomus epigaeus*. Mycologia 72:457–471
Davies FT, Potter JR, Linderman RG (1992) Mycorrhiza and repeated drought exposure affect drought resistance and extraradical hyphae development of pepper plants independent of plant size and nutrient content. J Plant Physiol 139:289–294
El-Atrach F, Vierheilig H, Ocampo JA (1989) Influence of non-host plants on vesicular-arbuscular mycorrhizal infection of host plants and on spore germination. Soil Biol Biochem 21:161–163
Faber B, Zasoski RJ, Munns DN, Shackel K (1991) A method for measuring hyphal nutrient and water uptake in mycorrhizal plants. Can J Bot 69:87–94
Finlay R, Söderström B (1992) Mycorrhiza and carbon flow to the soil. In: Allen MF (ed) Mycorrhizal functioning: an integrative plant-fungal process. Chapman & Hall, New York, pp 134–160
Frey B, Schuepp H (1993) A role of vesicular-arbuscular (VA) mycorrhizal fungi in facilitating interplant nitrogen transfer. Soil Biol Biochem 25:651–658
Gemma JN, Koske RE (1988) Pre-infection interactions between roots and the mycorrhizal fungus *Gigaspora gigantea*: chemotropism of germ tubes and root growth response. Trans Br Mycol Soc 91:123–135
Gianinazzi-Pearson V, Gianinazzi S (1992) Influence of intergeneric grafts between host and non-host legumes on formation of vesicular-arbuscular mycorrhiza. New Phytol 120:505–508
Gianinazzi-Pearson V, Branzanti B, Gianinazzi S (1990) In vitro enhancement of spore germination and early hyphal growth of a vesicular-arbuscular mycorrhizal fungus by host root exudates and plant flavonoids. Symbiosis 7:243–255
Giovannetti M, Avio L, Sbrana C, Citernesi AS (1993) Factors affecting appressorium development in the vesicular-arbuscular mycorrhizal fungus *Glomus mosseae* (Nicol. & Gerd.) Gerd. & Trappe. New Phytol 123:115–122
Goodwin J (1992) The role of mycorrhizal fungi in competitive interactions among native bunchgrasses and alien weeds: a review and synthesis. Northwest Sci 66:251–260
Graham JH (1982) Effects of citrus root exudates on germination of chlamydospores of the vesicular-arbuscular mycorrhizal fungus, *Glomus epigaeum*. Mycologia 74:831–835
Graham JH (1987) Non-nutritional benefits of VAM fungi – do they exist? In: Sylvia DM, Hung LL, Graham JH (eds) Mycorrhizae in the next decade. Practical applica-

tions and research priorities. Proc 7th N A Conf on Mycorrhizae, Gainesville, pp 237–239

Graham JH, Leonard RT, Menge JA (1981) Membrane-mediated decrease in root exudation responsible for phosphorus inhibition of vesicular-arbuscular mycorrhiza formation. Plant Physiol 68:548–552

Hardie K (1985) The effect of removal of extraradical hyphae on water uptake by vesicular-arbuscular mycorrhizal plants. New Phytol 101:677–684

Harley JL, Smith SE (1983) Mycorrhizal symbiosis. Academic Press, New York

Ianson DC, Linderman RG (1993) Variation in the response of nodulating pigeonpea (*Cajanus cajan*) to different isolates of mycorrhizal fungi. Symbiosis 15: 105–119

Jakobsen I, Rosendahl L (1990) Carbon flow into soil and external hyphae from roots of mycorrhizal cucumber plants. New Phytol 115:77–83

Janos DP (1993) Vesicular-arbuscular mycorrhizae of epiphytes. Mycorrhiza 4:1–4

Klironomos JN, Kendrick WB (1993) Research on mycorrhizas: trends in the past 40 years as expressed in the MYCOLIT database. New Phytol 125:595–600

Koske RE (1982) Evidence for a volatile attractant from plant roots affecting germ tubes of a VA mycorrhizal fungus. Trans Br Mycol Soc 79:305–310

Koske RE, Gemma JN (1992) Fungal reactions to plants prior to mycorrhizal formation. In: Allen MF (ed) Mycorrhizal functioning: an integrative plant-fungal process. Chapman & Hall, New York, pp 3–36

Kough JL, Molina R, Linderman RG (1985) Mycorrhizal responsiveness of *Thuja*, *Calocedrus*, *Sequoia*, and *Sequoiadendron* species of western North America. Can J For Res 15:1049–1064

Lee P, Koske RE (1994) *Gigaspora gigantea*: parasitism of spores by fungi and actinomycetes. Mycol Res 98:458–466

LeTacon F, Skinner FA, Mosse B (1983) Spore germination and hyphal growth of a vesicular-arbuscular mycorrhizal fungus, *Glomus mosseae* (Gerdemann and Trappe), under decreased oxygen and increased carbon dioxide concentrations. Can J Microbiol 29:1280–1285

Linderman RG (1988a) Mycorrhizal interactions with the rhizosphere microflora: the mycorrhizosphere effect. Phytopathology 78:366–371

Linderman RG (1988b) The VA (vesicular-arbuscular) mycorrhizal symbiosis. ISI Atlas of Science. Animal and plant sciences section 1. Institute for Scientific Information, Philadelphia, pp 183–188

Linderman RG (1992) Vesicular-arbuscular mycorrhizae and soil microbial interactions. In: Bethlenfalvay GJ, Linderman RG (eds) Mycorrhizae in sustainable agriculture. ASA Spec Publ 54, American Society of Agronomy Inc, Crop Science Society of America Inc, and Soil Science Society of America Inc, Madison, Wisconsin, pp 45–70

Linderman RG (1994) Role of VAM fungi in biocontrol. In: Pfleger FL, Linderman RG (eds) Mycorrhizae and plant health. APS Press, St Paul, pp 1–26

Linderman RG, Gilbert RG (1975) Influence of volatiles of plant origin on soilborne plant pathogens. In: Bruehl GW (ed) Biology and control of soilborne plant pathogens. APS Press, St Paul, pp 90–99

Lynch JM, Whipps JM (1990) Substrate flow in the rhizosphere. Plant Soil 129:1–10

Malajczuk N, Linderman RG, Kough J, Trappe JM (1981) Presence of vesicular-arbuscular mycorrhizae in *Eucalyptus* spp. and *Acacia* sp., and their absence in *Banksia* sp. after inoculation with *Glomus fasciculatus*. New Phytol 87:567–572

Malloch D (1987) The evolution of mycorrhizae. Can J Plant Pathol 9:398–402

Martins MA (1993) The role of the external mycelium of arbuscular mycorrhizal fungi in the carbon transfer process between plants. Mycol Res 97:807–810

Marx DH, Schenck NC (1983) Potential of mycorrhizal symbiosis in agriculture and forest productivity. In: Kommedahl T, Williams PH (eds) Challenging problems in plant health. APS Press, St Paul, pp 334–347

Mayo K, Davis RE, Motta J (1986) Stimulation of germination of spores of *Glomus versiforme* by spore-associated bacteria. Mycologia 78:426–431

Meyer JR, Linderman RG (1986) Selective influence on populations of rhizosphere or rhizoplane bacteria and actinomycetes by mycorrhizas formed by *Glomus fasciculatum*. Soil Biol Biochem 18:191–196

Miller RM, Jastrow JD (1992) The role of mycorrhizal fungi in soil conservation. In: Bethlenfalvay GJ, Linderman RG (eds) Mycorrhizae in sustainable agriculture. ASA Special Publ 54, American Society of Agronomy Inc, Crop Science Society of America Inc, and Soil Science Society of America Inc, Madison, Wisconsin, pp 29–44

Miller RM, Jastrow JD (1994) Vesicular-arbuscular mycorrhizae and biogeochemical cycling. In: Pfleger FL, Linderman RG (eds) Mycorrhizae and plant health. APS Press, St Paul, pp 283–308

Morandi D, Bailey JA, Gianinazzi-Pearson V (1984) Isoflavonoid accumulation in soybean roots infected with vesicular-arbuscular mycorrhizal fungi. Physiol Plant Pathol 24:357–364

Mugnier J, Mosse B (1987) Vesicular-arbuscular mycorrhizal infection in transformed root-inducing T-DNA roots grown axenically. Phytopathology 77:1045–1050

Nair MG, Safir GR, Siqueira JO (1991) Isolation and identification of vesicular-arbuscular mycorrhiza-stimulatory compounds from clover (*Trifolium repens*) roots. Appl Environ Microbiol 57:434–439

Nelson CE, Safir GR (1982) Increased drought tolerance of mycorrhizal onion plants caused by improved phosphorus nutrition. Planta 154:407–413

Parke JL, Linderman RG (1980) Association of vesicular-arbuscular mycorrhizal fungi with the moss *Funaria hygometrica*. Can J Bot 58:1898–1904

Parra-Garcia MD, Giudice VL, Ocampo JA (1992) Absence of VA colonization in *Oxalis pes-caprae* inoculated with *Glomus mosseae*. Plant Soil 145:298–300

Paulitz TC, Linderman RG (1991) Mycorrhizal interactions with soil organisms. In: Arora DK, Rai B, Mukerji KG, Knudsen GR (eds) Advances in applied mycology vol 1. soil and plants. Marcel Dekker, New York, pp 77–129

Pirozynski KA, Dalpé Y (1989) Geologic history of the Glomaceae with particular reference to mycorrhizal symbiosis. Symbiosis 7:1–36

Rabatin SC, Stinner BR (1991) Vesicular-arbuscular mycorrhizae, plant and invertebrate interactions in soil. In: Barbosa P, Krishik VA, Jones CG (eds) Microbial mediation of plant herbivore interactions. John Wiley, New York, pp 141–168

Ratnayake M, Leonard RT, Menge JA (1978) Root exudation in relation to supply of phosphorus and its possible relevance to mycorrhizal formation. New Phytol 81:543–552

Read DJ (1992) The mycorrhizal mycelium. In: Allen MF (ed) Mycorrhizal functioning: an integrative plant-fungal process. Chapman & Hall, New York, pp 102–133

Rhodes LH, Gerdemann JW (1975) Phosphate uptake zones of mycorrhizal and non-mycorrhizal onions. New Phytol 75:555–561

Rhodes LH, Gerdemann JW (1978) Translocation of calcium and phosphate by external hyphae of vesicular-arbuscular mycorrhizae. Soil Sci 126:125–126

Schwab SM, Menge JA, Leonard RT (1983) Quantitative and qualitative effects of phosphorus on extracts and exudates of sudangrass in relation to vesicular-arbuscular mycorrhiza formation. Plant Physiol 73:761–765

Secilia J, Bagyaraj DJ (1987) Bacteria and actinomycetes associated with pot cultures of vesicular-arbuscular mycorrhizas. Can J Microbiol 33:1067–1073

Simon L, Bousquet J, Levesque RC, Lalonde M (1993) Origin and diversification of endomycorrhizal fungi and coincidence with vascular land plants. Nature 363:67–69

Siqueira JO, Hubbell DH, Kimbrough JW, Schenck NC (1984) *Stachybotrys chartarum* antagonistic to azygospores of *Gigaspora margarita* in vitro. Soil Biol Biochem 16:679–681

Siqueira JO, Sylvia DM, Gibson J, Hubbell DH (1985) Spores, germination, and germ tubes of vesicular-arbuscular mycorrhizal fungi. Can J Microbiol 31:965–972

St John TV, Hays RI, Reid CPP (1983) Influence of a volatile compound on formation of vesicular-arbuscular mycorrhizae. Trans Br Mycol Soc 81:153–154

Sutton JC, Sheppard BR (1976) Aggregation of sand-dune soil by endomycorrhizal fungi. Can J Bot 54:326–333

Tisdall JM (1991) Fungal hyphae and structural stability of soil. Aust J Soil Res 29:729–743

Tommerup IC (1985) Inhibition of spore germination of vesicular-arbuscular mycorrhizal fungi. Trans Br Mycol Soc 85:267–278

Tsai SM, Phillips DA (1991) Flavonoids released naturally from alfalfa promote development of symbiotic *Glomus* spores in vitro. Appl Environ Microbiol 57:1485–1488

Tylka GL, Hussey RS, Roncadori RW (1991) Axenic germination of vesicular-arbuscular mycorrhizal fungi: effects of selected *Streptomyces* species. Phytopathology 81:754–759

Vancura V, Orozco MO, Grauova O, Prikryl Z (1989) Properties of bacteria in the hyphosphere of a vesicular-arbuscular mycorrhizal fungus. Agric Ecosyst Environ 29:421–427

Viera A, Glenn MG (1990) DNA content of vesicular-arbuscular mycorrhizal fungal spores. Mycologia 82: 263–267

Whitfield FB, Shea SR, Gillen KJ, Shaw KJ (1981) Volatile components from the root of *Acacia pulchella* R. Br. and their effect of *Phytophthora cinnamomi* Rands. Aust J Bot 29:195–208

Wilson GWT, Hetrick BAD, Kitt DG (1989) Suppression of vesicular-arbuscular mycorrhizal fungus spore germination by nonsterile soil. Can J Bot 67:18–23

Young JL (1982) Survival of endomycorrhizal spores/propagules during long-term air-dry storage. Agron Abstr 1982:199

8 Endophytes of Forest Trees: a Model for Fungus-Plant Interactions

J. Stone[1] and O. Petrini[2]

CONTENTS

I.	Introduction	129
II.	Endophytes as Models for Ecological Studies	131
III.	Endophytes as Models for the Study of Fungus-Plant Interactions	132
A.	Infection Processes	132
	1. Recognition Mechanisms	133
	a. *Hypoxylon fragiforme-Fagus sylvatica*	133
	b. Attachment by *Discula umbrinella*	133
	2. Infection, Penetration, and Colonization	134
	a. *Rhabdocline parkeri*	134
	b. *Discula umbrinella*	135
B.	Population Biology Studies	135
IV.	Conclusions	136
	References	138

I. Introduction

The tissues of higher plants support complex, species-rich assemblages of microorganisms. Microfungi colonizing foliar and twig surfaces (epiphytes), internal foliage (foliar endophytes), young and old bark (bark endophytes), and wood (xylotropic endophytes and wood decomposers) are recognized as dominant components of these assemblages. Symptomless internal colonization of healthy plant tissues by microfungi is a widespread and well-documented phenomenon, and increasing interest in the ecological roles of these fungi has stimulated research in recent years. Fungi causing such infections in certain grasses, as well as those colonizing a variety of nongrass hosts, are both often referred to as endophytes or endophytic fungi, although the two groups have little in common. A historical account of the early use of the terms "endophyte" and "endophytic fungi" in relation to fungal pathogens and plant disease is given in Large (1940). In recent times, these terms have come to be used synonymously to refer to symptomless occupation of apparently healthy host plants by a variety of nonpathogenic fungi (Carroll 1986), although they are also sometimes used in a broader sense to include host colonization by weakly pathogenic species that may be symptomless only during a portion of the infection cycle. The term "quiescent infection" has been proposed for infections by latent pathogens when host-imposed mechanisms inhibit colonization by the pathogen until a particular stage of maturation (e.g., ripening, senescence) has been accomplished (Williamson 1994). In the broadest sense, endophytes are fungi colonizing living plant tissue without causing any immediate, overt negative effects (Hirsch and Braun 1992).

With reference to grass hosts (primarily Poaceae), "endophyte" denotes a particular type of systemic, nonpathogenic symbiosis. Grass endophytes provide their host a number of benefits that increase host fitness; in particular, production of fungal alkaloids that inhibit insect herbivory forms the basis of a protective mutualism (Chap. 10, Vol. V, Part A; Chaps. 12, 14, this Vol.). These fungi are taxonomically restricted primarily to the Clavicipitaceae and are derived from pathogenic species (Schardl et al. 1994; Tsai et al. 1994; Chap. 14, this Vol.). Grass endophytes cause systemic infections, extensively colonize host leaf, culm, and root tissue, and are seed-transmitted. Fungal sporulation on the host is supressed, and host and fungus together function essentially as a single organism.

Endophytes of nongrass hosts, on the other hand, represent a broad range of genera from several orders and families of ascomycetes or anamorphic form genera. Several genera and/or species are often recovered concurrently infecting the same host tissue (Carroll 1995). Although there are a few reports of systemic endophyte colonization of nongrass hosts (Bose 1947; Boursnell 1950), host colonization by these fungi is frequently very restricted; the endophyte thallus may consist of only a few cells (Fig. 1A,B). Colonization is

[1] Department of Botany and Plant Pathology, Oregon State University, Corvallis, Oregon 97330, USA
[2] Tera d'Sott 5, 6949 Comano, Switzerland

The Mycota V Part B
Plant Relationships
Carroll/Tudzynski (Eds.)
© Springer-Verlag Berlin Heidelberg 1997

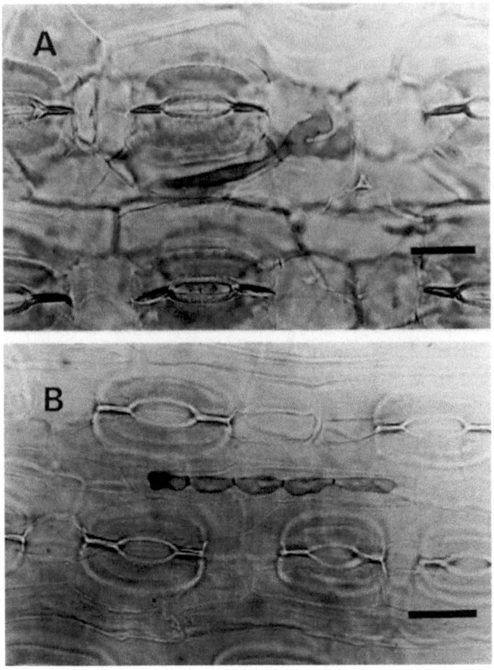

Fig. 1A,B. Endophytic fungal infections in conifer needles. **A** Intercellular infection of *Phyllosticta abietis* in *Abies grandis*. **B** Intracellular infection of *Rhabdocline parkeri* in Douglas-fir. *Bars* 25 μm

more often nonsystemic, and new or successive infections within hosts occurs by spores. Endophyte infections of this type are especially well documented for woody hosts, perennials, evergreen conifers, and hosts with long-lived or evergreen foliage. Similar infections are also found in graminaceous hosts, sometimes designated "p-endophytes" (Siegel et al. 1995).

Because of their symptomless nature, endophytic infections are often presumed to form mutualistic associations with their hosts in a manner similar to the endophytes of grasses. Metabolites that may inhibit herbivory by insects or colonization by other fungi, and thus form the basis for a mutualistic symbiosis, are produced in vitro by some endophytes (Noble et al. 1991; Johnson and Whitney 1994). It is clear from floristic and histological studies, however, that symptomless fungi indentified as endophytic represent a heterogeneous ecological group that encompasses a variety of symbiotic interactions with their hosts.

Most studies of endophytes have dealt with natural infections occurring in natural host populations. Such fungi represent endemic symbionts that have evolved in association with their hosts, and possess specialized adaptations to elude the general resistance mechanisms which serve to exclude most opportunistic fungi in the natural environment. Nevertheless, fungi regarded as pathogens of particular hosts are often isolated as endophytes, although these are usually not among the most abundant isolates from a given host. Strains of pathogens with impaired virulence may be considered endophytes (Freeman and Rodriguez 1993), as well as a number of commensal saprobic fungi and mutualistic species that have cryptic colonization patterns. The frequent occurrence of phytopathogenic fungi in endophyte associations emphasizes the heterogeneous ecology and the sometimes ambiguous distinction between latent pathogens and symptomless endophytes.

Some endophytic fungi also have evolved host recognition and infection mechanisms similar to those of the better-known host-specific biotrophic or necrotrophic fungi. These include host-specific attachment, spore germination in response to specific elicitors, and host-induced differentiation of infection structures. In contrast to typical pathogenic fungi, endophytes characteristically exhibit a prolonged, inconspicuous period of interrupted growth where colonization temporarily ceases and resumes following a physiological or maturational change in the host. Coadaptation and growth regulation between host and fungus are thus suggested, but their mechanisms remain unknown. This period of interrupted growth is a defining feature of endophytes, and may occur during any stage of infection, as with latent pathogens (Swinburne 1983). Indeed, the differences between many fungi considered as endophytes and those considered latent pathogens are slight and may simply reflect the duration of the latent or quiescent phase and the degree of injury sustained by the host during active growth of the fungus. Interruption of fungal growth may be under host control, fungal control, or some equilibrium between both. The long duration of quiescent phases of some endophytes suggests that active host defenses are involved and that the main difference between endophytic and pathogenic fungi lies in differential sensitivity to specific host resistance mechanisms. Active colonization and sporulation of many endophytes resumes following the onset of natural host senescence or induced necrosis, either in response to nutrient availability or removal of specific inhibitors.

II. Endophytes as Models for Ecological Studies

Endophytes are ideally suited for ecological investigations for a number of reasons. Firstly, they are extremely common in the plant kingdom and comparative ecological investigations on endophytic assemblages can be performed at several hierarchical levels, e.g., tissue, host, site, community, and ecosytem; secondly, a large number of them can be easily cultivated, which enables coordinated physiological, biochemical, and genetic investigations to be carried out. Thirdly, since many endophytic fungi are not of direct economic importance, they have rarely been subjected to any artificial selection pressure, yet many are apparently closely adapted to their hosts. Certain endophtye species may be highly host-specific while others are more generally distributed. Endophytes as a group thereby provide excellent subjects for comparative studies of population biology, evolutionary biology, and speciation of symbiotic fungi.

Detection and enumeration of endophyte infections have most often involved isolation of fungi from surface-sterilized tissue samples incubated on culture media (Petrini 1986). In fewer cases, a combination of microscopy and fungal isolations has been used to correlate histological colonization patterns with fungal taxa (Stone 1987, 1988). Biochemical methods could be a promising alternative for quantification of endophyte infections that may overcome some of the inherent problems associated with cultural or histological methods. Selective amplification of specific fungal nucleic acid sequences from infected host tissue may become a preferred approach, as has been unintentionally demonstrated (Smith and Klein 1994, 1996). The ITS1 region of the ribsomal DNA of gymnosperms is much larger (ca. 3000 bp), than that of most angiosperms and fungi (ca. 250 bp; Liston et al. 1996), which permits electrophoretic separation of host and fungal ITS from direct PCR amplification of conifer host tissue. Specific primers are useful for PCR amplification of conserved sequence motifs in certain fungal taxa and might be used to selectively amplify fungal DNA in mixed populations. Alternatively, techniques developed specifically to detect and quantify plant pathogens can also be readily adapted to the study of endophytes. Talbot et al. (1993) described a method to quantitatively determine the biomass of *Magnaporthe grisea* in planta during rice infection, based on mass-dependent differential DNA extraction from rice tissues and from *M. grisea*. This technique might be used to detect endophytes and to study fluctuations in endophyte infections in host tissues.

Because of the diffuse, sporadic nature of many endophyte infections, microscopic detection of hyphae in symptomless tissue requires either a clearing/staining procedure to scan large portions of host tissue (Stone 1988; Cabral et al. 1993), or tissue sectioning, Clearing procedures, however, may also remove host cuticle and also some fungal infection structures, making this approach unsuitable for some hosts (Viret 1993). Combined thick-thin sectioning, and combinations of SEM and TEM can be used to locate endophyte infections by light microscopy and subsequent ultrastructural examination (Stone 1988; Suske and Acker 1989). Understanding the structural and physiological aspects of the host-endophyte interaction also often depends on the ability to culture axenically both symbiont and host. Under natural conditions host tissue can be concurrently colonized by several different endophytic species, whereas colonization of axenic seedlings by single species can be experimentally controlled. This approach can then be useful in differentiating anatomical colonization patterns for comparison with samples from the field. Using field-collected material, immunological methods have been used to differentiate specific fungi in host tissue, such as *Naemacyclus minor* (= *Cyclaneusma minus*), *Lophodermium seditiosum*, and *Cenangium ferruginosum* in *Pinus sylvestris* (Franz et al. 1993) and *Lophodermium piceae* in tissue of *Picea abies* (Suske and Acker 1989). Observations of the complete infection cycle, from host penetration to sporulation, from either controlled inoculation studies or field-collected material in which all stages can be related to an identified stage, however, are known for only few endophyte species (Stone 1987, 1988; Cabral et al. 1993; Viret et al. 1993; Viret and Petrini 1994).

In phytocoenological mycofloristic surveys, the internal fungi of various hosts have been sampled methodically with the aim of describing the spatial distributions and temporal sequences of these fungal assemblages (Hirsch and Braun 1992). Investigators have tried to ascertain the underlying factors influencing colonization patterns of endophytic fungi in a manner analagous to synecological investigations of plant communities (Petrini 1991; Carroll 1995). Mycofloristic investigations of a variety of hosts by means of culture methods have been published (reviewed in Petrini 1991; Petrini et al. 1992; Carroll 1995). These ecological studies have generally emphasized endophyte distribution patterns according to host genera and families, specific habitat types (Petrini and Carroll 1981; Petrini et al. 1982; Petrini 1985), or of fungal taxa as endophytes (Petrini and Petrini 1985; Petrini 1986). Unfortunately, the importance of the microscopic colonization domain of endophytes with respect to sampling units and frequency estimates has only recently begun to be fully appreciated (Carroll 1995).

Infection frequencies for specific hosts have been related to foliage age (Stone 1987; Espinosa-Garcia and Langenheim 1990), host distribution (Petrini 1991; Bills and Polishook 1992; Rollinger and Langenheim 1993), and temporal and spatial variation patterns (Halmschlager et al. 1993; Wilson and Carroll 1994). Several studies have investigated the role of endophytic fungi in complex symbioses involving hosts, fungi, and insects (Todd 1988; Butin 1992; Wilson 1992), temporal successional patterns among endophytes (Petrini 1991; Halmschlager et al. 1993), and the influence of anthropogenic factors on endophytic assemblages (Petrini 1991). Helander et al. (1993, 1994) recently reported that air pollutants apparently affect the frequency of colonization of birch leaves and pine needles by epiphytic fungi, whereas the species composition of endophyte assemblages seems not to be significantly influenced.

Populations of endophytic fungi have been studied using morphological characters such as spore size (Toti et al. 1992a) and cultural characters including growth on different media and spot tests to detect extracellular enzyme production (Sieber et al. 1991; Sieber-Canavesi et al. 1991). Although informative, these methods have failed to fully explain population structures in endophyte taxa. More recently, biochemical and DNA amplification methods have been applied to studies of endophytes as an alternative to culturing and direct microscopy. Isozyme analyses of both extracellular (pectinase, amylase) and intracellular enzymes have been carried out with selected endophytic fungi to detect taxonomic affinities such as biological species (Sieber-Canavesi et al. 1991; Brunner and Petrini 1992; Leuchtmann et al. 1992; Rodrigues et al. 1993) or host-specific strain formation (Toti 1993). Detailed studies on the structure and composition of endophytic populations in beech have revealed patterns of geographic differentiation of strains within the endophyte *Discula umbrinella* (Berk. et Br.) Morelet [teleomorph: *Apiognomonia errabunda* (Rob.) Hoehnel], the most common beech (*Fagus sylvatica* L.) leaf endophyte. Shamoun and Sieber (1993), using isoenzyme analysis, have detected no difference between endophytic and disease syndrome associated (DSA) isolates of *Melanconium marginale*. Similar studies have been conducted with grass endophytes (Chap. 12, this Vol.). Recent studies on population biology of endophytes have used DNA amplification methods such as random amplified polymorphic DNA (RAPD) and RFLP (Haemmerli et al. 1992; McCutcheon et al. 1993) or somatic incompatibility and isozyme analyses (Rodrigues et al. 1993, 1995) to observe genetic variation in endophyte populations from individual hosts. These studies have demonstrated the occurrence of multiple genotypes of the same endophyte species concurrently infecting the same tissue.

III. Endophytes as Models for the Study of Fungus-Plant Interactions

A. Infection Processes

Colonization of plant tissue by fungi depends on a complex sequence of events, including recognition of the host surface by spores, germination and penetration. These processes are based on a physical or chemical signal transfer between host and fungus that commences at spore germination (Chap. 2, Vol. V, Part A). Chemical signals that elicit fungal responses typical of the infection process, such as germination or development of infection structures, are often present in plant extracts (Stockwell and Hanchey 1983; Stewart et al. 1989). The cell walls of plants contain, in addition to low-molecular weight compounds and proteins, also complex molecules such as glycoproteins and many polysaccharides (Brett and Waldron 1990) that may be recognized by fungi. Compatibility and defense mechanisms both may be mediated by substances present on the cell surfaces of both plant and microorganism, and such compounds may influence directly the success of the infection.

Espinosa-Garcia and Langenheim (1991a,b) investigated the effects of leaf essential oils of coastal redwood (*Sequoia sempervirens*) on the growth of several leaf-inhabiting fungi. They detected differential growth effects and hypothesized that leaf essential oil components may be responsible for controlling endophyte growth.

1. Recognition Mechanisms

a. Hypoxylon fragiforme – Fagus sylvatica

Fungi in the Xylariaceae have consistently been found as asymptomatic endophytes of many plants (Petrini 1986). The genus *Hypoxylon*, for example, contains both pathogenic (antagonistic) and neutral symbionts (Rogers 1979) and provides useful models for the study of fungus-host specific recognition mechanisms. In *Hypoxylon fragiforme*, the initial step of infection is mediated by a complex sequence of events collectively termed eclosion (Fig. 2; Chapela et al. 1990). Chapela et al. (1991) have demonstrated that the eclosion of ascospores is mediated by two monolignol glucosides extracted from the bark of the host. Consistent with the presence of a host signal, higher levels of eclosion of *H. fragiforme* ascospores were obtained with beech leaves compared to 16 other plants and inorganic materials. Eclosion could be mediated by a Na^+-sensitive receptor to host-derived monolignol glucosides, and could be related to a permeability change in the wall/membrane complex (Chapela et al. 1993). As soon as the

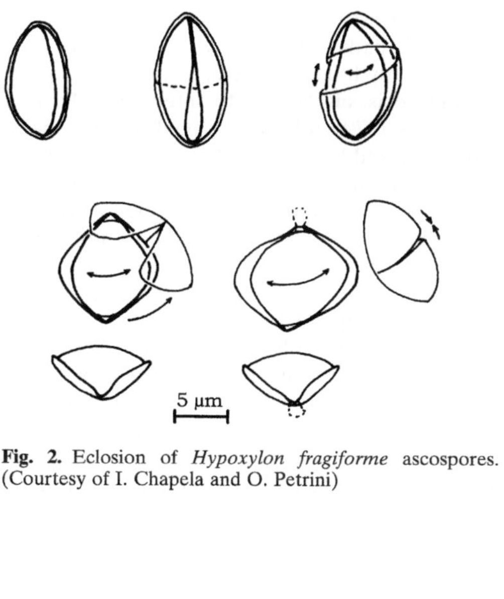

Fig. 2. Eclosion of *Hypoxylon fragiforme* ascospores. (Courtesy of I. Chapela and O. Petrini)

spore body is released from the exosporium, it swells and exposes an inner, plastic cell wall layer through the germ-fissure. In contrast to the dormant ascospore, the surface exposed upon eclosion binds readily both chitin and glucose/mannose-specific lectins (Chapela et al. 1993). Lectin-like molecules are known to mediate recognition and binding in a number of microbe-plant symbioses (Hamer et al. 1988; Diaz et al. 1989). Thus, the exosporial shell of *Hypoxylon* prevents its ascospores from being recognized by nonhost surfaces, and eclosion renders the ascospores recognizable only when the appropriate host is found. In addition, the relatively thick, pigmented outer wall may permit a prolonged period of ascospore dormancy on the appropriate host surface until threshold levels of host monolignol glucosides are present to trigger eclosion, indicating suitable conditions for infection to take place.

b. Attachment by Discula umbrinella

Adhesion of many fungi to hydrophobic host surfaces is related to the presence around spores and/or mycelium of a conspicuous extracellular matrix, mainly composed of glycoprotein and carbohydrates (Nicholson and Epstein 1991; Chap. 1, Vol. V, Part A). The role of an extracellular matrix in adhesion has been demonstrated by TonThat and Epstein (1991) with adhesive-reduced mutants of *Nectria haematococca*. Hamer et al. (1988) have reported that a concanavalin A-binding mucilage present on the spore tip is responsible for the attachment of *Magnaporthe grisea* conidia to the host surface. Conidia of *Rhabdocline parkeri* (an endophyte of Douglas-fir) are surrounded by a viscous sheath, visible as an attachment mucilage under TEM during early infection (Stone 1988).

The role of cellular adhesives in the infection process of a foliar endophyte has been investigated in detail in only one model to date. Toti et al. (1992b) showed that the conidia of *Discula umbrinella* attached differentially to the surfaces of various hosts and showed a marked specificity for beech. The adhesion was both pH- and temperature-dependent (Toti 1993). Electron microscopy investigations confirmed the presence on the conidial surface of a conspicuous sheath in which polysaccharides, as well as proteins (probably in their glycosylated form), are present (Toti et al. 1992b; Toti 1993). Adhesion of conidia to the host surface could be effectively inhibited by different lectins, and enzymatic treatment of the conidial sheath resulted in sharply reduced attachment and changes in the binding of fluorescence-labeled lectins. Digestion with helicase modified only slightly the structure of the sheath, while proteases completely dissolved the fibrillar sheath leaving a comparatively smooth cell wall. Viret et al. (1994) concluded that the extracellular sheath is responsible for the adhesion of *Discula umbrinella* conidia to the host and that glycoproteins may be involved in the recognition and attachment process.

The attachment mechanism of *D. umbrinella* is similar, with respect to host specificity, to the recognition system described for the xylotropic endophyte *Hypoxylon fragiforme* (Chapela et al. 1990), and also parallels many fungal pathogens that have been investigated. Esterases, glucosidases, peroxidases (Louis and Cooke 1985), and cutinases (Kolattukudy 1985) have been shown to be produced by fungal pathogens during infection and are presumed to be involved in the digestion of the host cell walls. Toti (1993) similarly demonstrated the presence of nonspecific esterases in the extracellular sheath of *D. umbrinella*, which may function as cutinases. Recently, Nicholson and Epstein (1991; Chap. 1, Vol. V, Part A) have postulated that the adhesion of fungal spores to the host surface may be triggered by the activity of small amounts of cutinase present in the spore matrix and in the conidial sheaths. This "sensing" cutinase may liberate small amounts of cutin monomers, which in turn could act as transcriptional inducers of the bulk synthesis of cutinase required for enzymatic penetration of the cuticle (Koeller 1991). Cuticular degradation could, in turn, enhance adhesion by providing a physical cast or by freeing new sites for chemical bonding. Whether this model applies broadly to endophytic species remains to be demonstrated.

2. Infection, Penetration, and Colonization

a. Rhabdocline parkeri

Rhabdocline parkeri is a common and widespread endophyte in the foliage of Douglas-fir, its only known host, in northwestern North America. Upon germination, conidia of the *Meria* anamorph of *R. parkeri* become two-celled by the formation of a median septum. In one of the cells, additional wall layers are formed, which become melanized; the second cell, which is apparently void of contents, typically degenerates and collapses as the melanized cell develops. Development proceeds with the formation of a small lateral germ hypha, which also becomes melanized and differentiates to form a short, lateral appressorium. From this appressorium, a fine penetration hypha with an outer wall layer contiguous with the inner layer of the appressorium wall and an inner layer contiguous with a newly synthesized inner wall near the base of the appressorium emerges. The outer wall layer terminates just below the penetration point in the host cuticle. A fine penetration hypha with a single wall layer continues to grow below this point, directly penetrating the epidermal cell wall (Stone 1988). Colonization of needles proceeds with the penetration of epidermal cells, which are killed as a result of invasion. The infection hypha eventually enlarges to occupy the entire lumen of the epidermal cell. Further colonization is interrupted at this point in healthy foliage. Infections are visible in cleared tissue as separate, discreet intracellular infections of single epidermal cells (Fig. 1; Stone 1987).

Differentiation of *R. parkeri* conidia apparently occurs in response to a chemical signal, but it is unclear whether this is triggered by specific host recognition or is a more generalized phenomenon. Conidia germinate on culture media without the initial cell division and melanization that is observed in conidia germinating on host needle surfaces. Aqueous extracts of host needles, as well as metal ion chelators, reported as stimulants of germination and appressorium formation in *Colletotrichum musae* by Swinburne (1986), increase the percentages of differentiated conidia on artificial media (Stone 1986).

In contrast to the pattern of infection of the congeneric pathogen *Meria laricis* on *Larix* spp., in which only foliage younger than 4 weeks postemergence is susceptible to infection (McBride and Hayes 1979), foliage of Douglas-fir is repeatedly reinfected by conidia of *R. parkeri* as it ages. Needles are initially infected during their first growing season and density of infected epidermal cells increases logarithmically on both upper and lower surfaces with needle age (Stone 1987). Thus, needles become increasingly more susceptible to infection with age. A relatively minor proportion, less than 5%, of the total needle epidermis is infected however, even in older foliage of highly infected trees (Fig. 3). The discreet nature of these infections was confirmed by McCutcheon et al. (1993), who demonstrated that the accumulated infections in a single older needle may contain a

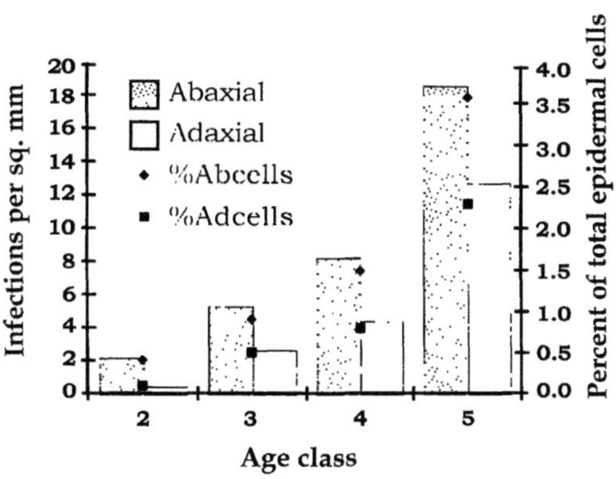

Fig. 3. Relationship between needle age and density of infection sites of *Rhabdocline parkeri* as infections per mm^2 and percent of total epidermal cells. (Data from Stone 1987)

number of unique genotypes. Similar infections restricted to single epidemal cells has also been described for *Stagonospora innumerosa* infecting *Juncus effusus* and an unidentified endophyte of *Juncus bufonius* (Cabral et al. 1993), and an unidentified endophyte of *Sequoia sempervirens* (J.K. Stone, unpubl.).

At the onset of senescence, i.e., in attached but yellowed needles, active growth of *R. parkeri* resumes with the invasion of adjacent cells by haustoria. Colonization proceeds to contiguous cells until needle abscission, when conidiophores of the *Meria* anamorph emerge from the stomata of abscised needles. The normal life span of Douglas-fir needles may exceed 7 years, and thus the quiescent phase of *R. parkeri* may commonly exceed 5 years (Stone 1987).

b. Discula umbrinella

Viret et al. (1993) investigated the sequence of events leading to infection by *Discula umbrinella* on leaves taken from axenically grown beech seedlings. Conidial germination on the host or on other surfaces varied from approx. 70 to 100%, whereas the germination rate of conidia suspended in water was never higher than 5%. Incubation of conidia in beech leaf extracts induced higher rates of germination, as compared to water control, even if no physical contact with a support was established, suggesting the presence of a diffusable inducer of germination in the host. Experiments carried out with extracts of other hosts (Viret et al. 1993, 1994) similarly suggest that induction of germination is a host-specific event. On beech leaves, four stages of germination could be observed between 16 and 24 h after infection. Ungerminated conidia, germinating conidia producing long superficial hyphae, conidia with distinct, appressorium-like swellings at the end of the germ tube, and germinated conidia with a halo around the tip of the germ tube were present. Transmission electron microscopy showed, however, that only the germ hyphae with halos were associated with sites of subculticular penetration. While conidial germ tubes were able to penetrate the cuticular layers of the outer edges of the guard cells, direct penetration through the stomata was rarely observed.

Penetration and colonization of the host tissue by *D. umbrinella* was often preceded by the formation of an appressorium at the end of the germ tube, but direct penetration without appressorium formation, with subsequent establishment of the fungus under the cuticle, was also observed (Viret and Petrini 1994). When appressoria were seen, a fine penetration hypha originating from the appressorium pierced the cuticle and the epidermal cell wall. The fungus established itself in the reticulate part of the cuticle, beneath the amorphous layer of the cuticle. At this stage the underlying parenchyma cells did not show signs of degradation, and Viret and Petrini (1994) postulated that *D. umbrinella* could remain as a quiescent subcuticular infection until appropriate conditions such as senescence of the tissues permit further development of the endophyte.

B. Population Biology Studies

The almost ubiquitous and abundant presence of selected endophyte taxa on different hosts has attracted the attention of mycologists interested in taxonomy and population biology. Endophytic fungi are often difficult to identify not only because of the lack of adequate literature but also because cultures often fail to develop diagnostic features, making the use of morphological and cultural characters alone often inadequate.

In the past few years, combined cultural, morphological and biochemical analyses have been used to study genetic variation in endophyte populations with respect to host and geographic distributions, and also apparent speciation in endophytes. Sieber-Canavesi et al. (1991) used growth on different media, substrate utilization patterns, water-soluble protein electrophoretic patterns, as well as isoenzyme analysis of extracellular enzymes to characterize a population of *Leptostroma* isolates morphologically almost indistinguishable from each other. These authors concluded that the population studied was composed of several distinct biological species of *Leptostroma*. Differences in electrophoretic profiles were indicative of an ongoing process of host-specific differentiation similar to that described for selected pathogens (e.g., Micales et al. 1987). Similar results were reported for *Melanconium* spp. isolated from *Alnus* spp. in Europe and Canada (Sieber et al. 1991). Rodrigues et al. (1993) examined 13 species of *Xylaria* by isozyme analysis and detected a high degree of intra- and interspecific diversity among them. In addition, using a combination of biochemical and cultural data, as well as vegetative compatibility group analysis Rodrigues (1992) characterized 35 isolates of *Xylaria cubensis*, and demonstrated that several genotypes of this taxon can be detected in a single tree. Leuchtmann et al. (1992) found high levels of infraspecific variation in isozymes of *Phyllosticta* spp. from conifers that reflected host-specific strains. Using RFLP analysis, Wilson et al. (1994) identified 27 different genotypes of morphologically indistinguishable *Lophodermium pinastri* (*Leptostroma* anamorph) isolates from red pine from three sites in eastern Canada.

Differences among *D. umbrinella* isolates derived from different hosts were detected using pectic enzyme electrophoretic patterns (Toti 1993) and corroborated the results obtained with morphometric studies (Toti et al. 1992a). Isozyme analysis of *Discula umbrinella* isolates showed a remarkably high homogeneity, and thus suggested a low gene flow, even within populations derived from the same host (Toti 1993). Haemmerli et al. (1992), using randomly amplified polymorphic DNA (RAPD) and restriction fragment length polymorphisms (RFLP), detected up to four different genotypes infecting the same leaf. McCutcheon et al. (1993) similarly used RAPD markers to compare genetic variation in *R. parkeri* isolates from old-growth and young trees. Four to 7 different genotypes were found among 9 isolates from single needles of old growth Douglas-firs. Fewer genotypes were found from isolated mature trees and young trees regenerating in clearcuts far from mature stands than in mature old growth trees and trees regenerating near mature stands, demonstrating the importance of mature trees in maintaining a reservoir of genetic heterogeneity in this species.

IV. Conclusions

Endophytic microfungi form symbiotic associations with higher plants. Endophytic fungi are characterized by having a phase of interrupted growth following infection, or at least in having a cryptic phase where asymptomatic colonization of living host tissue occurs. Their presence on host tissue ordinarily becomes apparent only after the onset of natural senescence or induced necrosis. Endophyte species assemblages are comprised of a heterogeneous ecological and taxonomic group in which species typical of phylloplane or cauloplane mycobiota, latent pathogens, and early establishing saprotrophic species occur together with host specific, nonpathogenic colonists. Some of the more host-specific endophytes possess specialized mechanisms for host recognition, attachment, and colonization. Accumulated evidence suggests that many of these fungi have evolved mechanisms for coordinated communication and regulation with their hosts. To date, however, most studies have concentrated primarily on identifying factors affecting abundance and distribution of endophytes. Few studies have examined the structural and physiological aspects of these symbioses.

The structural aspects of colonization of host tissue by most endophytes are largely unknown, since most investigations have relied on cultural isolations. Structural aspects of foliar endophyte-host associations of several hosts indicate that host colonization is very limited in healthy host tissue, but the mechanisms of growth limitation remain unclear (Carroll 1986; Stone 1986). Structural studies have covered only a few of the potential hosts and much more research is needed before firm generalizations on colonization patterns can be made.

Of the studies published on the histology of tree endophytes, one has been carried out in a natural system (Stone 1987, 1988), while others have used detached leaf systems in which the endophyte was stimulated to cause symptoms in order to detect it (Toti 1993; Viret 1993). The latter studies have thus emphasized the transition phase between quiescent and active colonization phase.

Understanding the phenomena that control the transition between quiescent and active phases will probably disclose information that will be helpful in understanding general aspects of fungal infection processes and mechanisms of host symbiont recognition-communication by asymptomatic endophytes as well as latent pathogens. Techniques such as those described by Talbot et al. (1993) and DNA amplification technologies should be useful in detecting endophytic fungi in host tissues, understanding their population dynamics, and assessing species diversity.

Expression specificity, the formation of fruiting structures or other identifiable morphological features only on a restricted number of plant taxa, has now been described for a number of endophytes, particularly in the Xylariaceae (Petrini and Petrini 1985; Whalley 1993). This important component of host specificity, often overlooked in pathogen-plant associations, is widespread and evident in endophytic systems. Anamorphic *Xylaria* and *Hypoxylon* spp., for example, are commonly reported as endophytic colonists of hosts from which the teleomorphs are entirely unknown. *Hypoxylon fragiforme*, until recently only known from its saprotrophic phase, establishes in an inconspicuous form within the wood of practically every beech tree (*Fagus sylvatica*) sampled in Europe (Chapela et al. 1988).

The ecological roles of endophytes are diverse and varied. Endophytic fungi have been described as playing a protective role against insect herbivory not only in grasses (Clay 1990) but also in conifers (Carroll 1991) and many are potential producers of novel antimicrobial secondary metabolites (Fisher et al. 1984a; Polishook et al. 1993). Fisher et al. (1984b) have reported antibacterial or antifungal activity for more than 30% of endophytic isolates from ericaceous plants and Dreyfuss (1986) has reported antibiotic activity from an endophytic isolate of *Pleurophomopsis* and of *Cryptosporiopsis* sp., as well as from a sterile endophyte from *Abies alba.* Dreyfuss (1986) also described the production of a new family of cytochalasines in endophytic *Xylaria* spp. from South America and Mexico. High frequencies of biological activities from endophytic isolates of *Xylaria* spp. have been reported by Brunner and Petrini (1992). Production of antifungal or antibacterial substances as well as compounds active against insects has been demonstrated in vitro (e.g. Johnson and Whitney 1994) but to date no investigation has been undertaken to detect the importance of such metabolites in planta. Future investigations might include studies aiming at detecting the production of antibiotics and pest deterrents in planta as a first step toward evaluating the ecological significance of secondary metabolite production by endophytes and the potential use of endophytes in biological control. An important outstanding question concerning the expression of these novel substances by endophytic fungi is whether they are produced during the quiescent phase of host occupation, or only during active growth, which generally occurs after the onset of host senescence and would thus be of minimal utility for biological control.

The early occupation of host tissue prior to either natural senescence or induced necrosis gives endophytes an advantage over saprobes normally excluded from healthy tissue. Endophytes that are quiescent during the normal lifespan of deciduous host organs are immediately able to intercept and utilize host metabolites mobilized during early senescence. Competitive interactions with later invading saprobic fungi may account for the apparent widespread production of antifungal metabolites by endophytic fungi. Other metabolites produced by some endophytes can modify host growth responses and accelerate or delay senescence (Petrini et al. 1992). Endophytes might therefore prove interesting subjects for the study of coordinated control of plant growth regulating and anti-fungal metabolites.

The high degree of diversity of fungal saprobes and epiphytes described up to now suggests that a large number of endophytic species can also be expected from detailed investigations of host plants. Additional survey work is needed and will augment our knowledge of endophyte biota and of fungal diversity. Investigations employing biochemical detection and differentiation methods will probably reveal a greater biodiversity within endophyte taxa that are either apparently ubiquitous or which can infect only selected hosts, as indicated by a number of studies showing that host-selective strain formation is a common trait of endophytes (Petrini et al. 1992). Population genetics will probably be an area in which endophyte research will flourish in the next decade. As noted above, endophytic fungi present an excellent group of symbiotic organisms that have not been subjected to artificial selection pressures, and thus are excellent subjects for testing hypotheses about population genetics and evolutionary biology in natural populations. Studies on the distribution of individual genotypes involving either vegetative compatibility testing, isozyme or DNA analysis can easily be carried out using forest endophytes. The presence in forests of large plant populations and natural sources of inoculum can be exploited to plan experiments otherwise difficult to perform in agricultural systems, where often the influence of anthropogenic factors is a limiting factor.

Finally, forest-endophyte systems are ideally suited to carry out investigations on the stability of endophytic symbioses. Recent studies suggest that the same host placed outside its native range may well be colonized by different endophyte assemblages. For instance, an investigation carried out with *Eucalyptus nitens* introduced in Britain and native in Australia showed that *Cytospora eucalypticola*, the main colonizer of this host in Australia, was not present to any extent in the British samples (Fisher et al. 1993). In addition, the endophyte assemblages of the trees planted in the United Kingdom appeared to be depauperate and consisted of species different from those found in trees collected in the native habitats. These results confirm earlier observations by Carroll et al. (1977) and Espinosa-Garcia and Langenheim (1990). The comparison of hosts growing in their native habitats with their counterparts present outside their native range could help to understand the dynamics of endophyte transmission and distribution.

References

Bills GF, Polishook JD (1992) Recovery of endophytic fungi from *Chamaecyparis thyoides*. Sydowia 44:1–12

Bose SR (1947) Hereditary (seed-borne) symbiosis in *Casuarina equisetifolia* Forst. Nature 159:512–514

Boursnell JG (1950) The symbiotic seed-borne fungus in the Cistaceae. I. Distribution and function of the fungus in the seedling and in the tissues of the mature plant. Ann Bot 14:217–242

Brett CT, Waldron K (1990) Physiology and biochemistry of plant cell walls. Topics in plant physiology 2. Unwin Hyman, London, 194 pp

Brunner F, Petrini O (1992) Taxonomy of some *Xylaria* species and xylariaceous endophytes by isozyme electrophoresis. Mycol Res 96:723–733

Butin H (1992) Effect of endophytic fungi from oak (*Quercus robur* L.) on mortality of leaf-inhabiting gall insects. Eur J For Pathol 22:237–246

Cabral D, Stone JK, Carroll GC (1993) The internal mycobiota of *Juncus* spp. microscopic and cultural observations of infection patterns. Mycol Res 97:367–376

Carroll FE, Müller E, Sutton BC (1977) Preliminary studies on the incidence of needle endophytes in some European conifers. Sydowia 29:87–103

Carroll GC (1986) The biology of endophytism in plants with particular reference to woody perennials. In: Fokkema NJ, van den Heuvel J (eds) Microbiology of the phyllosphere. Cambridge University Press, Cambridge, pp 205–222

Carroll GC (1991) Fungal associates of woody plants as insect antagonists in leaves and stems. In: Barbosa P, Krischik VA, Jones CG (eds) Microbial mediation of plant-herbivore interactions. Wiley, New York, pp 253–271

Carroll GC (1995) Forest endophytes: pattern and process. Can J Bot 73:1316–1324

Chapela IH, Boddy L, Rayner ADM (1988) Structure and development of fungal communities in beech logs four and a half years after felling. FEMS Microbiol Ecol 53:59–70

Chapela IH, Petrini O, Petrini LE (1990) Unusual ascospore germination in *Hypoxylon fragiforme*: first steps in the establishment of an endophytic symbiosis. Can J Bot 68:2571–2575

Chapela IH, Petrini O, Hagmann L (1991) Monolignol glucosides as specific recognition messengers in fungus-plant symbioses. Physiol Mol Plant Pathol 39:289–298

Chapela IH, Petrini O, Bielser G (1993) The physiology of ascospore eclosion in *Hypoxylon fragiforme*: mechanisms in the early recognition and establishment of an endophytic symbiosis. Mycol Res 97:157–162

Clay K (1990) Fungal endophytes of grasses. Annu Rev Ecol Syst 21:275–295

Diaz CL, Melchers LS, Hooykaas PJJ, Lugtenberg BJJ, Kijne JW (1989) Root lectin as a determinant of host-plant specificity in the *Rhizobium*-legume symbiosis. Nature 338:579–581

Dreyfuss MM (1986) Neue Erkenntnisse aus einem pharmakologischen Pilz-screening. Sydowia 39:22–36

Espinosa-Garcia FJ, Langenheim J (1990) The endophytic fungal community in leaves of a coastal redwood population – diversity and spatial patterns. New Phytol 116:89–97

Espinosa-Garcia FJ, Langenheim J (1991a) Effect of some essential leaf oil phenotypes from coastal redwood on growth of its predominant endophytic fungus, *Pleuroplaconema* sp. J Chem Ecol 17:1837–1857

Espinosa-Garcia FJ, Langenheim J (1991b) Effect of some leaf essential oil phenotypes in coastal redwood on the growth of several fungi with endophytic stages. Biochem Syst Ecol 19:629–642

Fisher PJ, Anson AE, Petrini O (1984a) Novel antibiotic activity of an endophytic *Cryptosporiopsis* sp. isolated from *Vaccinium myrtillus*. Trans Br Mycol Soc 83:145–187

Fisher PJ, Anson AE, Petrini O (1984b) Antibiotic activity of some endophytic fungi from ericaceous plants. Bot Helv 94:249–253

Fisher PJ, Petrini O, Sutton BC (1993) A comparative study of fungal endophytes in leaves, xylem, and bark of *Eucalyptus nitens* in Australia and England. Sydowia 45:338–345

Franz F, Grotjahn R, Acker G (1993) Identification of *Naemacyclus minor* hyphae within needle tissues of *Pinus sylvestris* by immunoelectron microscopy. Arch Microbiol 160:265–272

Freeman S, Rodriguez RJ (1993) Genetic conversion of a

fungal plant pathogen to a nonpathogenic, endophytic mutualist. Science 260:75–78

Haemmerli UA, Braendle UE, Petrini O, McDermott JM (1992) Differentiation of isolates of *Discula umbrinella* (teleomorph: *Apiognomonia errabunda*) from beech, chestnut and oak using RAPD markers. Mol Plant-Microbe Interact 5:479–483

Halmschlager E, Butin H, Donaubauer E (1993) Endophytische Pilze in Blättern und Zweigen von *Quercus petraea*. Eur J For Pathol 23:51–63

Hamer JE, Howard RJ, Chumley FG, Valent B (1988) A mechanism for surface attachment in spores of a plant pathogenic fungus. Science 239:288–290

Helander ML, Neuvonen S, Sieber T, Petrini O (1993) Simulated acid rain affects birch leaf endophyte populations. Microb Ecol 26:227–234

Helander ML, Sieber TN, Petrini O, Neuvonen S (1994) Endophytic fungi in Scots pine needles: spatial variation and consequences of simulated acid rain. Can J Bot 72:1108–1113

Hirsch GU, Braun U (1992) Communities of parasitic microfungi. In: Winterhoff W (ed) Handbook of vegetation science, vol 19, Fungi in vegetation science. Kluwer, Dordrecht, pp 225–250

Johnson JA, Whitney NJ (1994) Cytotoxicity and insecticidal activity of endophytic fungi from black spruce (*Picea mariana*) needles. Can J Microbiol 40:24–27

Koeller W (1991) The plant cuticle: a barrier to be overcome by fungal plant pathogens. In: Cole GT, Hoch HC (eds) The fungal spore and disease initiation in plants and animals. Plenum Press, New York, pp 219–246

Kolattukudy PE (1985) Enzymatic penetration of the plant cuticle by fungal pathogens. Annu Rev Phytopathol 23:223–250

Large EC (1940) The advance of the fungi. Holt, New York

Leuchtmann A, Petrini O, Petrini LE, Carroll GC (1992) Isozyme polymorphism in six endophytic *Phyllosticta species*. Mycol Res 96:287–294

Liston A, Robinson WA, Oliphant JM, Alvarez-Buylla ER (1996) Length variation in the nuclear ribosomal DNA internal transcribed spacer of non-flowering seed plants. Syst Bot 21:1–12

Louis I, Cooke RC (1985) Conidial matrix and spore germination in some plant pathogens. Trans Br Mycol Soc 84:661–667

McBride RP, Hays AJ (1979) Interactions of the leaf pathogen *Meria laricis* with nutrients and microorganisms of the larch phylloplane. Nova Hedwigia 31:507–517

McCutcheon TL, Carroll GC, Schwab S (1993) Genotypic diversity in populations of a fungal endophyte from Douglas-fir. Mycologia 85:180–186

Micales JA, Stipes RJ, Bonde MR (1987) On the conspecificity of *Endothia eugeniae* and *Cryptosphaeria cubensis*. Mycologia 79:707–720

Nicholson RL, Epstein L (1991) Adhesion of fungi to the plant surface: prerequisite for pathogenesis. In: Cole GT, Hoch HC (eds) The fungal spore and disease initiation in plants and animals. Plenum Press, New York, pp 3–23

Noble HM, Langley D, Sidebottom PJ, Lane SJ, Fisher PJ (1991) An echinocandin from an endophytic *Cryptosporiopsis* sp. and *Pezicula* sp. in *Pinus sylvestris* and *Fagus sylvatica*. Mycol Res 95:1439–1440

Petrini L, Petrini O (1985) Xylariaceous fungi as endophytes. Sydowia 38:216–234

Petrini O (1985) Wirtsspezifität endophytischer Pilze bei einheimischen Ericaceae. Bot Helv 95:213–238

Petrini O (1986) Taxonomy of endophytic fungi of aerial plant tissues. In: Fokkema NJ, van den Heuvel J (eds) Microbiology of the phyllosphere. Cambridge University Press, Cambridge, pp 175–187

Petrini O (1991) Fungal endophytes of tree leaves. In: Andrews JH, Hirano SS (eds) Microbial ecology of leaves. Springer, Berlin Heidelberg New York, pp 179–197

Petrini O, Carroll GC (1981) Endophytic fungi in foliage of some Cupressaceae in Oregon. Can J Bot 59:629–636

Petrini O, Stone J, Carroll FE (1982) Endophytic fungi in evergreen shrubs in western Oregon: a preliminary study. Can J Bot 60:789–796

Petrini O, Sieber TN, Toti L, Viret O (1992) Ecology, metabolite production, and substrate utilization in endophytic fungi. Nat Toxins 1:185–196

Polishook JD, Dombrowski AW, Tsou NN, Salituro GM, Curotto JE (1993) Preussomerin D from the endophyte *Hormonema dematiodes*. Mycologia 85:62–64

Rodrigues KF (1992) Endophytic fungi in the tropical palm *Euterpe oleracea* Mart. PhD Thesis, City University of New York, New York

Rodrigues KF, Leuchtmann A, Petrini O (1993) Endophytic species of *Xylaria*: cultural and isozymic studies. Sydowia 45:116–138

Rodrigues KF, Petrini O, Leuchtmann A (1995) Variability among isolates of *Xylaria cubensis* as determined by isozyme analysis and somatic incompatibility tests. Mycologia 87:592–596

Rogers JD (1979) The Xylariaceae: systematic, biological and evolutionary aspects. Mycologia 71:1–42

Rollinger JL, Langenheim JH (1993) Geographic survey of fungal endophyte community composition in leaves of coastal redwood. Mycologia 85:149–156

Schardl CL, Leuchtmann A, Tsai H-F, Collett MA, Watt DM, Scott DB (1994) Origin of a fungal symbiont of perennial ryegrass by interspecific hybridization of a mutualist with the ryegrass choke pathogen, *Epichloë typhina*. Genetics 136:1307–1317

Shamoun S, Sieber TN (1993) Isozyme and protein patterns of endophytic and disease syndrome associated isolates of *Melanconium apiocarpum* and *Melanconium marginale* collected from alder. Mycotaxon 49:151–166

Sieber TN, Sieber-Canavesi F, Petrini O, Ekramoddoullah AKM, Dorworth CE (1991) Characterization of Canadian and European *Melanconium* from some *Alnus* species by morphological, cultural, and biochemical studies. Can J Bot 69:2170–2176

Sieber-Canavesi F, Petrini O, Sieber TN (1991) Endophytic *Leptostroma* species on *Picea abies*, *Abies alba*, and *Abies balsamea*: a cultural, biochemical, and numerical study. Mycologia 83:89–96

Siegel MR, Schardl CL, Phillips TD (1995) Incidence and compatibility of nonclavicipitaceous fungal endophytes in *Festuca* and *Lolium* grass species. Mycologia 87:196–202

Smith DE, Klein AS (1994) Phylogenetic inferences on the relationship of North American and European *Picea* species based on nuclear ribosoimal 18S sequences and the 18S internal transcribed spacer 1 region. Mol Phylogenet Evol 3:17–26

Smith DE, Klein AS (1996) Erratum. Mol Phylogenet Evol 5:286–287

Stewart A, Backhouse D, Sutherland PW, Fullerton RA (1989) The development of infection structures of *Sclerotium cepivorum* on onion. J Phytopathol 126:22–32

Stockwell V, Hanchey P (1983) The role of the cuticle in resistance of beans to *Rhizoctonia solani*. Phytopathology 73:1640–1642

Stone JK (1986) Foliar endophytes of Douglas-fir: cytology and physiology of the host-endophyte relationship. PhD Thesis, University of Oregon, Eugene

Stone JK (1987) Initiation and development of latent infections by *Rhabdocline parkeri* on Douglas-fir. Can J Bot 65:2614–2621

Stone JK (1988) Fine structure of latent infections by *Rhabdocline parkeri* on Douglas-fir, with observations on uninfected epidermal cells. Can J Bot 66:45–54

Suske J, Acker G (1989) Endophytic needle fungi: culture, ultrastructural and immunocytochemical studies. In: Schulze E-D, Lange OL, Oren R (eds) Ecological studies, vol 77. Springer, Berlin Heidelberg New York, pp 121–136

Swinburne TR (1983) Quiescent infections in post-harvest diseases. In: Dennis C (ed) Post-harvest pathology of fruits and vegetables. Academic Press, New York, pp 1–21

Swinburne TR (1986) Stimulation of disease development by siderophores and inhibition by chelated iron. In: Swinburne TR (ed) Iron, siderophores and plant disease. Plenum Press, New York, pp 217–226

Talbot NJ, Ebbole DJ, Hamer JE (1993) Identification and characterization of MPG1, a gene involved in pathogenicity from the rice blast fungus *Magnaporthe grisea*. Plant Cell 5:1575–1590

Todd D (1988) The effects of host genotype, grwoth rate, and needle age on the distribution of a mutualistic, endophytic fungus in Douglas-fir plantations. Can J For Res 18:601–605

TonThat CC, Epstein L (1991) Adhesion-reduced mutants and the wild-type *Nectria haematococca*: an ultrastructural comparison of the macroconidial walls. Exp Mycol 15:193–205

Toti L (1993) The symbiosis *Discula umbrinella/Fagus sylvatica*: biochemical, ecological and morphological studies of the host endophyte relationship. PhD Thesis, Swiss Federal Institute of Technology, Zürich

Toti L, Chapela IH, Petrini O (1992a) Morphometric evidence for host-specific strain formation in *Discula umbrinella*. Mycol Res 96:420–424

Toti L, Viret O, Chapela IH, Petrini O (1992b) Differential attachment by conidia of the endophyte, *Discula umbrinella* (Berk. & Br.) Morelet, to host and non-host surfaces. New Phytol 121:469–475

Tsai H-F, Liu J-S, Staben C, Christensen MJ, Latch GCM, Siegel MR, Schardl CL (1994) Evolutionary diversification of fungal endophytes of tall fescue grass by hybridization with *Epichloë* species. Proc Natl Acad Sci USA 91:2542–2546

Viret O (1993) Infection of beech leaves by the endophyte *Discula umbrinella* (Berk. et Br.) Morelet [teleomorph: *Apiognomonia errabunda* (Rob.) Hoehn.]: an ultrastructural study. PhD Thesis, Swiss Federal Institute of Technology, Zürich

Viret O, Petrini O (1994) Colonisation of beech leaves (*Fagus sylvatica*) by the endophyte *Discula umbrinella* (Teleomorph: *Apiognomonia errabunda*). Mycol Res 98:423–432

Viret O, Scheidegger C, Petrini O (1993) Infection of beech leaves (*Fagus sylvatica*) by the endophyte *Discula umbrinella* (teleomorph: *Apiognomonia errabunda*). Low temperature scanning electron microscopy studies. Can J Bot 71:1520–1527

Viret O, Toti L, Chapela IH, Petrini O (1994) The role of the extracellular sheath in recognition and attachment of conidia of *Discula umbrinella* (Berk. & Br.) Morelet to the host surface. New Phytol 127:123–131

Whalley AJS (1993) Tropical Xylariaceae: their distribution and ecological characteristics. In: Isaac S, Frankland JC, Watling R, Whalley AJS (eds) Aspects of tropical mycology. Cambridge University Press, Cambridge, pp 103–120

Wilson D (1992) On endophyte-insect-plant interactions. PhD Thesis, University of Oregon, Eugene

Wilson D, Carroll GC (1994) Infection studies of *Discula quercina*, an endophyte of *Quercus garryana*. Mycologia 86:635–647

Wilson R, Wheatcroft R, Miller JD, Whitney NJ (1994) Genetic diversity among natural populations of endophytic *Lophodermium pinastri* from *Pinus resinosa*. Mycol Res 98:740–744

Williamson B (1994) Latency and quiescence in survival and success of fungal plant pathogens. In: Blakeman JP, Williamson B (eds) Ecology of plant pathogens. CAB International, Oxford, pp 187–207

The Consequences of Fungal Associations in Plant Populations

9 Plant Pathogenic Fungi in Agroecosystems

J.T. English[1] and J.J. Marois[2]

CONTENTS

I.	Introduction	143
II.	Roles of Pathogenic Fungi	144
III.	Advancements and Contributions of Epidemiology	146
A.	Characterization of the Epidemic Process	146
B.	Characterization of the Environment	146
C.	Quantification of Epidemics and Component Processes	147
D.	Characterization of Pathogen Populations	150
E.	Characterization of Plant Populations	151
IV.	Considerations for the Future	152
V.	Conclusions	154
	References	154

I. Introduction

Epidemics induced by plant pathogenic fungi have had dramatic impacts on human populations worldwide. The earliest records of epidemics appeared well before 500 B.C., but it was not until the mid-19th century that fungi were recognized as disease-causing agents (Whetzel 1918; Ainsworth 1981). Since that time, a massive amount of information has accumulated to provide insights into the biology of fungi as pathogens and processes of disease development. Despite this knowledge base, pathogenic fungi still limit the production of agronomically important plant species, and disease management still requires tremendous inputs of labor and money.

The impacts of pathogenic fungi on agroecosystems surpass mere reductions in food and fiber production. In many instances, epidemics have caused dramatic shifts in economic, social, and political patterns (Table 1). One such epidemic, that of potato late blight, caused by the soilborne fungus, *Phytophthora infestans*, contributed greatly to changes in the human demographic patterns of Ireland. The disease first appeared in continental Europe in the early summer of 1845, and by mid-September it had spread to Ireland. The rapid spread of the disease coincided with frequent, cool rains that, in combination with winds, rapidly dispersed the infective zoosporangia produced on diseased plants. Although the biology of *P. infestans* is now well characterized, at the time it was questionable if the fungus could cause disease. Heated debates among scientists, agriculturalists, politicians, and journalists throughout Europe continued for many years as to whether late blight was a disease caused by a microorganism or whether it represented an interaction of the genetic constitution of potato plants with unusual environmental conditions (Bourke 1991). It was not until 1861 that de Bary (1861, 1863) demonstrated the pathogenic nature of *P. infestans*. During the debate, late blight and societal factors in Ireland combined to cause widespread starvation that, over a 10-year period, led to a drop in population of 3 million, 1 million of whom died, and 2 million of whom emigrated to the United States, Canada, and elsewhere. After more than a century, this epidemic continues to stimulate studies of the interactions of disease and societal function (e.g., Willigan 1977). Additionally, studies on the population biology of *P. infestans* (Goodwin et al. 1994; Chap. 2, this Vol.) are ongoing, with current emphasis on the basis of genetic changes in populations across broad geographic areas.

The debate on the nature of potato late blight focused attention not only on the damage that can be caused by a pathogenic fungus, but also on the importance of disease as a population phenomenon and on the complexities of interactions among components of agroecosystems that lead to epidemic development.

As plant pathology has developed as a scientific discipline, understanding of cellular and genetic aspects of pathogenesis has become progressively clearer. However, numerous unan-

[1] Department of Plant Pathology, 108 Waters Hall, University of Missouri, Columbia, Missouri 65211, USA
[2] North Florida Research and Education Center, University of Florida, Quincy, Florida 32351-9529, USA

The Mycota V Part B
Plant Relationships
Carroll/Tudzynski (Eds.)
© Springer-Verlag Berlin Heidelberg 1997

Table 1. Historically important epidemics. (After Campbell and Madden 1990)

Year(s)	Epidemic and consequence	Reference
857	First recorded epidemic of ergotism: thousands died in the Rhine Valley	Carefoot and Sprott (1967)
1039	Ergotism in France: monks of the Order of St. Anthony relieved many symptoms	Carefoot and Sprott (1967)
1722	Ergotism at Astrakhan: aided in defeat of Peter the Great of Russia	Carefoot and Sprott (1967)
1845–1846	Late blight of potato: Irish potato famine; 1 million Irishmen died of starvation and related maladies, another 2 million emigrated	Bourke (1964), Carefoot and Sprott (1967), Woodham-Smith (1962)
1845–1860	Powdery mildew of grape in England and France: financial loss and importation of *Phylloxera* aphid from North America	Large (1940), Carefoot and Sprott (1967)
1882–1885	Downy mildew of grape in France: financial loss and discovery of Bordeaux mixture	Large (1940), Carefoot and Sprott (1967)
1870–1880	Coffee rust in Ceylon: financial ruin for planters; English people became primarily tea drinkers	Large (1940), Carefoot and Sprott (1967)
1904-present	Chestnut blight in the United States: destruction of American chestnut as a forest tree species in eastern United States; financial loss	Hepting (1974)
1913	Leaf spot on banana cultivar Gros Michaels in the Sigatoka Valley in Fiji: financial loss	Carefoot and Sprott (1967)
1915–1923 and 1930–1935	Panama disease of bananas in Costa Rica, Panama, Colombia, and Guatemala: financial loss	Carefoot and Sprott (1967)
1916–1917	Late blight of potato in Germany: food shortages in civilian population; contributing factor to demoralization of German troops in World War I	Carefoot and Sprott (1967)
1930-present	Dutch elm disease in the United States: loss of American elm as a shade tree species in many areas; loss of property value	Carefoot and Sprott (1967)
1942–1943	Leaf blight of rice in Bengal: great Bengal famine – nearly 2 million people died from starvation	Padmanabhan (1973)
1951	Ergotism at Pont-St.-Esprit, France: 4 deaths, 32 cases of insanity, numerous cases of hallucinations	Fuller (1968)
1970	Southern maize leaf blight in the United States: 15% of US maize crop lost	Horsfall (1972)
1977–1978	Ergotism in Ethiopia: hallucinations, human suffering, and death	Demeke et al. (1979)
1979–1980	Blue mold of tobacco in eastern United States and Canada: financial loss	Lucas (1980)

swered questions about the pathways by which discrete plant-fungal interactions translate into impacts on agroecosystems remain. Are pathogenesis and the defense responses of a plant examined in the laboratory the same as those of a plant in the midst of a dense population in an agricultural field or forest? If not, what factors contribute to the differences? And finally, what are the implications of studying plant-fungal interactions within the framework of populations?

Botanical epidemiology is the subdiscipline of plant pathology in which these and related questions are addressed. Epidemiological concepts and methodologies enable the establishment of linkages between fungal influences on individual plants and subsequent temporal and spatial behaviors of diseased plant populations. Inquiry at this level is a logical step in the hierarchy of investigative approaches of plant-fungus interactions.

This chapter focuses on three topics: the effects of pathogenic fungi on growth and development of individual plants; insights into epidemic development and agroecosystem function; and finally, future directions of epidemiological research and study of agroecosystem behavior.

II. Roles of Pathogenic Fungi

Fungal diseases affect plants that occur in natural and managed communities; in both situations pathogenic fungi affect plant life histories. The biotic and abiotic conditions of these two environments can differ dramatically, and the patterns of disease development and their impacts on plant life histories may also differ.

Several authors have suggested conceptual frameworks for characterizing pathogenic fungi and their influences on plant birth, growth, and reproduction. For instance, Burdon (1993) categorized pathogenic fungi and other microorganisms according to their modes of influence on host plants. Categories included castrators, that reduce fecundity of individual flowers, and killers, that eliminate seedlings before or after emerging from the soil surface. Debilitators included pathogens that cause discrete lesions which affect plant development to only a limited extent unless they occur in sufficiently great numbers. Jarosz and Davelos (1995) similarly characterized pathogens in relation to influence on plant life histories, relating influences to specific categories of disease symptom.

Within any of these categories, there can occur fungi that differ in life history and degree of specificity for plant interaction. A specific pathogen can also be included in more than one category, depending on the stage of plant development, stress, and prevailing environmental conditions. *Phytophthora sojae*, for instance, can function as a killer of emerging soybean seedlings, but the fungus also functions as a debilitator when it infects plants at later stages of maturity.

Boote et al. (1983) characterized pathogens and their influences on plant growth and development in a manner that is more appropriate for quantitative modeling of the plant-fungus interaction. These influences were categorized relative to specific plant tissues and plant growth processes, and most fungal influences were considered at the level of individual plants. For example, pathogenic fungi can affect plant growth and reproduction by reducing rates of photosynthesis, accelerating leaf senescence, "stealing" light, sapping tissues of assimilates, consuming plant tissues, or reducing plant cell turgor. However, pathogenic fungi also can reduce the productivity of a plant population by killing young seedlings and reducing plant densities.

This view of plant-pathogen interaction shifted attention to the quantification of very specific processes. Boote et al. (1983) used this approach, for instance, to estimate impacts of leaf spot necrosis, caused by *Cercospora* spp., on photosynthesis within peanut canopies. These fungi reduced photosynthetic rates through acceleration of leaf senescence rates, light stealing, and direct effects of fungal toxins on photosynthetic reactions. More recently, Pinnschmidt et al. (1995) used this system of classification to characterize the combined impacts of multiple pests, including pathogenic fungi, on rice growth and production.

These systems of categorization assume that environmental conditions are appropriate for infection and disease development. Generally, however, the behavior of a pathogen, its host, and their interactions are modified by less-than-ideal conditions. For instance, the behavior of a pathogen that comes into contact with an individual of a susceptible host species can be constrained directly by the condition of the plant. The developmental pattern of the plant determines whether specific tissues have been formed and whether the tissues are susceptible to infection at any point in time. This is the case for emerging seedling radicles and hypocotyls of numerous plants that are susceptible to infection by species of the soilborne fungi *Rhizoctonia*, *Pythium*, and *Fusarium*. Infection of seedlings usually results in death; however, after plant tissues have matured, debilitation is more common. Similarly, within the canopy of a plant, species of *Botrytis* and other pathogenic fungi cannot infect and kill shoots that are beyond some point of maturity; nor can they infect floral tissues that do not exist at the time that infective spores are dispersed in the plant population.

Of particular importance in managed plant systems, the potential for infection by a specific pathogenic fungus also is constrained by the genetic structure of the established plant population. This post-plant contribution is encountered for diseases of both roots (Schmitthenner et al. 1994) and foliage (Shaner 1983).

A plant or a population of plants also may influence the occurrence of disease indirectly through its influence on conditions, complex in structure, within root systems or within the canopy. Although numerous scenarios have been suggested for modes of influence of plant canopy structure on microenvironment and, subsequently, disease, it is difficult to quantify these effects under conditions in the field. One of the few examples in which the microenvironment has been quantified in relation to canopy structure and disease development was for grape vineyards in California. In these vine canopies, grape bunch rot, caused by *Botrytis cinerea*, could be managed effectively by removing leaves from around fruit clusters (Gubler et al. 1987). Leaf removal significantly altered conditions of air movement, humidity, and temperature to reduce disease to a point at which fungicides were not needed (English et al. 1989).

The soil environment is similarly complex in environmental structure. In recent studies multivariate approaches have proven useful in characterizing the combined influences of biotic and abiotic factors on the abilities of specific soilborne fungi to function effectively as biological control agents against a range of diseases (Fravel and Marois 1986; Eastburn and Butler 1988).

III. Advancements and Contributions in Epidemiology

Epidemiologists view disease in plant populations within a framework of interrelated processes and subprocesses. It is this hierarchical view of biology that has made epidemiological approaches important. As might be expected, much of the conceptual and technical evolution in this area of study has occurred in relation to disease in managed plant systems. Ainsworth (1981), Campbell and Madden (1990), and Whetzel (1918) summarized significant aspects of the history of epidemiology that emphasize the sequential recognition of host plants, pathogens, and the environment as components of epidemic development. The remainder of this section considers examples of significant contributions of epidemiology.

A. Characterization of Epidemic Processes

In the early development of the discipline of plant pathology, pathologists focused their efforts on establishing that microorganisms of various types were capable of causing diseases of plants. This coincided with the evolution of the germ theory in animal systems. However, pathologists rapidly moved beyond a focus on etiology and became interested in pathogenesis. Pathogenesis can be examined at the level of individual plants or as a population phenomenon. Examination of pathogenesis at the population-level presupposed that the sequence of events required for epidemic development were well defined. One of the earliest formal schemes for describing these events was presented by Gäumann (1950), who described the infection chain, the chain of events that link an infected plant to spread of disease to surrounding healthy plants. Gäumann's descriptive system included survival of a pathogen in the absence of its host plant, initial transmission of the pathogen to a susceptible plant, and infection with subsequent disease development and further transmission.

This line of thought has evolved such that contemporary introductory texts on plant pathology consider that epidemics require a succession of stages that make up the disease cycle. These stages, as defined by Agrios (1988), include inoculation, penetration of a host plant, establishment of infection, colonization of host tissue, pathogen growth and reproduction, pathogen dissemination, and survival in the absence of host plants. The manner in which the stages of a disease cycle proceed is reflected in the pattern of disease that develops in a population of susceptible plants.

The concept of a disease cycle now seems obvious to plant pathologists. However, it remains important because it provides a central focus for evaluating the myriad factors that influence epidemics.

B. Characterization of the Environment

The environment in which a pathogenic fungus contacts a susceptible plant constrains the extent to which any pathogenic interaction takes place. Ainsworth (1981) indicated that the limiting effects of environment on disease development were not always appreciated. The environment was the last component of disease to be considered by pathologists, and this recognition did not occur until the early part of the 20th century. Until this time, investigations had, to a great degree, emphasized attributes of plants and pathogens in relation to disease development.

Some of the earliest studies of the environment focused on its relationships to diseases caused by pathogenic fungi. For example, Jones et al. (1926) characterized the relationship of soil temperature to root diseases caused by several soilborne fungi including species of *Fusarium*, *Rhizoctonia*, and *Plasmodiophora*. During that same year, Keitt and Jones (1926) summarized 5 years of investigation of the relationships of climatic factors to the biology of *Venturia inaequalis* and its ability to initiate apple scab epidemics, relating temperature and moisture duration to release of ascospores from perithecia, germination efficiency of ascospores, and plant infection.

There is still much to learn about the influences of the environment on disease, and the level of examination of these influences has become increasingly sophisticated. In early work with soilborne fungi, Jones et al. (1926) acknowledged that when they manipulated soil temperature, soil moisture and aeration also changed. Beyond statements recognizing environmental complexity, few attempts were made to incorporate it into experimental design. In fact this was difficult to do, considering the technological constraints to replicating more than a small number of controllable environmental variables in factorially designed experiments. In addition, considering the vast array of potentially important biotic and abiotic variables in a given situation, it is not necessarily obvious which ones should be experimentally manipulated.

The availability of increased computing power and sophisticated analytical methodologies has enabled investigators to begin characterizing and integrating the range of biotic and abiotic factors that may be important in understanding plant-fungal interactions in complex environments. An example of this was provided by Fravel and Marois (1986), who investigated the influences of environmental factors on the ability of a fungus, *Talaromyces flavus*, to become established in soil. Interest in this fungus arose from its known efficacy as a biological control agent against a variety of soilborne diseases. Using multivariate statistical approaches, these authors sampled populations of the fungus from a number of different soils and related them to 23 chemical, physical, and biotic attributes of each environment. Of these attributes, they found that survival and proliferation of *T. flavus* in any soil was related to cation exchange capacity, concentrations of potassium, sodium, and zinc, and total population densities of soilborne bacteria. Eastburn and Butler (1988) also used multivariate approaches to characterize soil microenvironmental factors in relation to the distribution of *Trichoderma harzianum*, another fungus with biological control potential.

In the case of diseases that occur above ground, English et al. (1989) were able to distinguish the microclimatic conditions of grapevine canopies that were conducive or inhibitory for development of grape bunch rot. However, differences between the canopy microenvironments were not discernible when variables known to affect the biology of *B. cinerea* were compared individually; rather, differences were detected only by considering simultaneous fluctuations in wind, humidity, temperature, and leaf wetness. Although complex in interpretation, the requirement of multiple factors to distinguish environments was more realistic. This information collected from the field, in combination with controlled laboratory experiments (Thomas et al. 1988; English et al. 1990), suggested mechanisms by which these environmental factors interact to affect the reproductive ability of *B. cinerea* and, thus, its ability to disperse and cause further disease in vineyards.

In many cases, a single or small group of environmental factors is known to regulate the development of epidemics caused by a specific pathogen. Mathematical characterization of these relationships has become very important in the development of models that predict outbreaks of disease (Gutierrez and DeVay 1986; Campbell and Madden 1990). Predictive models are now routinely incorporated into field-based computers that continuously record microclimatic conditions and provide instantaneous risk assessments for multiple diseases specific to a crop at any one site.

C. Quantification of Epidemics and Component Processes

Many factors contribute to the way in which an epidemic develops in a population of plants, including the pattern of initial dispersal of fungal propagules in the population and the environmental conditions at sites of initial pathogen contact and plant infection. Variability in these and other factors makes any epidemic unique. However, there are some common temporal and spatial attributes of epidemics that are useful for understanding pathogenesis. The characterization of these attributes represents one of the major areas of advancement in pathology in recent years.

In the early years of epidemiological research, spatial and temporal patterns of disease were considered independently. Many of the initial studies of temporal aspects of epidemics focused on diseases incited by pathogenic fungi. For example, Large (1952) characterized temporal patterns of potato late blight development during seasonal epidemics. He faced difficulties in comparing epidemics because disease progress curves, representing patterns of disease increase, typically were nonlinear. He found, however, that it was possible to linearize these curves by plotting amounts of disease over time on probability paper (Fig. 1). Once linearized, the rate of an epidemic's development could be evaluated by examining the slope of the transformed data. By this approach, Large was able to compare epidemics among regions and to evaluate the impacts of various disease management strategies.

Descriptions of spatial attributes of plant diseases appeared at least as early as did temporal descriptions. Van der Plank (1947), for instance, evaluated several methods that had been suggested for assessing the randomness of diseased plants in a population. He emphasized that detection of nonrandomness in patterns of diseased plants would be indicative of the contagiousness of any disease that resulted from cyclic dispersal of pathogen propagules within the plant population.

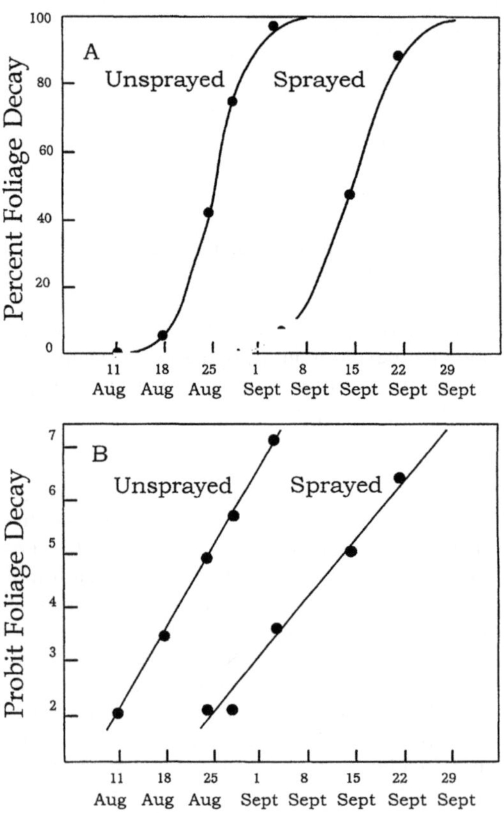

Fig. 1A,B. Progression of potato late blight with and without the application of copper fungicides as foliar protectants. **A** Temporal increase of diseased foliar tissue. **B** Linearized form of disease progression achieved by applying the probit transformation. (After Large 1945)

At about this same time, mechanisms of fungal spore liberation and patterns of dispersal were being examined. Ingold (1971) summarized important aspects of both short- and long-distance dispersal that were addressed in the early literature, with particular reference to pathogenic fungi including *Puccinia graminis tritici* (Craigie 1945) and *Cronartium ribicola* (Buchanan and Kimmey 1938), both hetero basidiomycetous fungi that cause rust diseases in plants, and *P. infestans* (Bonde and Schultz 1943). In addition to their economic importance, pathogenic fungi proved to be useful model organisms because their patterns of dispersal were reflected as gradients of diseased plants at increasing distances from sources of inoculum. Gregory (1945) provided a means for applying a mathematical power law to describe the typical pattern of declining spore density or disease from a source of inoculum. Although this descriptive approach was empirical, he provided biological interpretations for various gradient patterns.

Beyond these and a limited number of other initial efforts, described by Campbell and Madden (1990), interest in quantitative analyses of spatial and temporal attributes of disease did not increase significantly until the early 1960s, at which time Van der Plank (1963) published the first book on theoretical plant disease epidemiology entitled *Plant Diseases: Epidemics and Control.* In this book, Van der Plank provided a theoretical treatment of the relationship of pathogen biology and pattern of disease increase in a plant population. Examples used in the elaboration of this theory were based, in part, on epidemics caused by pathogenic fungi, including *P. infestans*, cause of potato late blight, *P. graminis tritici*, cause of wheat stem rust, and *Fusarium oxysporum* f. sp.*vasinfectum*, cause of cotton wilt. The inclusion of this group of pathogens in the development of initial epidemiological theory demonstrated that organisms exhibiting diverse modes of survival, dispersal, and reproduction could be evaluated within a common conceptual framework.

Van der Plank (1963) considered a number of important aspects of epidemics, including the relative rates of disease increase in a population, the relationships of growth and reproduction of pathogenic fungi and the likelihood of epidemic development, and host disease resistance. Zadoks and Schein (1979) also considered these and other

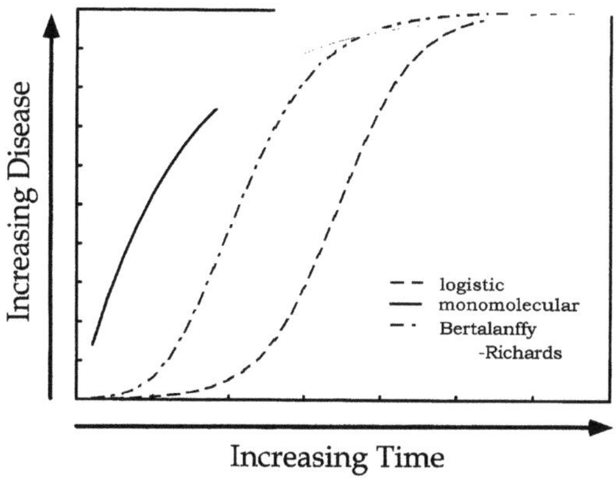

Fig. 2. Cumulative disease described by logistic, monomolecular, and Bertalanffy-Richards models. Flexible models, such as the Bertalanffy-Richards, describe a variety of patterns of disease progression, including those described by the logistic or monomolecular models

topics in a later text that provided an expanded ecological orientation to epidemic analysis. Van der Plank (1963) developed two mathematical models that proved to be very important in stimulating interest in analysis of disease progression over time. The first of these, the logistic model (Fig. 2), described polycyclic diseases that are incited by pathogens that reproduce on infected plant tissue and disperse to propagate additional disease in the plant population. The monomolecular model (Fig. 2) described diseases that result from infection derived from a single generation of a pathogen. By the application of the appropriate model, it was possible to linearize disease progress curves and obtain estimates of relative rates of disease increase.

The models proved useful in describing and comparing epidemics caused by pathogenic fungi in a variety of plant systems, and they proved useful in evaluating biological factors that contribute to specific patterns of disease. As pathologists gained experience with temporal analysis, they found that epidemics caused by any specific pathogenic fungus might not be characterized consistently by the logistic or monomolecular model (Campbell et al. 1984). Consequently, additional models were developed that provided greater flexibility for representation and comparison of epidemics (Fig. 2). Madden (1986) and Campbell and Madden (1990) provided detailed overviews of the considerations involved in using and developing models.

Coincident with the development of temporal analysis, there occurred renewed emphasis on the quantification of pathogen dispersal and the development of disease foci from initial infections in a population. For example, McCartney and Fitt (1985) and Minogue (1989) developed mechanistically based models to describe fungal dispersal on the basis of physical laws that characterize small-particle movement. Davis and Main (1986) described the application of atmospheric trajectory analysis to evaluate and predict the annual dispersal of *Peronospora tabacina* from the Caribbean to northern areas of the United States, where it incites blue mold on tobacco. On a much reduced scale, Ferrandino (1993), Aylor and Ferrandino (1990), and van den Bosch et al. (1988) developed methods of evaluating the expansion of disease from initial points of occurrence in a population.

The pattern of disease occurrence in a plant population reflects not only the pattern of fungal dispersal, but also the interaction of factors that enhance or inhibit disease development once inoculum has arrived. Increasingly sophisticated methods have been developed to characterize the degree of aggregation that disease often exhibits. Madden (1989) and Campbell and Benson (1994) provided an overview of concepts of random or other distributions of disease that are the basis of various indices of aggregation. These authors also described the more recent use of spatial-temporal autocorrelation methods to provide a view of the dynamics of simultaneous changes in disease over time and space. These and related methods have seen increasing application, in particular to diseases incited by soilborne pathogenic fungi, including *Cylindrocladium crotalariae*, cause of root rot on peanut (Griffin and Tomimatsu 1983), *Phytophthora nicotianae*, cause of pineapple heart rot (Chellemi et al. 1988), and *Phytophthora cactorum*, cause of leather rot of strawberry (Reynolds and Madden 1988). Most recently, two-dimensional distance analysis has been introduced as a means of characterizing spatial dimensions and arrangements of diseased plants. Based on binomial data for presence or absence of disease symptoms, this approach has proven useful for characterizing spatial patterns of foliar disease of clover caused by a complex of pathogens, including fungi (Nelson and Campbell 1993), and for characterizing disease symptom expression by bell pepper plants during epidemics incited by *Phytophthora capsici* (Ristaino et al. 1994).

D. Characterization of Pathogen Populations

Fungal populations can be characterized in relation to several attributes that affect their potential to incite disease. For example, the genetic structure of a population of a fungal species at a location obviously is important in deciding the outcome of plant-fungus interactions and how best to deploy appropriate resistance genes in plants. Also of importance is the spatial distribution of a pathogen population, which can affect the pattern of disease that develops within a plant population.

The spatial distribution of aerially dispersed fungi is dynamic because of the ease with which these organisms can be moved. However, a different situation exists in soil. Here, with the exception of oomycetous fungi, populations of soilborne pathogenic fungi consist of nonmotile propagules. Thus, patterns of plant infection and disease development strongly reflect the pattern of root system development in relation to pathogen distribution. Baker et al. (1967) stimulated interest in this subject area by developing several models to represent possible interactions between plant roots and pathogen populations. These models included scenarios in which susceptible infection sites on roots were either immovable or moved through soil that contained either motile or nonmotile propagules of a specific pathogen. For simplicity, these models assumed that the infective propagules of any fungus were distributed uniformly in the soil volume. From this and other assumptions, various families of curves were derived to represent the relationship of pathogen density and plant infection.

The relationships of infection to pathogen or inoculum density, as described in Baker's models, were used to interpret the biology of interactions of fungal populations with their host plants in terms of infection process and the influence of soil edaphic factors and competing microorganisms on the interaction (Rouse and Baker 1978). Of particular interest was the debate that these models stimulated concerning the relationship of rhizosphere size to infection by soilborne pathogens (Gilligan 1979; Grogan et al. 1980; Leonard 1980; Ferris 1981).

Many of these initial population descriptions focused on spatial attributes at a fine scale with a view towards elucidating microsite-level interactions of plant and pathogen within the rhizosphere. With time, however, attention shifted to considerations of spatial distributions at broader scales within the two-dimensional framework of agriculture fields. Using the methodologies de-

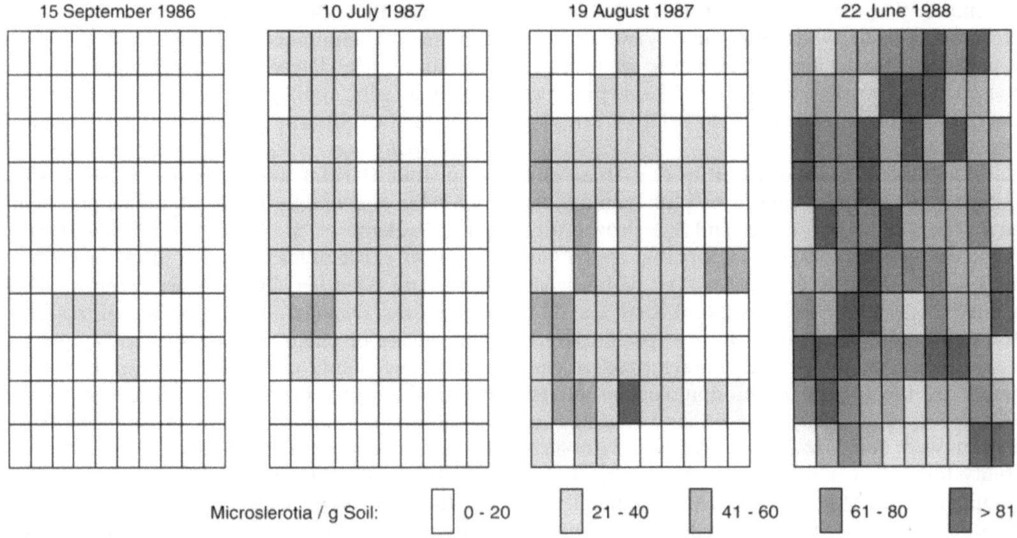

Fig. 3. Changes in spatial patterns of sclerotial populations of *Macrophomina phaseolina* during two seasons of growth of *Euphorbia lathyris* in contiguous plots. (After Mihail 1989)

scribed previously for characterizing aggregation of disease in plant populations, recent investigations have provided insights into the relationship of pathogen distribution and disease occurrence. For example, Mihail (1989) used spatio-temporal approaches to characterize changes in the aggregation of microsclerotial populations of *Macrophomina phaseolina* in relation to growth and development of *Euphorbia lathyris* (Fig. 3).

The life-history attributes of a pathogenic fungus also affect the pattern of disease that it incites. For example, the intrinsic ability of a fungus to reproduce after infection and to colonize plant tissue determines the potential for secondary dispersal and further disease development in a population. The importance of these events in relation to disease prediction was mentioned previously with reference to aerially dispersed fungi. Reproduction of soilborne fungi and its relationship to repeated plant infection and epidemic progression are more poorly understood. For example, Kuan and Erwin (1982) demonstrated that reproduction of *Phytophthora megasperma* f. sp. *medicaginis* increased significantly with increasing soil matric potential. The rate of mortality of alfalfa plants grown in soils infested with this pathogen coincidentally increased with increasing fungal reproduction.

In many cases, it is difficult to conduct experiments that clearly establish relationships between variable soil environmental factors and fungal reproduction. Consequently, mathematical modeling has become an important tool for evaluation of fungal reproduction. For example, Gilligan (1994) has recently described theoretical models of root disease epidemics that considered the likelihood of reproduction of pathogenic fungi in soil and subsequent initial or secondary infection of plants in a population.

E. Characterization of Plant Populations

It was mentioned previously that plants contribute to epidemic development through their expression of some level of disease resistance and through their influences on microclimatic conditions. Numerous approaches are used to assess such plant contributions. It has been difficult to visualize and assess how pathogenic fungi affect the size and structure of plant populations. The difficulty may be related to the variable impacts that fungi can have on plant populations, depending on the growth stage at which interaction with a pathogen occurs. Burdon (1987) described four phases of plant development that provide a framework for considering impacts of infection. A dormant seed phase, or seed bank, reflects the occurrence of seed populations established at the beginning of a growing season or that develop over successive years of reproduction of perennial plant species. The length of time of the dormant phase varies with environmental conditions at specific locations. At some point in time, seeds germinate and enter the early recruitment phase. The extent to which early recruitment depletes the dormant seed bank varies with plant species. Those seedlings that survive the early recruitment phase become part of the established plant phase that continues through the time that plants flower and produce seed. The final, predispersal seed phase encompasses the period from seed maturation to dispersal.

In any plant population, one or more pathogenic fungi can infect plants at each phase of development, and the effects of infection can carry beyond the phase where a plant-fungus interaction initially occurs. For example, fungi that infect and kill seed in the soil seed bank reduce the initial densities of plants that are established at a site. In contrast, fungi that infect during the establishment phase may not kill plants, but they may affect plant vigor and final fecundity at the predispersal phase. For example, Antonovics and Alexander (1989) described possible relationships of infection at different phases of plant development and plant reproductive fitness, with emphasis on annual and perennial plant species in natural systems.

Numbers and sizes of plants or their structural components can affect plant-fungus interactions at any phase of plant development. The density of plants at any stage may have impacts on the speed at which disease develops. Burdon and Chilvers (1975), for instance, demonstrated a positive relationship between density of garden cress seedlings at the early recruitment stage and rate of disease transmission by *Pythium irregulare*. In contrast, Melendez and Ackerman (1994) observed inconsistent positive correlations of the incidence of rust, caused by *Sphenospora saphena* and density of an epiphytic orchid *Tolumnia variegata*. They suggested that at low plant densities, clustering patterns of plants may have influenced the relationship of plant density and disease, and that plants with larger leaves were more likely to become infected than were plants with smaller leaves.

The impacts that infection may have on the structure or architecture of plant root systems or canopies have, until recently, also been ignored. It is important, however, to understand such impacts because the structure of these plant components will affect the efficiency of plant performance. The competitive ability of a plant depends on the efficiency of its root system in uptake of water and nutrients from soil, and the efficiency of its canopy in capture of light and transport of resources. Fitter (1982, 1986) developed morphometric and topological systems that have proven useful for characterizing root system structure. Larkin et al. (1995, 1996) recently used these systems to characterize alterations of alfalfa root systems that were brought about by infection with multiple species of *Pythium*. These and related systems are not restricted to descriptions of root systems; rather, they should prove useful in characterizing subtle structural changes in any plant component.

IV. Considerations for the Future

One of the strengths of epidemiology is the relevant biological framework that it provides for examining relationships of pathogenic fungi and plants as components of natural and managed systems. Epidemiological methods will prove important for evaluating the utilities of specific disease management strategies. For example, emphasis is being placed on the identification and introduction of specific disease resistance genes into agronomically important plant species. These molecular manipulations are technically complex but conceptually simple; it is not understood how manipulations of transformed plants will influence the behavior of any plant-pathogen system. Placement of molecular genetic approaches into a meaningful biological context will require greater emphasis on epidemiology and ecology, with an orientation to clearer definition of the biological phenomena for which genes are being isolated and manipulated. This background is needed to ask even simple questions such as: does the incorporation of a specific resistance gene confer advantages on a plant genotype in all management systems and under all environmental conditions?

Questions also are arising as to the best methods of deploying novel, resistant plant genotypes. Should they be deployed as monocultures or in mixtures with other plant genotypes of varying resistance to specific pathogenic fungi? Studies such as those of Mundt and Leonard (1985, 1986) are beginning to address this question. In one study, these authors found that the number of rust pustules caused by *Puccinia sorghi* on susceptible maize genotypes was reduced by 50% when the genotypes were mixed in any proportion of ground cover with resistant genotypes (Fig. 4).

The implications of manipulating specific genes in microorganisms also bears further evaluation within epidemiological and ecological contexts. As beneficial and virulence genes are

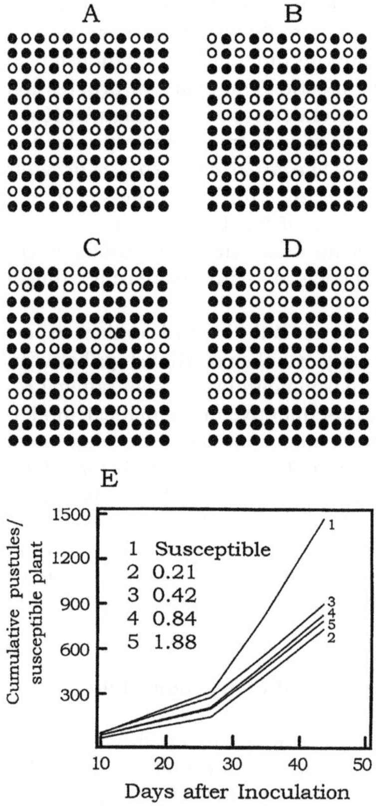

Fig. 4A–E. Patterns of temporal increase of maize rust in a mixture of 25% susceptible and 75% resistant genotypes as affected by the percentage of ground cover by maize genotypes. Pattern of susceptible genotype established in ground covers of **A** 0.21 m^2, **B** 0.42 m^2, **C** 0.84 m^2, **D** 1.88 m^2. **E** Progress of disease in relation to the percentage of ground cover and in a pure-line susceptible population. (After Mundt and Leonard 1986)

defined and introduced into fungi, it is logical to examine mechanisms that affect the successful introduction of novel fungal genotypes into populations of existing genotypes. What are the minimum requirements for stable introduction? Insights into this question may be gained by using experimental designs developed to assess mechanisms of microbial competition. For example, Adee et al. (1990) recently employed de Wit replacement series to assess the relative competitive abilities of *Pyrenophora tritici-repentis* and *Septoria nodorum* for utilization of wheat foliar tissues (Fig. 5). Zitko and Timmer (1994) used these same methods to evaluate competition between two

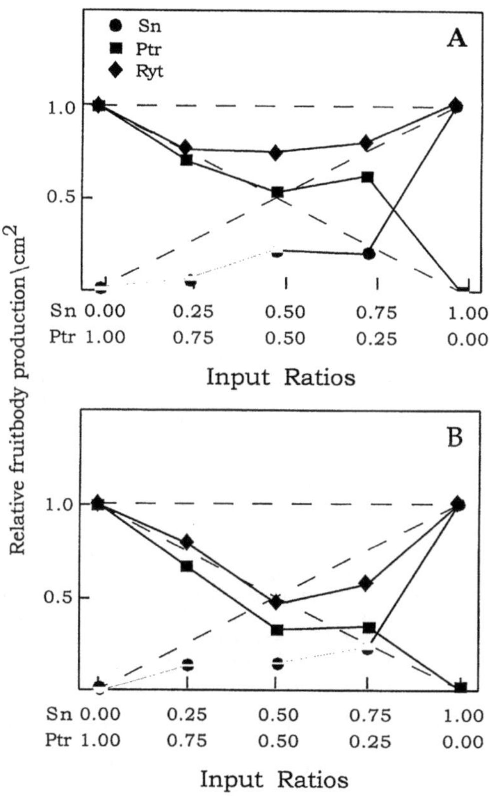

Fig. 5A,B. Comparative sporocarp production for *Pyrenophora tritici-repentis* (*Ptr*) and *Septoria nodorum* (*Sn*) cocolonizing the flag leaf of wheat. **A** Relative sporocarp production and total sporocarp yield (*Ryt*) for wheat inoculated simultaneously at various input ratios. **B** Sporocarp production and yield for wheat inoculated first with *S. nodorum*, followed 4 days later by *P. tritici-repentis*. (After Adee et al. 1990)

soilborne fungi, *Phytophthora parasitica* and *P. palmivora*, for occupation of root tissues of *Citrus jambhiri*.

Genetic manipulations are not the only means of managing plant disease. In many instances, disease can be controlled by cultural manipulations of plants and their environments. Cultural management will receive even greater emphasis with the increasing restrictions that are being imposed on the use of fungicides and soil fumigants (Civerolo et al. 1993). To meet the challenge of developing innovative cultural practices, pathologists need to grasp more clearly the impacts of environmental and plant modifications on system behavior. For example, there needs to be better characterization of environmental influences on plant-microbe interactions at the microsites where they occur. The approaches of Fravel and Marois (1986) and Eastburn and Butler (1988), described previously, provide a preliminary view of possible factors of importance in understanding microsite interactions in soil ecosystems. However, these selected factors have yet to be evaluated critically for their individual or combined effects on definable population processes within specific plant-pathogen systems. Elucidation of the contribution of plant modifications to epidemic development requires more creative approaches to plant assessment. For example, demographic methods have been used to assess population dynamics in animal systems, and they should be equally applicable to description of plant structural components (English 1992). Additionally, topological or morphometric systems should continue to be useful for characterizing the structural dynamics of both plant canopies and root systems in relation to disease (English 1993).

In any natural or managed plant population, multiple fungal species may function as pathogens. However, very few studies have considered the impacts of multiple pathogenic fungi on plant growth and development. Of these studies, Pinnschmidt et al. (1995) have begun to incorporate multiple pathogen effects in their physiological models of diseases on rice, and Nelson and Campbell (1993) have considered the temporal and spatial progession of foliar disease of white clover caused by a complex of pathogens including eight pathogenic fungi. Obviously, the reality of multiple pathogens and diseases needs to be addressed, not only to develop a more complete understanding of the impacts of disease on agroecosystems, but, from a more practical standpoint, to assess the extent to which predictive disease models for a crop species need to consider all or a subset of potential pathogens.

Many of the suggestions for future research in the area of epidemiology border on the areas of ecology, population biology, and population genetics. This is as it should be. Additionally, epidemiological studies do not stand alone; they require continued collaboration with geneticists, agronomists, and horticulturists in order to deal with the diversity of plant-pathogen systems that exist.

V. Conclusions

Pathogenic fungi affect plants at all stages of their development, and subsequently, their behaviors in natural and agroecosystems. The impacts of these microorganisms are strongly influenced by the environment in which plant-fungus interactions occur. Advancements in understanding the roles that pathogenic fungi play in agroecosystems, in particular, have coincided with the development of the discipline of epidemiology. The process-based framework of this discipline has been important in establishing the contributions of life-history and population traits of a plant species and its specific fungal pathogens to disease development. Epidemiological approaches have also been very important in elucidating the impacts that environment can have on plant-pathogen interactions and subsequent epidemic development.

Epidemiological research has provided the concepts and tools to evaluate not only basic biological processes, but it has provided relevant information to managers of agroecosystems. This will be increasingly important because of growing concerns for our abilities to manage disease within the constraints of environmental and economic sustainability.

References

Adee SR, Pfender WF, Hartnett DC (1990) Competition between *Pyrenophora tritici-repentis* and *Septoria nodorum* in the wheat leaf as measured by de Wit replacement series. Phytopathology 80:1177–1182

Agrios GN (1988) Plant pathology. Academic Press, New York

Ainsworth GC (1981) Introduction to the history of plant pathology. Cambridge University Press, Cambridge

Antonivics J, Alexander HM (1989) The concept of fitness in plant-fungal pathogen systems. In: Leonard KJ, Fry WE (eds) Plant disease epidemiology, vol 2. Genetics, resistance, and management. McGraw-Hill, New York, pp 185–214

Aylor DE, Ferrandino FJ (1990) Initial spread of bean rust close to an inoculated bean leaf. Phytopathology 80: 1469–1476

Baker R, Maurer CL, Maurer RA (1967) Ecology of plant pathogens in soil. VII. Mathematical models and inoculum density. Phytopathology 57:662–666

Bonde R, Schultz ES (1943) Potato refuse piles as a factor in the dissemination of late blight. Am Potato J 20:112–118

Boote KJ, Jones JW, Mishoe JW, Berger RD (1983) Coupling pests to crop growth simulators to predict yield reductions. Phytopathology 73:1581–1587

Bourke A (1991) Potato blight in Europe in 1845: the scientific controversy. In: Lucas JA, Shatttock RC, Shaw DS, Cooke LR (eds) *Phytophthora*. Cambridge University Press, Cambridge, pp 12–24

Bourke PMA (1964) Emergence of potato blight, 1843–1846. Nature (Lond) 203:805–808

Buchanan TS, Kimmey JW (1938) Initial tests of the distance of spread to and intensity of infection on *Pinus monticola* by *Cronartium ribicola* from *Ribes lacustre* and *R. viscosissimus*. J Agric Res 56:9–30

Burdon JJ (1987) Diseases and plant population biology. Cambridge University Press, Cambridge

Burdon JJ (1993) The structure of pathogen populations in natural plant communities. Annu Rev Phytopathol 31:305–323

Burdon JJ, Chilvers GA (1975) Epidemiology of damping-off disease (*Pythium irregulare*) in relation to density of *Lepidium sativum* seedlings. Ann Appl Biol 81:135–143

Campbell CL, Benson DM (1994) Spatial aspects of the development of root disease epidemics. In: Campbell CL, Benson DM (eds) Epidemiology and management of root diseases. Springer, Berlin Heidelberg New York, pp 195–243

Campbell CL, Madden LV (1990) Introduction to plant disease epidemiology. Wiley, New York

Campbell CL, Jacobi WR, Powell NT, Main CE (1984) Analysis of disease progression and the randomness of occurrence of infected plants during tobacco black shank epidemics. Phytopathology 74:230–235

Carefoot GL, Sprott ER (1967) Famine on the wind. Rand McNally, Chicago

Chellemi DO, Rohrbach KG, Yost RS, Sonoda RM (1988) Analysis of the spatial pattern of plant pathogens and diseased plants using geostatistics. Phytopathology 78: 221–226

Civerolo EL, Narang SK, Ross R, Vick KW, Greczy L (1993) Alternatives to methyl bromide: assessment of research needs and priorities. Proc USDA Worksh on Alternatives to methyl bromide, 29 June–1 July 1993, Arlington, Virginia

Craigie JH (1945) Epidemiology of stem rust in Western Canada. Sci Agric 25:285–401

Davis JM, Main CE (1986) Applying atmospheric trajectory analysis to problems in epidemiology. Plant Dis 70:490–497

de Bary A (1861) Die gegenwärtig herrschende Kartoffelkrankheit, ihre Ursache und ihre Verhütung. Felix, Leipzig

de Bary A (1863) Recherches sur le dévéloppement de quelques champignons parasites, etc. Ann Sci Nat 4e Ser Bot 20:1–148

Demeke T, Kidane Y, Wuhih E (1979) Ergotism, a report on the epidemic, 1977–1978. Ethiopian Med J 17:107–114

Eastburn DM, Butler EE (1988) Microhabitat characterization of *Trichoderma harzianum* in natural soil; evaluation of factors affecting distribution. Soil Biol Biochem 20:547–553

English JT (1992) Modular demography of *Lotus corniculatus* infected by *Rhizoctonia* spp. Phytopathology 82:1104

English JT (1993) Architecture of shoots of birdsfoot trefoil infected with *Rhizoctonia* spp. and *Stemphylium loti*. Phytopathology 83:1358

English JT, Thomas CS, Marois JJ, Gubler WD (1989) Microclimate of grape vine canopies associated with leaf removal and control of *Botrytis* bunch rot. Phytopathology 79:395–401

English JT, Bledsoe AM, Marois JJ (1990) Influence of leaf removal from the fruit cluster zone on the components of evaporative potential within grapevine canopies. Agric Ecosys Environ 31:49–61

Ferrandino FJ (1993) Dispersive epidemic waves. I. Focus expansion within a linear planting. Phytopathology 83:795–802

Ferris RS (1981) Calculating rhizosphere size. Phytopathology 71:1229–1231

Fitter AH (1982) Morphometric analysis of root systems: application of the technique and influence of soil fertility on root system development in two herbaceous species. Plant Cell Environ 5:313–322

Fitter AH (1986) The topology and geometry of plant root systems: influence of watering rate on root system topology in *Trifolium pratense*. Ann Bot 58:91–101

Fravel DR, Marois JJ (1986) Edaphic parameters associated with establishment of the biocontrol agent *Talaromyces flavus*. Phytopathology 76:643–646

Fuller JG (1968) The day of St. Anthony's Fire. Macmillan, New York

Gäumann E (1950) Principles of plant infection. Hafner, New York

Gilligan CA (1979) Modeling rhizosphere infection. Phytopathology 69:782–784

Gilligan CA (1985) Introduction. In: Gilligan CA (ed) Advances in plant pathology, vol 3. Academic Press, New York

Gilligan CA (1994) Temporal aspects of the development of root disease epidemics. In: Campbell CL, Benson DM (eds) Epidemiology and management of root diseases. Springer, Berlin Heidelberg New York, pp 148–194

Goodwin SB, Cohen BA, Deahl KL, Fry WE (1994) Migration from northern Mexico as the probable cause of recent genetic changes in populations of *Phytophthora infestans* in the United States and Canada. Phytopathology 84:553–558

Gregory PH (1945) The dispersion of air-borne spores. Transact Br Mycol Soc 28:26–72

Gregory PH (1968) Interpreting plant disease dispersal gradients. Annu Rev Plant Pathol 6:189–212

Griffin GJ, Tomimatsu GS (1983) Root infection pattern, infection efficiency, and infection density-disease incidence relationships of *Cylindrocladium crotalariae* on peanut in field soil. Can J Plant Pathol 5:81–88

Grogan RG, Sall MA, Punja ZK (1980) Concepts for modeling root infection by soilborne fungi. Phytopathology 70:361–363

Gubler WD, Marois JJ, Bettiga LF (1987) Control of *Botrytis* bunch rot of grape with canopy management. Plant Dis 81:599–601

Gutierrez AP, DeVay JE (1986) Studies of plant-pathogen-weather interactions: cotton and verticillium wilt. In: Leonard KJ, Fry WE (eds) Plant disease epidemiology, vol 1. Macmillan, New York, pp 205–231

Hepting GH (1974) Death of the American chestnut. For Hist 18:60–67

Horsfall JG (1972) Genetic vulnerability of major crops. National Academy of Science, Washington, DC

Ingold CT (1971) Fungal spores. Their liberation and dispersal. Clarendon Press, Oxford

Jarosz AM, Davelos AL (1995) Effects of disease in wild populations and the evolution of pathogen aggressiveness. New Phytol 129:371–387

Jeger JF (1985) The spatial component of plant disease epidemics. In: Jeger MJ (ed) Spatial components of plant disease epidemics. Prentice Hall, Englewood Cliffs, pp 1–13

Jones LR, Johnson J, Dickson JG (1926) Wisconsin studies on the relation of soil temperature to plant disease. Bull Wisc Agric Exp Stn 71, University of Wisconsin at Madison, Wisconsin, 144 pp

Keitt GW, Jones LK (1926) Studies of the epidemiology and control of apple scab. Res Bull Wisc Agric Exp Stn 73, University of Wisconsin at Madison, Wisconsin, 104 pp

Kuan TL, Erwin DC (1982) Effect of soil matric potential on *Phytophthora* root rot of alfalfa. Phytopathology 72:543–548

Large EC (1940) The advance of the fungi. Holt, New York

Large EC (1945) Field trials of copper fungicides for the control of potato blight. I. Foliage protection and yield. Ann Appl Biol 32:319–329

Large EC (1952) The interpretation of progress curves for potato blight and other plant diseases. Plant Pathol 1:109–117

Larkin RP, English JT, Mihail JD (1995) Effects of infection by *Pythium* spp. on root system morphology of alfalfa seedlings. Phytopathology 85:430–435

Larkin RP, English JT, Mihail JD (1996) Relationship of infection by *Pythium* spp. to root system morphology of alfalfa seedlings in the field. Plant Dis 80:281–285

Leonard KJ (1980) A reinterpretation of the mathematical analysis of rhizoplane and rhizosphere effects. Phytopathology 70:695–696

Lucas GB (1980) The war against blue mold. Science 210:147–153

Madden LV (1986) Statistical analysis and comparison of disease progress curves. In: Leonard KJ, Fry WE (eds) Plant disease epidemiology, vol 1. Population dynamics and management. Macmillan, New York, pp 55–84

Madden LV (1989) Dynamic nature of within-field disease and pathogen distributions. In: Jeger MJ (ed) Spatial components of plant disease epidemics. Prentice Hall, Englewood Cliffs, pp 96–126

McCartney HA, Fitt BD (1985) Construction of dispersal models. In: Gilligan CA (ed) Advances in plant pathology, vol 3. Academic Press, London

Melendez EJ, Ackerman JD (1994) Factors associated with a rust infection (*Sphenospora saphena*) in an epiphytic orchid (*Tolumnia variegata*). Am J Bot 81:287–293

Mihail JD (1989) *Macrophomina phaseolina*: spatio-temporal dynamics of inoculum and of disease in a highly susceptible crop. Phytopathology 79:848–855

Minogue KP (1989) Diffusion and spatial probability models for disease spread. In: Jeger MJ (ed) Spatial components of plant disease epidemics. Prentice Hall, Englewood Cliffs, pp 127–143

Mundt CC, Browning JA (1985) Development of crown rust epidemics in genetically diverse oat populations: effect of genotype unit area. Phytopathology 75:607–610

Mundt CC, Leonard KJ (1986) Effect of host genotype unit area on development of focal epidemics of bean rust and common maize rust in mixtures of resistant and susceptible plants. Phytopathology 76:895–900

Nelson SC, Campbell CL (1993) Comparative spatial analysis of foliar epidemics on white clover caused by viruses, fungi, and a bacterium. Phytopathology 83:288–301

Padmanabhan SY (1973) The great Bengal famine. Annu Rev Phytopathol 11:11–26

Pinnschmidt HO, Batchelor WD, Teng PS (1995) Simulation of multiple species pest damage in rice using CERES-rice. Agric Syst 48:193–222

Reynolds KM, Madden LV (1988) Analysis of epidemics using spatio-temporal autocorrelation. Phytopathology 78:240–246

Ristaino JB, Larkin RP, Campbell CL (1994) Spatial dynamics of disease symptom expression during *Phytophthora* epidemics in bell pepper. Phytopathology 84: 1015–1024

Rouse DI, Baker R (1978) Modeling and quantitative analysis of biological control mechanisms. Phytopathology 68:1297–1302

Schmitthenner AF, Hobe M, Bhat RG (1994) *Phytophthora sojae* races in Ohio over a 10-year interval. Plant Dis 78:269–276

Shaner G (1983) Growth of uredinia of *Puccinia recondita* in leaves of slow and fast-rusting wheat cultivars. Phytopathology 73:931–935

Thal WM, Campbell CL, Madden LV (1984) Sensitivity of Weibull model parameter estimates to variation in simulated disease progression data. Phytopathology 74: 1425–1430

Thomas CS, Marois JJ, English JT (1988) The effects of wind speed, temperature, and relative humidity on development of aerial mycelium and conidia of *Botrytis cinerea* on grape. Phytopathology 78:260–265

Van den Bosch F, Frinking HD, Metz JAJ, Zadoks JC (1988) Focus expansion in plant disease III: two experimental examples. Phytopathology 78:919–925

Van der Plank JE (1947) A method for estimating the number of random groups of adjacent diseased plants in a homogeneous field. Trans R Soc S Afr 31:269–278

Van der Plank JE (1963) Plant diseases: epidemics and control. Academic Press, New York

Whetzel HJ (1918) An outline of the history of phytopathology. WB Saunders, Philadelphia

Willigan J (1977) Famine and structural adaptation in mid-nineteenth-century Ireland. PhD Dissertation, University of North Carolina, Chapel Hill

Woodham-Smith C (1962) The Great Hunger, Ireland 1845–1849. Harper and Row, New York

Zadoks JC, Schein RD (1979) Epidemiology and plant disease management. Oxford University Press, New York

Zitko SE, Timmer LW (1994) Competitive parasitic abilities of *Phytophthora parasitica* and *P. palmivora* on fibrous roots of *Citrus*. Phytopathology 84:1000–1004

10 Speciation and Population Biology in *Colletotrichum*

D.O. TeBeest, J.C. Correll, and G.J. Weidemann

CONTENTS

I. Introduction 157
II. Pathogenesis in *Colletotrichum* 158
III. Genetics of Sexuality and Pathogenesis 159
A. Parasexuality 159
B. Sexual Systems 159
C. Genetics of Parasitism 161
IV. Population Biology of *Colletotrichum* Species 162
V. Summary and Conclusions 165
References 165

I. Introduction

The fungal genus *Colletotrichum* is composed of a number of saprophytic and plant pathogenic species of worldwide importance on a wide range of economic crops and ornamentals. In addition to economically important crop plants, several *Colletotrichum* species attack a variety of wild and weedy plants, and several host-specific strains have been utilized for biological weed control (Templeton 1992). Several species are noted for causing latent infections, particularly on tropical fruit crops (Simmonds 1963; Muirhead 1981; Jeffries et al. 1990). Most *Colletotrichum* species cause diseases known as anthracnoses, characterized by sunken necrotic lesions on leaves, stems, or fruit. Several species appear to exist exclusively as an asexual population and produce only the characteristic *Colletotrichum* anamorph. Other species also produce a *Glomerella* teleomorph and can reproduce both sexually and asexually (Fig. 1).

Considerable taxonomic uncertainty exists in this genus due to the lack of good morphological characters, the existence of both anamorphic and teleomorphic forms in some taxa, and uncertain host relationships. Additionally, several similar taxa may occur on the same host species and cause similar or identical symptoms. The taxonomic uncertainty has made accurate identification difficult and complicated efforts to understand host relationships, diagnose diseases accurately, develop effective control strategies, and establish cost-effective quarantine programs.

Since the 1950s, three major taxonomic treatments of the genus have served to reduce the number of described and to examine the value of various morphological and cultural features for the identification of taxa. Von Arx (1957, 1970) reduced over 900 described species to 21 species and several subspecies. In particular, nearly 600 species were placed in synonomy with *C. gloeosporioides* (Penz.) Sacc. In a later treatment, von Arx (1981) added two additional taxa. Sutton (1980) accepted 22 species although he considered *C. gloeosporioides*, *C. dematium* (Pers.:Fr.) Grove, *C. capsici* (H. Syd.) E. Butl. and *C. graminicola.* (Ces.) Wilson as heterogeneous group-species in need of further revision. Baxter et al. (1983, 1985) examined 12 South African *Colletotrichum* species and provided a thorough discussion of useful characters for each taxon. Most recently, Sutton (1992) reviewed previous work and described 38 currently accepted species.

Although previous authors have placed greater or lesser emphasis on different morphological characters, several features have been utilized by most taxonomists including conidial size and shape, appressorial features, presence/absence and appearance of setae, appearance of the conidiomata, presence or absence of sclerotia, and the development and morphology of the *Glomerella* teleomorph. In addition to the primary morphological features, taxonomists also have utilized cultural appearance and host relationships to differentiate taxa.

Recent studies utilizing biochemical and molecular approaches for differentiating taxa (Sreenivasaprasad et al. 1992, 1993; Sherriff et al. 1994) and host-specialized populations (Braith-

Department of Plant Pathology, University of Arkansas, Fayetteville, Arkansas 72701, USA

The Mycota V Part B
Plant Relationships
Carroll/Tudzynski (Eds.)
© Springer-Verlag Berlin Heidelberg 1997

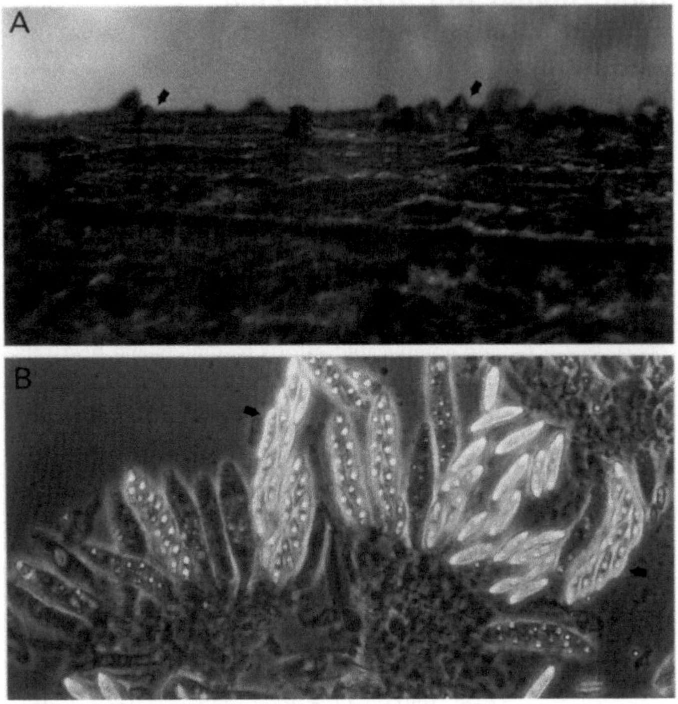

Fig. 1A,B. The sexual stage of *Glomerella cingulata*. **A** Perithecia (*arrows*) of *G. cingulata* in mated cultures of *Colletotrichum gloeosporioides* f. sp. *aeschynomene* and a heterothallic isolate of *G. cingulata* from pecan. **B** A fan of asci and ascospores produced from paired heterothallic isolates. Note the presence of eight ascospores in various developmental stages within several of the asci (*arrows*)

waite et al. 1990; Guthrie et al. 1992; Hodson et al. 1992; Mills et al. 1992a,b) have contributed to the understanding of taxonomic and biological relationships within the genus. However, the existence of numerous host-specialized populations within some species, lack of easily distinquishable morphological characters, and worldwide distribution has limited the development of a synthetic treatment incorporating morphological, molecular and biological information.

II. Pathogenesis in *Colletotrichum*

The basic disease cycles for anthracnose diseases caused by the several species of *Colletotrichum* are very similar (Agrios 1988). The fungus usually overwinters in diseased tissues as mycelium, spores, or perithecia (if perithecia are formed). Conidia produced on lesions are generally responsible for dispersal of the fungus and increase of the disease.

The infection processes in *Colletotrichum* are remarkably similar across species and isolates. The infection events following deposition of spores on susceptible hosts include adhesion, germination, appressoria formation and penetration. The events following deposition of spores on host tissues through penetration generally occur within 24 to 48h. Following penetration, events leading to visualization of disease generally can be separated into a biotrophic phase and a necrotrophic phase (Bailey et al. 1992). The biotrophic phase generally is short-lived and includes all events in which infection develops without visible disruption of host systems. In the necrotrophic phase, the host-pathogen relationship changes and leads to the death of cells and tissues normally associated with symptom development. Anthracnose diseases also can be broadly differentiated on the basis of latent period. Anthracnose diseases of tropical fruits generally have a long latent period of several weeks or months during which disease development following initial penetration and infection is arrested (Swinburne 1983; Jeffries et al. 1990), whereas stem or leaf infections generally have a latent period of only several hours or days.

Recent work has begun to elucidate the mechanisms responsible for the delayed infection of tropical fruits by *C. gloeosporioides*. Mechanisms for induction of appressoria formation were suggested that included both physical or topographical (Miehle and Lukezic 1972; Lapp and Skoropad 1978; Suzuki et al. 1982; Staples and Macko 1984) and chemical signals (Swinburne 1976; Parberry and Blakeman 1978). Podila et al. (1993) recently showed that chemical signals from the wax triggers both germination and appressoria formation on avocado. Waxes from nonhost plants did not induce appressorium formation in the avocado pathogen, and avocado wax did not induce appressorium formation in most *Colletotrichum* species from other hosts. Further testing indicated that fatty alcohols greater than 24 carbons in length induced appressoria to form. Waxes from nonhost plants inhibited appressorium induction by avocado wax.

Conida of *Colletotrichum* species are produced in acervuli and the mass of conidia is encased in a sticky, water-soluble matrix. The composition and function of the matrix of *C. graminicola* has been studied in detail. The primary components of the mucilage are high molecular weight glycoproteins which are thought to be important as antidessicants for enhancing conidial survival within an acervulus (Nicholson and Moraes 1980; Ramadoss et al. 1985). Proline-rich proteins in the matrix detoxify phenols produced by plants within lesions (Nicholson et al. 1986, 1989). The matrix of *C. graminicola* also contains several enzymes (Nicholson and Epstein 1991) in addition to cutinase (Pascholati et al. 1992b). Cutinase, produced by many fungal pathogens, is believed to be important for penetration of the cuticule of many plants (Kolattukudy 1985). Cutinases have been described for several *Colletotrichum* species (Kolattakudy 1985; Dickman et al. 1982, 1989; Liyanage et al. 1993; Pascholati et al. 1992b). However, the role of cutinases in plant pathogenesis remains unclear (Stahl and Schafer 1992; Sweigard et al. 1992a,b; Pascholati et al. 1992a).

III. Genetics of Sexuality and Pathogenesis

A. Parasexuality

Because several *Colletotrichum* species appear to exist exclusively as asexually reproducing populations, heterokaryosis and parasexuality have been investigated as mechanisms for inducing genetic variability and as a means for asexual gene exchange between different host-specific populations. Dutta and Garber (1960) produced heterokaryons of *C. orbiculare* (Berk. and Mont.) von Arx but were unable to demonstrate recombination. In contrast, Stephan (1968) reported several recombinants from heterokaryotic colonies of *C. gloeosporioides* but was unable to demonstrate conclusively that the recombinants resulted from mitotic recombination rather than mutation.

Chacko et al. (1994) used nitrate nonutilizing and auxotrophic mutants to characterize vegetative compatibility, heterokaryosis, and mitotic recombination within *C. gloeosporioides* f. sp. *aeschynomene* (Cga) and isolates of *C. gloeosporioides* from other hosts. Vegetative compatibility analysis using nitrate nonutilizing mutants (Puhalla 1985; Correll et al. 1987) demonstrated that all isolates of Cga were vegetatively compatible but were incompatible with *C. gloeosporioides* isolates obtained from other hosts. Cultural studies using auxotrophic mutants showed that several paired auxotrophic strains of Cga were able to produce heterokaryons, but no recombinant phenotypes were detected. Likewise, no recombinant phenotypes were recovered from anthracnose lesions on the legume host, *Aeschynomene virginica* (L.) BSP, coinoculated with auxotrophic strains of Cga. Vegetative incompatibility, lack of a demonstrated parasexual cycle, and host specialization may serve to genetically isolate *Colletotrichum* species that lack a sexual state and limit the potential for gene exchange within the species.

B. Sexual Systems

A significant number of *Colletotrichum* species including *C. gloeosporioides* (von Arx 1957; Brenneman and Reilly 1989), *C. gossypii* South. (Farr et al. 1989), *C. graminicola* (Politis 1975), *C. lindemuthianum* (Sacc. and Magnus) Briosi and Cav. (Kimati and Galli 1970), *C. destructivum* O'Gara (Manandhar et al. 1986), *C. falcatum* Went. (Mordue 1967), *C. orbiculare* (Jenkins and Winstead 1964; McClean 1966), and *C. musae* (Berk. and Curtis) von Arx (Rodriquez and Owen 1992) produce a *Glomerella* teleomorph. However, many host-specialized strains within these species have not been reported to produce the sexual state in nature or in culture. The remaining species in *Colletotrichum* appear to lack a perfect state.

The ubiquitous pathogen *C. gloeosporioides* and its associated teleomorph, *Glomerella cingulata* (Stonem.) Spauld. and Schrenk, has been reported to cause diseases on many different host species, particularly in the tropics and subtropics (Fig. 1) Von Arx (1957) placed nearly 600 previously described species into synonymy with *C. gloeosporioides* based on similar morphology, although most species was described from different hosts. Likewise, 42 *Glomerella* species were placed in synonymy with *Glomerella cingulata* (von Arx and Muller 1954). Several forma speciales have been retained for both *C. gloeosporioides* (Daniel et al. 1969; Templeton and Heiny 1989; Sutton 1992) and *C. cingulata*, based on host relationships. Not all special forms and strains of *C. gloeosporioides* are known to produce a *Glomerella* state. The ontogeny of the ascoma of *G. cingulata* has most recently been described by Uecker (1994).

Mating systems in fungi can be broadly divided into heterothallic (self-incompatible) or homothallic (self-compatible) forms. The mating system in ascomycetes has been intensively studied in *Saccharomyces cerevisiae* Hansen and *Neurospora crassa* Shear and Dodge. Both fungi have fairly simple bipolar mating systems (one locus, two alleles) and their mating type genes have been cloned (Astell et al. 1981; Glass et al. 1990a,b; Staben and Yanofsky 1990; see also Vol. I, this Series).

Genetics studies have been conducted on three *Colletotrichum* species (Wheeler 1956; Bryson 1990; Vailancourt and Hanau 1991). Bipolar mating was first reported in *Glomerella* by Edgerton (1912, 1914). In the 1940s and 1950s, several studies on the genetics of mating of *G. cingulata* isolated from morning glory, apple, and tomato were conducted (Huttig 1935; Lucas et al. 1994; Edgerton et al. 1945; Chilton et al. 1945; Chilton and Wheeler 1949; Struble and Keitt 1950; Wheeler and McGahan 1952). Edgerton (1914) described the production of a ridge of fertile perithecia forming between two isolates when coinoculated on the same petri dish. This reaction was attributed to a sexual interaction of two mating types. This work has been confirmed by other studies and is considered a general phenomenon of *G. cingulata*. However, self-fertility prevents designation of specific plus and minus mating types.

A knowledge of the genetic basis for the difference between + and − strains resulted from work published from 1945 to 1952. Ascospores collected from asci of perithecia from plus cultures or from the ridge at the interface of paired plus and minus isolates segregated as 4 plus: 4 minus or contained either 8 plus or 8 minus ascospores, suggesting that a single gene with two alleles (B+ or B1) was present (Wheeler and McGahan 1952). However, in plus cultures, B+ regularly mutated to B1 (Olive 1951). As a result, B+ nuclei are essential for fertility but rarely self-fertilize, while B1 nuclei are self-sterile, unless in the presence of a B+ nucleus. The involvement of the B1 allele accounts for the ridge of perithecia at the interface of plus and minus strains. Chilton et al. (1945) reported a third allele at the B locus, differing from B1 only in the fertility of perithecia produced in single cultures.

However, Chilton et al. (1945) demonstrated that plus cultures not only gave rise to minus cultures, but also to cultural variants that were sterile but produced large masses of conidia. These variants crossed with + and − isolates, and results of the crosses indicated a single gene A with two alleles (A+ and A1) determining the switching from a perithecial to conidial form. A third allele, A2, described by Wheeler and McGahan (1952), affected the size of the perithecial clumps, pigmentations, and self-fertility in B+ and B1 genotypes. Chilton and Wheeler (1949) noted that clumped conidial genotypes (A1B+) are very similar to the nonascogenous *Colletotrichum* forms isolated from many host plants, and Bryson et al. (1992) concluded that many *Colletotrichum* species are naturally mutated at a gene equivalent to A.

The studies described above were conducted with isolates collected from the same hosts. Many workers have assumed that the lack of a known sexual stage in culture or on plants is both proof and evidence that a sexual stage does not exist in the many anamorphic forms of *C. gloeosporioides* that produce only conidia in culture. However, Cisar et al. (1994) have recently described a cross of a self-sterile isolate of *G. cingulata* from pecan and self-sterile isolates of *C. gloeosporioides* isolated from several unrelated host species. The fertility of this cross would lend support to the contention that conidial isolates of *C. gloeosporioides* from these hosts may be mutants at a gene equivalent to A or perhaps even may represent an undefined gene equivalent in function to A.

Vaillancourt and Hanau (1991) reported that a single strain of *C. graminicola* produced fertile perithecia when grown alone on maize stem pieces, suggesting homothallism. However, two monoascosporic siblings of this isolate, each capable of producing fertile perithecia, produced a greater number of perithecia when two monoascosporic siblings were grown on maize leaves together. In pairings, perithecia were concentrated at the junction of the two colonies. Vaillancourt and Hanau (1991) questioned whether *C. graminicola* is sterile in nature and suggest that sterility might be a result of mutation and selection during subculturing. The sexual stage of *C. graminicola* has never been observed in nature and was observed only once previously in culture (Politis 1975).

C. lindemuthianum strains are usually conidial in nature. *Glomerella lindemuthianum* Shear was first identified by Shear and Wood (1913). Kimati and Galli (1970) later reported the formation of fertile perithecia in specific pairings of conidial strains (Bryson 1990). Batista and Chaves (1982) and Bryson (1990) extended studies to additional strains, and results indicated that all strains are self-sterile when grown alone but will develop fertile perithecia when grown in paired combinations with some other strains. Bryson (1990) and Bryson et al. (1992) concluded that the genetic control of sexuality in *C. lindemuthianum*, as in many other *Glomerella* species, is complex.

Colletotrichum musae causes a postharvest disease of banana, damaging crowns, pedicels, stalks, and fingers. The species is conidial when grown in culture and had never been known to produce a sexual stage until Rodriguez and Owen (1992) reported the production of perithecia in controlled cultural conditions. Development of the sexual stage required light and suboptimal growth conditions. Perithecia were fertile, containing 50 to 80 asci and up to eight ascospores in each ascus. Results of experiments with chlorate-resistant mutants indicated that this species was heterothallic, since no isolate produced perithecia when grown alone or paired with itself. The complexity of the sexual compatibility was analyzed by crossing 15 monoascosporic progeny with four additional progeny. All of the progeny screened mated with at least one of the four test progeny. They concluded that *Glomerella musae* Petch possessed at least four mating types.

Bryson et al. (1992) concluded that the mating system in *Colletotrichum graminicola*, *Colletotrichum lindemuthianum*, and *Glomerella cingulata* is fundamentaly homothallic. First, self-fertile genotypes exist in all three species; second, there is evidence for induced selfing of apparently self-sterile genotypes; and third, the failure of sexual compatibility to fit any simple model of heterothallism. The reports by Rodriguez and Owen (1992) and Manandhar et al. (1986) suggest, however, that *G. musae* and *G. glycines* Lehman and Wolf are heterothallic. The discovery of additional sexual compatibilities within and between species of *Colletotrichum* increases the importance of the phenomenon and warrants further study.

C. Genetics of Parasitism

Few investigations have studied the genetics of parasitism in *Glomerella*, perhaps due to difficulties in obtaining fertile crosses (Meiners 1981). Nevertheless, a few recent studies have been reported. Physiological specialization of a pathogen to differential host genotypes was demonstrated by Barrus (1911) for *C. lindemuthianum* and subsequently many races have been identifed. Although evidence from host plant genetic studies suggested that a gene-for-gene relationship existed, no information was available describing the inheritance of pathogenicity in *C. lindemuthianum*. Bryson et al. (1992) crossed Brazilian strains of this fungus pathogenic to a set of differential hosts. One strain was virulent to both cultivars, while the other strain was avirulent to both. Twenty-five ascosporic progeny from a cross of these strains were analyzed for virulence to the differential cultivars. The data suggested that two independent genes existed, each determining avirulence to one cultivar. Also, a significant number recombinants were isolated from the cross with new virulence patterns to the two cultivars. Batista and Chaves (1982) also reported the formation of new races following sexual crosses. Host resistance in bean to *C. lindemuthianum* appears to be conferred by single genes, consistent with the gene-for-gene system reported for many pathogens.

A limited number of progeny from a cross of *C. gloeosporioides* f. sp. *aeschynomene*, a pathogen of the legume northern jointvetch (*Aeschynomene virginica*), and *G. cingulata* from pecan also have been analyzed (Fig. 1; Cisar et al. 1994, 1996). *C. gloeosporioides* f. sp. *aeschynomene* infects only northern jointvetch and a few other legumes, but not pecan or wounded apple fruit, while *G. cingulata* from pecan infects wounded apple fruit, causing a characteristic lesion. The results from Southern hybridization experiments show that these progeny exhibit a variety of parental characters demonstrating that sexual recombination has occurred. However, while all progeny from a single cross were pathogenic to apple fruit, none is pathogenic to northern jointvetch. Pathogenicity of these progeny to other hosts of either parental isolate have not been reported. Recently, Cisar et al. (1996) showed that crosses of the isolates from *Aeschynomene* and pecan could be crossed on living northern jointvetch plants.

IV. Population Biology of *Colletotrichum* Species

The use of genetic and molecular markers in recent years has greatly improved our understanding of inter- and intraspecies variation within *Colletotrichum*. Although species concepts within this genus remain uncertain, recent work has helped to more clearly define genotypes within and between species. As a result, a more thorough understanding of the relationship between species, subspecific taxa, populations, and their biology is beginning to emerge. It is clear from many of these recent molecular studies that some species and subspecies of *Colletotrichum* are more clearly delineated than others. Also, some species and subspecies may be composed of many diverse genotypes, whereas others may be more genetically homogeneous.

Many genetic and molecular techniques are available for characterizing closely related taxa and populations within a given taxon for plant pathogenic fungi. The utility of these various techniques and their relative level of resolution has been reviewed elsewhere (Kohn 1992). Genetic techniques, such as sexual and vegetative compatibility, recently have proven to be very useful for characterizing diversity in several species of *Colletotrichum* (Brooker et al. 1991; Correll et al. 1993a,b,c; Vaillancourt and Hanau 1994). Various molecular studies on inter- and intraspecies diversity have included the use of genomic and mitochondrial restriction fragment length polymorphisms (RFLPs), randomly amplified polymorphic DNA (RAPDs), DNA sequencing of specific regions of the *Colletotrichum* genome, and electrophoretic karyotyping. The molecular genotype diversity which has been identified both within and between species varies considerably. Overall, many of these molecular techniques have been very useful for characterizing subspecific population structure among isolates from a particular host.

C. gloeosporioides is recognized as a group species because of the morphological and virulence diversity observed among isolates from various plant hosts. One of the first studies to utilize RFLP analysis of *Colletotrichum* examined isolates of *C. gloeosporioides* from *Stylosanthes* spp. (Braithwaite and Manners 1989; Braithwaite et al. 1990; Manners et al. 1992). Examination of genomic RFLPs detected with human minisatellite and rDNA probes revealed two distinct populations of *C. gloeosporioides*, designated type A and type B, in Australia. Limited RFLP variation was observed within each population which indicated that each probably represented a distinct clonally reproducing population. Examination of isolates of *C. gloeosporioides* from *Stylosanthes* spp. from outside of Australia has revealed that additional genotypes occur throughout the world. Within the two populations in Australia, four major physiological races have been identified among type B isolates. Examination of electrophoretic karyotypes also has revealed a considerable amount of diversity in chromosome size and number among isolates within type A and B populations (Masel et al. 1990, 1993; Manners et al. 1992). The electrophoretic karyotypes of the isolates examined, however, apparently are mitotically stable both in vitro and after multiple infection cycles. A unique 1.2 Mb minichromosome also has been found associated with race 3 isolates in the type B population. However, it remains to be determined if indeed this represents a race-specific marker (Masel et al. 1993).

Several studies have examined the molecular population diversity among isolates of *C. gloeosporioides* from tropical fruits (Hodson et al. 1992; Mills et al. 1992a,b; Freeman et al. 1996; McCormick et al. 1995). Examination of rDNA and mtDNA RFLPs, RAPDs, and sequence comparisons of the internal transcribed spacer region (ITS1) among isolates from various tropical fruits has revealed a considerable degree of genotype variation among isolates from avocado and papaya. However, there was some correspondence between genotype and geographical origin within these host populations. In contrast, isolates of *C. gloeosporioides* recovered from mango from throughout the world showed very little molecular genotypic diversity. In general, all the molecular techniques employed yielded similar interpretations of the level of genotype diversity observed from these various host populations.

Anthracnose of strawberry and bitter-rot of apple are disease complexes caused by *Colletotrichum* spp. Anthracnose of strawberry is caused by *C. gloeosporioides*, *C. fragariae*, and *C. acutatum*. The taxonomic identity and genetic variation among isolates in this disease complex has been the focus of considerable debate (Bonde et al. 1991; Gunnell and Gubler 1992). Recently, *Colletotrichum* isolates recovered from strawberry and several other hosts were examined for genotype diversity using rDNA and mtDNA RFLPs, RAPDs, and sequence diversity of the internally transcribed spacer region (Sreenivasaprasad et al. 1992; Freeman and Rodrigues 1995). These data indicated that *C. gloeosporioides* and *C. fragariae* are genetically distinct from *C. acutatum*, but similar to one another. Although a very limited sample size was examined, isolates of *C. gloeosporioides* from strawberry represented a single genotype, whereas isolates of *C. fragariae* belonged to two distinct genotypes. Considerable diversity was identified among the isolates of *C. acutatum* from diverse hosts. European isolates of *C. acutatum* from strawberry fell into a discrete group, whereas isolates from strawberry from the US belonged to a different molecular group that included isolates from several other hosts. Again, there was a close correspondence between the groupings based upon either rDNA RFLPs, mtDNA RFLPs, or RAPDs.

Examination of over 800 *Colletotrichum* isolates from apple bitter-rot lesions in the US has revealed that three taxa, *C. acutatum*, *C. gloeosporioides*, and *Glomerella cingulata*, are involved in this disease complex (Correll et al. 1993c, 1994; Shi et al. 1993, 1996). Growth of monoconidial isolates indicated that isolates designated as *G. cingulata* produced perithecia and ascospores, indicating that they were homothallic. Isolates of *C. gloeosporioides*, *G. cingulata*, and *C. acutatum* from apple each belonged to a distinct mtDNA RFLP group. Thus, the three taxa apparently represent three genetically distinct populations (Fig. 2; Correll et al. 1993; Shi et al. 1996). Several DNA fingerprinting techniques also have recently been developed for examining nuclear RFLP diversity in these *Colletotrichum* spp. (Correll et al. 1994; Dyson et al. 1993; Cawthon et al. 1995).

Postbloom fruit drop of citrus (*Citrus sinensis* and *C. aurantifolia*) is an economically important disease caused by *C. gloeosporioides* (Agostini et al. 1992). However, many isolates of *C. gloeosporioides* which vary in conidial morphology, colony morphology, and pathogenicity have been recovered from citrus. Liyanage et al. (1992) examined

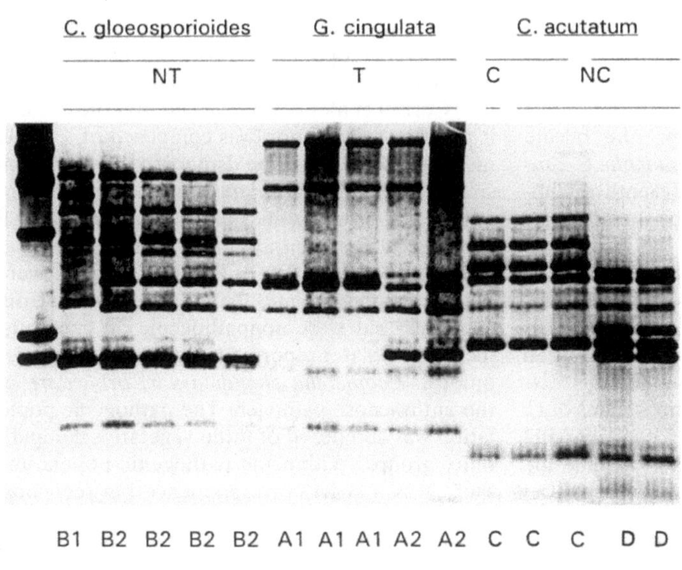

Fig. 2. Mitochondrial DNA RFLP haplotypes identified among nonteleomorphic (*NT*) and teleomorphic (*T*) isolates of *Colletotrichum gloeosporioides* and chromogenic (*C*) and nonchromogenic (*NC*) isolates of *C. acutatum*. Note mtDNA haplotypes at the bottom of the gel. Genomic DNA was cut with the restriction enzymes PalI and probed with a 23.8 mtDNA clone from *Colletotrichum orbiculare*

a collection of citrus isolates using rDNA RFLP comparisons, heterologous hybridization to several genes from *Neurospora crassa*, and electrophoretic karyotypes, and concluded there were two genetically distinct populations of *C. gloeosporioides* on citrus. The two populations were designated slow-growing orange (SGO) and fast-growing gray (FGG) corresponding to the type 1 and type 2 populations, respectively, identified using molecular criteria (Agostini et al. 1992; Liyanage et al. 1992).

Several recent studies have utilized molecular techniques to compare inter- and intraspecies genotype diversity in *Colletotrichum*. Freeman et al. (1993) used a repetitive DNA element recovered from *C. lindemuthianum* and RAPDs to examine diversity of isolates of *C. acutatum*, *C. coccodes*, *C. fragariae*, *C. lindemuthianum*, *C. magna*, *C. orbiculare*, and *C. graminicola*. The study revealed a close correspondence between the molecular groupings identified and the taxonomic delineations although several exceptions were observed. Interspecies similarity was low (<34%), whereas intraspecies similarity was greater than 85% on the basis of RAPD comparisons. Genotype variation was observed among several isolates of *C. gloeosporioides* from strawberry and several isolates of *C. musae* from banana.

A study by Sherriff et al. (1994) utilized ribosomal sequence analysis of the ITS2 region to examine the relatedness between nine species of *Colletotrichum*. Based on the sequence data, the authors concluded that the similarity between isolates of *C. lindemuthianum*, *C. malvarum*, *C. orbiculare*, and *C. trifolii* warrant lumping them under a single species designation, *C. orbiculare*. Considerable heterogeneity was identified among six isolates of *C. gloeosporioides*. The falcate conidial species, *C. capsici*, *C. graminicola*, *C. caudatum*, and *C. truncatum*, were sufficiently dissimilar to warrant delineation as a separate species.

Sexual compatibility recently has been used to characterize mating populations within *C. graminicola* (Vaillancourt and Hanau 1992). Isolates of *C. gloeosporioides* from maize and sorghum belong to a sibling species but represented two genetically isolated mating populations. Examination of the maize and sorghum isolates of *C. graminicola* using mtDNA RFLP and RAPD analysis also supported the conclusion that the maize and sorghum host populations each represent distinct genetic lineages (Vaillancourt and Hanau 1992). Guthrie et al. (1992) also examined isolates of *C. graminicola* from sorghum and johnsongrass from different geographical locations. Although a limited sample size was examined, all isolates from Puerto Rico were quite similar, as were isolates from johnsongrass from the US, indicating that these populations may indeed be clonal. Several polymorphic loci were detected among sorghum isolates of *C. graminicola* from the US.

Although the teleomorph of several species of *Colletotrichum* has been reported, distinct mating populations, like those observed for *C. graminicola*, have not been reported (Bryson et al. 1992; Rodriguez and Owen 1992). As previously discussed, Cisar et al. (1994) demonstrated that, under laboratory conditions, apparently genetically diverse isolates of *C. gloeosporioides* from different hosts may be sexually compatible.

Vegetative compatibility has been examined in several species of *Colletotrichum* including *C. gloeosporioides*, *C. destructivum*, *C. fragariae*, *C. malvarum*, and *C. trifolii* (Brooker et al. 1991). Vegetative compatibility has been very useful in characterizing genetic diversity within populations of *C. dematium* on spinach (Correll et al. 1993) and *C. orbiculare* on cucurbits (Correll et al. 1993a) and *C. gloeosporioides* on almond (Katan and Shabi 1996). In studies with other plant pathogenic fungi, vegetative compatibility groupings have served as a useful framework for guiding investigations of molecular diversity (Leslie 1993).

Anthracnose, caused by *C. orbiculare*, is an important disease of cucurbits worldwide. Examination of a worldwide collection of isolates using vegetative compatibility (Wasilwa et al. 1993), mtDNA RFLPs, RAPDs, and DNA fingerprinting (Correll et al. 1993b) revealed that the cucurbit pathogen population was composed of a single mtDNA RFLP genotype. Isolates recovered from spiny cocklebur (*Xanthium spinosum*) were nonpathogenic on cucurbits, but had a mtDNA RFLP genotype almost identical to the pathogen population. Two other mtDNA RFLP genotypes were detected among isolates that were recovered from cucurbits, but were nonpathogenic on cucurbits, including isolates reported to be the putative teleomorph *Glomerella cingulata* var. *orbiculare* of the anthracnose pathogen. The pathogenic population was composed of three vegetative compatibility groups. Within the pathogenic population, there was a strict correspondence between host origin, vegetative compatibility group, and virulence (race) phenotype. Also, a single RAPD primer effectively differentiated two distinct races of the anthracnose pathogen. Overall, there was a strict correspondence between mtDNA RFLP genotype, DNA fingerprint group, and RAPD group.

A significant limitation to many studies on the genetic population diversity of *Colletotrichum* species is the lack of virulence data provided for many isolates being examined. *Colletotrichum* isolates recovered from a given host may or may not be pathogens of that host. Efforts to more thoroughly understand virulence among isolates from a given host will help in understanding the degree of genotype diversity among host populations of *Colletotrichum*. Although the mechanism of parasexuality has been invoked to explain the extreme genetic diversity found within populations of *C. gloeosporioides* (Alahakoon et al. 1992, 1994), alternative hypotheses to explain the genetic diversity in natural populations of *C. gloeosporioides* have not been adequately explored.

V. Summary and Conclusions

Over the past several decades, significant advances have been made in our understanding of the biology and genetics of *Colletotrichum*. The huge number of described species have been reduced to a manageable number of taxa, and the use of moleular and genetic analyses is beginning to provide a clearer picture of genetic relationships between and within taxa. This work, in combination with population genetic studies, has greatly improved the understanding of host plant relationships of several important species that will enhance our ability to diagnose several important crop diseases accurately. Our understanding of the physiology of infection and pathogenesis is greatly improved and will facilitate the development of crop-resistant cultivars and other disease control strategies. *Colletotrichum* and its *Glomerella* teleomorph could remain a widely useful model system for studies of facultative parasitism, host relationships, latent infectivity, genetic variability, and population biology.

References

Agostini JP, Timmer LW, Mitchell DJ (1992) Morphological and pathological characteristics of strains of *Colletotrichum gloeosporioides* from Citrus. Phytopathology 82:1377–1382

Agrios G (1988) Plant pathology, 3rd edn. Academic Press, New York, pp 381–390

Alahakoon PW, Sreenivasaprasad S, Brown AE, Mills PR (1992) Selection of a genetic variant within *Colletotrichum gloeosporioides* isolates pathogenic on mango by passage through wounded tomato fruits. Physiol Mol Plant Pathol 41:227–240

Alahakoon PW, Brown AE, Sreenivasaprasad S (1994) Cross-infection potential of genetic groups of *Colletotrichum gloeosporioides* on tropical fruits. Physiol Mol Plant Pathol 44:93–103

Astell CR, Ahlstrom-Jonasson L, Smith M, Tatchell K, Nasmyth KA, Hall BD (1981) The sequence of the DNAs coding for the mating-type loci of *Saccharomyces cerevisiae*. Cell 27:15–23

Bailey JA, O'Connell RJ, Pring RJ, Nash C (1992) Infection strategies of *Colletotrichum* species. In: Bailey JA, Jeger MJ (eds) *Colletotrichum*: biology, pathology and control. Commonwealth Agricultural Bureau, Wallingford, pp 88–120

Barrus MF (1911) Variations of varieties of beans in their susceptibility to anthracnose. Phytopathology 1:190–195

Batista UG, Chaves MG (1982) Pathogenicity of monoascosporic cultures of crosses between races of *Colletotrichum lindemuthianum* (Sacc. et Magn) Scrib. Fitopatol Bras 7:15–18

Baxter AP, van der Westhuizen CGA, Eicker A (1983) Morphology and taxonomy of South African isolates of *Colletotrichum*. S Afr J Bot 2:259–289

Baxter AP, van der Westhuizen GCA, Eicker A (1985) A review of the literature on the taxonomy, morphology and biology of the fungal genus *Colletotrichum*. Phytophylactica 17:15–18

Bernstein B, Zehr EI, Dean RA, Shabi E (1995) Characteristics of *Colletotrichum* from peach, apple, pecan, and other hosts. Plant Dis 79:478–782

Bonde MR, Peterson GL, Maas JL (1991) Isozyme comparisons for identification of *Colletotrichum* species pathogenic to strawberry. Phytopathology 81:1523–1528

Braithwaite KS, Manners JM (1989) Human hypervariable minisatellite probes detect DNA polymorphisms in the fungus *Colletotrichum gloeosporioides*. Curr Genet 16:473–475

Braithwaite KS, Irwin JAG, Manners JM (1990) Restriction fragment length polymorphisms in *Colletotrichum gloeosporioides* infecting *Stylosanthes* species in Australia. Mycol Res 94:1129–1137

Brenneman TB, Reilly CC (1989) Recent occurrence of pecan anthracnose caused by *Glomerella cingulata*. Plant Dis 73:775

Brooker NL, Leslie JF, Dickman MB (1991) Nitrate non-utilizing mutants of *Colletotrichum* and their uses in studies of vegetative compatibility and genetic relatedness. Phytopathology 81:672–677

Bryson RJ (1990) Sexual hybridization and the genetics of pathogenic specificity in *Colletotrichum lindemuthianum*. PhD Thesis, University of Birmingham, Birmingham

Bryson RJ, Caten CE, Hollomon DW, Bailey JA (1992) Sexuality and genetics in *Colletotrichum*. In: Bailey JA, Jeger MJ (eds) *Colletotrichum*: biology, pathology and control. Commonwealth Agricultural Bureau, Wallingford, pp 27–46

Cawthon D, Rhoads DD, Correll JC (1995) RAPD isolation of fingerprint probes for analysis of *Colletotrichum* spp. 18th Fungal Genetics Conf, Asilomar

Chacko RJ, Weidemann GJ, TeBeest DO, Correll JC (1994) The use of vegetative compatibility and heterokaryosis to determine potential asexual gene exchange in *Colletotrichum gloeosporioides*. Biol Control 4:382–389

Chilton SJP, Wheeler HE (1949) Genetics of *Glomerella*. VII. Mutation and segregation in plus cultures. Am J Bot 32:549–554

Chilton SJP, Lucas GB, Wheeler HE (1945) Genetics of *Glomerella*. III. Crosses with a conidial strain. Am J Bot 36:549–554

Cisar CR, Spiegel FW, TeBeest DO, Trout C (1994) Evidence for mating between isolates of *Colletotrichum gloeosporioides* with different host specificities. Curr Genet 25:330–335

Cisar CR, Thornton AB, TeBeest DO (1996) Isolates of *Colletotrichum gloeosporioides* (Teleomorph: *Glomerella cingulata* with different host specificities mate on northern jointvetch. Biol Control 7:75–83

Correll JC, Klittich CJR, Leslie JF (1987) Nitrate nonutilizing mutants of *Fusarium oxysporum* and their use in vegetative compatibility tests. Phytopathology 77:1640–1646

Correll JC, Morelock TE, Guerber JC (1993a) Vegetative compatibility and virulence of the spinach anthracnose pathogen, *Colletotrichum dematium*. Plant Dis 77:688–691

Correll JC, Rhoads DD, Guerber JC (1993b) Examination of mitochondrial DNA restriction fragment length polymorphism, DNA fingerprints, and randomly amplified polymorphic DNA of *Colletotrichum orbiculare*. Phytopathology 83:1199–1204

Correll JC, Rhoads DD, Guerber JC (1993c) Taxonomic, mtDNA haplotype, VCG, and morphological diversity of *Colletotrichum* spp. causing fruit-rot of apples. (Abstr.) Phytopathology 83:1412–1413

Correll JC, Rhoads DD, Guerber JC (1994) Genetic and molecular diversity of populations of *Glomerella cingulata*, *Colletotrichum gloeosporioides*, and *Colletotrichum acutatum* from apple fruit. 5th Int Mycology Congr, Vancouver, BC

Daniel JT, Templeton GE, Smith RJ Jr, Fox WT (1969) Biological control of northern jointvetch in rice with an endemic fungal disease. Weed Sci 21:303–307

Dickman MB, Patil SS, Kolattukudy PE (1982) Purification, characterization and role in infection of an extracellular cutinolytic enzyme from *Colletotrichum gloeosporioides* Penz. on *Carica papaya* L. Physiol Plant Pathol 20:333–347

Dickman MB, Podila GK, Kolattukudy PE (1989) Insertion of cutinase gene into a wound pathogen enables it to infect intact host. Nature 342:446–448

Dutta SK, Garber ED (1960) Genetics of phytopathogenic fungi. III. An attempt to demonstrate the parasexual cycle in *Colletotrichum lagenarium*. Bot Gaz 122:118–121

Dyson WC, Rhoads DD, Correll JC (1993) RAPD examination of phylogenetic relationships in *Colletotrichum gloeosporioides*. Phytopathology 83:1413

Edgerton CW (1912) Plus and minus strains in an ascomycete. Science 35:151 (Abstr)

Edgerton CW (1914) Plus and minus strains in the genus *Glomerella*. Am J Bot 1:244–254

Edgerton CW, Chilton SJP, Lucas GB (1945) Genetics of *Glomerella*. II. Fertilization between strains. Am J Bot 32:115–118

Farr DF, Bills CF, Chamuris GP, Rossman AY (1989) Fungi on plant and plant products in the United States. APS Press, St Paul

Freeman S, Rodriguez RJ (1995) Differentiation of *Colletotrichum* species responsible for anthracnose of strawberry by arbitrarily primed PCR. Mycol Res 99:501–504

Freeman S, Pham M, Rodriguez RJ (1993) Molecular genotyping of *Colletotrichum* species based on arbitrarily primed PCR, A + T-rich DNA, and nuclear DNA analyses. Exp Mycol 17:309–322

Freeman S, Katan T, Shabi E (1996) Characterization of *Colletotrichum gloeosporioides* isolates from avocado and almond fruits with molecular and pathogenicity tests. Appl Environ Microbiol 62:1014–1020

Glass NL, Metzenberg RL, Raju NB (1990a) Homothallic Sordariaceae from nature: the absence of strains containing only the a mating type sequence. Exp Mycol 14:274–289

Glass NL, Grotelueschen J, Metzenberg RL (1990b) *Neurospora crassa* A mating-type region. Proc Natl Acad Sci USA 87:4912–4916

Gunnell PS, Gubler WD (1992) Taxonomy and morphology of *Colletotrichum* species pathogenic to strawberry. Mycologia 84:157–165

Guthrie PAI, Magill CW, Fredrickson RA, Odvody GN (1992) Random amplified polymorphic DNA markers-a system for identifying and differentiating isolates of *Colletotrichum graminicola*. Phytopathology 82:832–835

Hodson A, Mills PR, Brown AE (1992) Ribosomal and mitochondrial DNA polymorphisms in *Colletotrichum gloeosporioides* isolated from tropical fruits. Mycol Res 97:329–335

Huttig W (1935) Die Sexualitat bei *Glomerella lycopersici* Kruger und ihre Vererbung. Biol Zentralbl 55:74–83

Jeffries P, Dodd JC, Jeger MJ, Plumbley RA (1990) The biology and control of *Colletotrichum* species on tropical fruit crops. Plant Pathol 39:343–366

Jenkins SF, Winstead NN (1964) *Glomerella magna*, cause of a new anthracnose of cucurbits. Phytopathology 54:452–454

Katan T, Shabi E (1996) Vegetative compatibility among isolates of *Colletotrichum gloeosprorioides* from almond in Israel. Eur J Plant Pathol (in press)

Kimati H, Galli F (1970) *Glomerella cingulata* (Stonem) Spaul et v Schrenk f. sp. *phaseioli*, fase ascogena do agente causal da antracnose do feijoeiro. An Esc Super Agric "Luis de Queiroz" 27:411–437

Kohn LM (1992) Developing new characteristics for fungal systematics: an experimental approach for determining the rank of resolution. Mycologia 84:139–153

Kolattukudy PE (1985) Enzymatic penetration of the plant cuticle by fungal pathogens. Annu Rev Phytopathol 23:223–250

Lapp MS, Skoropad WP (1978) Location of appressoria of *Colletotrichum graminicola* on natural and artificial barley leaf surfaces. Trans Br Mycol Soc 70:225–228

Leslie JF (1993) Fungal vegetative compatibility. Annu Rev Phytopathol 31:127–150

Liyanage HD, McMillan RT Jr, Kistler HC (1992) Two genetically distinct populations of *Colletotrichum gloeosporioides* citrus. Phytopathology 82:1371–1376

Liyanage HD, Koller W, McMillan RT Jr, Kistler HC (1993) Variation in cutinase from two populations of *Colletotrichum gloeosporioides* from citrus. Phytopathology 93:113–116

Lucas GB, Chilton SJP, Edgerton CW (1944) Genetics of *Glomerella*. I. Studies on the behavior of certain strains. Am J Bot 31:233–239

Manandhar JB, Hartman GL, Sinclair JB (1986) *Colletotrichum destructivum*, the anamorph of *Glomerella glycines*. Phytopathology 76:282–285

Manners JM, Masel A, Braithwaite KS, Irwin JAG (1992) Molecular analysis of *Colletotrichum gloeosporioides* pathogenic on the tropical pasture legume *Stylosanthes*. In: Bailey JA, Jager MJ (eds) *Colletotrichum*: biology, pathology, and control, CAB International, Wallingford

Masel A, Braithwaite KS, Irwin JAG, Manners JM (1990) Highly variable molecular karyotypes in the plant pathogen *Colletotrichum gloeosporioides*. Curr Genet 18:81–86

Masel AM, Irwin JAG, Manners JM (1993) DNA addition or deletion is associated with a major karyotype polymorphism in the fungal phytopathogen *Colletotrichum gloeosporioides*. Mol Gen Genet 237:73–80

McClean DM (1966) Sexual expression in *Colletotrichum orbiculare* Plant Dis Rep 50:871–873

McCormick TW, Correll JC, Rhoads DD (1994) Variation of ITS1 among isolates of *Colletotrichum orbiculare*, *C. magna*, and *Glomerella cingulata* var. *orbiculare*. Phytopathology 84:1097 (Abstr)

McCormick TW, Correll JC, Rhoads DD (1995) RFLP analysis of the intergenic spacer region (IGS) of selected isolates of *Colletotrichum* spp. Phytopathology 85: 1118 (Abstr)

Meiners JP (1981) Genetics of disease resistance in edible legumes. Annu Rev Phytopathol 21:189–209

Miehle BR, Lukezic FL (1972) Studies on conidial germination and appressorium formation by *Colletotrichum trifolii* Bain and Essary. Can J Microbiol 18:1263–1269

Mills PR, Hodson A, Brown AE (1992a) Molecular differentiation of *Colletotrichum gloeosporioides* isolates infecting tropical fruits. In: Bailey JA, Jager MJ (eds) *Colletotrichum*: biology, pathology, and control. CAB International, Wallingford

Mills PR, Sreenivasaprasad S, Brown AE (1992b) Detection and differentiation of *Colletotrichum gloeosporioides* isolates using PCR. FEMS Microbiol Lett 98:137–144

Mordue JEM (1967) *Glomerella tucumanensis*. CMI Descriptions of pathogenic fungi and bacteria, No 133. Easter Press, London

Muirhead IF (1981) The role of appressorial dormancy in latent infection. In: Blakeman JP (ed) Microbial ecology of the phylloplane. Academic Press, New York, pp 155–167

Nicholson RL, Epstein L (1991) Adhesion of fungi to the plant cuticle. In: Roberts DW, Aist JR (eds) The fungal spore and disease initiation in plants and animals. Plenum Press, New York, pp 3–23

Nicholson RL, Moraes WBC (1980) Survival of *Colletotrichum graminicola*: importance of the spore matrix. Phytopathology 70:255–261

Nicholson RL, Butler LG, Asquith TN (1986) Glycoproteins from *Colletotrichum graminicola* that bind phenols: implications for survival and virulence of phytopathogenic fungi. Phytopathology 76:1315–1318

Nicholson RL, Hipskind J, Hanau RM (1989) Protection against phenol toxicity by the spore mucilage of *Colletotrichum graminicola*, an aid to secondary spread. Physiol Mol Plant Pathol 35:243–252

Olive LS (1951) Homothallism and heterothallism in *Glomerella*. Trans NY Acad Sci 13:238–242

Parberry DG, Blakeman JP (1978) Effects of substances associated with leaf surfaces on appressorium by *Colletotrichum acutatum*. Trans Br Mycol Soc 70:7–19

Pascholati SF, Yoshioka H, Kunoh H, Nicholson RL (1992a) Preparation of the infection court by *Erysiphe graminis* f.sp. *hordei*: cutinase is a component of the conidial exudate. Physiol Mol Plant Pathol 41:53–59

Pascholati SF, Deising H, Leite B, Anderson D, Nicholson RL (1992b) Cutinase and non-specific esterase activities in the conidial mucilage of *Colletotrichum graminicola*. Physiol Mol Plant Pathol 42:37–51

Podila GK, Rogers LM, Kolattukudy PE (1993) Chemicals signals from avocado surface wax trigger germination and appressorium formation in *Colletotrichum gloeosporioides*. Plant Physiol 103:267–272

Politis DJ (1975) The identity of the perfect state of *Colletotrichum graminicola*. Mycologia 67:56–62

Puhalla JE (1985) Classification of strains of *Fusarium oxysporum* on the basis of vegetative compatibility. Can J Bot 63:179–183

Ramadoss CS, Uhlig J, Carlson DM, Butler LG, Nicholson RL (1985) Composition of the mucilaginous spore matrix of *Colletotrichum graminicola*, a pathogen of corn, sorghum and other grasses. J Agric Food Chem 33:728–732

Rodriquez RJ, Owen JL (1992) Isolation of *Glomerella musae* [Teleomorph of *Colletotrichum musae* (Berk & Curt.)Arx] and [segregation analysis of ascospore progeny. Exp Mycol 16:291–301

Sreenivasaprasad S, Mills PR, Brown AE (1994) Nucleotide sequence of the rDNA spacer 1 enables identification of isolates of *Colletotrichum* as *C. Acutatum*. Mycol Res 98(2):186–188

Shear CL, Wood AK (1913) Studies of fungus parasites belonging to the genus *Glomerella*. US Dep Agric, Bureau of Plant Industry, Bull 252, pp 1–110

Sherriff C, Whelan MJ, Arnold GM, LaFay JF, Brygoo Y, Bailey JA (1994) Ribosomal DNA sequence analysis reveals new species groupings in the genus *Colletotrichum*. Exp Mycol 18:121–138

Shi Y, Rom C, Corrrell JC (1993) Examination of virulence of isolates of the bitter rot pathogen on apple fruit. HortScience 28:94–95

Shi Y, Correll JC, Guerber JC, Ron CR (1996) Frequency of *Colletotrichum* species causing bitter rot of apple in the Southeastern United States. Plant Dis 80:692–696

Simmonds JH (1963) Studies in the latent phase of *Colletotrichum* species causing ripe rots of tropical fruits. Queensl J Agric Sci 20:373–424

Sreenivasaprasad S, Brown AE, Mills PR (1992) DNA sequence variation and interrelationships among *Colletotrichum* species causing strawberry anthracnose. Physiol Mol Plant Pathol 41:265–281

Sreenivasaprasad S, Brown AE, Mills PR (1993) Coffee berry disease pathogen in Africa: genetic structure and relationship to the group species *Colletotrichum gloeosporioides*. Mycol Res 97:995–1000

Staben C, Yanofsky C (1990) *Neurospora crassa* a mating-type region. Proc Natl Acad Sci USA 87:4917–4921

Stahl DJ, Schafer W (1992) Cutinase is not required for fungal pathogenicity on pea. Plant Cell 4:621–629

Staples RC, Macko V (1984) Germination of urediospores and differentiation of infection structures. In: Bushnell WR, Roelfs AR (eds) The cereal rusts, vol I. Origins, specificity, structure and physiology. Academic Press, New York, pp 255–289

Stephan BR (1968) Untersuchungen zum Nachweis der Heterokaryose bei *Colletotrichum gloeosporioides* Penzig unter Verwendung auxotropher Mutanten. Zentralbl Bakteriol Parasitenkd Abt II 122:420–435

Struble FB, Keitt GW (1950) Variability and inheritance in *Glomerella cingulata* (Stonem.) S. and v. S. from apple. Am J Bot 37:563–574

Sutton BC (1980) The Coelomycetes: fungi imperfecti with pycnidia, acervuli and stromata. Commonwealth Mycological Institute, Kew, Surrey, pp 523–539

Sutton BC (1992) The genus *Glomerella* and its anamorph *Colletotrichum*. In: Bailey JA, Jeger MJ (eds) *Colletotrichum*: biology, pathology and control. CAB International, Wallingford, pp 1–26

Suzuki K, Furasawa I, Ishida N, Yamamoto M (1982) Chemical dissolution of cellulose membranes as a prerequisite for penetration from appressorium of *Colletotrichum lagenarium*. J Gen Microbiol 128:1035–1039

Sweigard JA, Chumley FG, Valent B (1992a) Cloning and analysis of **CUT1**, a cutinase gene from *Magnaporthe grisea*. Mol Gen Genet 232:174–182

Sweigard JA, Chumley FG, Valent B (1992b) Disruption of a *Magnaporthe grisea* cutinase gene. Mol Gen Genet 232:183–190

Swinburne TR (1976) Stimulants of germination and appressorium formation by *Colletotrichum musae* (Berk. & Curt.) Arx in banana leachate. Phytopathol Z 87:74–90

Swinburne TR (1983) Quiescent infections in post-harvest diseases. In: Dennis C (ed) Post-harvest pathology of fruits and vegetables. Academic Press, London, pp 1–21

Templeton GE (1992) Use of *Colletotrichum* strains as mycoherbicides. In: Bailey JA, Jeger MJ (eds) *Colletotrichum*: biology, pathology and control. CAB International, Wallingford, pp 358–380

Templeton GE, Heiny DK (1989) Improvement of fungi to enhance mycoherbicide potential. In: Whipps JM, Lumsden RD (eds) Biotechnology of fungi for improving plant growth. Cambridge University Press, Cambridge, pp 127–151

Uecker FA (1994) Ontogeny of the ascoma of *Glomerella cingulata*. Mycologia 86:82–88

Vaillancourt LJ, Hanau RM (1991) A method for genetic analysis of *Glomerella graminicola* (*Colletotrichum graminicola*) from maize. Phytopathology 81:530–534

Vailancourt LJ, Hanau RM (1992) Genetic and morphological comparison of *Glomerella* (*Colletotrichum*) isolates from maize and sorghum. Exp Mycol 16:219–229

Vaillancourt LJ, Hanau RM (1994) Nitrate-nonutilizing mutants used to study heterokaryosis and vegetative compatability in *Glomerella graminicola* (*Colletotrichum graminicola*). Exp Mycol 18:311–319

von Arx JA (1957) Die Arten der Gattung *Colletotrichum* Cda. Phytopathol Z 29:413–468

von Arx JA (1981) The genera of fungi sporulating in pure culture. Strauss and Cramer, Vaduz

von Arx JA, Muller E (1954) Die Gattungen der amerosporen Pyrenomyceten. Beitr Kryptogamenflora Schweiz 11:1–434

von Arx JA, Muller E (1970) A revision of the fungi classified as *Gloeosporium*. Bibl Mycol 24:1–203

Wasilwa LA, Correll JC, Morelock TE, McNew RE (1993) Reexamination of races of the cucurbit anthracnose pathogen *Colletotrichum orbiculare*. Phytopathology 83:1190–1198

Wheeler HE (1956) Linkage groups in *Glomerella*. Am J Bot 43:1–6

Wheeler HE, McGahan JW (1952) Genetics of *Glomerella*. X. Genes affecting sexual reproduction. Am J Bot 39: 110–119

11 The Epidemiology of Mycorrhizal Fungal Infection During Succession

M.F. Allen, J.N. Klironomos, and S. Harney

CONTENTS

I.	Introduction	169
II.	Linkage and Epidemiology and Succession	170
III.	Disturbances and Mycorrhizal Survival	171
A.	Aboveground disruption	171
B.	Minor Soil Perturbations	172
C.	Severe Soil Disturbance	172
D.	Bootstrapping and Survival of Residuals	172
IV.	Immigration	173
A.	Biotic Vectors	173
B.	Physical Vectors	174
C.	Interactions of Biotic and Abotic Vectors	175
D.	Artificial Inoculation	176
E.	Source Areas, Landscape Structure, and Spatial Models	176
	1. Mount St. Helens: a Case Study	176
V.	Establishment	177
A.	Physiological/Environmental Constraints	178
B.	Genetic Constraints	178
C.	Spatial Constraints	179
VI.	Community Diversity and Mycorrhizal Fungal Epidemiology	180
VII.	Conclusions	181
	References	181

I. Introduction

Mycorrhizae are essential symbioses regulating the trajectory of succession following severe perturbation of a site (E.B. Allen and M.F. Allen 1990). As Pickett and McDonnell (1989) noted, succession is probably the single law of ecology. However, the course of succession is not a predetermined one; rather, it depends on the composition of residuals surviving a disturbance (a function of the degree and type of perturbation), the activity of vectors, and the changing chemical/physical environment (E.B. Allen 1988). Because a mycorrhiza is a symbiosis of fungus and plant, conceptual models that describe mycorrhizal development and the effects of mycorrhizae on ecosystems require an understanding of the ecology of both organisms independently, together with information on their interactions.

In ecological time, the only function of an organism is to maintain itself until it can reproduce. To do this, a single organism must first find a suitable habitat and then survive until reproduction, thus maximizing fitness. Similarly, with mutualistic symbioses, two organisms must maximize fitnesses jointly, although not necessarily at the same time. In such situations, physiological interdependence and reproductive separateness can mitigate effective dispersal of the association.

Mycorrhizal fungi and their host plants are sessile organisms which reproduce independently. One strategy for the invasion of disturbed habitats by sessile organisms might involve ingrowth from contiguous surrounding areas. A few plants, such as those with rhizomes, including *Ammophila* and *Ipomea*, are able to grow across open dunes. Hyphae can expand into open habitats and along roots in a similar manner. However, invasion by such means proceeds slowly, and the two symbionts in mycorrhizal associations often advance at different rates. Warner and Mosse (1980) estimated that *Glomus mosseae* invaded an onion patch at about 10 cm $year^{-1}$, while the host plants could invade sites at rates up to a few meters per annum. Since most disturbed sites are colonized more rapidly than this, in the face of a severe disturbance both plant and fungus must persist dormantly in the absence of a compatible symbiont or reinvade independently across a matrix of unfavorable conditions.

Often agriculture and restoration of native vegetation are viewed as distinctly different systems with little commonality in underlying ecological processes. However, techniques for sustainable agriculture depend on the establishment of a stable soil environment that supports the growth and production of desirable crops (Bethlenfalvay and Linderman 1992). The desired crop

Department of Biology, Soil Ecology and Restoration Group, San Diego State University, San Diego, California 92182-4614, USA

The Mycota V Part B
Plant Relationships
Carroll/Tudzynski (Eds.)
© Springer-Verlag Berlin Heidelberg 1997

can be viewed as an intermediate successional state that is maintained through tillage or other combination of practices. From the standpoint of a mycorrhizal fungus, agriculture may be no different from restoration practices that seek to establish a desirable mix of native vegetation.

During the remainder of this chapter, we will discuss the mechanisms whereby organisms survive a disturbance, reinvade a disturbed habitat, and successfully reestablish in the face of continually changing conditions. To organize thoughts about succession, we will use Clements' (1916) conceptual model of processes (nudation, residual survival, immigration, establishment) that comprise succession. Further details on these processes and additional biotic interactions can be found elsewhere (Miller and Allen 1982; M.F. Allen 1991).

II. Linkage of Epidemiology and Succession

Most studies of fungal epidemiology are focused on preventing a pathogen from invading a particular tract of cropland. However, mycorrhizal fungi are deemed desirable, and thus their establishment and persistence are important. We generally assume that mycorrhizal fungi are already present at a site prior to a perturbation. Therefore, most efforts at studying mycorrhizal fungi involve reestablishing these fungi in sites where they are missing, or changing the species composition where less desirable mycorrhizal species are present. Thus, the emphasis in studying the epidemiology of mycorrhizal fungi differs from the focus of more traditional work on plant pathogens. With plant pathogens in agroecosystems, one studies a small patch of fungal colonization within the larger matrix where the fungus does not occur. With mycorrhizae, one usually deals with a larger matrix with the fungus and concentrates on the smaller patch with few or no mycorrhizae (Fig. 1).

For plant pathogens, most efforts at describing epidemiological processes are based on mathematical models of dispersal and deposition (one patch into another patch). These models are based on physical principles that have been verified by experimental observations; usually they are adequate because the conditions prevailing in the area from which a plant pathogen migrates are similar to those of the habitat which it invades.

When dealing with mycorrhizae, the epidemiological situation is reversed, with dispersal from matrix to patch. To date, the observational database is insufficient for comparisons between these two inverse epidemiological circumstances. Further, after disturbance, large differences in the

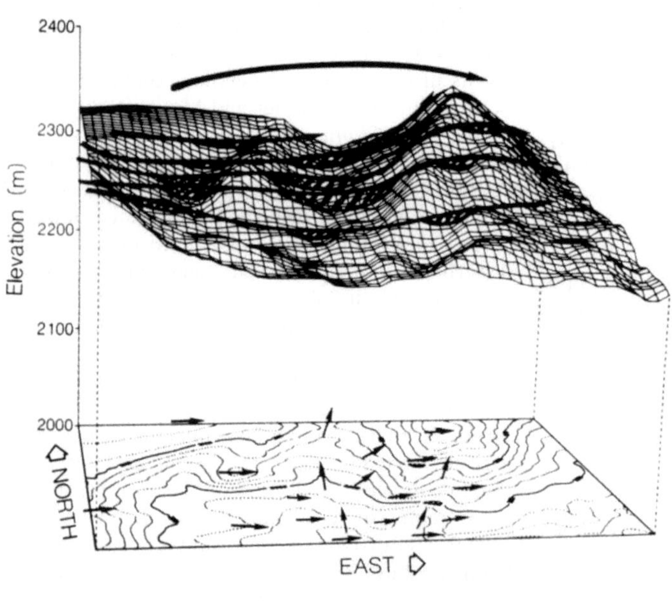

Fig. 1. Structure of the landscape from which and to which mycorrhizal fungi are expected to migrate. Shown is the Kemmerer restoration project with the wind structure across the complex terrain. The disturbed area is the small valley to the *right*. All else is undisturbed matrix with mycorrhizae. More information can be found in M.F. Allen (1991) and M.F. Allen et al. (1989c)

physical environment of patch and matrix may prevail, a fact which led Marx (1975) to suggest that the use of exotic inoculum would prove essential for the reintroduction of mycorrhizae into severely disturbed habitats.

Considerable effort has been made to introduce exotic or native mycorrhizal fungi into disturbed sites, either to enhance production or restore natural vegetation. Where enhanced production is the goal, either short-term large increases in annual production or longer-term moderate increases in sustained yield may be envisioned. Often, this may entail introducing an inoculum into an existing plant community. Where site restoration is the goal, mycorrhizae are introduced to manage native ecosystems for preservation of endangered organisms or habitats. In this case, the primary effort is to reintroduce mycorrhizal fungi (although not necessarily the same species) into conditions where the existing inoculum has declined or disappeared.

Although these two goals differ considerably, the underlying situations which they seek to alter are the same. Individual hyphae of mycorrhizal fungi are microscopic and comprise only small fractions of the soil volume or root surface. (Nevertheless, that may be adequate to generate some interesting patterns of competition; Miller and Allen 1992.) Moreover, even in the most stable of environments, soils are subjected to small disturbances such as gopher and ant diggings that create new resources for occupation and can be considered in a successional context (M.F. Allen 1991). Thus, the models for mycorrhizal fungal epidemiology may be applicable to both situations.

III. Disturbances and Mycorrhizal Survival

Disturbance is a broadly applicable term. It applies to minor perturbations that reduce aboveground plant cover and also to events which affect large areas such as a volcanic eruption or a glacial moraine. The types of disturbances that affect mycorrhizae also run a gamut from minor to catastrophic. Rangeland fires or herbicides may selectively kill off trees, reducing ectomycorrhizae and favoring arbuscular mycorrhizal plants, thereby indirectly affecting the fungal soil communities. In contrast, a large-scale volcanic eruption such as the explosion of Mt. St. Helens can completely sterilize large areas. We will discuss three predominant types of disturbance (above ground only, minor soil perturbations, and major soil disturbances) and show how these affect the ability of mycorrhizal fungi to survive.

A. Aboveground Disruption

Aboveground disruption tends to affect mycorrhizal fungi indirectly and to involve a wide array of disturbances that reduce, alter, or eliminate leafy and standing dead plant material. This type of disturbance includes fire, pesticides, and herbivory. In such cases, elimination of carbon fixation by host plants deprives the fungi of energy. In nature, this is a common and natural occurrence, and mycorrhizal fungi have made some remarkable adaptations for surviving such disturbances. In other cases, these disruptions may impose regulatory constraints on the ecosystem.

In general, the influence of simple aboveground disruptions on belowground organisms and mycorrhizal fungi is complicated, although of minimal long-term significance where production is not limited by available sunlight. Examples include habitats with high-intensity sunlight dominated by grasses with buried meristems that are adapted to grazing. Thus, M.F. Allen et al. (1989b) noted that, depending on the year and sampling regime, mycorrhizal activity increased, decreased, or showed no change in response to clipping of grasses. They concluded that the effect in this grassland ecosystem was probably transitory. Gibson and Hetrick (1988) reported that fire had little effect on mycorrhizal infection but did have some effect on mycorrhizal species composition.

In contrast, leaf removal can significantly reduce mycorrhizae in shrublands and forests where shading or low light levels limit production somewhere in the community. In a mixed shrub-grassland, E.B. Allen and West (1994) found that herbicide removal of the dominant shrub, sagebrush, reduced spore counts of VA mycorrhizae and total root infection, a change to be expected, since eliminating the shrub which constituted the major plant cover and source of fixed carbon also eliminated carbon input to the soil. Mycorrhizae associated with the grasses appeared to be unaffected by the treatment. Similarly, C. Klopatek et al. (unpubl.) found that the mycorrhizae of trees were virtually eliminated following fire but that the interspace grasses in pinyon-juniper woodlands were unaffected. This suggests that the total mycorrhizal inoculum was reduced by reducing carbon inputs from the dominant plant but that at the scale of individual plants, grass biomass supported the remainder. Further, Gehring and Whitham (1994) demonstrated that ectomycorrhizae clearly declined in pinyon pine in response to heavy insect herbivory and that mycorrhizal activity declined about a year following the Yellowstone fire after fungal carbon reserves had been exhausted (S. Miller, pers. comm.). On Mount St. Helens, arbuscular mycorrhizal spores persisted for about 2 to 4 years following the eruption in buried, intact soils (M.F. Allen et al. 1992b); basidiospores and sclerotia persisted about 2 years following fire in the Grand Teton Park (Miller et al. 1994).

B. Minor Soil Perturbations

Minor soil perturbations are common in "undisturbed" conditions. These range from natural disturbances such as animal diggings and animal fertilizations, to small-patch perturbations by humans for roads, trampling by grazing animals, and tillage for small agricultural plots. In most cases, mycorrhizal inoculum is reduced, although minimally. For example, gopher mounds in undisturbed habitats have been shown to contain reduced inoculum (M.F. Allen et al. 1984; Koide and Mooney 1987); Wallace (1987) suggested that soil compaction by grazing animals was responsible for inoculum decline more than the grazing itself. In small pasture plots established in tropical seasonal forests, the plant infection rates were unaffected by the conversion from forest to introduced grasses, but the fungal species diversity was reduced dramatically (M. Allen, unpubl.). In some cases, such disturbances actually increase local inoculum density. Examples include badger mounds in which the badgers preferentially dig into shrub root systems and move the soil to the interspace regions (M. Allen, unpubl.) and ant mounds in which harvester ants use mycorrhizal root tips to construct the seed chambers, resulting in high-density patches of mycorrhizal inoculum (Friese and Allen 1993).

Smaller-scale human disturbances can also reduce mycorrhizal inoculum. Reeves et al. (1979) reported that highway cuts had lower inoculum densities than the surrounding undisturbed matrix.

C. Severe Soil Disturbance

Severe disturbance of the soil eliminates or substantially reduces mycorrhizal inoculum. When topsoil is retained, some mycorrhizal propagules survive. Examples include strip mines where the soils are disrupted but retained (e.g., Miller 1979; E.B. Allen and M.F. Allen 1980; Rives et al. 1980), volcanic eruptions in which the topsoil is simply preserved, but not too deeply, under the ashfall (M.F. Allen et al. 1984), and tillage agriculture, where the soil is disrupted seasonally (M.F. Allen and Boosalis 1983; McGonigle and Miller 1990). When the topsoil is not preserved, the mycorrhizal inoculum approaches nil. Examples include clearcutting and slash burning (Pilz and Perry 1984), strip mining where the topsoil is not retained (Waaland and Allen 1987), volcanic eruptions under pyroclastic flows and deep ashfall (M.F. Allen et al. 1984), and glacial retreats (M.F. Allen et al. 1994; Cazares and Trappe 1994a).

D. Bootstrapping and Survival of Residuals

In most instances, mycorrhizal fungi are not completely eliminated from a disturbed site. Mycelium, infected roots, and spores all survive most disturbances and can reinitiate new infections with surviving or invading plants. The surviving mycorrhizal fungi can dictate successional trajectories since they preferentially support some plant species over others (E.B. Allen and M.F. Allen 1990). For example, Molina and Trappe (1982) noted that specificity existed among mycorrhizal fungi, and if *Alpova* survived a disturbance, then *Alnus* would preferentially receive mycorrhizal benefits. E.B. Allen and M.F. Allen (1984) found clear competitive advantages for nonmycotrophic species following a disturbance when mycorrhizal fungi were eliminated, but when the fungi were present, the mycorrhizal plants outcompeted the nonmycotrophs. At the chaparral/conifer forest boundary, mycorrhizal fungi could be found associated with madrone after fire; the same fungi also rapidly formed mycorrhizae with invading seedlings of Douglas fir which persisted into the mature forest (Amaranthus and Perry 1989).

These observations led Perry et al. (1989) to suggest that plant communities survive disturbances by "bootstrapping". That is, mycorrhizal fungi and other microorganisms survive many disturbances and initiate new soil communities with the surviving vegetation. This, in turn, allows the ecosystems to "pull themselves up by their bootstraps". Clearly, residual survival is important. When topsoil is protected from disturbance (e.g., mining with retention of mycorrhizae), the subsequent inoculum and plant species composition more nearly resembles that of the native vegetation. Alternatively, in large severe disturbances, when no inoculum is retained, both the plant and mycorrhizal fungal communities do not resemble the native vegetation, and the mycorrhizal inoculum density may not recover, even after several decades (Waaland and Allen 1987).

IV. Immigration

In any situation where mycorrhizal inoculum is reduced or eliminated, immigration by natural vectors or artificial inoculation is important for further development of the plant community. However, the patterns of immigration depend on the vectors and on the spatial relationships of the disturbance to source areas. Teasing apart these relationships is crucial to determining if and when inoculation is important and what reinvasion patterns will emerge.

A. Biotic Vectors

A wide array of biotic vectors disperse mycorrhizal fungi. These include animals that intentionally feed on mycorrhizal fungi and transport inoculum, as well as those that incidentally transport fungi on body parts or unintentionally ingest infected root fragments and spores while feeding on other items. Not all vectors transport viable inoculum. For example, we have found spores of arbuscular mycorrhizal fungi in the guts of grasshoppers, but have been unable to germinate them.

Many rodents ingest mycorrhizal fungi and transport these fungi along their travel routes. Examples include gophers in subalpine and alpine meadows (Maser et al. 1978), *Prochimys* in Peruvian tropical forests (Emmons 1982), rats in the Australian *Eucalyptus* forests (McGee and Baczocha 1994), squirrels in pine forests (Kotter and Farentinos 1984; States 1985), and deer mice in shrub steppes (Warner et al. 1987). Large mammals may also serve as vectors. These include moose, deer, elk, and antelope (M.F. Allen 1987; Warner et al. 1987; Cazares and Trappe 1994a). Between 75 and 100% of all grizzly bear feces contained roots of legumes such as *Hedysarum* (Murie 1962). Depending on the location, such plants can have arbuscular or ectomycorrhizae. Even invertebrates have been found to serve as inoculum vectors (e.g., McIlveen and Cole 1976; Ponder 1980; Harinikumar and Bagyaraj 1994).

Some of these animals may move considerable distances into disturbed areas. For example, deer mice have been found to move up to 1 km in a single night (Broome 1987) and often become a dominant animal moving from undisturbed vegetation onto reclaimed strip mines and agricultural fields. The home range of a grizzly bear is several hundred square km.

Other animals burrow through soils or walk across soil surfaces and carry bits of soil containing inoculum with them as they migrate. Gophers were found to move considerable amounts of soil and could inoculate newly established vegetation patches on the Mount St. Helens Pumice Plain (M.F. Allen and MacMahon 1988). Wapati fed in undisturbed meadows by tearing up grasses with soil and infected roots attached. They then transported the inoculum in the roots across the pumice disturbances inoculating plant seedlings up to several km distant (Fig. 2). Even cicadas burrowing back up out of the soil contained spores attached to the outsides of their bodies (M. Allen, unpubl. observ.).

Understanding the effectiveness of biotic vectors depends not only on listing the vectors but also on estimating their effective dispersal rate. For animals, the limits to dispersal are twofold. The first is the effective distance over which a small animal is likely to move a spore, and the second is the number of propagules carried by the animal.

For example, we found a high density of deer mice on shrub-steppe disturbed site in one of our studies. However, assuming that a mouse ingested a sporocarp at the edge of the undisturbed site, it would take only a few hours before the spore was egested, because of the animal's small size. In a 24-h period, deer mice have been found to move over 1 km, but in a nonlinear fashion, searching for food. We predict that spores from the source area would migrate at most about 20–50 m before being deposited. Subsequent migration would require germination, establishment, resporulation, reingestion. During a limited growing season, mycorrhizal fungi migrate at most only a few m.

Fig. 2. Emergence of mycorrhizal seedlings from elk feces on the Pumice Plain of Mount St. Helens compared to the naturally invading nonmycorrhizal adjacent seedlings. (Data discussed in M.F. Allen 1987)

At that rate, we would expect that recolonization of an area the size of our Wyoming restoration site (500 ha) would take a decade. However, within a narrow corridor, such as along a roadside, colonization across the entire disturbance could occur within a single growing season.

Intermediate-sized animals should move spores farther, but their density is lower. We found that jack rabbits ingested mycorrhizal root fragments and spores when they grazed on grasses in loose soils. We also found pellets of jackrabbits with mycorrhizal fungi scattered at low densities all across the shrub-steppe site during the first year (M.F. Allen 1988). Those pellets may have been important in moving viable inoculum onto the larger sites.

For larger animals, the low density of animals makes high density dispersal unlikely. Thus, while wapiti can migrate several km in a 24-h period, an individual may deposit inoculum at only three or four points per trip. Thus, only widely dispersed sites would be inoculated, although these would be numerous across a very large area. At Mount St. Helens, in an area about 5 km in diameter, we found only widely dispersed patches of plants that were mycorrhizal during the first 5 years (M.F. Allen 1987).

B. Physical Vectors

Physical vectors, including both wind and water, are also effective at dispersing propagules of mycorrhizal fungi. Wind is probably the best known of these vectors. Ingold (1971) described wind movement around a mushroom that could entrain spores of epigeous fungi into the prevailing winds for dispersal. Warner et al. (1987) and M.F. Allen et al. (1989c), using trapped spores and spore mimics, further demonstrated that arbuscular mycorrhizal fungi could be dispersed up to 2 km by wind. It is often assumed that epigeous fungi are primarily dispersed by wind, while hypogeous fungi are dispersed by animals. However, wind remains a tricky and highly unpredictable factor, and wind dispersal is equally difficult to predict. These difficulties arise because wind usually exhibits not laminar but turbulent flow, with all of the associated problems with calculation.

Wind dispersal of mycorrhizal fungi can be divided into three crucial processes: entrainment, migration, and deposition. During entrainment, fungal propagules enter into the large-scale wind-flow patterns. Once entrained, the propagules are dispersed as long as the turbulence maintains them within the atmosphere. Spores are deposited when the vertical velocity drops below that of the terminal velocity of the propagule.

Generally, one can assume that the vertical velocities of wind approximate 10% of the horizontal velocity. At the Mount St. Helens site, the wind velocities of the forest rarely exceeded $100\,cm\,s^{-1}$ and were much lower at the forest floor where the epigeous mycorrhizal fungal sporocarps predominated (M.F. Allen 1987). This suggests a vertical wind velocity of only $1–10\,cm\,s^{-1}$. The range of the settling velocities of many ectomycorrhizal fungal spores is around 5 cm s^{-1}. Indeed, when we experimentally measured the horizontal dispersal of spores from a *Russula emetica* sporocarp, we found that dispersal of spores beyond 20 cm from the mushroom occurred only when we placed a fan on the forest floor (M.F. Allen et al. 1993). In the field, one often finds that an intense deposition of spores occurs directly beneath mushroom sporocarps, equivalent to spore prints in the lab.

However, production of fruiting bodies is not limited to the middle of the forest. *Thelephora* spores are commonly found in wind traps, located some distance from the nearest forest, 5 km in the case of one of ours (M.F. Allen 1987). Many fungi fruit in canopy gaps and forest margins. These habitats are likely to have turbulent bursts that are capable of entraining spores from epigeous fungi. In another study, we measured the light at the point of a single sporocarp of a *Russula emetica* compared with a 1-m light bar that integrated the surrounding light environment. Through a daily period, we found that the light was consistently higher than for the surrounding area, indicating that there may have been preferential fruiting in a gap in the forest. Potentially, fruiting can occur across a wide area, since an individual mycelium can extend several m across the forest floor. Further observations are needed with a wide array of fungi, but these observations suggest that some fungi may preferentially fruit in gaps where dispersal is most likely.

Other fungi may preferentially fruit along forested margins or in small patches, such as pine or oak savannas, where wind dispersal could be high. At our shrub-steppe habitats in southern California, we have observed that *Pisolithus tinctorius* always fruited well beyond the canopy edge when associated with *Quercus englemannii* or *Chrysolepis sempervirens*. Near the glaciers at the Kenai Fjords in southern Alaska, sporocarps of *Inocybe* spp. could be found fruiting along the edge of alder thickets where turbulence is probably adequate to entrain spores, but were rarely found well under the canopy at the base of the plant. Clearly then, wind dispersal alone is not as common a mechanism for dispersal of all epigeous fungi as has generally been thought in forested habitats. Dispersal by wind appears to be a function of both fruiting body structure and habitat architecture.

In arid to semiarid regions, wind becomes of greater importance as an agent of dispersal. At a sagebrush steppe site in western Wyoming, we commonly measured horizontal wind velocities of 1000–1500 cm s^{-1} with vertical velocities (measured using a sonic anemometer at 20 Hz) of 100–300 cm s^{-1}. Interestingly, because of the low settling velocities of most ectomycorrhizal fungal spores, fungi in these habitats have an almost indefinite dispersal capacity once their spores are entrained in wind currents. Indeed, at this site we often found unknown basidiomycete spores along with pine pollen even though the nearest ectomycorrhizal trees were over 30 km distant. The smaller *Glomus* spores, such as *G. occultum*, *G. deserticola*, and *G. albidum*, have terminal velocities of about 20 cm s^{-1} (M.F. Allen et al. 1993). These fungi have wide distributions across the Great Basin. Many larger-spored species, such as *Acaulospora elegans*, *A. gerdemannii*, and *Gigaspora margarita* have terminal velocities of 200 to 400 cm s^{-1} and localized distribution patterns within this region. We were never able to collect the larger-spored arbuscular mycorrhizal fungi on our wind traps from Wyoming, Utah, Nevada, or California sites.

Another less-well documented vector is water. Friese (1984) and Gemma and Koske (1990) noted that strands of mycorrhizal *Ammophila* were found that had washed out to sea and been redeposited back on beach fronts. The inoculum remained viable and could serve to reinoculate beach-front plant communities. There are many rhizomotous species living on primary dunes, and in these habitats mycorrhizal inoculum appears act as one constraint on succession (E.B. Allen and M.F. Allen 1990; Gemma and Koske 1990). In contrast, we could not find spores in erosion depositional areas on Mount St. Helens. There may be important differences in the survival of spores versus root inoculum in materials immersed in water.

C. Interactions of Biotic and Abiotic Vectors

Generally, studies of mycorrhizal fungal dispersal concentrate on describing either biotic or abiotic variables alone. However, dispersal is often dependent on an interaction of variables. Murie (1962) noted that squirrels cached sporocarps in the crotches of trees and allowed them to dry in the air. Both epigeous and hypogeous mycorrhizal fungi are commonly collected by squirrels, which carry them into tree branches before feeding

(M.F. Allen 1991). In either case, spores dropping from these sporocarps would be entrained in wind currents and thus be capable of dispersal over very long distances.

We also demonstrated that *Glomus* spores could be dispersed over 2 km across a shrub steppe in Wyoming (Warner et al. 1987). At one site on Mount St. Helens, we were also able to trap *Glomus* spores, but not at others. However, how such spores reached the soil surface where they might be available for wind dispersal proved a major question. The Mount St. Helens site provided some clues. No spores were trapped in dense canopy sites or in the midst of the Pumice Plain, a long distance from the source area. However, at the margins of the blow-down areas, where there was virtually no plant canopy, but where pocket gophers piled buried soil on the ash surface, *Glomus* spores were entrained and dispersed locally (M.F. Allen 1987). At the Wyoming site, we found badger mounds that were up to 30 cm high, nearly the height of some of the shrubs. Importantly, the badgers preferentially dug into a shrub root system and deposited that soil in a mound in the interspace. The spore counts of this soil equaled those of the shrub rhizosphere. Thus, the badger mounds, composed of soil from the shrub rhizosphere, received the full impact of the 5–20 $m s^{-1}$ winds with vertical speeds of up to 200 $cm s^{-1}$, well above the 20 to 100 $cm s^{-1}$ settling velocities of these fungi. We suggest that disruption of the soil surface by animals may be essential for the process of spore dispersal, because they bring inoculum to the soil surface where winds can carry it into the atmosphere for long-distance transport.

D. Artificial Inoculation

We do not wish to discuss artificial inoculation techniques or approaches in detail here. However, intentional inoculation has clearly played a key role in the global dispersal of some mycorrhizal fungi. Examples include the inoculation of *Pisolithus tinctorius* into many disturbed forest types (Marx et al. 1984).

E. Source Areas, Landscape Structure, and Spatial Models

Linking animal and physical vectors to mycorrhizal dispersal is crucial for predicting mycorrhizal reestablishment in disturbed sites. These linkages depend on understanding animal behavior and physiology as well as the dynamics of turbulent flow and landscape ecology. Intensive studies on a single site have provided interesting and important insights.

1. Mount St. Helens: a Case Study

On the Mount St. Helens Pumice Plain, the migration of arbuscular mycorrhizal fungi occurred exclusively by means of animal dispersal. This area proved especially interesting since it contained both animal- and wind-dispersed plants (e.g., lupine and fireweed, respectively) which had established there by the second growing season following the deposition of the sterile material, none of which was initially mycorrhizal. By 1982, individuals of the wind-dispersed species were widely spaced, whereas the lupines occurred in two distinct patches that appeared to have come from single individuals that survived and set seed. By 1983, a few more of these lupine patches were initiated downstream from the original patches, while the fireweed remained scattered. Through 1990, the only arbuscular mycorrhizal plants were restricted to the lupine patches (M.F. Allen et al. 1992b). We presume that the animals carrying inoculum did not stop and defecate at individual plants, but moved directly to the larger plant patches with additional food and cover resources.

The spacing of the initial patches and inoculum was also interesting. Through 1985 the patches of mycorrhizal plants were restricted to the edge of a second eruption that deposited material in 1981 about 2 km from the crater mouth (Fig. 3). This created a distinct topographic and substrate boundary. The majority of animal trails could be found skirting this edge, and feces of wapati could frequently be found at this interface. The early plant patches also could be found at this interface (followed by the mycorrhizal plants 2 years later). The mycorrhizal patches then moved downstream as animals increasingly crossed the pumice plain and migrated down to Spirit Lake or as propagules washed downstream with the rapidly eroding pumice.

In addition, the highly disturbed pumice areas were the only sites with low enough plant cover to allow wind to reach the soil surface. The gopher mounds containing chopped plant roots and spores from inoculated plants were raised 10–15 cm above the soil surface, enough to allow the

Fig. 3. Landscape structure of Mount St. Helens 2 years following the initial eruption which occurred in 1980. Shown is the Pumice Plain (*foreground*) and the material from a second eruption in 1981. The animals appeared to follow the steep edge of the secondary eruption around the mountain. This landscape feature may have contributed to the deposition of plant and mycorrhizal fungal diaspores across the eruption zone

wind to strike the inoculum material directly. These were the only locations from which spores of arbuscular mycorrhizal fungi were trapped (M.F. Allen 1987).

Thus, we suggested that the animals were moving across the Pumice Plain in search of food and shelter. They moved along an edge created by the second smaller eruption and deposited patches of plant and fungal diaspores. These served as the first patches from which inoculum could spread. Those plants that were widely scattered (fireweed) provided few resources for animals and appeared not to serve as focal points for mycorrhizal establishment. Thus, succession at this site appears to occur in distinct patches, triggered by animal migration along routes structured by topography. Such patches serve as distinct foci for the spread and colonization of the site as a whole. Succession, in this case, does not follow an old-field model of distinct successional seres initiated by nonmycotrophic species as predicted by earlier models, but occurs as disjunct patches linked by animal migration across the landscape (M.F. Allen 1988).

V. Establishment

Establishment acts as a major limitation to the recovery of populations of mycorrhizal plants. Propagules of both plants and fungi disperse across great distances. However, the ability of propagules to find a safe site in which to germinate, grow, and encounter a compatible symbiont is no trivial task for a sessile organism. Many have argued that a mycorrhizae should be studied as a symbiosis, not as two individual organisms. However, both symbionts must disperse **independently** and survive in a hostile environment until a suitable partner can be found. This makes it crucial to understand both organisms as individuals before the ecology of the symbiosis can be understood.

Plants and mycorrhizal fungi typically occupy a limited zone of overlap in the surface soil and litter layers where they encounter each other as mature individuals; both seeds and spores must germinate and establish initial contact in this zone. Thus, an understanding of the characteristics of surface soil and litter may be necessary for describing mycorrhizal development. Nicolson (1960) noted that formation of a mycorrhiza required an appropriate propagule density, chemical/physical environment, and spatial distribution of both symbionts. Therefore, the constraints include the physical/chemical characteristics of the environment, which affect the physiological status of both fungus and plant, the genetic structure of the fungus, and the compatibility of fungal and plant genomes, and the spatial relationship between plant and fungus.

A. Physiological/Environmental Constraints

The establishment of seedling plants is well described and will not be reevaluated in detail here. However, seedling growth is generally described according to agronomic principles which dictate maximum growth as the most important measure of fitness. Importantly, abundant information indicates that high levels of soil fertility, conducive to maximum seedling growth, are detrimental to mycorrhizal formation. Despite the intensity of studies on the mechanisms by which mycorrhizae enhance nutrient uptake, few studies document the conditions under which a plant root becomes or fails to become mycorrhizal. Even when such studies are undertaken, results in the field often prove inconclusive, variable (E.B. Allen et al. 1987), or even counter-intuitive (M.F. Allen et al. 1987).

In glasshouse experiments, Mosse et al. (1981) showed that perturbing the N/P ratios from an optimum value would stimulate arbuscular mycorrhizal formation. In agar culture, M.F. Allen et al. (1981) noted that changing the form of P from an inorganic to an organic form would stimulate arbuscular mycorrhizal formation. Cold, heat, salinity, heavy metals, waterlogging, and drought have all been shown to inhibit mycorrhizal formation (Harley and Smith 1983). Since it is virtually impossible to keep ectomycorrhizae from forming in glasshouse studies, it is apparent that in ectomycorrhizal plants, the environmental limits for mycorrhizal formation are very broad.

Field studies indicate, however, that formation of a mycorrhiza is rarely limited by soil physical/chemical characteristics that did not also limit the plants themselves. We have never seen a plant seedling capable of forming mycorrhizae to be nonmycorrhizal in the field if inoculum was present, even under a wide variety of temperature, salinity, fertility, pH, and soil moisture conditions (M.F. Allen 1991). In addition, the limits of establishment for mycorrhizal fungi in exotic habitats are largely unknown. Marx et al. (1984) inoculated *P. tinctorius* worldwide in a wide array of habitats. Weinbaum et al. (1996) found that arbuscular mycorrhizal fungi could survive up to 2 years following transplanting into exotic habitats although the survival rates were lower in exotic habitats than when reintroduced into the habitat of origin.

However, while mycorrhizal formation may occur under unfavorable circumstances, the amount of mycorrhizal infection may very well depend on the physiological status of the plant and fungus. Thus, if a host plant grows in a favorable physical environment with high levels of inorganic nutrients, the root appears to grow faster than the fungus is capable of infecting. Buwalda et al. (1982) noted that under high P conditions, the total root infection was not diminished, but that infection units were spread farther apart resulting in a decreasing number of contacts between host and fungus. M.F. Allen et al. (1989b) found that the total amount of mycorrhizal infection and spore counts was more related to variation in root length than to percent infected roots.

B. Genetic Constraints

The role of fungal and plant genetics on mycorrhizal establishment has not received the level of careful study that is deserved, but is real, and occurs at two levels. First, the fungus must have the appropriate genetic constitution, and second, the host and fungus must be compatible.

The fungal genetic composition is unusual compared with that of many complex animals and plants. Spores of mycorrhizal fungi may be sexually or asexually produced. The information to date suggests that the spores of Glomales are asexually produced but that each contains multiple nuclei. Presumably, these are haploid although we have not seen data to prove this. The ability of hyphae in this group to anastomose is controversial and has not been documented from the field. However, given the vast numbers of nuclei per hypha and spore, a single "individual" fungus is probably quite capable of forming mycorrhizae with all compatible plants in the neighborhood. Moreover, following dispersal by wind or animals, a complete set of genetic information is available.

Some of the ectomycorrhizae may be different, but this depends on the type of fungus. The Endogonales form suspensor cells and diploid zygospores which probably undergo meiosis upon germination, returning to the multinucleate haploid state. They presumably disperse as diploid zygospores. The remaining mycorrhizal fungi (Dikaryomycota) live most of their lives as a mycelium with individual cells containing haploid nuclei. Following karyogamy and meiosis in the ascus or basidium, the resulting spores also tend to be haploid. After dispersal and germination, the fungus must find a compatible partner to become heterokaryotic. When these fungi disperse long distances into new sites, finding a compatible partner becomes less likely with increasing distance. This creates some interesting dynamics. *Laccaria bicolor*, an early seral mycorrhizal fungus, when monokaryotic, often forms a poorly developed mycorrhiza without a Hartig net (Kropp et al. 1987). Even when a "normal" mycorrhiza is formed by a monokaryon, the physiological variation among progeny (e.g., Kropp and Fortin 1988; Wagner et al. 1989; Kropp 1990; Chap. 6, this Vol.) suggests that a single individual may have a more difficult time surviving in a rapidly changing early seral site or the highly diverse environment of a later successional site than would a genetically more diverse heterokaryon. This may account for the poor ectomycorrhizal development in early-colonizing conifers on Mount St. Helens (M.F. Allen et al. 1992b).

Compatibility between mycorrhizal fungus and plant is also an important and little-studied consideration. Many early-colonizing plant species do not form mycorrhizae (e.g., Brassicaceae; M.F. Allen et al. 1992a) or form a pathogenic relationship between host and fungus (e.g., *Salsola kali*; M.F. Allen et al. 1989a). Most plants form only one type of mycorrhiza, although there are a few switchers such as *Alnus*, *Populus*, and *Carya*. However, recent observations suggest that this notion may deserve closer scrutiny. Cazares and Trappe (1994b) reported arbuscular mycorrhizal vesicles in the seedling roots of species in the Pinaceae; at the same site, more mature representatives of the same species were ectomycorrhizal. More detailed observations on seedlings and mature plants of the same host species need to be undertaken.

C. Spatial Constraints

We generally study establishment as an interaction between fungus, plant, and environment. In all published models, the development of the symbiosis assumes random or equitable spacing of inoculum and root growth. Rarely in undisturbed sites are either root growth or inoculum equitably dispersed. However, following a severe disturbance, inoculum is either eliminated or, if surviving, tends to become more evenly distributed (M.F. Allen and MacMahon 1985). Thus, vectors become important in initiating the establishment of a mycorrhizal symbiosis after all other factors are in place. Since the fungal hypha cannot grow long distances to invade a new host, it must encounter that host based on the spatial arrangement in which it is placed.

In undisturbed systems, roots and inoculum tend to be spatially aggregated. Much of this aggregation is caused by either physical constraints on production (in deserts) or by animals (in all habitats). In arid ecosystems, plants are widely dispersed, although their roots may be intertwined. In these systems, the fungus must be present where new roots are initiated following drought or cold periods (M.F. Allen and MacMahon 1985) or near the water table, where conditions are more conducive to growth and survival (Virginia et al. 1986). Indeed, some evidence suggests that mycorrhizal inoculum will concentrate in old root channels that are reexplored on an annual basis (Friese and Allen 1991). In more mesic systems such as grasslands or tropical forests, we have not been able to find microscale spatial differentiation in inoculum dispersion (Anderson and Liberta 1987; M.F. Allen and Friese 1992). Animal vectors also affect the spatial dispersion of inoculum. Inoculum was found to be concentrated in ant hills in sagebrush shrub-steppes (Friese and Allen 1993), under shrub patches in a successional shrub-steppe (M.F. Allen 1988), and in gopher- and ant-worked soils in the pumice on Mount St. Helens (M.F. Allen et al. 1984). These were preferred sites for the reestablishment of mycorrhizae following disturbance (M.F. Allen et al. 1992b; Friese and Allen 1993).

Spatial constraints are also regulated by aboveground architecture. Wind velocity is dramatically reduced by individual shrubs, which leads to the deposition of spores under and immediately leeward of shrubs (M.F. Allen 1988). These are "safe" sites in which the fungus can establish because of a continuous input of organic matter and soils from leaf deposition and deposition of erosional soils. Importantly, two or more shrubs, if spaced closely enough, will enhance that deposition even more (Fig. 4). This occurs because two shrubs spaced within ten times their height will reduce turbulence, resulting in dead-air spaces. Thus, a patch of plants will trap spores of mycorrhizal fungi better than individuals, which in turn, traps spores better than open soil.

Clearly, the survival of and dispersal of mycorrhizal fungi to "safe" sites where there is an appropriate host is crucial to the establishment of a mycorrhiza and survival of both the host and fungus. Plant patches provide shelter for animals which disperse inoculum. They also trap wind-dispersed propagules by spore impaction. Thus, the architecture and distribution of plants can regulate the types of mycorrhiza formed and the rate at which disturbed sites revegetate.

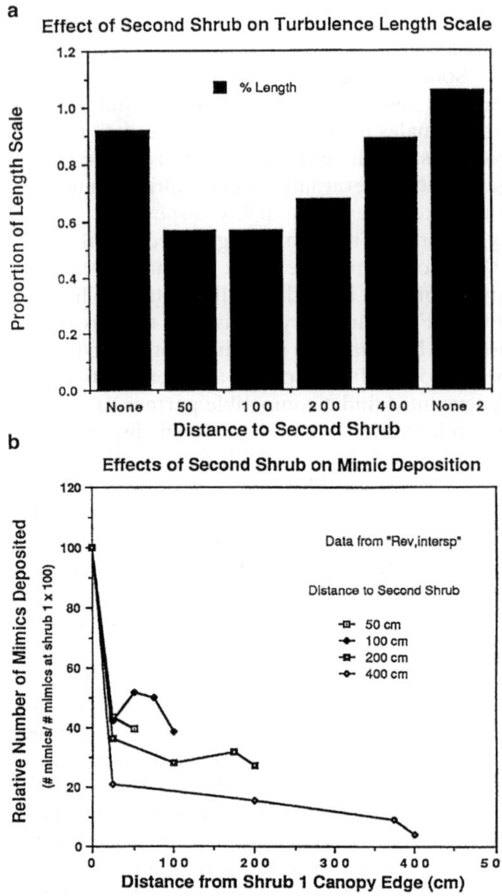

Fig. 4a,b. Deposition and wind structure between shrubs placed at varying distances apart. Note that the size of the turbulent eddies was reduced as the shrubs were moved closer together (**a**) and the spore mimic deposition increased in between the shrubs (**b**). The length scale with no second shrub was about 1 m. Thus, the spatial arrangement of the shrubs can dictate the wind fields of a site and thereby regulate the harvest of wind-dispersed propagules

VI. Community Diversity and Mycorrhizal Fungal Epidemiology

Most of this chapter has been devoted to describing the general process whereby a mycorrhiza reestablishes, without regard to the plant or fungal species involved. However, an equally important topic is the importance of fungal epidemiology in the face of the growing loss of biodiversity. There is a growing recognition that biodiversity is linked to ecosystem process (Tilman and Downing 1994) but the exact relationship is probably unknown or currently not even measurable.

Mycorrhizae play an important role both in regulating plant community diversity; further, the fungi that form mycorrhizae can be diverse. While the actual level of diversity is highly debated (Hawksworth 1993; E.B. Allen et al. 1995), clearly the number of species to be described is large.

Much of this debate centers on the relationship between environmental heterogeneity, plant diversity, and structure of mycorrhizal fungal communities. A key question is the structure of a mycorrhizal fungus. Since the mycorrhizal fungi are microscopic in structure, there is always "open" space for continuous colonization. In an arbuscular mycorrhizal system, Friese and Allen (1993) demonstrated that *Gigaspora margarita* could be inoculated into a small patch of soil and that it survived through a second growing season despite an already existing inoculum of a different species. We know of no similar experiment using other mycorrhizal fungi in late seral ecosystems, although inoculation of disturbed systems is common. This type of experiment would allow for the calculation of maximal fungal diversity in different habitats and the invasibility of different locations. Epidemiological models could aid in the understanding of fungal migration and establishment and, thus, biodiversity of mycorrhizae in ecosystems.

Much of the concern regarding biodiversity centers around the possibility of a catastrophic decline in the face of global climate change. Clearly, CO_2 and NO_x concentrations are increasing dramatically. Global climate change in response to increases in these greenhouse gases is uncertain. However, climatic change might plausibly accelerate local extinction in native plant communities. In that case, the ability of mycorrhizal fungi to disperse and reestablish in new habitats might become as crucial to maintaining native communities as to restoring disturbed lands.

VII. Conclusions

Time has been recognized as an important variable in the reestablishment of mycorrhizae for many years. However, space has been rarely considered. We believe that this neglect arises from assumptions inherent in the use of successional seres as the primary unit of study in succession, the "old-field" model of succession. This model assumes that site characteristics do not differ among sites and that "safe sites" crucial for establishment are unimportant. Rather, the model holds that establishment is a function only of the composition of the preceding community. This idea was an important step in understanding the mechanisms of succession and the role of mycorrhizae. For example, Janos (1980) used the old-field model of succession in conjunction with a knowledge of plant mycotrophy to provide an important advance in recognizing the increasing need for mycorrhizae during older successional seres.

As biologists have begun to regard succession as a landscape-level process instead of a plot-level process, we begin to see that space is as crucial as time in regulating mycorrhizal dynamics and plant succession. Succession generally occurs within many small patches advancing at different rates depending on the many nonlinear processes occurring in those patches (M.F. Allen 1988; M.F. Allen and E.B. Allen 1992). Some patches are initiated from a passing mammal that deposits both mycorrhizal inoculum and plant propagules, others from a single seed. Wind entrains inoculum from forest patches if a turbulent burst breaks through an open patch, but not if the forest is continuous. Forest edges probably provide a great deal more inoculum than forest understory, despite the fact that there are probably more fruiting bodies in the forest understory than along a single edge. Wind preferentially entrains small spores and deposits large spores in the lee edges of various structures such as plants. These are only a few of the qualities of space that must be considered in explaining the reestablishment of mycorrhizal symbioses.

These and the myriad of other interacting factors make the epidemiology of mycorrhizae both exciting from a scientific perspective and crucial from a land management perspective. Developing accurate predictive models of entrainment, dispersal, and inoculum sources and sinks will be critical to the management of sustainable agriculture and the restoration of disturbed lands.

Acknowledgments. This research was funded by grants from the US Department of Energy and the National Science Foundation.

References

Allen EB (1988) Some trajectories of succession in Wyoming sagebrush-grassland: implications for restoration. In: Allen EB (ed) The reconstruction of disturbed arid lands: an ecological approach. Westview Press, Boulder, pp 89–112

Allen EB, Allen MF (1980) Natural re-establishment of vescucular-arbuscular mycorrhizae following stripmine reclamation in Wyoming. J Appl Ecol 17:139–147

Allen EB, Allen MF (1984) Competition between plants of different successional stages: mycorrhizae as regulators. Can J Bot 62:2625–2629

Allen EB, Allen MF (1990) The mediation of competition by mycorrhizal fungi in successional and patchy environments. In: Grace J, Tilman D (eds) Perspectives in plant competition. Academic Press, New York, pp 367–390

Allen EB, West N (1994) Nontarget effects of the herbicide tebuthiuron in sagebrush-semidesert. Mycorrhiza 3:75–78

Allen EB, Chambers JC, Conner KF, Allen MF, Brown RW (1987) Natural re-establishment of mycorrhizae in disturbed alpine ecosystems. Arct Alp Res 19:11–20

Allen EB, Allen MF, Helm DJ, Trappe JM, Molina R, Rincon E (1995) Patterns and regulations of arbuscular and ectomycorrhizal plant and fungal diversity: a hypothesis. Plant Soil 170:47–62

Allen MF (1987) Re-establishment of mycorrhizae on Mount St. Helens: migration vectors. Trans Br Mycol Soc 88:413–417

Allen MF (1988) Re-establishment VA of mycorrhizae following severe disturbance: comparative patch dynamics of a shrub desert and a subalpine volcano. Proc R Soc Edinb 94B:63–71

Allen MF (1991) The ecology of mycorrhizae. Cambridge University Press, Cambridge, 184 pp

Allen MF, Allen EB (1992) Development of mycorrhizal patches in a successional arid ecosystem. In: Read DJ, Lewis DH, Fitter AH, Alexander IJ (eds) Mycorrhizae in ecosystems. CAB International Press, Cambridge, pp 164–170

Allen MF, Boosalis MG (1983) Effects if two species of VA mycorrhizal fungi on drought tolerance of winter wheat. New Phytol 93:67–76

Allen MF, Friese CF (1992) Mycorrhizal and reclamation success: importance and measurement. In: Chambers JC, Wade GL (eds) Evaluating reclamation success. USDA Forest Service General Technical Report, Northeastern Forest Experiment Station, Radnor, Pennsylvania, pp 17–25

Allen MF, MacMahon JA (1985) Importance of disturbance on cold desert fungi: comparative microscale dispersion patterns. Pedobiologia 28:215–224

Allen MF, MacMahon JA (1988) Direct VA mcyorrhizal inoculation of colonizing plants by pocket gophers (*Thomomys talpoides*) on Mount St. Helens. Mycologia 80:754–756

Allen MF, Sexton JC, Moore TS Jr, Christensen M (1981) Influence of phosphate source on vesicular-arbuscular mycorrhizae of *Bouteloua gracilis*. New Phytol 87:687–694

Allen MF, MacMahon JA, Andersen DC (1984) Reestablishment of Endogonaceae on Mount St. Helens: survival of residuals. Mycologia 76:1031–1038

Allen MF, Allen EB, West NE (1987) Influence of parasitic and mutualistic fungi of *Artemesia tridentata* during high precipitation years. Bull Torrey Bot Club 114:272–279

Allen MF, Allen EB, Friese CF (1989a) Responses of the non-mycotrophic plant *Salsola kali* to invasion by vesicular-arbuscular mycorrhizal fungi. New Phytol 111:45–49

Allen MF, Richards JH, Busso CA (1989b) Influence of clipping and soil water status on vesicular-arbuscular mycorrhizae of two semiarid tussock grasses. Biol Fert Soils 8:285–289

Allen MF, Hipps LE, Wooldridge GE (1989c) Wind dispersal and subsequent establishment of VA mycorrhizal fungi across a successional a rid landscape. Landscape Ecol 2:165–167

Allen MF, Clouse SD, Weinbaum BS, Jeakins S, Friese CF, Allen EB (1992a) Mycorrhizae and the integration of scales: from molecular to ecosystems. In: Allen MF (ed) Mycorrhizal functioning. Chapman and Hall, New York, pp 488–515

Allen MF, Crisafulli C, Friese CF, Jeakins S (1992b) Reformation of mycorrhizal symbiosis on Mount St. Helens, 1980–1990: interactions of rodents and mycorrhizal fungi. Mycol Res 96:447–453

Allen MF, Allen EB, Dahm CN, Edwards FS (1993) Preservation of biological diversity in mycorrhizal fungi: importance and human impacts. In: Sundnes G (ed) Human impacts on self-recruiting populations. 3rd Int Knogsvoll Symp. Tapir Press, Trondheim, pp 81–108

Amaranthus MP, Perry DA (1989) Interaction between vegetation type and madrone soil inocula in the growth, survival and mycorrhizal formation of Douglas fir. Can J For Res 19:550–556

Anderson RC, Liberta A (1987) Variation in vesicular-arbuscular mycorrhizal relationships of two sand prairie species. Am Midl Nat 118:56–63

Bethlenfalvay G, Linderman R (1992) Mycorrhizae in sustainable agriculture. American Society of Agronomy Spec Publ 54, Madison, Wisconsin, 124 pp

Broome LS (1987) Determinants of small rodent distribution and abundance in a shrub-steppe ecosystem: influence of seeds, ants, and shrubs. PhD Diss, Utah State University, Logan

Buwalda JG, Ross GJS, Stribley DP, Tinker PB (1982) The development of endomycorrhizal root systems. IV. The mathematical analysis of effects of phosphorus on the spread of vesicular arbuscular mycorrhizal infection in root systems. New Phytol 92:391–399

Cazares E, Trappe JM (1994a) Spore dispersal of ectomycorrhizal fungi on a glacial forefront by mammal mycophagy. Mycologia 86:507–510

Cazares E, Trappe JM (1994b) Vesicular endophytes in roots of the Pinaceae. Mycorrhiza 2:153–156

Clements FE (1916) Plant succession: an analysis of the development of vegetation. Carnegie Institute of Washington Publ No 242:1–512

Emmons LH (1982) Ecology of Prochimys, Rodentia, Echimyidae, in southeastern Peru. Trop Ecol 23:280–290

Friese (1984) The distribution of VAM fungi in a sand dune. MS Thesis, University of Rhode Island, Kingston

Friese CF, Allen MF (1991) The spread of VA mycorrhizal fungal hyphae in the soil: inoculum type and external hyphal architecture. Mycologia 83:409–418

Friese CF, Allen MF (1993) The interaction of harvester ants and vesicular-arbuscular mycorrhizal fungi in a patchy semi-arid environment: the effects of mound structure on fungal dispersion and establishment. Funct Ecol 7:13–20

Gehring CA, Whitham TG (1991) Herbivore-driven mycorrhizal mutualism in insect-susceptible pinyon pine. Nature 353:556–557

Gehring CA, Whitham TG (1994) Interactions between aboveground herbivores and the mycorrhizal mutualists of plants. Trends Ecol Evol 9:251–255

Gemma JN, Koske RE (1990) Mycorrhizae in recent volcanic substrates in Hawaii. Am J Bot 77:1193–1200

Gibson DJ, Hetrick BAD (1988) Topographic and fire effects on the composition and abundance of VA-mycorrhizal fungi in tallgrass prairie. Mycologia 80: 433–441

Harinikumar KM, Bagyaraj DJ (1994) Potential of earthworms, ants, millipedes, and termites for dissemination of vesicular-arbuscular mycorrhizal fungi in soil. Biol Fertil Soils 18:115–118

Harley JL, Smith SE (1983) Mycorrhizal symbiosis. Academic Press, London

Hawksworth DL (1993) The tropical fungal biota: census, pertinence, prophylaxis, and prognosis. In: Issac S, Frankland JC, Watling R, Whalley ASS (eds) Aspects of tropical mycology. Cambridge University Press, Cambridge, pp 265–293

Ingold CT (1971) Fungal spores: their liberation and dispersal. Clarendon Press, Oxford

Janos DP (1980) Mycorrhizae influence tropical succession. Biotropica 12:56–64

Koide RT, Mooney HA (1987) Spatial variation in inoculum potential of vesicular-arbuscular mycorrhizal fungi caused by formation of gopher mounds. New Phytol 107:173–182

Kotter MM, Farentinos RC (1984) Tassel-eared squirrels *Sciurus aberti ferreus* as spore dispersal agents of hypogeous mycorrhizal fungi. J Mammal 65:684–687

Kropp BR (1990) Variation in acid phosphatase activity among progeny from controlled crosses in the ectomycorrhizal fungus *Laccaria bicolor*. Can J Bot 68:864–866

Kropp BR, Fortin JA (1988) The incompatibility system and relative ectomycorrhizal performance of monokaryons and reconstituted dikaryons of *Laccaria bicolor*. Can J Bot 66:289–294

Kropp BR, McAfee BJ, Fortin JA (1987) Variable loss of ectomycorrhizal ability in monokaryotic and dikaryotic cultures of *Laccaria bicolor*. Can J Bot 65:500–504

Marx DH (1975) Mycorrhizae and establishment of trees on strip mined lands. Ohio J Sci 75:288–297

Marx DH, Cordell CE, Kenney DS, Mexal JG, Artman JD, Riffle JW, Molina RJ (1984) Commercial vegetative inoculum of *Pisolithus tinctorius* and inoculation techniques for development of ectomycorrhizae on bareroot seedlings. For Sci Monogr 25:1–101

Maser C, Trappe JM, Nussbaum RA (1978) Fungal-small mammal interrelationships with emphasis on Oregon forests. Ecology 59:799–809

McGee PA, Baczocha N (1994) Sporocarpic Endogonales and Glomales in the scats of *Rattus* and *Perameles*. Mycol Res 98:246–249

McGonigle TP, Miller MM (1990) Effect of degree of soil disturbance on mycorrhizal colonization and phosphorus absorption by maize in growth chamber and field experiments. New Phytol 116:629–636

McIlveen WD, Cole H Jr (1976) Spore dispersal of Endogonaceae by worms, ants, wasps and birds. Can J Bot 54:1486–1489

Miller RM (1979) Some occurrences of vesicular-arbuscular mycorrhizae in natural and disturbed ecosystems of the Red Desert. Can J Bot 57:619–623

Miller SL, Allen EB (1992) Mycorrhizae, nutrient translocation, and interactions between plants. In: Allen MF (ed) Mycorrhizal functioning, an integrative plant-fungal process. Chapman and Hall, New York, pp 301–332

Miller SL, Torres P, McClean TM (1994) Persistence of basidiospores and sclerotia of ectomycorrhizal fungi and *Morchella* in soil. Mycologia 86:89–95

Molina RJ, Trappe JM (1982) Patterns of ectomycorrhizal host specificity and potential amongst Pacific Northwest conifers and fungi. For Sci 28:423–457

Mosse B, Stribley DP, LeTacon F (1981) Ecology of mycorrhizae and mycorrhizal fungi. Adv Microb Ecol 5:137–210

Murie A (1962) Mammals of Denali. Alaska Natural History Association. Denali National Park, Alaska

Nicolson TH (1960) Mycorrhizae in the Gramineae. II. Development in different habitats particularly sand dunes. Trans Br Mycol Soc 43:132–145

Perry DA, Amaranthus MP, Borchers JG, Borchers SL, Brainerd RE (1989) Bootstrapping in ecosystems. BioScience 39:230–237

Pickett STA, McDonnell MJ (1989) Changing perspectives in community dynamics: a theory of successional forces. Trends Ecol Evol 4:241–245

Pilz DP, Perry DA (1984) Impact of clearcutting and slash burning on ectomycorrhizal associations of Douglas-fir seedlings. Can J For Res 14:94–100

Ponder F Jr (1980) Rabbits and grasshoppers: vectors of endomycorrhizal fungi on new coal mine spoil. North Central For Exp Stn Res Note NC 250:1–2

Reeves FB, Wagner DW, Moorman T, Kiel J (1979) The role of endomycorrhizae in revegetation practices in the semi-arid west. I. A comparison of incidence of mycorrhizae in severely disturbed vs. natural environments. Am J Bot 66:1–13

Rives CS, Bajwa MI, Liberta AE, Miller RM (1980) Effects of topsoil storage during surface mining on the viability of VA mycorrhiza. Soil Sci 129:253–257

States J (1985) Hypogeous mycorrhizal fungi associated with ponderosa pine: sporocarp phenology. In: Molina R (ed) Proc 6th North American Conf on Mycorrhizae. Forest Research Laboratory, Corvallis, 271 pp

Tilman D, Downing JA (1994) Biodiversity and stability in grasslands. Nature (Lond) 367:363–365

Virginia RA, Jenkins MB, Jarrell WM (1986) Depth of root symbiont occurrence in soil. Biol Fertil Soils 2:127–130

Waaland M, Allen EB (1987) Relationships between VA mycorrhizal fungi and plant cover following surface mining in Wyoming. J Range Manage 40:271–276

Wagner F, Gay G, Debaud JC (1989) Genetic variation of nitrate reductase activity in mono- and dikaryotic populations of the ectomycorrhizal fungus, *Hebeloma cylindrosporum* Romagnesi. New Phytol 113:259–264

Wallace LL (1987) Effects of clipping and soil compaction of *Schizachyrium scoparium*, a C_4 bunchgrass. Oecologia (Berl) 72: 423–428

Warner A, Mosse B (1980) Independent spread of vesicular-arbuscular mycorrhizal fungi in soil. Trans Br Mycol Soc 74:407–410

Warner NJ, Allen MF, MacMahon JA (1987) Dispersal agents of vesicular-arbuscular mycorrhizal fungi in a disturbed arid ecosystem. Mycologia 79:721–730

Weinbaum B, Allen MF, Allen EB (1996) Survival of arbuscular mycorrhizal fungi following reciprocal transplanting across the Great Basin, U.S.A. Ecol Appl 6:1365–1372

12 The Population Biology of Grass Endophytes

A. LEUCHTMANN[1] and K. CLAY[2]

CONTENTS

I.	Introduction	185
II.	Life Cycle of the Fungi	186
A.	Vegetative Growth and Seed Transmission	186
B.	Stroma Formation and Sporulation	186
III.	Reproductive Biology of the Fungi	187
A.	Horizontal Transmission of Spores	187
B.	Vertical Transmission	188
C.	Dispersal of Endophytes in Different Host Groups	189
IV.	Population Genetics of Endophytes	190
A.	Genetic Variation in Endophyte Populations	190
B.	Mode of Transmission and Genetic Variation	192
C.	Other Mechanisms of Variability	193
V.	Specificity of Endophyte Associations	193
A.	Evidence for Genetic Differences in Host Range	193
B.	Evidence for Host Resistance to Endophyte Infection	195
C.	Taxonomic Correlation of Specificity	195
VI.	Population Biology of Infected and Uninfected Plants	196
A.	Morphology and Physiology of Infected Plants	196
B.	Interactions with Competitors, Herbivores, and Pathogens	196
C.	Comparative Demography	197
D.	Community Changes	197
VII.	Experimental Approaches to Unanswered Questions	198
A.	Long-term Dynamics of Endophyte Populations	198
B.	Mechanisms of Mutualism	199
C.	Contagious Spread	199
D.	The Genetic Basis of Compatibility	199
VIII.	Conclusions	199
	References	200

I. Introduction

The population biology of grass endophytes is uniquely different from that in other groups of fungi given their status (primarily) as mutualistic symbionts, their effects on organisms interacting with hosts, and their variable means of horizontal and/or vertical transmission. Because grass endophytes are biotrophic and form perennial associations with their hosts, population biology of the fungi is tightly connected with that of the host. While the infection of grasses by endophytic fungi has been known for some time, their ecological role was seldom considered in early investigations. The first study that explicitly considered the population biology of endophytes was that of Bradshaw (1959) on *Agrostis* infected by *Epichloë typhina* (Pers.) Tul. However, most research has been conducted in the past 10 years, only after the ecological consequences of endophyte infection became widely appreciated.

The term endophyte has been used widely in different contexts and for many groups of organisms. In this chapter we use the term specifically in reference to systemic clavicipitaceous fungi (Ascomycotina; Balansieae) infecting grasses that are completely asymptomatic, or show no external signs of infection prior to flowering. These fungi include species of the sexual genera *Atkinsonella*, *Balansia*, *Epichloë*, and *Myriogenospora*, as well as their asexual relatives referred to *Acremonium* sect. *Albo-lanosa* and *Ephelis*. We recognize that many members of the tribe Balansieae are actually epiphytic on host plants (Leuchtmann and Clay 1988a; Clay and Frentz 1993), and that members of other plant families (e.g., Cyperaceae) also serve as hosts (Clay 1986b). In the narrowest definition, grass endophyte has been used only in reference to seed-borne *Acremonium* endophytes which are asexual relatives of the choke fungus *Epichloë typhina*. However, because of the close phylogenetic relationship and obvious functional similarity of the whole group, and because of variations in sexual expression of the fungus in many associations, we do not adhere to such a restricted definition.

[1] Geobotanisches Institut, ETH-Zürich, Zollikerstrasse 107, CH-8008 Zürich, Switzerland
[2] Department of Biology, Indiana University, Bloomington, Indiana 47405, USA

The Mycota V Part A
Plant Relationships
Carroll/Tudzynski (Eds.)
© Springer-Verlag Berlin Heidelberg 1997

The purpose of this chapter is to consider several aspects of the population biology of grass endophytes. We focus on their reproductive biology and patterns of genetic variation, their specificity with different grass hosts, and their effects on the ecology of host plants. Recent research has revealed that endophyte associations are extremely common and widespread, and that they may exert cascading effects through communities by virtue of their impact on host plants, plant communities, and the organisms that feed on plants. Much past research on grasses is incomplete, since the infection status of plants was not considered; future research on grasses requires that these symbionts be considered and incorporated into experiments. The ecological and economic importance of the Poaceae insures that the biology of endophytes will continue to be a major research area.

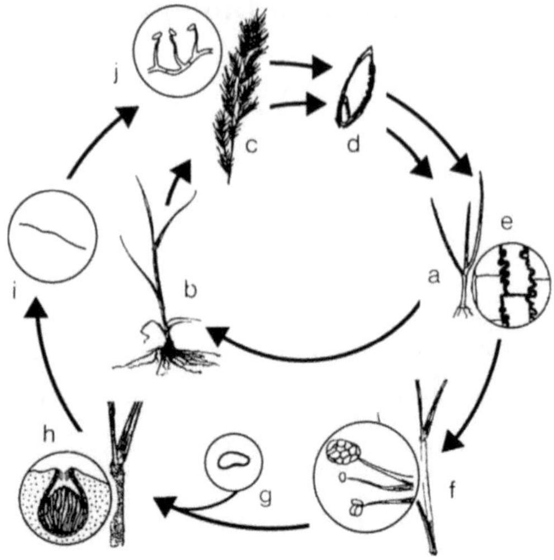

Fig. 1. Alternative sexual and asexual life cycles of *Epichloë* species. Magnified fungal structures are shown in *circles*. In the asexual cycle, convoluted hyphae grow intercellularly in leaf sheath (*a*), shoot meristems (*b*), and in the ovules of the florets (*c*), later becoming seed-transmitted (*d*). In the sexual cycle, the fungus also grows intercellularly in leaf sheaths (*e*) and meristems, but then forms external stromata with conidia around developing inflorescences aborting them in the process (*f*). Fertilization occurs by transfer of conidia (spermatia) between stromata of opposite mating types (*g*), upon which perithecia containing asci develop (*h*), and filamentous ascospores are ejected (*i*). Ascospores, germinating with iterative cycles of asexual sporulation, are postulated to cause infection of host florets (*j*) and then seed (*k*). (Schardl et al. 1994)

II. Life Cycle of the Fungi

Grass endophytes may have two alternative modes of reproduction. The sexual or asexual life cycles of an *Epichloë* species (with its asexual *Acremonium* anamorph) are depicted in Fig. 1 in association with the grass host.

A. Vegetative Growth and Seed Transmission

During the vegetative phase of grass development, endophytes grow intercellularly in stems, leaves, and sheaths forming straight or typically convoluted, sparsely branched hyphae parallel to the long axis of plant cells (Fig. 2A). In epiphytic species mycelium develops in leaf axils and on the surface of young leaves (Leuchtmann and Clay 1988a; Clay and Frentz 1993). Endophyte hyphae are most abundant in nodal or basal leaf meristems and adjacent tissues, but are sparse in leaf blades, and absent in roots (Hinton and Bacon 1985; White et al. 1993). No haustorial structures are known from grass endophytes; therefore growth of these fungi depends on nutrients available in the intercellular spaces of the apoplast. Nutritional supply may be greatest in meristems and young tissues. Mechanisms to overcome physical barriers to endophytic growth between cell walls must exist. Cell wall-degrading enzymes and plant hormones produced by the fungus were suggested to be involved during colonization of plant tissues (Hill 1993).

Endophytes are systemic in above-ground plant parts and invade all vegetatively formed tillers or stolons. As inflorescence are developing, asymptomatic endophytes grow into ovules and proliferate in the nucellus tissues, where they become incorporated into seeds, forming a characteristic layer of mycelium beneath the seed coat (Fig. 2B). Symptomless endophytes may also invade stamen filaments and anther walls, but have never been found in pollen grains, and apparently are not disseminated via paternal structures (Sampson 1933; Hinton and Bacon 1985).

B. Stroma Formation and Sporulation

Sexually reproducing endophytes form external fruiting structures (stromata) on elongating culms or on leaves of the host. In *Epichloë*, fungal hyphae proliferate on the surface of inflorescence

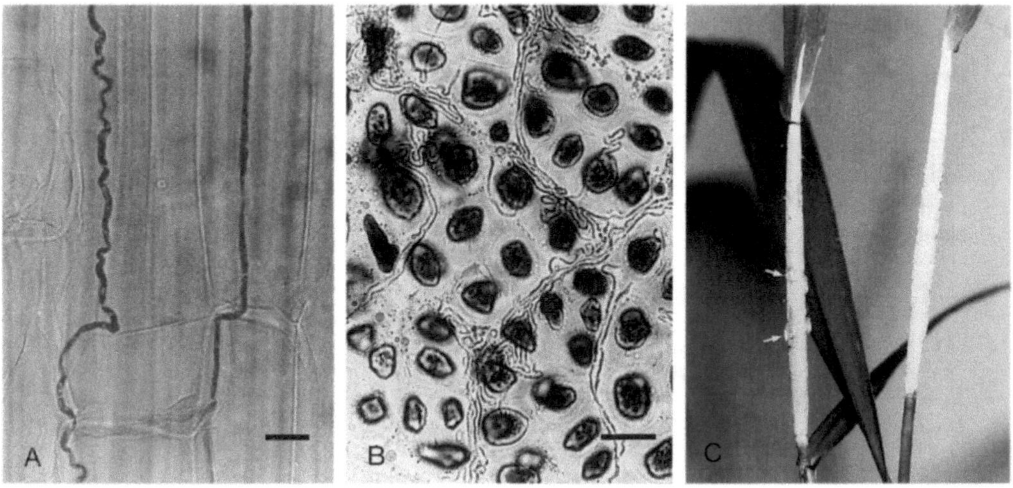

Fig. 2A–C. Endophytic fungi of grasses. **A** Typically convoluted, intercellular hyphae of *Epichloë clarkii* in leaf sheath of *Holcus lanatus*. *Bar* 10 μm. **B** Endophytic mycelium of *Acremonium lolii* among aleuron cells of *Lolium perenne* seed. *Bar* 40 μm. **C** Stromata of *Epichloë typhina* on aborted flowering culms of *Elymus virginicus*; *left* stroma with cases (*arrows*) of fly larvae adapted to feed on fungal tissue (natural size)

during early stages of development and eventually form a dense mycelial mat that closely binds inflorescence and young leaves together. This process effectively prevents further development of inflorescence into flowers and seed. At the same time, external mycelium gives rise to the white primary conidial stroma from which numerous *Acremonium* conidia are produced (White et al. 1991; Fig. 2C). Following the conidial state, perithecia with asci and ascospores develop in a thick mycelium layer, which then overgrows the conidial stroma. As perithecia mature the stroma turns yellow and then orange from carotenes formed in the perithecial wall. Filamentous, multiseptate ascospores are actively discharged from the asci either entire or as part-spores and are wind-transmitted (Webster 1980; Welch and Bultman 1993; White 1993). Ascospores may serve as inoculum to infect other plants either directly or indirectly through iterative germination. Bacon and Hinton (1988) suggest that iterative germination of ascospores could result in infective conidia.

Most sexual endophytes, including *Epichloë*, are heterothallic, requiring the transfer of conidia from one stroma to another of opposite mating type for fertilization (White and Bultman 1987; Leuchtmann and Clay 1989c; White and Owens 1992; Leuchtmann et al. 1994). The conidia therefore act as spermatia in sexual reproduction. Observations in the field and experimental evidence indicate that the anthomyiid fly, *Phorbia phrenione* Séguy, is the main vector transporting conidia between stromata of *Epichloë* (Kohlmeyer and Kohlmeyer 1974; Bultman and White 1988). Adults of parasitic *Ph. phrenione* visit infected plants to lay eggs on the conidial stromata. Conidia are then deposited in the process. The method of conidial transport by the fly may involve adherence of conidia to hairs and spines, or excretion of viable conidia after passage through the alimentary tract (Kohlmeyer and Kohlmeyer 1974; Bultman et al. 1995). Hatching larvae are adapted to feed and develop on the mature perithecial stroma and thus may substantially reduce potential ascospore release by the fungus (Welch and Bultman 1993).

III. Reproductive Biology of the Fungi

A. Horizontal Transmission of Spores

In sexually reproducing endophytes, both conidia and ascospores can potentially serve as propagules for horizontal transmission (Fig. 3). At least three mechanisms of contagious spread have been suggested in grass endophytes, but none is well docu-

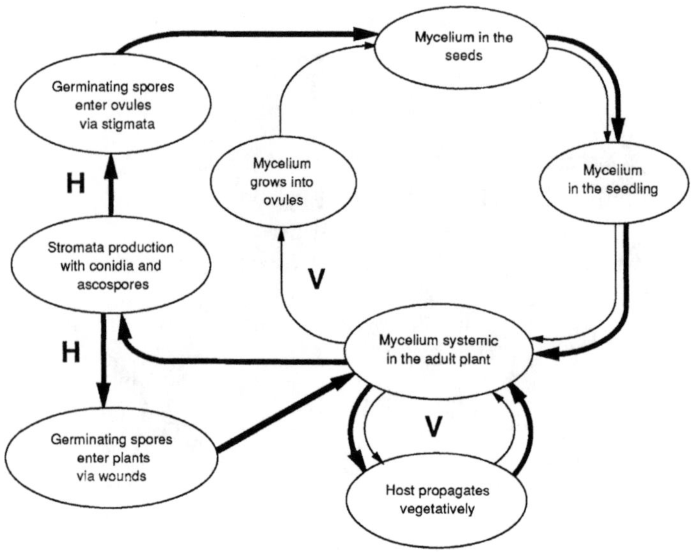

Fig. 3. Flow chart depicting vertical (*V*) and horizontal (*H*) transmission in stroma-forming (*thick-lined arrows*) and nonstroma-forming (*thin-lined arrows*) endophytes

mented, and they are supported experimentally only in single endophyte/host associations: (1) germinating spores enter ovules through stigmata, resulting in infected seed, (2) germinating spores enter the plant through wounds, and (3) spores contaminate seed, leading to seedling infection. Stigma infection has not been proven experimentally in *Epichloë* spp., but was shown with conidia of *Balansia obtecta* Diehl infecting *Cenchrus echinatus* L. seeds (Diehl 1950). Western and Cavett (1959) demonstrated invasion of *E. typhina* by growth of hyphae down the pith through freshly cut stubble of *Dactylis glomerata* L. inoculated with ascospores and conidia. Infection by contaminated seed was tested but has never been confirmed (Western and Cavett 1959). This seems an unlikely mechanism because of the ephemeral nature of thin-walled spores.

Contagious spread of sexual species is not epidemic, as in many plant pathogens, and infection of previously uninfected grass individuals may be rare in most species. For example, no spread of choke disease from infected to healthy *Dactylis glomerata* individuals occurred when grown as spaced plants (Western and Cavett 1959). However, Large (1954), in an extensive survey of *D. glomerata* seed crops in England and Wales, documented a dramatic increase in percentage of infected crops and severity of infection, which was correlated with the age of the stand. In individual crops where infection was established, the percentage of choked tillers increased by up to 30% in 2 consecutive years. This high rate of contagious spread may be connected with the particular crop management and the favorable environmental conditions for spore dispersal and germination on cut stubble following crop harvest. Ascospore discharge, as well as germination, depends on high relative humidity and may occur predominantly during the night (Raynal 1991). In wild grasses under natural conditions, rates are expected to be much lower. However, in related clavicipitaceous fungi, rates comparable to those seen in *E. typhina* under crop conditions have been observed in the field (Rykard et al. 1985; Clay 1990a).

B. Vertical Transmission

In infected, symptomless plants, ovules and seeds are usually invaded by the endophyte from internally borne hyphae such that the endophyte becomes seed-transmitted (Fig. 3). This was first reported by Sampson (1993) in *Festuca rubra* L. plants, and later confirmed in a number of studies on symptomless *Acremonium* endophytes of *Festuca* and *Lolium* species (Latch and Christensen 1982; Siegel et al. 1984; White and Cole 1986; White et al. 1991). However, viable seeds which are free of endophyte may be produced by plants infected by stroma-forming *E. typhina*, when shoot meristems of individual tillers escape infection and consequently will give rise to seed which bear no hyphae (Sampson 1933; Kirby 1961; A. Leuchtmann pers. observ.). The capacity of individual tillers to escape infection may be controlled by the particular fungal strain, the plant genotype, as well as by environmental factors. Successful dissemination of endophytes by seeds depends on viability of endophytic mycelium in seed and embryo until seed germination. Experiments with infected seed lots of *Lolium perenne* L. showed that endophyte viability declined rapidly after 12 months from an initial 84% to less than 15% when seed moisture content was above 11.5% at ambient temperatures (Rolston et al. 1986). Similar observations were made with infected seed of *Festuca arundinacea* Schreber stored under high relative humidity at room temperature (Welty et al. 1987). However, when stored at low humidity and at 0 to 5 °C, viable endophytes were reported to persist for up to 15 years in *L. perenne* seed (Latch 1983). These observations suggest that, under certain circumstances, seeds germinating in the season following their production may have lost the endophyte.

Another mode of vertical transmission in both sexual and asexual endophytes occurs when host plants propagate vegetatively through rhizomes, tillers, tubers, or viviparous plantlets (Diehl 1950; Bradshaw 1959; Clay 1986a,b).

C. Dispersal of Endophytes in Different Host Groups

The reproductive capability of both host plant and fungus greatly affects the potential rate of horizontal or vertical transmission of the endophyte. Considerable variation in the degree of disease symptoms preventing flowering and seed formation exists in different grass populations and host species. In grasses infected by *Epichloë* and related asexual *Acremonium* endophytes, three main categories (or types) of host groups were proposed based on life-cycle variations (White 1988). In the first host group, referred to as type I, stroma formation is obligatory on all infected grass individuals so that seed formation of host plants is completely suppressed (Table 1). Here, spore transmission is the only means of aerial dispersal, although empirical evidence suggests that plant populations with this type of infection tend to remain small, spreading predominantly as clonally propagating individuals (Bradshaw 1959; Clay and Leuchtmann 1989). This host group includes numerous members of the subfamily Pooideae such as *Agrostis tenuis* Sibth. and *Dactylis glomerata* as well as nonpooid grasses (for a complete list of genera see Clay 1989).

In the second host group (type II), stromata are formed on only a subset of infected host individuals of a population, and not all culms of an infected plant may be sterilized. The endophyte may therefore be both seed-transmitted and dispersed by fungal propagules. Grass populations forming this type of association with an endophyte typically show moderate levels of infection (40–75%), and stroma-forming plants tend to occur in clusters within the population (Sampson 1933; White and Chambless 1991; Leuchtmann and Clay 1993). It is possible that stroma-forming plants contain endophyte strains which are genetically different from asymptomatic strains, and are reproductively isolated from one another within a population. Allozyme data from an infected *Brachypodium sylvaticum* (Huds.) P.B. population seem to support this view (Bucheli and Leuchtmann 1996). Maintenance of type II associations may occur as a result of a balance between seed transmission and contagious spread by spores on the one hand and loss of endophytes in seeds be-

Table 1. Dispersal in endophyte genera and different host groups

Fungal genus	Host group	Symptoms	Seed transmission	Spore transmission
Epichloë	Type I	Stroma obligatory on infected plants	–	+
Epichloë	Type II	Stroma optional	+	+
Acremonium	Type III	No stroma formed	+	–
Atkinsonella	*Danthonia*, *Stipa* (Poaceae)	Stroma obligatory	+	+
Balansia	Nonpooid grasses or sedges	Stroma obligatory	–	+
Myriogenospora	Nonpooid grasses	Stroma obligatory	–	+

fore germination and the survival of uninfected seedlings on the other. Based on evidence described earlier, it is assumed that seed transmission is much more efficient than contagious spread, which would explain the higher proportion of plant individuals without stromata in type II associations. This association has been found exclusively in the subfamily Pooideae (White 1988).

In the third host group (type III), no stromata are formed on any of the infected plants, and seed transmission is generally the only means of dispersal. However, seedlings or pieces of tissues of many of these host grasses produce epiphytic mycelium with conidia on roots or cut surfaces when kept on an agar medium under axenic conditions (Clay and Leuchtmann 1989; Azevedo and Welty 1995; pers. observ.). Thus, there is a potential for dissemination of the endophyte during seed germination or when host plants are wounded in nature which should be investigated. Infection rates of type III associations are usually very high throughout the range of a grass species, reaching 100% in many natural populations. This host group, which is restricted to pooid grasses, includes the agronomically important grass species *Festuca arundinacea* and *Lolium perenne*, as well as many typical woodland grasses infected by symptomless *Acremonium* endophytes (Latch and Christensen 1982; Siegel et al. 1984; Clay and Leuchtmann 1989; Leuchtmann 1992). Endophytes of this category completely lack sexual reproduction and therefore exhibit limited genetic variability, often resulting in clonally structured endophyte populations with wide distribution (Leuchtmann and Clay 1990; Leuchtmann 1992, 1994a).

By comparison, species of the other genera in the tribe, *Atkinsonella*, *Balansia*, and *Myriogenospora*, always form fruiting structures on infected hosts (Diehl 1950; Table 1). In *Atkinsonella*-infected species of *Danthonia* and *Stipa*, vertical and horizontal transmission may occur simultaneously. Fertile stromata suppress flowering of terminal panicles, whereas cleistogamous flowers enclosed by the sheaths in the lower portion of the culms are not affected and produce viable seed containing the fungus (Clay and Jones 1984; Clay 1994). Species of *Balansia* and *Myriogenospora* completely sterilize their hosts, and seeds are never infected. Consequently, dispersal of this group relies exclusively on contagious spread by fungal spores (Rykard et al. 1985; Clay 1986b; Clay et al. 1989). However, in some species of *Balansia*, dissemination through vegetative structures of the host such as rhizomes or viviparous plantlets may be of greater importance (Clay 1986b).

IV. Population Genetics of Endophytes

Population genetic studies of endophytes are limited by their restricted number of morphological characters, their difficulty of isolation and culture, and their often clandestine existence within asymptomatic plants. The advent of molecular techniques has allowed the examination of fungal population structure within and among populations and host species, levels of variation in endophyte groups characterized by different means of reproduction, differentiation among host-specific races, and the concordance of endophyte phylogenies with host phylogenies. Most data are from allozyme electrophoresis, a conservative measure of genetic variation, but more recently DNA RFLP and sequence data have become available.

A. Genetic Variation in Endophyte Populations

The first population genetic study of systemic clavicipitaceous fungi examined variation of *Atkinsonella* isolates from several *Danthonia* grass hosts and *Stipa leucotricha* Trin. & Rupr. (Table 2; Leuchtmann and Clay 1989b). Considerable variation was found within and among populations from *D. compressa* Austin and *D. spicata* (L.) Beauv., but no variation was found in populations from *D. sericea* Nutt. and *Stipa*. More recently, intensive sampling from southwestern North Carolina, where the greatest concentration of *Danthonia* hosts occurs, has revealed a major split in *Atkinsonella* lineages infecting *D. compressa* and *D. spicata* vs. *D. epilis* Scrib. and *D. sericea* (Leuchtmann and Clay 1996), corroborating mtDNA analyses of isolates from the same host complexes (Van Horn and Clay 1995).

Most research has focused on *Acremonium* or *Epichloë*. Wide-ranging samples of isolates from many host species and genera revealed no distinct clustering of *Acremonium* vs. *Epichloë* isolates, confirming that sexual vs. asexual reproduction does not correspond to genetic differentiation (Leuchtmann and Clay 1990). There was also only weak correlation between host taxonomy and pat-

Table 2. Isozyme diversity in grass endophytes of the genera *Atkinsonella*, *Epichloë*, and *Acremonium*

Fungus Host species	Polymorphic loci (%)	No. of genotypes	No. of isolates	Ratio	Reference[a]
Atkinsonella hypoxylon	76.9	26	345	0.075	Combined
Danthonia compressa	46.2	9	122	0.074	(3)
Danthonia epilis	50.0	5	58	0.086	(5)
Danthonia sericea	23.1	3	22	0.136	(5)
Danthonia spicata	76.9	11	143	0.077	(3)
Atkinsonella texensis	–	1	20	0.050	
Stipa leucotricha	–	1	20	0.050	(3)
Epichloë (type I)	100	19	60	0.317	Combined
Agrostis stolonifera	33.3	2	14	0.143	(7)
Anthoxanthum odoratum	8.3	2	3	0.667	(7)
Bromus erectus	8.3	2	11	0.182	(7)
Calamagrostis villosa	–	1	3	0.333	(7)
Dactylis glomerata	41.7	8	18	0.444	(7)
Glyceria striata	–	1	7	0.143	(4)
Holcus lanatus	–	1	1	–	(4)
Poa trivialis	8.3	2	3	0.667	(7)
Epichloë (type II)	100	51	256	0.199	Combined
Agrostis hiemalis	50.0	9	13	0.692	(4)
Agrostis perennans	50.0	6	6	1.000	(4)
Brachyelytrum erectum	30.0	3	18	0.167	(4)
Brachypodium sylvaticum	41.7	14	96	0.146	(1)
Elymus villosus	–	1	1	–	(4)
Elymus virginicus	80.0	4	40	0.100	(4)
Festuca rubra	41.7	6	29	0.207	(2)
Hystrix patula	60.0	6	40	0.150	(4)
Poa sylvestris	70.0	3	13	0.231	(4)
Acremonium (type III)	100	60	343	0.175	Combined
Agropyron caninum	66.7	2	6	0.333	(1)
Bromus benekenii	41.7	4	22	0.182	(1)
Bromus ramosus	–	1	15	0.067	(1)
Elymus europaeus	58.3	2	41	0.049	(1)
Festuca altissima	–	1	13	0.077	(2)
Festuca arundinacea	90.9	9	69	0.130	(4, 6)
Festuca gigantea	16.7	3	21	0.143	(2)
Festuca glauca	–	1	1	–	(2)
Festuca heterophylla	16.7	2	2	1.000	(2)
Festuca longifolia	16.7	3	8	0.375	(2)
Festuca obtusa	90.0	5	14	0.357	(4)
Festuca ovina	41.7	2	9	0.222	(2)
Festuca pratensis	9.1	2	17	0.118	(4, 6)
Festuca pulchella	8.3	2	20	0.100	(2)
Festuca valesiaca	–	1	1	–	(2)
Lolium perenne	63.6	8	28	0.286	(4, 6)
Melica ciliata	41.7	5	48	0.104	(7)
Poa autumnalis	–	1	1	–	(4)
Poa wolfii	–	1	1	–	(4)
Sphenopholis nitida	70.0	2	2	1.000	(4)
Sphenopholis pallens	10.1	2	3	0.667	(4)
Stipa robusta	–	1	1	–	(4)

[a] (1) Leuchtmann (1992); (2) Leuchtmann (1994a); (3) Leuchtmann and Clay (1989b); (4) Leuchtmann and Clay (1990); (5) Leuchtmann and Clay (1996); (6) Christensen et al. (1993); (7) A. Leuchtmann (unpubl.).

terns of relatedness. Low levels of variation were found in endophytes from cultivated grass varieties; 90% of all isolates from *Festuca arundinacea* (N = 52) had identical multilocus genotypes (Leuchmann and Clay 1990). More recently, Leuchtmann (1994a) examined in more detail endophyte isolates from *Festuca* species and found they formed six distinct groupings that corresponded to a large extent to groupings within the host genera, suggesting some degree of coevolution. A study of endophytes from five different genera of cool-season grasses suggested that host identity played a more important role in determining patterns of endophyte variation than did geographical proximity (Fig. 4; Leuchtmann 1992), again supporting a coevolutionary or cophylogenetic interpretation of endophyte evolution.

Other molecular techniques have recently been applied to endophytes and related fungi. Abundant mitochondrial DNA (mtDNA) restriction fragment length polymorphisms (RFLPs) have been found in *Atkinsonella hypoxylon* (Pk.) Diehl that indicate two major clades within the species, each corresponding to different pairs of *Danthonia* hosts (Van Horn and Clay 1995). Considerable variation was also found within populations with up to seven mtDNA haplotypes found on a single host species. DNA sequence data have also been collected for phylogenetic investigations (Schardl et al. 1991; Chap. 14, this Vol.), but these data have limited value for revealing population genetic variation, given the small number of isolates that have been processed.

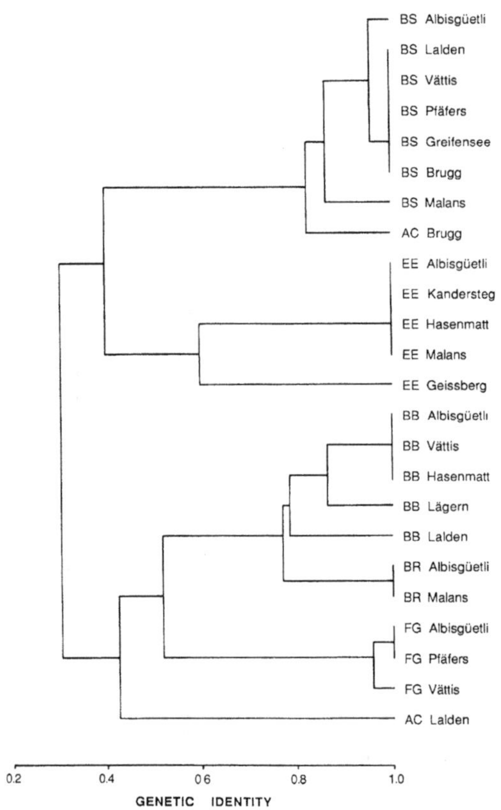

Fig. 4. Cluster diagram (UPGMA) showing the genetic relationship among 24 *Acremonium* endophyte populations from six different host grasses of beech forest communities in Switzerland. The distance measure is based on Nei's genetic identity using 14 isozyme loci. AC = *Agropyron caninum*; BB = *Bromus benekenii*; BR = *B. ramosus*; BS = *Brachypodium sylvaticum*; EE = *Elymus europaeus*; FG = *Festuca gigantea*. (Leuchtmann 1992)

B. Mode of Transmission and Genetic Variation

The grass endophytes as a group have three basic modes of reproduction (Sect. II). Clonal growth in conjunction with vegetative reproduction of the host results in the propagation of the same host and endophyte genotype (Bradshaw 1959; Clay 1986b; Stovall and Clay 1988). Single genotypic combinations may be geographically widespread and long-lived, but to date there have been no studies examining both endophyte and host population structure in parallel. Vegetative growth of the fungus into developing seeds in type II and type III associations results in the propagation of single fungal genotypes in a broader range of host genotypes produced by sexual reproduction. For example, the large majority of *Acremonium coenophialum* Morgan-Jones & Gams isolates collected from *Festuca arundinacea* from throughout eastern North America had only a single allozyme genotype (Leuchtmann and Clay 1990). The third mechanism of transmission is by contagious spread of spores. Hosts infected by ascospores, the product of meiosis, are each likely to possess a unique endophyte genotype. Asexual conidia might also be a source of contagious spread but, at least in *Epichloë typhina*, it appears that conidia act as spermatia in a heterothallic mating system and not as infective propagules (White and Bultman 1987).

The allozyme data bases from *Atkinsonella* and *Epichloë* (including *Acremonium*) are not directly comparable, given their difference in growth habit, host specificity, and phylogenetic relationships (Leuchtmann and Clay 1988a, 1989a; Schardl et al. 1991). Data from several studies are summarized in Table 2. For *Epichloë* there is a clear prediction that genetic variation in type I associations (new infections propagated only by contagious spread) should be greater than in type II associations which, in turn, should be greater than in strictly seedborne type III associations. Considerable genotypic variation exists among endophytes within each group, but the predicted pattern holds true; endophytes from type I associations exhibited the highest genotype/ isolate ratio on average, while endophytes from type III associations had the lowest value (Table 2). This ratio should approach zero, where endophytes reproduce asexually and approach one, where all reproduction is sexual. The allozyme data leave unanswered, however, the question of how diversity is generated in type III associations (e.g., *Melica ciliata* L.), and how genetic uniformity is maintained in some sexual Type I associations [e.g., *Glyceria striata* (Lam.) Hitchc.]. It should be noted that allozyme analysis underestimates the actual number of genotypes, since only a limited number of loci can be assessed, and not all alleles can be resolved; these may also be under the force of stabilizing selection.

In *Atkinsonella* the observed pattern is the reverse of that predicted. Due to the differences in basal seed production, through which the fungus is seed-transmitted, it was expected that *Danthonia epilis* and *D. sericea* should have the lowest rate of seed transmission (and asexual reproduction), while the other two *Danthonia* species should have higher rates of seed transmission (Clay 1994). Yet *D. compressa* and *D. spicata* had substantially lower genotype/isolate ratios than the former two species (Table 2). It should be noted that the number of isolates and the geographical range of collections differed between the two groups of *Danthonia* species.

C. Other Mechanisms of Variability

Other sources of variability may exist in endophytes besides meiosis (present only in type I and type II associations). Mutation is an important source of variability in asexual organisms. In endophytes, the probability of a mutation persisting in a population is limited both by the systemic growth of the fungus in the plant and the reproduction of the fungus via seeds or spores, separate from any fitness consequences of the mutation. Even if competition between mutant and nonmutant hyphae in one plant should occur, only a fraction of the seeds produced will incorporate the mutant, which may not propagate over time. Competition among fungal genotypes within one plant is also possible, but may be rare. Detrimental mutations should be purged from an asexual population since their bearers will have lower survival and/or reproduction. For example, due to variation in probability of seed establishment, seed progeny of certain plants infected by mutant endophytes may not become established.

Parasexuality represents another possible source of genetic variation in asexual endophytes. It depends on the formation of heterokaryotic mycelium which, in turn, depends upon the coinfection of a single plant by two distinct endophyte genotypes. This process might explain the existence of *Acremonium* endophytes that appear to be hybrids derived from sexual endophytes (Leuchtmann and Clay 1990; Schardl et al. 1994; Chap. 14, this Vol.). At least one of the parental endophytes would need to sporulate for contagious infection of a previously infected host. An increasing amount of empirical data supports the importance of parasexual cycles in *Acremonium* endophytes (Schardl and Tsai 1992; Tsai et al. 1994) but as yet there is no direct evidence.

V. Specificity of Endophyte Associations

A. Evidence for Genetic Differences in Host Range

Host specificity of endophytes directly affects the epidemiology of infection in plant communities by constraining the contagious spread within and among hosts. Host range reports suggest that particular endophytes are restricted to certain host species or genera (Diehl 1950; White 1987; Farr et al. 1989; Leuchtmann 1992). Evidence for host specificity may also be derived from distribution of endophyte genotypes among different sympatric host species. For instance, multiple isolates of *E. typhina* from three sympatric, closely related *Elymus* or *Hystrix* species were found to have identical or very similar allozyme genotypes, whereas isolates from also sympatric *Brachyelytrum erectum* (Schreb.) Beauv. were genetically quite divergent (Leuchtmann and Clay 1990). When reciprocal inoculations were conducted, genetically similar strains were cross-compatible with all four hosts, while the divergent *B. erectum* strain was specific to the original host (Leuchtmann and Clay 1993). Host-specific genetic differentiation of several geographically distant endophyte populations was also evident in six sympatric woodland grass species in Switzerland (Fig. 4; Leuchtmann 1992). It should be noted, however, that stromata were observed only on *Brachypodium sylvaticum*, whereas endophytes of the other hosts appeared to be strictly seed-borne. Genotypes on these hosts are therefore reproductively isolated. Whether or not they are specifically adapted to their host (i.e., form host races) requires confirmation by experimental inoculations.

An increasing number of studies have used molecular techniques such as sequencing of particular genes or analysis of restriction fragment length polymorphism (Schardl et al. 1991; An et al. 1992; Van Horn and Clay 1995; Chap. 14, this Vol.). Although usually only one or a few isolates from a host were used, the studies clearly confirm host-specific genetic variation of endophytes.

A direct approach to explore specificity of endophyte associations is experimental inoculation. However, a reliable method to infect mature plants through stigmas or wounds is not available, and successful attempts in the past remained anecdotal (Diehl 1950; Western and Cavett 1959; Rykard et al. 1985). Therefore, techniques have been established to infect grasses with endophytes using young seedlings (Latch and Christensen 1985; Leuchtmann and Clay 1988b), callus cultures (Johnson et al. 1986), somatic embryos (Kearney et al. 1991), or plantlets derived from meristem cultures (O'Sullivan and Latch 1993). All of these techniques rely on undifferentiated or young plant tissues to establish infection, suggesting that plants are most susceptible to infection only during the initial stages of development.

For example, using the seedling inoculation technique differences in compatibility among strains of *Atkinsonella hypoxylon* naturally infecting four host species of *Danthonia* were found (Leuchtmann and Clay 1989a; Clay 1993). Isolates of *D. spicata* and *D. compressa* could reciprocally infect seedlings of those species but not *D. sericea* and *D. epilis*, whereas isolates from *D. sericea* and *D. epilis* could infect all four host species (Table 3; Clay 1993). Host races appeared to be genetically distinct with regard to both allozyme variation and polymorphism (RFLP) of mtDNA (Leuchtmann and Clay 1989b; Van Horn and Clay 1995), but experimental hybrid strains could infect both parental hosts at frequencies intermediate between those of the parental isolates (Clay 1993). These findings indicate that there is a genetic basis for host specificity in *A. hypoxylon* and that it is under polygenic control.

Table 3. Nonreciprocal compatibility interactions of *Atkinsonella hypoxylon* with hosts of *Danthonia* as indicated by seedling inoculations

Source of fungal isolate	Source of seedlings			
	D. compressa	*D. spicata*	*D. epilis*	*D. sericea*
D. compressa	+	+	−	−
D. spicata	+	+	−	−
D. epilis	+	+	+	+
D. sericea	+	+	+	+

(+) = compatibility, (−) = incompatibility.

Table 4. Nonreciprocal compatibility interactions of *Epichloë typhina* with hosts of *Brachyelytrum*, *Elymus*, and *Hystrix* as indicated by seedling inoculations

Source of fungal isolate	Source of seedlings			
	B. erectum	*E. villosus*	*E. virginicus*	*H. patula*
B. erectum	+	−	−	−
E. villosus	+	+	+	+
E. virginicus	+	+	+	+
H. patula	+	+	+	+

(+) = compatibility, (−) = incompatibility.

A similar case of nonreciprocal compatibility was observed in *E. typhina* infecting four grass species in southern Indiana (Table 4; Leuchtmann and Clay 1993). Isolates from two *Elymus* species and from the closely related *H. patula* Moench were compatible with seedlings of all four species, whereas *Epichloë* isolates from *B. erectum* could infect only their original host and so were strictly host-specific. The successful infection of *B. erectum* with *Elymus* isolates documented, for the first time, compatibility of endophytes across grass genera in different tribes.

Most successful cross-infections of other *Epichloë* or seed-borne *Acremonium* endophytes were accomplished within or between the closely related genera *Festuca* and *Lolium* (reviewed in Leuchtmann 1994b). However, in only few cases were long-term persistence and seed transmission of novel endophyte/host associations observed. Several endophytes from hosts other than *Festuca* or *Lolium* could infect seedlings of *F. arundinacea* and grow into newly formed tillers, but they did not persist for more than 8 months (Siegel et al. 1987; Leuchtmann 1992).

B. Evidence for Host Resistance to Endophyte Infection

As indicated by the narrow host range of most endophytes, host grasses are resistant to the majority of endophytes. This nonhost resistance has been called basic resistance and may involve constitutive and induced defenses which are nonspecific towards pathogens (Heath 1987; Chap. 16, this Vol.). Physical barriers or unsuitable nutritional compounds within host tissues may be important factors in constitutive resistance. Other mechanisms which could explain limited compatibility of most endophytes involve nonspecific plant defense responses. For instance, increased levels of chitinase, an antifungal hydrolase associated with disease resistance, has been found in endophyte-infected *F. arundinacea* (Roberts et al. 1992). Many of the unsuccessful inoculation attempts in *F. arundinacea* seedlings using isolates from unrelated host genera are likely to be the result of basic resistance (Leuchtmann 1992). Similar cases of incompatibility are known in *L. perenne* and *F. rubra* seedlings (Latch and Christensen 1985; Latch et al. 1988).

Another type of incompatibility reaction resulting in limited persistence of novel *Acremonium*/host associations has recently been described by Koga et al. (1993). Hyphae of *A. coenophialum* isolates inoculated in nonhost *L. perenne* lost viability and degenerated in the second and third leaf sheath of most young tillers. Transmission electron microscopy showed that the extracellular matrix differed in nonhost associations from that seen in natural associations and had become electron-dense. The authors concluded that this structure may indicate a change in the permeability to nutrients, which could limit the survival of endophyte mycelium. Incompatibility affecting host tissue was noticed in seedlings of *F. arundinacea* inoculated with certain *Acremonium* genotypes from either *L. perenne* or *Festuca pratensis*, where necrosis in the region of the apical meristem was observed (Christensen 1995).

Host genotype-specific resistance towards endophytes is not very well documented, but has been suggested in *Atkinsonella*, based on differences in infection rates found in inoculation experiments with isolates and seedlings of different host populations (Leuchtmann and Clay 1989a). Clay (1993) reported later that seedlings of *D. spicata* inoculated with isolates of *A. hypoxylon* from the same population showed an unusually high mortality rate compared to isolates from different populations. In many seedlings which later died, extensive lignification and cell death occurred in areas where fungal hyphae were found. It was suggested that this hypersensitive response may indicate race-specific or gene-for-gene resistance to endophyte infection, as seen in many typical plant pathogens.

C. Taxonomic Correlation of Specificity

Evidence from allozyme variability and inoculation experiments suggests that seedborne *Acremonium* endophytes exhibit a high degree of specificity for their original host and/or to a few congeneric or closely related host species. Absence of spore production and thus of contagious spread in seedborne endophytes favors close adaptation to a host and may lead to formation of host-specific physiological races. Stroma-forming endophytes, on the other hand, should tend to have wider host ranges, given the ability of conidia or ascospores to potentially infect new host species. Although prevalent within the Pooideae, the sexual genus *Epichloë* has spread among six subfamilies of the Poaceae (Table 5). However,

Table 5. Distribution of endophytes in host families and subfamilies with known number of host species

Host family / Subfamily	Endophyte genus				
	Epichloë	*Acremonium*	*Atkinsonella*	*Balansia*	*Myriogenospora*
Poaceae					
Pooideae	88	60	1	1	–
Panicoideae	2	–	–	66	15
Chloridoideae	3	–	–	35	1
Bambusoideae	–	–	–	7	–
Oryzoideae	1	–	–	4	–
Arundinoideae	2	–	4	1	–
Cyperaceae	–	–	–	10	–

with the taxonomic characterization of the genus only recently begun, correlation of species with grass hosts is very incompletely known and not conclusive.

The related spore-transmitted genera, *Balansia* and *Myriogenospora*, are also found in several grass subfamilies with predominance on panicoid grasses (Table 5). Many *Balansia* species have broad host ranges and are known to infect up to 33 different host species (Diehl 1950). However, as in *Epichloë*, unresolved taxonomic questions loom, since a single species with a broad host range may prove to consist of many host-specific species or races upon further study. Except for Diehl's (1950) monograph, there has been little systematic research conducted outside *Epichloë* and *Acremonium*.

VI. Population Biology of Infected and Uninfected Plants

Infection induces several important ecological changes in host plants. Because the population biology of the endophyte is primarily determined by that of its host, an understanding of the dynamics of endophyte populations can be best gained by examining the infected plants.

A. Morphology and Physiology of Infected Plants

Infected plants exhibit changes relative to uninfected plants that affect their performance in field situations. Fungi that produce stromata on inflorescences generally abort them so that seed production is eliminated or reduced (Diehl 1950; Clay 1984; Clay and Leuchtmann 1989). Aborted inflorescences tend to be shorter and are often produced in greater numbers than healthy inflorescences. This does not occur in type III associations with *Acremonium*, and only part of the time in type II associations (White 1988). In some *Balansia* species, stromata are produced on leaves and flowering is inhibited (Clay et al. 1989; Marks and Clay 1990).

In addition to effects on flowering, increased tillering and/or rhizomatous growth is often induced by infection. Both Diehl (1950) and Bradshaw (1959) described aggressive vegetative spread in *Agrostis* spp. infected by *E. typhina*, and in *Cenchrus echinatus* infected by *Balansia obtecta*. Increased tillering of caespitose (clump) grasses appears to be a widespread effect of infection in a broad range of hosts (Diehl 1950; Latch et al. 1985; Leuchtmann and Clay 1988b; Clay et al. 1989; Clay 1990a). These effects are not limited to grasses; enhanced vegetative spread of infected plants occurs in *Cyperus virens* Michx., where viviparous plantlets are induced by infection (Clay 1986b), and in *C. rotundus* L., where subterranean bulbil numbers are increased (Stovall and Clay 1988).

Physiological changes reported to occur in infected plants include increased photosynthetic rate, increased response to elevated CO_2, altered stomatal physiology and response to nutrient changes (Belesky et al. 1987; Arachevalta et al. 1989; Cheplick et al. 1989; Marks and Clay 1990; Richardson et al. 1993; West et al. 1993). In *F. arundinacea*, infected plants exhibit greater tolerance to drought as a result (Arachevaleta et al. 1989). Phytohormone production by endophytes has been documented, which could influence a variety of morphological and physiological responses (Porter et al. 1985; De Battista et al. 1990). The occurrence of endophyte alkaloids in infected plants (Chap. 10, Vol. V, Part A) also has important effects on the interaction with herbivores (Clay 1989). Physiological studies have mainly been conducted with *F. arundinacea* and *L. perenne* so the generality of these results awaits additional studies on other species.

B. Interactions with Competitors, Herbivores, and Pathogens

The physiological and morphological changes induced by endophyte infection will influence both the amount and mode of reproduction by infected plants and their ecological relationships with pests and competitors in ecological communities. Several studies investigating competitive ability have generally shown an advantage for infected vs. uninfected plants (Kelley and Clay 1987; Clay et al. 1993). The increasing rates of infections over time found in type III associations also support the hypothesis of increased competitive success (see below). A notable exception occurred in *L. perenne*, where infection suppressed intra- and interspecific competitive ability (Marks et al. 1991).

Interactions with pests and pathogens could also affect competition by altering the relative growth, survival, and reproduction of infected vs. uninfected plants. Much interest in endophytes has been stimulated by their well-known effects on herbivores. Alkaloid compounds confer resistance to a wide range of invertebrate and vertebrate herbivores, including seed predators (Hoveland et al. 1983; Bacon et al. 1986; Cheplick and Clay 1988; Clay 1991a,b; Chap. 10, Vol. V, Part A). There are some exceptions, where certain insects are oblivious to the infection status of their host plant (Clay 1991b), but the overwhelming number of studies support the notion that infected plants exhibit increased resistance to most herbivores. Greater damage to uninfected plants would help shift the competitive advantage to infected plants (Clay et al. 1993). Several examples are known of endophyte-infected species dominating heavily grazed areas due to their unpalatability (Shaw 1873; Bacon et al. 1986).

Data on resistance to pathogens are less frequent, but increased resistance of endophyte-infected plants to several fungal pathogens has been reported (Clay et al. 1989; Ford and Kirkpatrick 1989; Gwinn et al. 1991; Stovall and Clay 1991), although lack of resistance to other pathogens has also been reported (Welty et al. 1991). Pure culture studies have also revealed direct inhibition of endophytes or endophyte extracts on a range of fungal pathogens (White and Cole 1985; 1986; Siegel and Latch 1991; Stovall and Clay 1991). The potential ecological significance of pathogen resistance has not been examined in field studies.

C. Comparative Demography

A number of long-term demographic studies have been conducted in field situations comparing the relative survival, growth, and fecundity of infected vs. uninfected plants. In these realistic ecological settings, the specific mechanisms generating differences among plants have not generally been examined, but they are likely to arise from a variety of physiological, biochemical, and biotic sources. Only a limited number of species have been examined, but most infected plants exhibited greater survival and growth than uninfected plants (Clay 1984, 1990a). Effects on reproduction depend on the nature of the association between plant and fungus. In cases where the plants are partly or completely sterilized by infection, fecundity is reduced. Given that infection of seeds is not known to occur in *Balansia*-infected grasses and sedges, the fungi are dependent upon contagious spread or vegetative propagation of their hosts for their dissemination. In contrast, seed-transmitted species of *Acremonium* or *Epichloë* can reproduce via the sexual reproduction of their hosts. For example, in field studies with *F. arundinacea*, infected plants produced more inflorescences than uninfected plants (Clay 1990a). Taken together with increased survival rates, it was estimated that infected plants had an over twofold fitness advantage over uninfected plants. These differences would translate into an increasing proportion of infected plants in mixed populations over time. The high level of endophyte infection reported in many grasses is de facto evidence of a mutualistic interaction, since a purely type III association is not expected to persist if infected plants show reduced fitness.

Much research has been conducted with *F. arundinacea*, given its prevalence and economic importance, in both field and greenhouse environments. A large number of studies, summarized in Table 6, have consistently found increased survival, growth, and/or fecundity of infected vs. uninfected plants. While this represents only a single species among many endophyte hosts, there is no reason to believe that the results are atypical of other associations. With the exception of reduced fecundity in type I and II associations, negative consequences of endophyte infection have rarely been demonstrated in any species (Cheplick et al. 1989; Marks et al. 1991). Thus, the great majority of data support the hypothesis that infected plants exhibit increased vigor compared to uninfected plants. Even in cases where seed production is impaired, infected plants are still vigorous, and may persist and spread by vegetative growth.

D. Community Changes

In cases where individual plants have not been distinguished, the ecological effect of endophyte infection may still be inferred by changes in infection frequency within populations and by changes in species composition over time. Consistent with greenhouse and field studies, all studies which have followed vegetational changes over time indicate that the frequency of infected plants within populations or the frequency of infected species within communities increase over time (reviewed

Table 6. Summary of relative productivity of endophyte infected *Festuca arundinacea* in a variety of studies where the productivity of uninfected *F. arundinacea* is set at 1

Relative productivity	Variable measured	Environment	Reference
1.55	Available forage	Field	Read and Camp (1986)
1.20	Tiller number	Greenhouse	Clay (1987)
1.24	Herbage	Greenhouse	"
1.45	Tiller number	Field	Clay (1990a)
1.54	Herbage	Greenhouse	Arachevaleta et al. (1989)
1.18	Herbage	Greenhouse	De Battista et al. (1990)
1.05	Herbage	Greenhouse	Hill et al. (1990)
0.91	Tiller number	Greenhouse	"
1.41	Seed number	Field	Rice et al. (1990)
1.29	Herbage	Field	Hill et al. (1991)
1.26	Seed weight	Field	"
1.22	Herbage	Greenhouse	Clay et al. (1993)
1.42	Herbage	Greenhouse/herbivory	"
1.26	Herbage	Field	West et al. (1993)
1.18	Tiller density	Field	"
Average 1.31			

in Clay 1990b; see also Shaw 1873; Francis and Baird 1989). Where the endophytes sporulate on their hosts, these changes could result from contagious spread, but where the endophyte is strictly seed-borne, an increased fitness and/or competitive ability of infected plants is the only mechanism by which infection frequency can increase (Lewis and Clements 1986; Saha et al. 1987; Francis and Baird 1989; Clay 1990a). However, species infected by type III endophytes that have infection rates less than 100% suggest that there may be situations where infection is not mutualistic. Similarly, endophytes that sterilize their host may decline in frequency within populations where seed production is an important source of plant regeneration. There is a great need for additional long-term studies of endophyte population dynamics at the community level.

VII. Experimental Approaches to Unanswered Questions

A number of important questions about the population biology of endophytes still remain to be answered. Answering these questions will involve detailed and long-term field experiments as well as molecular genetic techniques to track endophyte genotypes in the field and to create hybrid genotypes that can be reciprocally inoculated into different hosts. Much basic research on the grass endophytes is being driven by their agricultural effects and biotechnological potential.

A. Long-Term Dynamics of Endophyte Populations

As discussed above (Sect. VI), there are very few data available on the long-term dynamics of endophyte infections in populations and communities. Regular sampling can reveal changes in infection frequency within host populations, the basis of those changes (i.e., seedling establishment, contagious spread), and the changing dominance of different plant species over time. It is not known how long-term changes in infection frequency in type I vs. type III associations compare, or whether shifts in infection frequency are slow and steady or correlated with major environmental disturbances (i.e., drought, fire). If, indeed, most endophyte associations are mutualistic, as argued here, then we predict that the frequency of infection should increase within host populations and the dominance of infected plants and species within the multispecies community should increase. However, managing practices in grassland such as heavy grazing or frequent cutting before seed set of grasses may disrupt this process. In natural, long-standing grass communities such as those found in woodlands, a large proportion of the grass species were shown to be infected with often 100% infection rate, whereas infection frequencies in agriculturally managed grasslands were generally much lower (Clay and Leuchtmann 1989; Bucheli and Leuchtmann 1996).

B. Mechanisms of Mutualism

Past research has shown that both intrinsic and extrinsic mechanisms favor infected plants. For example, direct physiological changes induced by infection can alter photosynthetic rate and drought tolerance while infection-induced resistance can affect interactions with insect pests and pathogens. Studies are needed where intrinsic vs. extrinsic mechanisms of mutualism can be separated and quantified (see Clay et al. 1993). For example, by using fencing and insecticide applications, it is possible to create plots with both vertebrate and insect herbivory, only one type of herbivory or the other, and with no herbivory. Superimposed upon these treatments could be different fertilization or watering regimes. The levels of significance of different factors and their interaction would provide insights into the major environmental factors affecting endophyte infections.

C. Contagious Spread

The mechanism and frequency of contagious spread are not known for any system (see Sect. II.A). This deficiency can be addressed by field experiments in natural setting where the arrays of known uninfected adults and seedlings surround infected individuals where the endophyte has been genetically typed (see Alexander 1989). Comparisons between seedlings and adults would reveal whether mature plants can become infected and whether chimeric plants (partly infected and partly uninfected) exist. This process could be examined indirectly by introducing plants infected by endophytes with unique genetic markers into natural populations and then sampling the endophyte population several years later to look for recombinants with the unique marker. The proposed role of parasexuality and hybridization in endophytes (Schardl et al. 1994) is dependent upon contagious spread. Detailed sampling of endophytes in type III associations growing in the midst of related type I associations probably offers the best opportunity for detecting dual infections of single plants by different endophyte genotypes.

D. The Genetic Basis of Compatibility

An understanding of the basis of compatibility between endophytes and their hosts is relevant to coevolutionary interactions, to the dynamics of infections in mixed host species communities, to the role of contagious spread, and to the artificial creation of new host/endophyte combinations. Two approaches offer great potential for elucidating the basis for host compatibility.

Hybridization of endophytes with different host ranges can provide insight into the number and natures of genes controlling compatibility. Three steps are required: (1) to identify host range variants, (2) to hybridize the variants and obtain progeny, and (3) to use the hybrid progeny to inoculate uninfected plants of both parental hosts. If compatibility was controlled by a single nuclear gene, then a 1:1 ratio of progeny compatible with the alternate hosts would be expected. Cosegregation of compatibility with specific genetic markers in hybrid progeny could reveal the approximate location of compatibility genes, allowing their characterization. More complex patterns would result from more complex genetic control of compatibility. A second approach is mutagenesis and/or selection on different hosts. Large numbers of ascospores or conidia could be subjected to UV or chemical mutagens and then be used to inoculate novel hosts. Successful (or unsuccessful) inoculants could be compared with parental strains to determine the nature of the mutational changes. A greater understanding of the physiological basis of host resistance responses would also provide insights into how the host recognizes compatible vs. incompatible endophytes.

VIII. Conclusions

Our knowledge of the population biology of endophytes has increased exponentially over recent years. However, new knowledge also raises new questions. Unique features of endophytes (e.g., seed transmission) provide an opportunity for investigating a range of basic problems in population biology not found in other fungi, while their abundance and ecological significance make them an excellent model system for investigating problems common to a much larger range of organisms.

References

Alexander HM (1989) An experimental field study of anther-smut disease of *Silene alba* caused by *Ustilago violacea*: genotypic variation and disease incidence. Evolution 43:835–847

An Z, Liu J, Siegel MR, Bunge G, Schardl CL (1992) Diversity and origins of endophytic fungal symbionts of the North American grass *Festuca arizonica*. Theor Appl Genet 85:366–371

Arachevaleta M, Bacon CW, Hoveland CS, Radcliffe DE (1989) Effect of tall fescue endophyte on plant response to environmental stress. Agron J 81:83–90

Azevedo MD, Welty RE (1995) A study of the fungal endophyte *Acremonium coenophialum* in the roots of tall fescue seedlings. Mycologia 87:289–297

Bacon CW, Hinton DM (1988) Ascosporic iterative germination in *Epichloë typhina*. Trans Br Mycol Soc 90:563–569

Bacon CW, Lyons PC, Porter JK, Robbins JD (1986) Ergot toxicity from endophyte-infected grasses: a review. Agron J 78:106–116

Belesky DP, Devine OJ, Pallas JE Jr, Stringer WC (1987) Photosynthetic activity of tall fescue as influenced by a fungal endophyte. Photosynthetica 21:82–87

Bradshaw AD (1959) Population differentiation in *Agrostis tenuis* Sibth. II. The incidence and significance of infection by *Epichloë typhina*. New Phytol 58:310–315

Bucheli E, Leuchtmann A (1996) Evidence for genetic differentiation between choke-inducing and asymptomatic strains of the *Epichloë* grass endophyte from *Brachypodium sylvaticum*. Evolution 50:1879–1887

Bultman TL, White JF Jr (1988) "Pollination" of a fungus by a fly. Oecologia 75:317–319

Bultman TL, White JF Jr, Bowdish TI, Welch AM, Johnston J (1995) Mutualistic transfer of *Epichloë* spermatia by *Phorbia* flies. Mycologia 87:182–189

Cheplick GP, Clay K (1988) Acquired chemical defenses of grasses: the role of fungal endophytes. Oikos 52:309–318

Cheplick GP, Clay K, Wray S (1989) Interactions between fungal endophyte infection and nutrient limitation in the grasses *Lolium perenne* and *Festuca arundinacea*. New Phytol 111:89–97

Christensen MJ (1995) Variation in the ability of *Acremonium* endophytes of perennial rye-grass (*Lolium perenne*), tall fescue (*Festuca arundinacea*) and meadow fescue (*F. pratensis*) to form compatible associations in the three grasses. Mycol Res 99:466–470

Christensen MJ, Leuchtmann A, Rowan DD, Tapper BA (1993) Taxonomy of *Acremonium* endophytes of tall fescue (*Festuca arundinacea*), meadow fescue (*F. pratensis*) and perennial rye-grass (*Lolium perenne*). Mycol Res 97:1083–1092

Clay K (1984) The effect of the fungus *Atkinsonella hypoxylon* (Clavicipitaceae) on the reproductive system and demography of the grass *Danthonia spicata*. New Phytol 98:165–175

Clay K (1986a) New disease (*Balansia cyperi*) of purple nutsedge (*Cyperus rotundus*). Plant Dis 70:597–599

Clay K (1986b) Induced vivipary in the sedge *Cyperus virens* and the transmission of the fungus *Balansia cyperi* (Clavicipitaceae). Can J Bot 64:2984–2988

Clay K (1987) Effects of fungal endophytes on the seed and seedling biology of *Lolium perenne* and *Festuca arundinacea*. Oecologia 73:358–362

Clay K (1989) Clavicipitaceous endophytes of grasses: their potential as biocontrol agents. Mycol Res 92:1–12

Clay K (1990a) Comparative demography of three graminoids infected by systemic, clavicipitaceous fungi. Ecology 71:558–570

Clay K (1990b) Insects, endophytic fungi and plants. In: Burdon JJ, Leather SR (eds) Pests, pathogens and plant communities. Blackwell, Oxford, pp 111–130

Clay K (1991a) Ecological interactions among fungal endophytes, grasses, and herbivores. In: Barbosa P, Krischik V, Jones C (eds) Multi-trophic interactions among microorganisms, plants and insects. Wiley, New York, pp 199–226

Clay K (1991b) Endophytes as antagonists of plant pests. In: Andrews JH, Hirano SS (eds) Microbial ecology of leaves. Springer, Berlin Heidelberg New York, pp 331–357

Clay K (1993) Recent advances in the ecology of endophytes. In: Hume DE, Latch GCM, Easton HS (eds) Proc 2nd Int Symp *Acremonium*/Grass Interactions: Plenary Papers. AgResearch, Grasslands Research Center, Palmerston North, New Zealand, pp 1–30

Clay K (1994) Hereditary symbiosis in the grass genus *Danthonia*. New Phytol 126:223–231

Clay K, Frentz IC (1993) *Balansia pilulaeformis*, an epiphytic species. Mycologia 85:527–534

Clay K, Jones JP (1984) Transmission of *Atkinsonella hypoxylon* (Clavicipitaceae) by cleistogamous seed of *Danthonia spicata* (Gramineae). Can J Bot 62:2893–2895

Clay K, Leuchtmann A (1989) Infection of woodland grasses by fungal endophytes. Mycologia 81:805–811

Clay K, Cheplick GP, Marks S (1989) Impact of the fungus *Balansia henningsiana* on *Panicum agrostoides*: frequency of infection, plant growth and reproduction, and resistance to pests. Oecologia 80:374–380

Clay K, Marks S, Cheplick GP (1993) Effects of insect herbivory and fungal endophyte infection on competitive interactions among grasses. Ecology 74:1767–1777

DeBattista JP, Bacon CW, Severson R, Plattner RD, Bouton JH (1990) Indole acetic acid production by the fungal endophyte of tall fescue. Agron J 82:878–880

Diehl WW (1950) *Balansia* and the Balansiae in America. US Department of Agriculture, Washington, DC

Farr DF, Bills GF, Chamuris GP, Rossman AY (1989) Fungi on plant and plant products in the United States. APS Press, St Paul

Ford VL, Kirkpatrick TL (1989) Effects of *Acremonium coenophialum* in tall fescue on host disease and insect resistance and allelopathy to *Pinus taeda* seedlings. Proc AR Fescue Tox Conf 140:29–34

Francis SM, Baird DB (1989) Increase in the proportion of endophyte-infected perennial ryegrass plants in overdrilled pastures. NZ J Agric Res 32:437–440

Gwinn KD, Blank CA, Gavin AM (1991) *Acremonium coenophialum* infestation levels of tall fescue seed lots affect seedling losses due to *Rhizoctonia* spp. Phytopathology 81:810

Heath MC (1987) Evolution of plant resistance and susceptibility to fungal invaders. Can J Plant Pathol 9:389–397

Hill NS (1993) Physiology of plant–endophyte interactions: implications and use of endophytes in plant breeding. In: Hume DE, Latch GCM, Easton HS (eds) Proc 2nd Int Symp *Acremonium*/Grass Interactions: Plenary Papers. AgResearch, Grasslands Research Center, Palmerston North, New Zealand, pp 161–169

Hill NS, Stringer WC, Rottinghaus GE, Belesky DP, Parrot WA, Pope DD (1990) Growth, morphological, and chemical component responses of tall fescue to *Acremonium coenophialum*. Crop Sci 30:156–161

Hill NS, Belesky DP, Stringer WC (1991) Competitiveness of tall fescue as influenced by *Acremonium coenophialum*. Crop Sci 31:185–190

Hinton DM, Bacon CW (1985) The distribution and ultrastructure of the endophyte of toxic tall fescue. Can J Bot 63:36–42

Hoveland CS, Schmidt SP, King CC, Odum JW, Clark EM, McGuire JA, Smith LA (1983) Steer performance and association of *Acremonium coenophialum* fungal endophyte of tall fescue. Agron J 75:821–824

Johnson MC, Bush LP, Siegel MR (1986) Infection of tall fescue with *Acremonium coenophialum* by means of callus cultures. Plant Dis 70:380–382

Kearney JF, Parrott WA, Hill NS (1991) Infection of somatic embryos of tall fescue with *Acremonium coenophialum*. Crop Sci 31:979–984

Kelley SE, Clay K (1987) Interspecific competitive interactions and the maintenance of genotypic variation within the populations of two perennial grasses. Evolution 41:92–103

Kirby JM (1961) Host–parasite relations in the choke disease of grasses. Trans Br Mycol Soc 44:493–503

Koga H, Christensen MJ, Bennett RJ (1993) Incompatibility of some grass-*Acremonium* endophyte associations. Mycol Res 97:1237–1244

Kohlmeyer J, Kohlmeyer E (1974) Distribution of *Epichloë typhina* (Ascomycetes) and its parasitic fly. Mycologia 66:77–85

Large EC (1954) Surveys for choke (*Epichloë typhina*) in cocksfoot seed crops, 1951–53. Plant Pathol 3:6–11

Latch GCM (1983) Incidence of endophytes in seed lines and their control with fungicides. Proc NZ Grassland Assoc 44:251–253

Latch GCM, Christensen MJ (1982) Ryegrass endophyte, incidence, and control. NZ J Agric Res 25:443–448

Latch GCM, Christensen MJ (1985) Artificial infection of grasses with endophytes. Ann Appl Biol 1907:17–24

Latch GCM, Hunt WF, Musgrave DR (1985) Endophytic fungi affect growth of perennial ryegrass. NZ J Agric Res 28:165–168

Latch GCM, Christensen MJ, Hickson RE (1988) Endophytes of annual and hybrid ryegrasses. NZ J Agric Res 31:57–63

Leuchtmann A (1992) Systematics, distribution, and host specificity of grass endophytes. Nat Toxins 1:150–162

Leuchtmann A (1994a) Isozyme relationships of *Acremonium* endophytes from twelve *Festuca* species. Mycol Res 98:25–33

Leuchtmann A (1994b) Isozyme characterization, persistence and compatibility of fungal and grass mutualists. In: Bacon CW, White J Jr (eds) Biotechnology of endophytic fungi of grasses. CRC Press, Boca Raton, pp 21–33

Leuchtmann A, Clay K (1988a) *Atkinsonella hypoxylon* and *Balansia cyperi*, epiphytic members of the Balansiae. Mycologia 80:192–199

Leuchtmann A, Clay K (1988b) Experimental infection of host grasses and sedges with *Atkinsonella hypoxylon* and *Balansia cyperi*. Mycologia 80:291–297

Leuchtmann A, Clay K (1989a) Experimental evidence for genetic variation in compatibility between the fungus *Atkinsonella hypoxylon* and its three host grasses. Evolution 43:825–834

Leuchtmann A, Clay K (1989b) Isozyme variation in the fungus *Atkinsonella hypoxylon* within and among populations of its host grasses. Can J Bot 67:2600–2607

Leuchtmann A, Clay K (1989c) Morphological, cultural and mating studies on *Atkinsonella*, including *A. texensis*. Mycologia 81:692–701

Leuchtmann A, Clay K (1990) Isozyme variation in the *Acremonium*/*Epichloë* fungal endophyte complex. Phytopathology 80:1133–1139

Leuchtmann A, Clay K (1993) Nonreciprocal compatibility between *Epichloë typhina* and four host grasses. Mycologia 85:157–163

Leuchtmann A, Clay K (1996) Isozyme evidence for host races of the fungus *Atkinsonella hypoxylon* (Clavicipitaceae) infecting the *Danthonia* (Poaceae) complex in the southern Appalachians. Am J Bot 83:1144–1152

Leuchtmann A, Schardl CL, Siegel MR (1994) Sexual compatibility and taxonomy of a new species of *Epichloë* symbiotic with fine fescue grasses. Mycologia 86:802–812

Lewis GC, Clements RO (1986) A survey of ryegrass endophyte (*Acremonium loliae*) in the U.K. and its apparent ineffectuality on a seedling pest. J Agric Sci 107:633–638

Marks S, Clay K (1990) Effects of CO_2 enrichment, nutrient addition, and fungal endophyte infection on the growth of two grasses. Oecologia 84:207–214

Marks S, Clay K, Cheplick GP (1991) Effects of fungal endophytes on interspecific and intraspecific competition in the grasses *Festuca arundinacea* and *Lolium perenne*. J Appl Ecol 28:194–204

O'Sullivan BD, Latch GCM (1993) Infection of plantlets, derived from ryegrass and tall fescue meristems, with *Acremonium* endophytes. In: Hume DE, Latch GCM, Easton HS (eds) Proc 2nd Int Symp *Acremonium*/Grass Interactions. AgResearch, Grasslands Research Center, Palmerston North, New Zealand, pp 16–17

Porter JK, Bacon CW, Cutler HG, Arrendale RF, Robbins JD (1985) In vitro auxin production by *Balansia epichloë*. Phytochemistry 24:1429–1431

Raynal G (1991) Libération des ascospores d'*Epichloë typhina*, agent de la quenouille du dactyle. Conséquences pour l'épidémiologie et la lutte. Fourrage 127:345–358

Read JC, Camp BJ (1986) The effect of fungal endophyte in tall fescue on animal performance, toxicity, and stand maintenance. Agron J 78:848–850

Rice JS, Pinkerton BW, Stringer WC, Undersander DJ (1990) Seed production in tall fescue as affected by fungal endophyte. Crop Sci 30:1303–1305

Richardson MD, Hoveland CS, Bacon CW (1993) Photosynthesis and stomatal conductance of symbiotic and nonsymbiotic tall fescue. Crop Sci 33:145–149

Roberts CA, Marek SM, Niblack TL, Karr AL (1992) Parasitic *Meloidogyne* and mutualistic *Acremonium* increase chitinase in tall fescue. J Chem Ecol 18:1107–1116

Rolston MP, Hare MD, Moore KK, Christensen MJ (1986) Viability of *Lolium* endophyte fungus in seed stored at different moisture contents and temperatures. NZ J Exp Agric 14:297–300

Rykard DM, Bacon CW, Luttrell ES (1985) Host relations of *Myriogenospora atramentosa* and *Balansia epichloë* (Clavicipitaceae). Phytopathology 75:950–956

Saha DC, Johnson-Cicalese JM, Halisky PM, Van Heemstra MI, Funk CR (1987) Occurrence and significance of endophytic fungi in the fine fescues. Plant Dis 71:1021–1024

Sampson K (1933) The systemic infection of grasses by *Epichloë typhina* (Pers.) Tul. Trans Br Mycol Soc 18: 30–47

Schardl CL, Tsai HF (1992) Molecular biology and evolution of the grass endophytes. Nat Toxins 1:171–184

Schardl CL, Liu J, White JF Jr, Finkel RA, An Z, Siegel MR (1991) Molecular phylogenetic relationships of nonpathogenic grass mycosymbionts and clavicipitaceous plant pathogens. Plant Syst Evol 178:27–41

Schardl CL, Leuchtmann A, Tsai HF, Collett MA, Watt DM, Scott DB (1994) Origin of a fungal symbiont of perennial ryegrass by interspecific hybridization of a mutualist with the ryegrass choke pathogen, *Epichloë typhina*. Genetics 136:1307–1317

Shaw J (1873) On the changes going on in the vegetation of South Africa. Bot J Linn Soc 14:202–208

Siegel MR, Latch GCM (1991) Expression of antifungal activity in agar culture by isolates of grass endophytes. Mycologia 83:529–537

Siegel MR, Johnson MC, Varney DR, Nesmith WC, Buckner RC, Bush LP, Burrus II PB, Jones TA, Boling JA (1984) A fungal endophyte in tall fescue: incidence and dissemination. Phytopathology 74:932–937

Siegel MR, Järlfors U, Latch GCM, Johnson MC (1987) Ultrastructure of *Acremonium coenophialum*, *Acremonium lolii*, and *Epichloë typhina* endophytes in host and nonhost *Festuca* and *Lolium* species of grasses. Can J Bot 65:2357–2367

Stovall ME, Clay K (1988) The effect of the fungus *Balansia cyperi* on the growth and reproduction of purple nutsedge, *Cyperus rotundus*. New Phytol 109:351–359

Stovall M, Clay K (1991) Fungitoxic effects of *Balansia cyperi* (Clavicipitaceae). Mycologia 83:288–295

Tsai HF, Liu JS, Staben C, Christensen MJ, Latch GCM, Siegel MR, Schardl CL (1994) Evolutionary diversification of fungal endophytes of tall fescue grass by hybridization with *Epichloë* species. Proc Natl Acad Sci USA 91:2542–2546

Van Horn R, Clay K (1995) Mitochondrial DNA variation in the fungus *Atkinsonella hypoxylon* infecting sympatric *Danthonia* grasses. Evolution 49:360–371

Webster J (1980) Introduction to fungi. Cambridge University Press, Cambridge

Welch AM, Bultman TL (1993) Natural release of *Epichloë typhina* ascospores and its temporal relationship to fly parasitism. Mycologia 85:756–763

Welty RE, Azevedo MD, Cooper TM (1987) Influence of moisture content, temperature, and length of storage on seed germination and survival of endophytic fungi in seeds of tall fescue and perennial ryegrass. Phytopathology 77:893–900

Welty CP, Barker RE, Azevedo MD (1991) Reaction of tall fescue infected and noninfected by *Acremonium coenophialum* to *Puccinia graminis* subsp. *graminicola*. Plant Dis 75:883–886

West CP, Izekor E, Turner KE, Elmi AA (1993) Endophyte effects on growth and persistence of tall fescue along a water-supply gradient. Agron J 85:264–270

Western JH, Cavett JJ (1959) The choke disease of cocksfoot (*Dactylis glomerata*) caused by *Epichloë typhina* (Fr.) Tul. Trans Br Mycol Soc 42:298–307

White JF Jr (1987) Widespread distribution of endophytes in the Poaceae. Plant Dis 71:340–342

White JF Jr (1988) Endophyte-host associations in forage grasses. XI. A proposal concerning origin and evolution. Mycologia 80:442–446

White JF Jr (1993) Endophyte-host associations in grasses. XIX. A systematic study of some sympatric species of *Epichloë* in England. Mycologia 85:444–455

White JF Jr, Bultman TL (1987) Endophyte-host associations in forage grasses. VIII. Heterothallism in *Epichloë typhina*. Am J Bot 74:1716–1721

White JF Jr, Chambless DA (1991) Endophyte–host associations in forage grasses. XV. Clustering of stromata-bearing individuals of *Agrostis hiemalis* infected by *Epichloë typhina*. Am J Bot 78:527–533

White JF, Cole GT (1985) Endophyte–host associations in forage grasses. III. In vitro inhibition of fungi by *Acremonium coenophialum*. Mycologia 77:487–489

White JF Jr, Cole GT (1986) Endophyte–host associations in forage grasses. IV. The endophyte of *Festuca versuta*. Mycologia 78:102–107

White JF Jr, Owens JR (1992) Stromal development and mating system of *Balansia epichloë*, a leaf-colonizing endophyte of warm-season grasses. Appl Environ Microbiol 58:513–519

White JF Jr, Morrow AC, Morgan-Jones G, Chambless DA (1991) Endophyte–host associations in forage grasses. XIV. Primary stromata formation and seed transmission in *Epichloë typhina*: developmental and regulatory aspects. Mycologia 83:72–81

White JF Jr, Morgan-Jones G, Morrow AC (1993) Taxonomy, life cycle, reproduction and detection of *Acremonium* endophytes. In: Joost R, Quisenberry S (eds) *Acremonium*/grass interactions. Agric Ecosyst Environ, vol 44. Elsevier, Amsterdam, pp 13–37

The Evolution of Fungus-Plant Associations

13 Genomic Structure and Genetic Flexibility in Pathogenic Fungi

J.W. Wöstemeyer

CONTENTS

I. Introduction ... 205
II. Coevolution Between Host and Pathogen ... 205
A. Interactions Between *Leptosphaeria maculans* and *Brassica napus* ... 206
B. Correlations with Genetic Complexity ... 207
III. Induction of Genetic Variation by Meiotic and Mitotic Recombination ... 207
IV. Observations at the Karyotype Level ... 208
A. Electrophoretic Karyotyping ... 208
B. Mapping Genes by Electrophoretic Karyotyping ... 208
C. Comparing Phytopathogens with Other Fungi ... 209
1. Karyotypes of Yeasts ... 209
2. Karyotypes of Filamentous Fungi ... 209
3. Correlations Between Karyotype Diversity and Pathogenicity ... 210
4. Comparing Parasites with Related Saprotrophs ... 210
5. Are Species Defined by Karyotypes? ... 211
V. Observations Involving RFLPs, RAPDs, and DNA Fingerprints ... 211
A. Techniques ... 211
B. Variability at the Subchromosomal Level ... 212
1. *Fusarium* sp. ... 212
2. *Leptosphaeria* sp. ... 213
VI. Has Horizontal Gene Transfer Influenced Fungal Adaptation and Evolution? ... 213
A. General Considerations ... 213
B. Experimental Fungal Systems ... 214
1. *Plasmodiophora brassicae* ... 214
2. *Aspergillus niger* ... 215
3. *Parasitella parasitica* ... 215
VII. Conclusions ... 215
References ... 216

I. Introduction

In order to live and survive as successful pathogens, fungi had to develop several fundamental capabilities during the course of evolution. A pathogen must recognize and penetrate potential hosts and must circumvent their various defense mechanisms. Finally, it must solve the problem of successful propagule formation within or on the host. Most fungi have not developed the genetic and physiological apparatus for colonization of a host. The prerequisites which define a successful pathogen are unclear. Wall-degrading enzymes produced by pathogens such as cutinases, the various cellulose and pectin degrading activities or proteases have been discussed as important pathogenicity factors (Chap. 4, Vol. V, Part A), but such enzymes are not specific for phytopathogens and are also found in similar combinations among saprotrophic fungi. Over time host plants have adapted to neutralize and control the degradative capabilities of fungi.

One reason why plants can cope with fungal attack is that they build up a hostile environment for the invader. Various cell wall-degrading exoenzymes of plant origin including glucanases and chitinases have been described as plant responses following the infection by a fungal pathogen (Pearce 1990; Isaac 1992; Chap. 6, Vol. V, Part A). The production of phytoalexins, elicited by a variety of metabolites or degradation products, as well as the cooperative action of several other defense mechanisms (Dean and Kuc 1990) help to slow the propagation of fungal mycelia within the plant (Bailey and Mansfield 1982; Chap. 6, Vol. V, Part A). Whether coevolution between plants and their pathogens correlates with flexibility in the fungal genome is currently controversial, and the present chapter will thus consider and interpret data from situations in which both pathogenic behavior and genomic structure are at least in part understood.

II. Coevolution Between Host and Pathogen

An important aspect of the control of pathogens is that many of the plant's carbon supplies are not readily degraded by fungal enzymes. Some simple

Institut für Mikrobiologie, Lehrstuhl für Allg. Mikrobiologie und Mikrobengenetik, Friedrich-Schiller-Universität, Neugasse 24, D-07743 Jena, Germany

The Mycota V Part B
Plant Relationships
Carroll/Tudzynski (Eds.)
© Springer-Verlag Berlin Heidelberg 1997

considerations based on the infection behavior of the rape seed (*Brassica napus*) pathogen *Leptosphaeria maculans* (anamorph: *Phoma lingam*) may illustrate this (Fig. 1). Although not extensively characterized at the genetic level, its infection cycle is known and data are available on the genomic structure of this fungus. As these latter aspects will form the central part of this chapter, the essentials of its life cycle will be pointed out.

A. Interactions Between *Leptosphaeria maculans* and *Brassica napus*

After inoculation of leaves with an aggressive isolate of *Leptosphaeria maculans*, the fungus grows only for a short time in the intercellular space of the mesophyll (Hammond et al. 1985; Hammond and Lewis 1987). During this stage, the fungus produces toxins, the sirodesmins (Badawy and Hoppe 1989), which probably render the plant cells permeable to nutrients which, in turn, feed the fungus. However, plant defense mechanisms normally control spread of the pathogen effectively, and thus lesions within the leaves remain small. The interpretation of this cycle from an evolutionary point of view is obvious. While the fungus can gain access to nutrients by the production of toxins, the host plant has adapted to control mycelial spread within its living tissue. The fungus pays for easily degradable, intracellular carbon and nitrogen supplies with limited growth and only rare formation of spores. During a second stage of growth in the plant, *Leptosphaeria maculans* withdraws from the plant's active defense mechanisms. The fungus enters the xylem vessels and grows through petioles and the stem towards the hypocotyl region. During this stage, the nutrient supply is very low and the fungus has essentially no access to primary products of photosynthesis although its spread is no longer controlled by the plant. In a final growth stage, in the lower stem region and the hypocotyl, *Leptosphaeria maculans* leaves the xylem and attacks the surrounding tissue through the production of toxins and a massive secretion of pectin- and cellulose-degrading enzymes (Hanacziwskyj and Drysdale 1984a,b; Easton and Rossall 1985). Only during this stage does the fungus sporulate, both sexually and asexually. Host plants tend to die at this stage, but after producing enough seeds to warrant its survival. As in many other plant-pathogen relationships, coevolution has led to a metastable equilibrium which allows for the persistence of both partners.

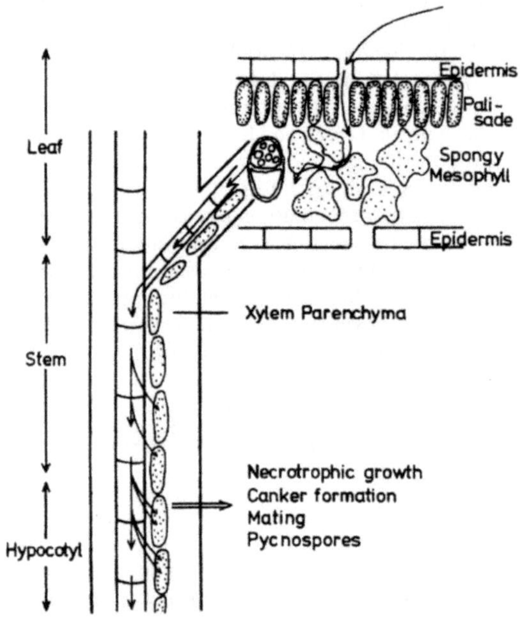

Fig. 1. Infection behaviour of a typical aggressive isolate of the rapeseed pathogen *Leptosphaeria maculans* (*Phoma lingam*; based on Hammond et al. 1985). Typically, the fungus invades its host plant on the leaves and grows mainly intercellularly without causing much harm. It enters xylem vessels and grows towards the lower stem and hypocotyl region, where it leaves the xylem and causes severe damage

B. Correlations with Genetic Complexity

Certainly the complex interdependence between host and parasite at many different physiological levels requires a certain mutability of both partners to maintain a balanced situation like that in *Leptosphaeria maculans*. It seems unlikely, however, that the evolutionary stress on the partners of a pathogenic relationship is substantially stronger than that on common saprotrophs which live under conditions of extreme competition for nutrients in soil and litter (Luttrell 1974). Given that residence within a host plant provides a fungus with a protected ecological niche once the defense mechanisms have been coped with, pathogens should not necessarily be expected to show a tendency towards higher mutability and greater genomic flexibility than saprotrophs.

The mere existence of fungal pathotypes, specified by many different avirulence genes, which correspond on a one-to-one basis with plant resistance genes, does not require huge overall genomic mutability or chromosomal flexibility. This relationship at the genetic level between hosts and pathogens was first described by Flor (1946) in the flax (*Linum usitatissimum*)/rust fungus (*Melampsora lini*) system. The general value of this gene-for-gene hypothesis has been demonstrated in a variety of host-pathogen systems and has also been verified at the molecular level by cloning and molecular characterization of avirulence genes from the tomato (*Lycopersicon esculentum*) pathogen *Cladosporium fulvum* (Van Kan et al. 1991; Marmeisse et al. 1993; Joosten et al. 1994; Chap. 1, this Vol.). However, the number of different avirulence genes within a given pathogen does not seem to be so high that genetic mechanisms for introducing a general genetic instability would be required. Moreover, even in systems where phenotypic variability is indeed realized by rearrangements at the genetic level, the recombination processes are likely to be restricted to specific regions of the genomes.

A good example is the transposition process leading to surface variation in trypanosomes (Van der Ploeg et al. 1984), a phenomenon which increases the pathogenicity of these flagellates by helping them to escape the host's immunoresponse.

Whether the complex relations between host plants and their fungal pathogens have promoted genetic instability and flexibility cannot be answered on the basis of present knowledge. We can, however, address the question whether plant pathogenic fungi are really genetically unstable and exhibit an increased genomic variability. The past years have provided us with several powerful methods which allow the physical analysis of fungal genomic DNA, independently of the availability of genetic markers and Mendelian approaches for characterizing the fungal genome.

III. Induction of Genetic Variation by Meiotic and Mitotic Recombination

Meiotic events in a heterothallic fungus may lead to two alternative consequences with regard to genetic variability. If the population is essentially homogeneous, meiosis will probably contribute to streamlining of the genome. Efficient pairing of the chromosomes presumably depends on homology of the chromosomes involved. Extended chromosomal deviations or aneuploidy, which arise from time to time by mutation, will lead to chromosomal losses, to the breakage of chromosomes, and other aberrations, which may prove lethal for a majority of the offspring. Sexual propagation will therefore select for individuals with reasonably homogeneous chromosomal complements. The situation is different in fungal populations which contain a wealth of diverse genotypes. Again, meiosis will tend to streamline the chromosomes with regard to map position of loci, to chromosome lengths, and to the degree of aneuploidy, but frequent recombination between the parents' genetically different genes will primarily favor the generation of novel genotypes.

The situation may be different in homothallic fungi. Although a limited RFLP analysis has shown less variation in homothallic species of some saprotrophic Sordariaceae than in heterothallic ones (Glass et al. 1990), other scenarios are possible under conditions where high frequencies of somatic deviations occur. Because under these conditions meiosis involves the pairing of chromosomes bearing identical aberrations, these should be tolerated more often. Thus, sexually propagating homothallic fungi might show a higher tendency towards variability at the chromosome level. Mathematical modeling shows that polymorphisms can be stably propagated under heterothallic as well as homothallic conditions (Nauta and Hoekstra 1992).

Because fungi which propagate asexually by mitotic sporulation lack the error-correction process active during meiotic chromosome pairing and symmetrical segregation by the spindle apparatus, the variability at the chromosomal level might be higher in imperfect fungi in comparison with sexual species. In principle, this hypothesis could be verified experimentally. Unfortunately, studies on the correlation of chromosomal polymorphisms with the predominant type of propagation (sexual or asexual) have not been carried out. Appropriate fungal phytopathogens are available, however, and species amenable to molecular karyotype analysis have recently been proposed (Kistler and Miao 1992).

An instructive example for the introduction of diversity into a fungal pathogen population by meiosis is the rapid acquisition of genetic variability in the oomycotan fungus *Phytophthora infestans*, the causal agent of late blight in potato (*Solanum tuberosum*; Drenth et al. 1993; Drenth and Govers 1994; Chap. 2, this Vol.) after the A2 genotype appeared in European potato fields after 1980 (Hohl and Iselin 1984; Tantius et al. 1986; Fry et al. 1992). New genotypes and many different and complex races have been shown to arise by recombination, some of which are evidently more fit than the asexual race previously resident in the fields, a race which has now been completely displaced. (Chap. 2, this Vol.).

IV. Observations at the Karyotype Level

A. Electrophoretic Karyotyping

Apart from very few organisms with relatively large nuclei like *Basidiobolus ranarum* (Robinow 1963) or *Neottiella rutilans* (Westergaard and von Wettstein 1970), fungi are poorly suited for cytogenetic studies due to the minuteness of their nuclei and chromosomes. While small genome size may argue for the use of fungi as good model systems for studying the molecular biology of eukaryotes, it also constitutes a major drawback when chromosome structure is considered. Some excellent studies on the chromosomes of *Neurospora crassa* (McClintock 1945; Perkins and Barry 1977) are no more than exceptions to this rule.

Investigations on the cytogenetics of fungi depend on physical separations of chromosomes combined with hybridization with molecular probes, not on microscope observations. During the last decade, a variety of technical tricks for the separation of DNA molecules in the size range between 0.2 and 12 Mb have been published (Schwartz and Cantor 1984; Orbach et al. 1988; Kayser and Wöstemeyer 1991). This range spans roughly the size of typical fungal chromosomes. All these techniques share two characteristics: (1) chromosomes are set free from embedded mycelia or from protoplasts, thus avoiding breakage of chromosomal DNA; (2) molecules of chromosome size are separated in normal agarose gels in a discontinuous electric field. The methods for generating the discontinuity in field direction are various and range from simply turning the gel between the electrodes, to turning the electrodes around a fixed gel and to the elaborate electronic control of many individual point-shaped electrodes around the gel. For reasons of simplicity, the term pulsed-field electrophoresis will be used throughout this review for all variants of this technique. The reader especially interested in techniques or in theory of chromosome separation should consult a recent review (Cantor et al. 1988; Mathew et al. 1988; Lai et al. 1989; Olson 1989; Walz 1995).

B. Mapping Genes by Electrophoretic Karyotyping

Genetic maps for fungi used to be based on segregation analysis of genetic markers in genetic crosses. This method depends on the availability of many clearly defined, nonlethal mutants with a visible phenotype and on the possibility for mating and efficient germination of meiospores under laboratory conditions. Alternatives for sexual recombination based on naturally occurring parasexual cycles (Pontecorvo and Käfer 1958) or protoplast fusions in the laboratory (Davis 1985; Manzinger and Ferenczy 1985; Harling et al. 1988; Wöstemeyer et al. 1990) have been established for a variety of fungi, but they are rather inefficient in most instances. Therefore, the precise routine assignment of genetic markers to fungal chromosomes has usually not been possible.

Under these circumstances pulsed-field electrophoresis is the method of choice. Even in cases where it has been used in conjunction with a well-established conventional genetic system, it has added considerably to our understanding of genome structure. A good example involves the causative agent of rice blast, *Magnaporthe grisea* (Skinner et al. 1993). The combination of conventional genetic markers with approximately 100 RFLP markers, which were assigned to linkage groups and chromosomes by crosses and hybridization of pulsed-field gels, allowed the construction of a detailed chromosome map. This work also allowed a precise determination of the number of chromosomes in *Magnaporthe grisea*. Genetic linkage groups (Sweigard et al. 1993) and cytological observations (Leung and Williams 1987) were streamlined and correlated with the number of electrophoretically resolvable chromosomal molecules (Skinner et al. 1993).

C. Comparing Phytopathogens with Other Fungi

1. Karyotypes of Yeasts

The first electrophoretic karyotype published for a fungal phytopathogen was that for the causative agent of corn smut, *Ustilago maydis* (Kinscherf and Leong 1988). *U. maydis* has at least 20 small chromosomes between 0.3 and 2–4 Mb. Even this first karyotype determination revealed a striking and unexpected degree of polymorphism when different field isolates and laboratory strains were analyzed. Southern hybridization of chromosome patterns with a probe for the gene for a particular heat shock protein (*hsp*70) suggested that a translocation had taken place. Similar experiments with *Ustilago hordei*, which causes covered smut of barley and oats, confirmed the impression that different isolates of a single species may have distinctive electrophoretic karyotypes (McCluskey and Mills 1990). Meiotic tetrads, however, exhibited a constant chromosome pattern in nine out of ten crosses. Variability may be due to aneuploidy or to chromosome length polymorphisms as a consequence of translocations, deletions, or insertions.

In view of the concern here with chromsomal variability in phytopathogenic fungi, these results should be compared with those obtained for nonpathogenic yeasts. The answer is clear-cut: laboratory strains of *Saccharomyces cerevisiae* have a fairly constant chromosome pattern, as might have been expected, since most current laboratory strains date back to a small selection of natural isolates (Lindegren and Lindegren 1943). The situation is different when naturally occurring strains of *Saccharomyces cerevisiae* from wine are analyzed. These isolates show a surprisingly high degree of aneuploidy (Bakalinsky and Snow 1990) and probably other chromosomal aberrations (Longo and Vezinhet 1993). Similarly large amounts of karyotype variation have been recorded in other ascomycetous and basidiomycetous yeasts, be they saprotrophs (Marri et al. 1993; Naumova et al. 1993), plant pathogens (Russell and Mills 1993), or animal pathogens (Lott et al. 1993).

2. Karyotypes of Filamentous Fungi

What is the situation in filamentous fungi with regard to the relative degree of karyotypic variation among phytopathogens? The chromosomes of many mycelial plant pathogens have been resolved by pulsed-field electrophoresis. Several examples demonstrate immediately that the situation is similar to that in the yeasts: natural isolates have a high degree of karyotype variability. As already mentioned, *Magnaporthe grisea* showed considerable chromosome length polymorphism (Talbot et al. 1993). Evidence for deletions, translocations, and other, more complex chromosomal rearrangements has been provided by chromosome separation and hybridization studies, but the overall chromosome number appears to be amazingly constant. In addition, a number of highly variable minichromosomes, characterized by meiotic instability, occur in many, but not all, isolates of this fungus. With respect to pathogenicity all of the observed rearrangements among rice-pathogenic strains are neutral; no evidence for the generation of new forms differing in virulence or aggressivity was provided. In addition, no correlations between individual pathotypes and specific karyotypes were observed (Talbot et al. 1993). In the pea wilt pathogen, *Nectria haematococca*, the situation may be different. Here, a family of cytochrome P-450 genes involved in the detoxification of the phytoalexin pisatin could be ascribed to a dispensable, meiotically unstable B-chromosome (Miao et al. 1991). Obviously, losing or acquiring this chromosome by sexual or parasexual processes will affect the behavior of the pathogen on pisatin-producing host plants.

The high chromosomal variability seen in *Magnaporthe grisea* is certainly not an unusual situation. Similar results have been obtained for the Dutch elm disease fungus *Ophiostoma ulmi* (Dewar and Bernier 1993), the wheat pathogen *Septoria nodorum* (Cooley and Caten 1991), and *Septoria tritici* (McDonald and Martinez 1991). This last case shows that polymorphisms seem to arise fairly frequently and are stable within a population, since all of the seven distinct isolates analyzed were sampled from a population infecting a single wheat field. A high degree of chromosomal polymorphism has also been found in the imperfect fungus *Colletotrichum gloeosporioides*, which causes anthracnose diseases in different pasture legumes in Australia (Masel et al. 1990; Chap. 10, this Vol.). This variability seems to be caused, at least in part, by DNA additions and deletions (Masel et al. 1993). As in other cases, it has been suggested that the plasticity of fungal chromosomes, which is observed at the level of electrophoretic karyotypes, might be the reason for the rapid adaptation of *Colletotrichum gloeosporioides* to new resistant host genotypes in the field. However, experimental data in support of this hypothesis are unavailable.

3. Correlations Between Karyotype Diversity and Pathogenicity

It is tempting to speculate that frequent chromosomal rearrangements change certain genes in such a way that dominant avirulence genes are inactivated. Such events would alter the host range of pathogenic fungi. On the other hand, it is hard to imagine that recombination points are situated predominantly within avirulence genes. Probably most of the recombination points involved in the generation of karyotypic diversity are situated within noncoding, repetitive elements, which are interspersed in the genome. In this case, the high variability visualized on chromosome gels will probably not mean very much. With the exception of position effects the expression and regulation of most genes would probably not be influenced. Based on our present knowledge we can, however, not totally exclude the possibility that highly elaborate recombination mechanisms exist, which specifically affect genes involved in pathogen/host recognition. At least in *Schizosaccharomyces pombe*, mutants have been isolated recently, which are recombination-deficient specifically around the *ade6* locus on chromosome III (De Veaux and Smith 1994). Thus, specific recombinational activators may exist for limited arrays of loci. With regard to phytopathogens, this is mere speculation, and we must wait for more detailed molecular analyses of recombination events in fungi.

4. Comparing Parasites with Related Saprotrophs

If indeed phytopathogenesis is strictly and directly correlated with high genomic variability, we might expect that saprotrophic species would exhibit more conserved chromosome patterns than their pathogenic relatives. A detailed analysis with this aim in mind has not been performed to my knowledge, but karyotypes have been established for several different species within the genus *Fusarium*. Six saprotrophic antagonistic strains and seven pathogenic formae speciales of *Fusarium oxysporum* were compared (Migheli et al. 1993). All of them exhibited high variability, and no indication of elevated chromosomal stability in any of the groups was found. These results are confirmed by the karyotype analysis with eight species from the *Fusarium* sections *Arthrosporiella* and *Sporotrichiella*, which are either associated with cereals as weak pathogens or saprotrophs (Fekete et al. 1993). Similar results have been obtained with *Trichoderma longibrachiatum* (Mäntilä et al. 1992), a species which ranges somewhere between a saprotroph and a weak soft-rot fungus. If we include mere saprotrophic species with no relations to pathogens in the analysis, the general picture already seen for the yeasts holds true. The electrophoretic karyotypes of field isolates have a strong tendency towards high variability. The number of saprotrophic species whose karyotypes have been compared is still rather limited. Examples are drawn from the Zygomycetes *Absidia glauca* (Kayser and Wöstemeyer 1991), *Mucor circinelloides* (Nagy et al. 1994), and *Parasitella parasitica* (Burmester and Wöstemeyer 1994), the latter being, however, a facultative parasite of other *Mucor*-like fungi.

The constant karyotypic patterns of many laboratory strains can be explained by their origins often in just a few initial isolates, or even a single one in the case of *Aspergillus nidulans* (Yuill 1939; Pontecorvo et al. 1953; Clutterbuck 1974). Consequently, the general picture emerges that electrophoretic karyotypes vary considerably within single species independently from their lifestyle as saprotroph or pathogen.

5. Are Species Defined by Karyotypes?

Apart from special situations, karyotyping is not generally suited for defining the boundaries of species at the molecular level. In this respect, the rape seed pathogen *Leptosphaeria maculans* (*Phoma lingam*) may serve as an interesting example. Although no generally expected pathotype structure has been established for this pathogen, all isolates of *Leptosphaeria maculans* can be ascribed consistently to one of two major pathotype groups, which are designated as virulent and avirulent (Gabrielson 1983) or aggressive and nonaggressive (Badawy and Hoppe 1989). Both groups can be differentiated by several biochemical and molecular characteristics (Hassan et al. 1991; Lim and Howlett 1994; Plummer et al. 1994). A current debate deals with whether these pathotype groups really belong to the same species, or if the many differences between them justify the delineation of new species (Hassan et al. 1991; Taylor et al. 1991; Schäfer and Wöstemeyer 1992; Plummer and Howlett 1993). As in all the other experimental systems, the electrophoretic karyograms show a high degree of diversity even within both of the pathotype groups. In addition, clear-cut differences separate both groups (Taylor et al. 1991; Morales et al. 1993a,b). This observation may be used as an argument for the separation of both pathotype groups into two species. However, karyotype patterns should generally be regarded with care in the context of species definition. We know that chromosome patterns have a strong tendency towards variation. Genetically, this must not necessarily mean very much, as few translocations may lead to a considerable amount of variability. We do not even know positively whether extensive, undisturbed chromosome pairing is really a necessary prerequisite for fungal meiosis. In the facultative mycoparasite *Parasitella parasitica* highly variable intraspecific karyotypes have been observed (Burmester and Wöstemeyer 1994), although mating occurs with comparable efficiency between all isolates. The effects of such variation on speciation are unclear.

Leptosphaeria maculans provides one of the few examples where chromosomal polymorphisms have not only been observed in randomly collected natural field isolates, but among the offspring of single meiotic events. Tetrads from a single pseudothecium gave rise to four distinct karyotypes in a 2:2:2:2 ratio (Plummer and Howlett 1993). The authors explain these patterns by assuming recombination between parental homologues of unequal size. Recombination would produce progeny with differing chromosomal sizes. Normally, sister ascospores have identical karyotypes. Howlett and Plummer propose that the crossing-over events leading to karyotypic variations take place within interspersed repetitive DNA segments which should be situated between coding regions. In order to prove this, much more work must be done. The presence of repetitive DNA has indeed been proven for many fungi. In Zygomycetes such sequences and foldback structures comprise roughly 35% of the genome (Wöstemeyer and Burmester 1986).

The wheat stem rust fungus *Puccinia graminis* has also been shown to have a very high amount (ca. 30%) of repetitive DNA (Backlund and Szabo 1993). However, repeated DNA elements have also been found to lie in the range of below 5% of total DNA in many phytopathogenic Ascomycotina (*Colletotrichum gloeosporioides*; Braithwaite et al. 1990; *Magnaporthe grisea*; Hamer et al. 1989; *Leptosphaeria maculans*; Schäfer and Wöstemeyer 1992), levels still sufficient to promote recombination. Inverted DNA repeats in particular have been proposed as a general feature of the eukaryotic genome which could serve as a source of genomic instability (Gordenin et al. 1993). However, cellular mechanisms may also control the frequency of recombination within repeated DNA elements, as with rDNA sequences of *Saccharomyces cerevisiae* (Petes et al. 1982) or *Coprinus cinereus* (Cassidy et al. 1984). Similar consequences would result from selection against the expression of repetitive DNA by RIP-like phenomena, which are based on the methylation of DNA and have been described in detail for *Neurospora crassa* (Selker 1990, 1991). It has also been demonstrated that defects in DNA methylation in *Neurospora* lead to aneuploidy, which, in this context, may be interpreted as the inability to control the degree of repetition at the level of single chromosomes (Foss et al. 1993).

V. Observations Involving RFLPs, RAPDs, and DNA Fingerprints

A. Techniques

Whereas electrophoretic karyotyping permits a view of variation at the level of whole chromosomes and, therefore, may be regarded as a surrogate for cytogenetics, the various molecular genetic fingerprint techniques provide some insight into the arrangement of DNA at the subchromosomal level. The term "fingerprinting" has never been defined unequivocally. In this chapter, we will use it to mean the characterization of DNAs from single fungal isolates by means of hybridization with synthetic DNA probes which correspond to hypervariable, minisatellite-like targets in chromosomal DNA (Jeffreys et al. 1985; Braithwaite and Manners 1989; Meyer et al. 1992). In a broader sense, fingerprinting is often used for techniques which allow the identification of defined clonal lines. The results obtained by the fingerprinting technique will be discussed together with those based on RFLP (restriction fragment length polymorphism) and RAPD (random amplified polymorphic DNA) experiments.

In principle, all these experimental approaches are suited to obtain a resolution down to the molecular identification of single lines or clones. This resolution is almost routinely obtained with most of the fingerprint probes, which are normally being used, whereas the RFLP and RAPD (Welsh and McClelland 1990; Williams et al. 1990) techniques usually require a more intensive search for adequate probes or primers to reach this level of resolution. On the other hand, an advantage of the RAPD approach to pathogen diagnosis is the relative ease with which primers can be found which detect either highly variable or reasonably conserved regions within a given segment of DNA. Thus, the choice of primers makes it possible to discriminate between defined groups within a species, which may correspond to pathotypes (Goodwin and Annis 1991; Wöstemeyer et al. 1992).

In addition, RAPD markers have proven their value for the construction of physical genomic maps. *Cladosporium fulvum* is a good example (Arnau et al. 1994) in which the combination of induced mitotic recombination and the RAPD technique has led to information on the genome structure. Especially because of its simplicity, the very low amount of fungal DNA required, and the speed of the analysis, the RAPD technique is often favored over RFLP or fingerprinting. The major considerations which may be of some help in deciding among molecular methods for the identification of phytopathogens have been discussed recently (Wöstemeyer et al. 1992; Schäfer and Wöstemeyer 1994).

B. Variability at the Subchromosomal Level

The results obtained for the genomes of phytopathogenic fungi by fingerprinting, RFLP, and RAPD are similar to those obtained by electrophoretic karyotyping. In most cases, a strong tendency towards high variability also occurs at the subchromosomal level. An exception to this general rule seems to occur in *Sclerotinia sclerotiorum*, where only a few chromosome length differences and an unusually low degree of RAPD polymorphisms have been observed among field isolates from different locations (Fraissinet-Tachet et al. 1996). A complementary approach involves tracing the fates of different fungal genotypes in populations when confronted with different varieties of host plants. The RFLP analysis of ten loci in field collections of *Mycosphaerella graminicola* over several years revealed no significant changes in overall genotypic diversity (Chen et al. 1994).

A complete list of fungal species with high degrees of genetic polymorphism is unnecessary in the present treatment. Instead, we will concentrate on two examples in the genera *Fusarium* and *Leptosphaeria*.

1. *Fusarium* sp.

Fungi from the anamorphic genus *Fusarium* can be found worldwide as important pathogens on a variety of different host plants. Many *Fusarium* species are poorly defined, and unequivocal recognition of formae speciales in particular can prove difficult and time-consuming. Often the systematic status of subgroups or mating populations, as determined through laboratory crosses, is unclear and controversial. An instructive example involves the different mating populations or compatibility groups of *Gibberella fujikuroi* (imperfect form: *Fusarium moniliforme*), which are treated as different species by several authors (Nelson et al. 1983; Leslie 1991).

Several approaches have been made to the use of molecular methods at the DNA level for differentiating different species and the various formae speciales within the genus *Fusarium*. Results obtained by electrophoretic karyotyping have been mentioned above (Migheli et al. 1993). Techniques based on the RFLP technique have been similarly successful for the differentiation of *Fusarium* isolates either at the level of mitochondrial DNA (Kuninaga and Yokosawa 1992) or at the level of nuclear DNA (Dickman and Partidge 1989; Correll et al. 1992; Whitehead et al. 1992; Elias et al. 1993). Comparable results have been obtained by RAPD analysis (Grajal-Martín et al. 1993; Voigt et al. 1995). These results show clearly that the sequence variability between the genomes is high enough to differentiate between different subspecies and between most of the mating populations by means of appropriate random primers in PCR assays (Voigt et al. 1995).

2. *Leptosphaeria* sp.

The classification of isolates of *Leptosphaeria maculans* is not as complicated as in the genus *Fusarium*. Investigations by means of PCR (Morales et al. 1993a), RAPD (Goodwin and Annis 1991; Schäfer and Wöstemeyer 1992; Plummer et al. 1994), RFLP (Johnson and Lewis 1990; Hassan et al. 1991; Koch et al. 1991), and minisatellite fingerprinting (Meyer et al. 1992) confirm the results that have been obtained by electrophoretic karyotyping and even by conventional enzymatic measurements (Hassan et al. 1991). The general conclusion from all these investigations is that the species *Leptosphaeria maculans* comprises many different genotypes even within both major pathotype groups, the aggressive and the nonaggressive isolates. The diversity at the DNA level is more pronounced between these groups than within. Thus, the DNA analysis supports measurements at the physiological level, e.g., the differential pigment formation (Koch et al. 1989) or toxin synthesis (Badawy and Hoppe 1989). Whether the observations at the DNA level are sufficient to define two species is unclear. Not all parts of a given genome mutate at the same rate. Conserved DNA stretches lie like islands within highly mutable DNA regions. RAPD primers or RFLP probes can be found which recognize any of these regions. The example in Fig. 2 (C. Schäfer and J. Wöstemeyer, unpubl.) illustrates this feature. The decamer primer sequence 5′-GGCATCGGCC-3′ separates clearly both major pathotype groups of *Leptosphaeria maculans* (Schäfer and Wöstemeyer 1992, 1994), whereas the nonamer primer sequence 5′-GTGAGCGTC-3′ reveals an enormous degree of genetic heterogeneity even within the aggressive isolates. This means that the detection of variability depends largely on the type of the primer employed. Comparable results have been found recently for the facultative mycoparasite *Parasitella parasitica*. Thus, one must be careful about defining species borders mainly on fingerprinting, RFLP, or RAPD.

VI. Has Horizontal Gene Transfer Influenced Fungal Adaptation and Evolution?

A. General Considerations

Horizontal gene transfer between different but related species is a phenomenon which is accepted as a widespread mechanism for parasexuality in bacteria. Also, the naturally occurring interking-

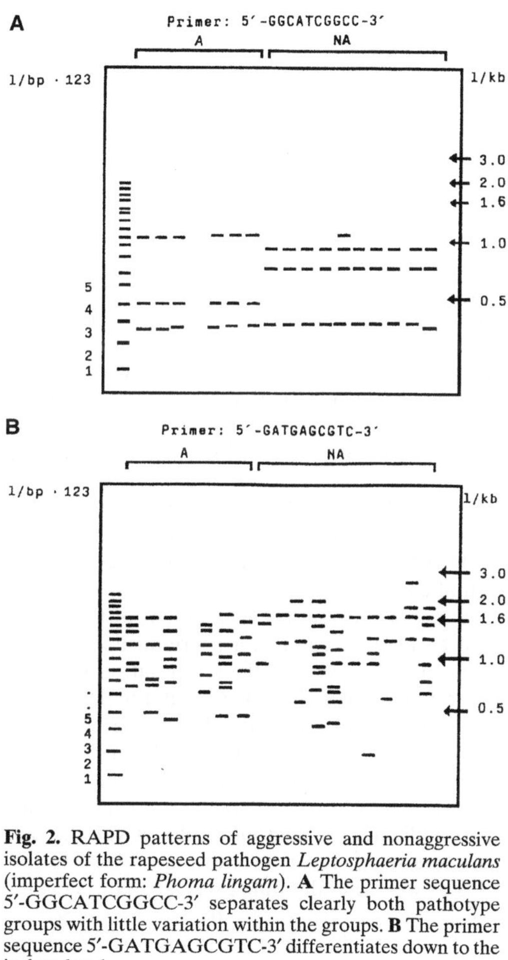

Fig. 2. RAPD patterns of aggressive and nonaggressive isolates of the rapeseed pathogen *Leptosphaeria maculans* (imperfect form: *Phoma lingam*). **A** The primer sequence 5′-GGCATCGGCC-3′ separates clearly both pathotype groups with little variation within the groups. **B** The primer sequence 5′-GATGAGCGTC-3′ differentiates down to the isolate level

dom gene transfer between *Agrobacterium tumefaciens* or *Agrobacterium rhizogenes* and its hosts (Binns and Thomashow 1988; Zambryski et al. 1989), which is functionally correlated with crown gall formation, is well known among biologists. Recently, it has been shown that a flux of genetic material via conjugative processes is also possible between *Escherichia coli* and baker's yeast (Heinemann and Sprague 1989). However, it remains unclear whether these examples are rare exceptions or whether alternatively, horizontal, interspecies gene transfer should be regarded as a major driving force in speciation and evolution.

It is tempting to speculate that fungal parasites or symbionts and their host plants, which exist in intimate contact, may be good candidates for demonstrating frequent interkingdom parasexuality. This idea is not new; some of the general prerequisites that allow the transfer of genetic information from fungi to their hosts were reported over 10 years ago (Lamboy 1984). Moreover, horizontal gene transfer has been proposed as one of the possible mechanisms for the evolutionary origin of parallelism in biology (Kubitzki et al. 1991). The discovery of unexpectedly similar genes for isopenicillin synthase in fungi, gram-positive and gram-negative bacteria may serve as an example in this respect (Landan et al. 1990). Horizontal gene transfer could indeed serve to explain these similarities. We should be aware, however, that this interpretation is merely hypothetical, and refers to a single or not more than a few transfer events some 2 billion years ago. A lucid review on the evolutionary aspects of horizontal or lateral gene transfer has been published recently (Kidwell 1993). We should now concentrate on experimental systems, which allow possibly the transfer of DNA between host and parasite under experimental conditions.

First, DNA donor and recipient of the DNA must be in close contact, a prerequisite which is always met in plant-pathogen relationships. A closer look shows, however, that the DNAs of both organisms are separated by the plasma membranes of the partners as well as by the nuclear envelopes. These membrane systems present an enormous barrier to the transport of the highly hydrophilic nucleic acid molecules. Even the nucleosome structure of chromosomal DNA, in the hosts as well as in the fungi, renders a direct gene transfer between healthy partners highly improbable. However, the chances for interspecific gene transfer may be regarded differently if we also consider situations where the fungal cell undergoes disintegration, either by autolytic processes or by direct plant attack with cell wall lytic enzymes. Degenerating fungal nuclei could release nucleic acids, which might enter into plant protoplasts. Occasionally, as in vesicular arbuscular mycorrhizal associations, older mycelial coils, arbuscules, or vesicles are routinely degraded, exposing much fungal material including the nuclear DNA directly to the host's cytoplasm.

B. Experimental Fungal Systems

1. *Plasmodiophora brassicae*

Few studies have looked for DNA transfer in fungi under these or similar conditions. A single exception has examined the transfer of high molecular weight DNA from the clubroot disease organism *Plasmodiophora brassicae* to its host, *Brassica napus* (oilseed rape). Bryngelsson et al. (1988) looked for possible gene transfer in the reverse direction, from the plant to the protist and analyzed the appearance of repetitive plant DNA in zoospores and in resting spores. By Southern hybridization with an isolated repetitive DNA of *Brassica napus* they could demonstrate the appearance of corresponding fragments in *Plasmodiophora*. Analogous results were obtained with the repeat units for ribosomal DNA, and the authors infer that generally all kinds of plant DNA may be transferred to the parasite. They state that this DNA has a high molecular weight in the recipient. Nevertheless, this DNA is neither covalently integrated into the chromosomal complement of the fungi nor is it established permanently by autonomous replication. A passage of the parasites on a different host species, *Sinapis alba*, removes the former plant DNA completely. Instead, a different pattern of repetitive DNA, specific for the new host plant can be found. Although far from representing a true parasexual system due to the lack of permanent DNA establishment, this phenomenon of transient DNA uptake deserves attention. Unfortunately, this observation has not been checked against experimental errors by all possible experimental approaches, and it remains possible that most of the signals obtained on the Southern blots are due to contamination of fungal spores by plant DNA.

2. *Aspergillus niger*

Recently, evidence has been provided for the uptake of plant DNA from *Brassica nigra* and *Brassica napus* by a saprotrophic fungus, *Aspergillus niger*. Very simple microcosm experiments show that a gene conferring hygromycin resistance appears in fungal single spore cultures after cocultivation of both partners. These results have been checked carefully at the DNA level by Southern hybridizations. In addition, recloning and analysis of the transferred piece of DNA showed that plant DNA had been cotransferred into the fungus (Hoffmann et al. 1994).

3. *Parasitella parasitica*

The transfer of DNA from the parasite to its host has been proven in a mycoparasitic system. *Parasitella parasitica*, a zygomycete which is a facultative parasite of several other *Mucor*-like fungi transfers several of its nuclei to the host at a defined stage during infection. By an unknown mechanism, part of the parasite's DNA appears in the host's genome. It has been shown for several markers that this DNA can be expressed (Kellner et al. 1993). This mycoparasitic relation certainly represents a special case, because the formation of a cytoplasmic continuum is an exception among most known host-parasite systems (Jeffries and Young 1994). However, it may represent a situation close to those, in which the fungal cell of an intracellular parasite disintegrates within its host cell. The infection of special fungal mutants, which are genetically engineered such that the transferred DNA can be expressed and easily recognized in the plant could shed some light on whether the transfer occurs frequently enough to play an important role in coevolution between fungal phytopathogens and their hosts.

VII. Conclusions

Our present knowledge, based on genomic structure and flexibility of fungi, is based mainly on electrophoretic karyotyping and the various molecular fingerprinting techniques (RFLP, RAPD, microsatellite fingerprinting). Many field isolates of plant pathogens have been analyzed. For saprotrophs the amount of data is considerably lower, and we have to be aware that many of these are laboratory strains maintained in culture for a long time. Although exceptions are known, wild-type isolates of fungi generally exhibit highly polymorphic genomes independent of their lifestyle. The question whether plant pathogens are endowed with a higher degree of variability than saprotrophs cannot be answered conclusively. More pronounced genetic variability and hence flexibility may be expected in fungi that have to cope with high selection pressure. However, the evolutionary stress on the partners of a pathogenic relationship is probably not stronger than on a common saprotroph. The challenges by competing microorganisms are comparable to those exerted by plant defense mechanisms.

Apart from genetic variability based on heterokaryosis, evidence is increasing that the uptake of foreign DNA into the fungal chromosomal complement adds to the evolutionary flexibility of fungal genomes. Model systems that have been analyzed by the transfer of genetically engineered resistance markers and at the molecular level are the uptake of plant DNA by *Aspergillus niger* in the rhizosphere of *Brassica*, and the highly efficient transfer of genetic information from the fusion biotroph *Parasitella parasitica* to its hosts.

Acknowledgments. I thank Barbara Howlett (Melbourne) for sharing and discussing with me many observations on *Leptosphaeria maculans* prior to publication and Torsten Hoffmann for valuable information on plant-fungus gene transfer. Anke Wöstemeyer (Berlin), Kerstin Voigt (Jena), Anke Burmester (Jena), and Christine Schäfer (Berlin) contributed helpful discussions and critical comments on the manuscript. I also thank our librarian, Uta Böhme, for the excellent job she did in providing me with many "difficult" papers and Angela Opitz for help with drawing the figures. The experimental work from my laboratory, which was cited in this chapter, was supported by grants from the Deutsche Forschungsgemeinschaft, Bundesministerium für Forschung und Technologie, and by Fonds der Chemischen Industrie.

I must apologize to all those colleagues whose publications are not cited in this chapter. Today, so many relevant papers on molecular characteristics of fungal genomes are available, pathogenic or saprotrophic, that mentioning all of them would go beyond the scope of these pages. No attempt has been made to give a complete account of the literature available on genomic variability in filamentous fungi. Most of the literature which was used during the preparation of this chapter is available as an ASCII file on request from the author by sending in a formatted diskette ($5^1/_4$ or $3^1/_2$in.) readable by IBM compatible computers.

References

Arnau J, Housego AP, Oliver RP (1994) The use of RAPD markers in the genetic analysis of the plant pathogenic fungus *Cladosporium fulvum*. Curr Genet 25: 438–444

Backlund JE, Szabo LJ (1993) Physical characteristics of the genome of the phytopathogenic fungus *Puccinia graminis*. Curr Genet 24:89–93

Badawy HMA, Hoppe HH (1989) Production of phytotoxic sirodesmins by aggressive strains of *Leptosphaeria maculans* differing in interactions with oil seed rape genotypes. J Phytopathol 127:146–157

Bailey JA, Mansfield JW (eds) (1982) Phytoalexins. Wiley, New York

Bakalinsky AT, Snow R (1990) The chromosomal constitution of wine strains of *Saccharomyces cerevisiae*. Yeast 6:367–382

Binns AN, Thomashow MF (1988) Cell biology of *Agrobacterium* infection and transformation of plants. Annu Rev Microbiol 42:575–606

Braithwaite KS, Manners JM (1989) Human hypervariable minisatellite probes detect DNA polymorphisms in the fungus *Colletotrichum gloeosporioides*. Curr Genet 16: 473–475

Braithwaite KS, Irwin JAG, Manners JM (1990) Restriction fragment length polymorphisms in *Colletotrichum gloeosporioides* infecting *Stylosanthes* spp. in Australia. Mycol Res 94:1129–1137

Brody H, Carbon J (1989) Electrophoretic karyotype of *Aspergillus nidulans*. Proc Natl Acad Sci USA 86:6260–6263

Bryngelsson T, Gustafsson M, Gréen B, Lind C (1988) Uptake of host DNA by the parasitic fungus *Plasmodiophora brassicae*. Physiol Mol Plant Pathol 33:163–171

Burmester A, Wöstemeyer J (1994) Variability in genome organization in the zygomycete *Parasitella parasitica*. Curr Genet 26:456–460

Cantor CR, Gaal A, Smith CL (1988) High-resolution separation and accurate size determination in pulsed-field gel electrophoresis of DNA. 3. Effect of electrical field shape. Biochemistry 27:9216–9221

Cassidy JR, Moore D, Lu BC, Pukkila PJ (1984) Unusual organization and lack of recombination in the ribosomal RNA genes of *Coprinus cinereus*. Curr Genet 8:607–613

Chen R-S, Boeger JM, McDonald BA (1994) Genetic stability in a population of a plant pathogenic fungus over time. Mol Ecol 3:209–218

Clutterbuck AJ (1974) *Aspergillus nidulans*. In: King RC (ed) Handbook of genetics, vol 1. Plenum Press, New York, pp 447–510

Cooley RN, Caten CE (1991) Variation in electrophoretic karyotype between strains of *Septoria nodorum*. Mol Gen Genet 228:17–23

Correll JC, Gordon TR, McCain AH (1992) Genetic diversity in California and Florida populations of the pitch canker fungus *Fusarium subglutinans* f. sp. *pini*. Phytopathology 82:415–420

Davis B (1985) Factors influencing protoplast isolation. In: Peberdy JF, Ferenczy L (eds) Fungal protoplasts. Marcel Dekker, New York, pp 45–71

Dean RA, Kuc JA (1990) Immunisation against disease: the plant fights back. In: Pegg GF, Ayres PG (eds) Fungal infections of plants. Cambridge University Press, Cambridge, pp 383–410

De Veaux LC, Smith GR (1994) Region-specific activators of meiotic recombination in *Schizosaccharomyces pombe*. Genes Dev 8:203–210

Dewar K, Bernier L (1993) Electrophoretic karyotypes of the elm tree pathogen *Ophiostoma ulmi*. Mol Gen Genet 238:43–48

Dickman MB, Partidge JE (1989) Use of molecular markers for monitoring fungi involved in stalk rot of corn. Theor Appl Genet 77:535–539

Drenth A, Govers F (1994) DNA fingerprinting of the potato late blight fungus *Phytophthora infestans*. In: Schots A, Dewey FM, Oliver R (eds) Modern assays for plant pathogenic fungi. CAB International, Oxford, pp 223–230

Drenth A, Goodwin SB, Fry WE, Davidse LC (1993) Genotypic diversity of *Phytophthora infestans* in The Netherlands revealed by DNA polymorphisms. Phytopathology 83:1087–1092

Drenth A, Tas ICQ, Govers F (1994) DNA fingerprinting uncovers a new sexually reproducing population of *Phytophthora infestans* in The Netherlands. Eur J Plant Pathol 100:97–108

Easton C, Rossall S (1985) The production of certain cell wall-degrading enzymes by *Leptosphaeria maculans* in culture and in stem canker lesions of oilseed rape. Physiol Plant Pathol 26:185–197

Elias KS, Zamir D, Lichtman-Pleban T, Katan T (1993) Population structure of *Fusarium oxysporum* f. sp. *lycopersici*: restriction fragment length polymorphisms provide genetic evidence that vegetative compatibility group is an indicator of evolutionary origin. Mol Plant-Microbe Interact 6:565–572

Fekete CS, Nagy R, Debets AJM, Hornok L (1993) Electrophoretic karyotypes and gene mapping in eight species of the *Fusarium Arthrosporiella* and *Sporotrichiella*. Curr Genet 24:500–504

Flor HH (1946) Genetics of pathogenicity in *Melampsora lini*. J Agric Res 73:335–357

Foss HM, Roberts CJ, Claeys KM, Selker EU (1993) Abnormal chromosome behavior in *Neurospora* mutants defective in DNA methylation. Science 262:1737–1741

Fraissinet-Tachet L, Reymond-Cotton P, Fèvre M (1996) Molecular karyotype of the phytopathogenic fungus *Sclerotinia sclerotiorum*. Curr Genet 29:496–501

Fry WE, Goodwin SB, Matuszak JM, Spielman LJ, Milgroom MG (1992) Population genetics and intercontinental migrations of *Phytophthora infestans*. Annu Rev Phytopathol 30:107–129

Gabrielson RL (1983) Blackleg disease of crucifers caused by *Leptosphaeria maculans* (*Phoma lingam*) and its control. Seed Sci Technol 11:749–780

Glass NL, Metzenberg RL, Raju NB (1990) Homothallic Sordariaceae from nature: the absence of strains containing only the a mating type sequence. Exp Mycol 14:274–289

Goodwin PH, Annis SL (1991) Rapid identification of genetic variation and pathotype of *Leptosphaeria maculans* by random amplified polymorphic DNA assay. Appl Environ Microbiol 57:2482–2486

Gordenin DA, Lobachev KS, Degtyareva NP, Malkova AL, Perkins E, Resnick MA (1993) Inverted DNA repeats: a source of eukaryotic genomic instability. Mol Cell Biol 13:5315–5322

Grajal-Martín MJ, Simon CJ, Muehlbauer FJ (1993) Use of random amplified polymorphic DNA (RAPD) to characterize race 2 of *Fusarium oxysporum* f. sp. *pisi*. Mol Plant Pathol 83:612–614

Hamer JE, Farrall L, Orbach MJ, Valent B, Chumley FG (1989) Host species-specific conservation of a family of repeated DNA sequences in the genome of a fungal plant pathogen. Proc Natl Acad Sci USA 86:9981–9985

Hammond KE, Lewis BG (1987) Variation in stem infections caused by aggressive and non-aggressive isolates of *Leptosphaeria maculans* on *Brassica napus* var. *oleifera*. Plant Pathol 63:53–65

Hammond KE, Lewis BG, Musa TM (1985) A systemic pathway in the infection of oilseed rape plants by *Leptosphaeria maculans*. Plant Pathol 34:557–565

Hanacziwskyj P, Drysdale RB (1984a) Cultural and biochemical characterization of isolates of *Leptosphaeria maculans* varying in pathogenicity. Aspects Appl Biol 6:387–397

Hanacziwskyj P, Drysdale RB (1984b) Polysaccharide-degrading enzymes produced in vitro by *L. maculans*. Cruciferae Newsl 9:77–78

Harling R, Kenyon L, Lewis BG, Oliver RP, Turner JG, Coddington A (1988) Conditions for efficient isolation and regeneration of protoplasts from *Fulvia fulva*. J Phytopathol 122:143–146

Hassan AK, Schulz C, Sacristan MD, Wöstemeyer J (1991) Biochemical and molecular tools for the differentiation of aggressive and nonaggressive isolates of the oilseed rape pathogen, *Phoma lingam*. J Phytopathol 131:120–136

Heath IB (1994) The cytoskeleton in hyphal growth, organelle movements, and mitosis. In: Wessels JGH, Meinhardt F (eds) The Mycota, vol 1. Springer, Berlin Heidelberg, pp 43–65

Heinemann JA, Sprague GF Jr (1989) Bacterial conjugative plasmids mobilize DNA transfer between bacteria and yeast. Nature 340:205–209

Hoffmann T, Golz C, Schieder O (1994) Foreign DNA sequences are received by a wild type strain of *Aspergillus niger* after co-culture with transgenic higher plants. Curr Genet 27:70–76

Hohl HR, Iselin K (1984) Strains of *Phytophthora infestans* with A2 mating type behavior. Trans Br Mycol Soc 83:529–530

Isaac S (1992) Fungal-plant interactions. Fungal-plant confrontation. Chapman and Hall, London, pp 147–207

Jeffreys AJ, Wilson V, Thein SL (1985) Hypervariable "minisatellite" regions in human DNA. Nature 314:67–73

Jeffries P, Young TWK (1994) Interfungal parasitic relationships. CAB International, Wallingford

Johnson RD, Lewis BG (1990) DNA polymorphism in *Leptosphaeria maculans*. Physiol Mol Plant Pathol 37:417–424

Joosten MHAJ, Cosijnsen TJ, De Wit PJGM (1994) Host resistance to a fungal tomato pathogen lost by a single base-pair change in an avirulence gene. Nature 367: 384–386

Kayser T, Wöstemeyer J (1991) Electrophoretic karyotype of the zygomycete *Absidia glauca*: evidence for differences between mating types. Curr Genet 19:279–284

Kellner M, Burmester A, Woestemeyer A, Woestemeyer J (1993) Transfer of genetic information from the mycoparasite *Parasitella parasitica* to its host *Absidia glauca*. Curr Genet 23:334–337

Kidwell MG (1993) Lateral transfer in natural populations of eukaryotes. Annu Rev Genet 27:235–256

Kinscherf TC, Leong SA (1988) Molecular analysis of the karyotype of *Ustilago maydis*. Chromosoma 96:427–433

Kistler HC, Miao VPW (1992) New modes of genetic change in filamentous fungi. Annu Rev Phytopathol 30:131–152

Koch E, Badawy HMA, Hoppe HH (1989) Differences between aggressive and non-aggressive single spore lines of *Leptosphaeria maculans* in cultural characteristics and phytotoxin production. J Phytopathol 124:52–62

Koch E, Song K, Osborn TC, Williams PH (1991) Relationship between pathogenicity based on restriction fragment length polymorphism in *Leptosphaeria maculans*. Mol Plant-Microbe Interact 4:341–349

Kubitzki K, Sengbusch P, Poppendieck HH (1991) Parallelism, its evolutionary origin and systematic significance. Aliso 13:191–206

Kuninaga S, Yokosawa R (1992) Genetic diversity of *Fusarium oxysporum* f. sp. *lycopersici* in restriction fragment length polymorphisms of mitochondrial DNA. Trans Mycol Soc Jpn 33:449–459

Lai E, Birren BW, Clark SM, Simon MI, Hood L (1989) Pulsed field gel electrophoresis. BioTechniques 7:34–42

Lamboy WF (1984) Evolution of flowering plants by fungus-to-host horizontal gene transfer. Evol Theor 7:45–51

Landan G, Cohen G, Aharonowitz Y, Shuali Y, Graur D, Shiffman D (1990) Evolution of isopenicillin N synthase genes may have involved horizontal gene transfer. Mol Biol Evol 7:399–406

Leslie JF (1991) Mating populations in *Gibberella fujikuroi* (*Fusarium* section *Liseola*). Phytopathology 81:1058–1060

Leung H, Williams PH (1987) Nuclear division and chromosome behavior during meiosis and ascosporogenesis in *Pyricularia oryzae*. Can J Bot 65:112–123

Lim L, Howlett BJ (1994) Linear plasmids, pLm9 and pLm10, can be isolated from the phytopathogenic ascomycete *Leptosphaeria maculans* by pulsed field gel electrophoresis. Curr Genet 26:276–280

Lindegren CC, Lindegren D (1943) A new method for hybridizing yeast. Proc Natl Acad Sci USA 29:306–308

Longo E, Vezinhet F (1993) Chromosomal rearrangements during vegetative growth of a wild strain of *Saccharomyces cerevisiae*. Appl Environ Microbiol 59:322–326

Lott TJ, Kuykendall RJ, Welbel SF, Pramanik A, Lasker BA (1993) Genomic heterogeneity in the yeast *Candida parapsilopsis*. Curr Genet 23:463–467

Luttrell ES (1974) Parasitism of fungi on vascular plants. Mycologia 66:1–15

MacPherson JM, Eckstein PE, Scoles GJ, Gajadhar AA (1993) Variability of the random amplified polymorphic DNA assay among thermal cyclers, and effect of primer and DNA concentration. Mol Cell Probes 7:293–299

Mäntilä AL, Rossi KH, Vanhanen SA, Penttilä ME, Suominen PL, Nevalainen KMH (1992) Electrophoretic karyotyping of wild-type and mutant *Trichoderma longibrachiatum* (*reesei*) strains. Curr Genet 21:471–477

Manzinger L, Ferenczy L (1985) Somatic cell fusion of *Trichoderma reesei* resulting in new genetic combinations. Appl Microbiol Biotechnol 22:72–76

Marmeisse R, Van den Ackerveken GFJM, Goosen T, De Wit PJGM, Van den Broek HWJ (1993) Disruption of the avirulence gene *avr*E9 in two races of the tomato pathogen *Cladosporium fulvum* causes virulence on tomato genotypes with the complementary resistance gene *Cf*E9. Mol Plant-Microbe Interact 6:412–417

Marri L, Rossolini GM, Satta G (1993) Chromosome polymorphisms among strains of *Hansenula polymorpha* (syn. *Pichia angusta*). Appl Environ Microbiol 59:939–941

Masel A, Braithwaite K, Irwin J, Manners J (1990) High variable molecular karyotypes in the plant pathogen *Colletotrichum gloeosporioides*. Curr Genet 18:81–86

Masel AM, Irwin JAG, Manners JM (1993) DNA addition or deletion is associated with a major karyotype polymorphism in the fungal phytopathogen *Colletotrichum gloeosporioides*. Mol Gen Genet 237:73–80

Mathew MK, Smith CL, Cantor CR (1988) High-resolution separation and accurate size determination in pulsed-field gel electrophoresis of DNA. 2. Effect of pulse time and electric field strength and implications for models of the separation process. Biochemistry 27:9210–9216

McClintock B (1945) *Neurospora*. I. Preliminary observations of the chromosomes of *Neurospora crassa*. Am J Bot 32:671–678

McCluskey K, Mills D (1990) Identification and characterization of chromosome length polymorphisms among strains representing fourteen races of *Ustilago hordei*. Mol Plant-Microbe Interact 3:366–373

McDonald BA, Martinez JP (1991) Chromosome length polymorphisms in a *Septoria tritici* population. Curr Genet 19:265–271

Meunier JR, Grimont PAD (1993) Factors affecting reproducibility of random amplified polymorphic DNA fingerprinting. Res Microbiol 144:373–379

Meyer W, Lieckfeldt E, Wöstemeyer J, Börner T (1992) DNA fingerprinting for differentiating aggressivity groups of the rape seed pathogen *Leptosphaeria maculans*. Mycol Res 96:651–657

Miao VP, Covert SF, Van Etten HD (1991) A fungal gene for antibiotic resistance on a dispensable ("B") chromosome. Science 254:1773–1776

Migheli Q, Berio T, Gullino ML (1993) Electrophoretic karyotypes of *Fusarium* spp. Exp Mycol 17:329–337

Morales VM, Pelcher LE, Taylor JL (1993a) Comparison of the 5.8S rDNA and internal transcribed spacer sequences of isolates of *Leptosphaeria maculans* from different pathogenicity groups. Curr Genet 23:490–495

Morales VM, Séguin-Swartz G, Taylor JL (1993b) Chromosome size polymorphism in *Leptosphaeria maculans*. Phytopathology 83:503–509

Nagy A, Vágvölgyi CS, Balla E, Ferenczy L (1994) Electrophoretic karyotype of *Mucor circinelloides*. Curr Genet 26:45–48

Naumova E, Naumov G, Fournier P, Nguyen H-V, Gaillardin C (1993) Chromosomal polymorphism of the yeast *Yarrowia lipolytica* and related species: electrophoretic karyotyping and hybridization with cloned genes. Curr Genet 23:450–454

Nauta MJ, Hoekstra RF (1992) Evolution of reproductive systems in filamentous ascomycetes. I. Evolution of mating types. Heredity 68:405–410

Nelson PE, Toussoun TA, Marasas WFO (1983) *Fusarium* species: an illustrated manual for identification. The Pennsylvania State University Press, University Park

Olson M (1989) Pulsed field electrophoresis. In: Setlow JK (ed) Genetic engineering. Plenum Press, New York, pp 183–227

Orbach M, Vollrath D, Davis RW, Yanofsky C (1988) An electrophoretic karyotype of *Neurospora crassa*. Mol Cell Biol 8:1469–1473

Perkins DD, Barry EG (1977) The cytogenetics of *Neurospora*. Adv Genet 19:133–285

Petes TD, Smolik-Utlaut S, McMahon M (1982) Recombination in yeast ribosomal DNA. Recent Adv Yeast Mol Biol 1:69–75

Pearce RB (1990) Antimicrobial defenses in secondary tissues of woody plants. In: Pegg GF, Ayres PG (eds) Fungal infections of plants. Cambridge University Press, Cambridge, pp 219–238

Plummer KM, Howlett BJ (1993) Major chromosomal length polymorphisms are evident after meiosis in the phytopathogenic fungus *Leptosphaeria maculans*. Curr Genet 24:107–113

Plummer KM, Dunse K, Howlett BJ (1994) Non-aggressive strains of the blackleg fungus, *Leptosphaeria maculans*, are present in Australia and can be distinguished from aggressive strains by molecular analysis. Aust J Bot 42:1–8

Pontecorvo G, Käfer E (1958) Genetic analysis based on mitotic recombination. Adv Genet 9:71–104

Pontecorvo G, Roper JA, Hemmons LM, MacDonald KD, Bufton AWJ (1953) The genetics of *Aspergillus nidulans*. Adv Genet 5:141–238

Robinow CF (1963) Observations on cell growth, mitosis, and division in the fungus *Basidiobolus ranarum*. J Cell Biol 17:123–152

Russell BW, Mills D (1993) Electrophoretic karyotypes of *Tilletia caries, T. controversa*, and their F1 progeny: further evidence for conspecific status. Mol Plant-Microbe Interact 6:66–74

Schäfer C, Wöstemeyer J (1992) Random primer dependent PCR differentiates aggressive from non-aggressive isolates of the oilseed rape pathogen *Phoma lingam* (*Leptosphaeria maculans*). J Phytopathol 136:124–136

Schäfer C, Wöstemeyer J (1994) Molecular diagnosis of the rapeseed pathogen *Leptosphaeria maculans* based on RAPD-PCR. In: Schots A, Dewey FM, Oliver R (eds) Modern assays for plant pathogenic fungi. CAB International, Oxford, pp 1–8

Schwartz DC, Cantor CR (1984) Separation of yeast chromosome-sized DNAs by pulsed field gradient gel electrophoresis. Cell 37:67–75

Selker E (1990) Premeiotic instability of repeated sequences in *Neurospora crassa*. Annu Rev Genet 24:579–613

Selker E (1991) Repeat-induced point mutation and DNA methylation. In: Bennett JW, Lasure LL (eds) More gene manipulations in fungi. Academic Press, San Diego, pp 258–265

Skinner DZ, Budde AD, Farman ML, Smith JR, Leung H, Leong SA (1993) Genome organization of *Magnaporthe grisea*: genetic map, electrophoretic karyotype, and occurrence of repeated DNAs. Theor Appl

Genet 87:545–557

Sweigard JA, Valent B, Orbach MJ, Walter AM, Rafalski A, Chumley FG (1993) Genetic map of the rice blast fungus *Magnaporthe grisea*. In: O'Brien SJ (ed) Genetic maps, 6th edn. Cold Spring Harbor Laboratory Press, Cold Spring Harbor, pp 3.112–3.115

Talbot NJ, Salch YP, Ma M, Hamer JE (1993) Karyotypic variation within cloning lineages of the rice blast fungus, *Magnaporthe grisea*. Appl Environ Microbiol 59:585–593

Tantius PM, Fyfe AM, Shaw DS, Shattock RC (1986) Occurrence of the A2 mating type and self fertile isolates of *Phytophthora infestans* in England and Wales. Plant Pathol 35:578–581

Taylor JL, Borgmann I, Séguin-Swartz G (1991) Electrophoretic karyotyping of *Leptosphaeria maculans* differentiates highly virulent from weakly virulent isolates. Curr Genet 19:273–277

Van der Ploeg LHT, Cornelissen AWCA, Michels PAM, Borst P (1984) Chromosome rearrangements in *Trypanosoma brucei*. Cell 39:213–221

Van Kan JAL, Van den Ackerveken GFJM, De Wit PJGM (1991) Cloning and characterization of cDNA of avirulence gene *avr*E9 of the fungal pathogen *Cladosporium fulvum*, causal agent of tomato leaf mold. Mol Plant-Microbe Interact 4:52–59

Voigt K, Schleier S, Brückner B (1995) Genetic variability in *Gibberella fujikuroi* and some related species of the genus *Fusarium* based on random amplification of polymorphic DNA (RAPD). Curr Genet 27:528–535

Walz M (1995) Electrophoretic karyotyping. In: Kück U (ed) The Mycota, vol 2. Springer, Berlin Heidelberg, pp 61–73

Welsh J, McClelland M (1990) Fingerprinting genomes using PCR with arbitrary primers. Nucleic Acids Res 18:7213–7218

Westergaard M, von Wettstein D (1970) Studies on the mechanism of crossing over. IV. The molecular organization of the synaptinemal complex in *Neottiella* (Cooke) *saccardo* (ascomycetes). C R Trav Lab Carlsberg 37:239–268

Whitehead DS, Coddington A, Lewis BG (1992) Classification of races by DNA polymorphism analysis and vegetative compatibility grouping in *Fusarium oxysporum* f. sp. *pisi*. Physiol Mol Plant Pathol 41:295–305

Williams JGK, Kubelik AR, Livak KJ, Rafalski JA, Tingey SV (1990) DNA polymorphisms amplified by arbitrary primers are useful as genetic markers. Nucleic Acids Res 18:6531–6535

Wöstemeyer J, Burmester A (1986) Structural organization of the genome of the zygomycete *Absidia glauca*: evidence for high repetitive DNA content. Curr Genet 10:903–907

Wöstemeyer A, Teepe H, Wöstemeyer J (1990) Genetic interactions in somatic inter-mating type hybrids of the zygomycete *Absidia glauca*. Curr Genet 17:163–168

Wöstemeyer J, Schäfer C, Kellner M, Weisfeld M (1992) DNA polymorphisms detected by random primer dependent PCR as a powerful tool for molecular diagnosis of plant pathogenic fungi. Adv Mol Gen 5:227–240

Yuill E (1939) Two new *Aspergillus* mutants. J Bot 77:174–175

Zambryski P, Tempe J, Schell J (1989) Transfer and function of T-DNA genes from *Agrobacterium* Ti and Ri plasmids. Cell 56:193–201

14 Evolution of Mutualistic Endophytes from Plant Pathogens

C.L. SCHARDL[1] and K. CLAY[2]

CONTENTS

I. Introduction . . . 221
II. Concepts and Definitions . . . 222
A. Types of Symbiosis . . . 222
B. Costs and Benefits . . . 222
III. Plant-Endophyte Systems . . . 224
A. Clavicipitaceous Endophytes of Grasses . . . 224
B. Fungal Characteristics . . . 224
C. Host Range and Specificity . . . 225
IV. The Symbiotic Continuum and the Clavicipitaceae . . . 225
A. *Claviceps* Species . . . 225
B. *Balansia obtecta* . . . 226
C. *Atkinsonella hypoxylon* . . . 226
D. The *Epichloë* Group . . . 227
V. Phylogenetic Relationships . . . 229
A. Family Clavicipitaceae, Tribe Balansieae . . . 229
B. *Epichloë* Species . . . 229
C. Asexual *Epichloë*-Type Endophytes . . . 230
D. *Atkinsonella* Species . . . 232
VI. Causes and Trends in Evolution of Endophytes . . . 232
A. Proximate Causes of Mutualism . . . 232
B. Selection on the Symbiotic Systems . . . 233
C. Selection on Asymptomatic Associations and Beneficial Endophytes . . . 234
VII. Conclusion . . . 235
References . . . 236

I. Introduction

The continuum of symbiotic interactions includes those that cause disease (antagonism) and others that enhance the survival and fitness of the hosts (beneficial symbiosis and mutualism). In this chapter we explore this symbiotic continuum as it applies to interactions of grasses with members of the ascomycete family, Clavicipitaceae, which includes both pathogens and mutualists. Mutualistic endophytes benefit their hosts by protecting against herbivores and abiotic stresses, and are probably important for the long-term survival of certain grass species.

A wide variety of eukaryotes enter beneficial or mutualistic symbiosis with photosynthetic organisms. Some examples are corals with zooxanthellate algae (Rowan and Powers 1991), lichens (Honegger 1991; Chap. 12, this Vol.), mycorrhizae (Simon et al. 1993; Chap. 12, Vol. V, Part A; Chaps. 6, 7, this Vol.), *Hydra* and *Paramecium* species with endosymbiotic *Chlorella* (Meints et al. 1985; Muscatine and McNeil 1989), and vascular plants with endophytic fungi (Law and Lewis 1983; Carroll 1988; Clay 1988a; Chap. 8, this Vol.). Many symbioses have great ecological importance because the photobionts are major primary producers. Among vascular plant symbioses, many are important for host nutrition (e.g., mycorrhizae and nitrogen-fixing nodules), whereas endophytic fungi (endophytes) principally protect their hosts from biotic and abiotic stresses (Carroll 1988).

Species of family Clavicipitaceae (Ascomycota) associate with grasses and sedges in a continuum that includes antagonism and mutualism. Many have host interactions with both antagonistic and mutualistic characteristics. Some are vertically transmitted in the seeds of infected plants, thus heritable in maternal lineages of their hosts; others produce contagious spores. Both characteristics typify life cycles of *Epichloë* species and their asexual relatives, as diagrammed in Fig. 1 (see also Bacon and de Battista 1990). The asexual clavicipitaceous endophytes lack an external infective stage and are strictly seed-disseminated (maternally transmitted), having essentially lost their identities as individual organisms. They are related to the holomorphic species of genus *Epichloë* (Pers.: Fr.) Tul. (*E. typhina* sensu lato), so we will refer to them as e-endophytes (*Epichloë*-type endophytes). The e-endophytes are also known as *Acremonium* endophytes because of their proposed removal from genus *Sphacelia* Sacc. and reclassification in *Acremonium* Link sect. *Albolanosa* G. Morgan-Jones et W. Gams (Morgan-

[1] Department of Plant Pathology, University of Kentucky, Lexington, Kentucky 40546, USA
[2] Department of Biology, Indiana University, Bloomington, Indiana 47405, USA

The Mycota V Part B
Plant Relationships
Carroll/Tudzynski (Eds.)
© Springer-Verlag Berlin Heidelberg 1997

Jones and Gams 1982) (= *Neotyphodium* Glenn, Bacon et Hanlin; Glenn et al. 1996). Since many *Epichloë* holomorphs have sexual (pathogenic and antagonistic) and asexual (benign or mutualistic) life cycles, evolution of some asexual e-endophytes may occur simply by loss of the pathogenic stage. In many cases, this simple scenario does not account for results of phylogenetic analysis, which indicate numerous interspecific hybridizations between e-endophytes and *Epichloë* species (Schardl et al. 1994; Tsai et al. 1994). Hybridization in endophyte evolution may have important implications for maintenance and diversification of the symbiotic systems, and for long-term fitness of host grasses.

In addition to clavicipitaceous endophytes, other systemic grass endophytes are known (Latch et al. 1984; White et al. 1990; An et al. 1992), but will not be discussed in this chapter. Endophytic fungi also appear in trees and dicotyledonous angiosperms (Carroll 1988; Chap. 8, this Vol.).

In this chapter we outline the current concepts and state of knowledge about the evolution of the plant-associated Clavicipitaceae and emphasize the diverse clavicipitaceous tribe Balansieae. The various associations of members of this tribe and their host plants provide an excellent model of the symbiotic continuum and some underlying causes of mutualism and benign parasitism. We describe the basic biology of the organisms and their host interactions, with a focus on several fungal species that span the range between pathogens and mutualists. Data on the evolutionary relationships between genera are reviewed, especially relationships between *Epichloë* species and their asexual relatives. Evidence is provided that many of these asexual, nonpathogenic endophytes are interspecific hybrids. Finally, we discuss selection and evolution of the symbiotic systems (plant-fungus coadaptation) and the maintenance of grass-endophyte relationships over time.

II. Concepts and Definitions

A. Types of Symbiosis

Symbiosis, as defined by be Bary, refers to intimate (long-term and physiological) interactions of individuals of distinct species (Lewis 1985). The symbiotic continuum includes antagonisms, mutualisms, and commensalisms. They also include pleiotropic symbioses, defined as those that vary spatially or temporally in relative antagonism or mutualism (Michalakis et al. 1992). The endophyte systems discussed here present nontrivial problems in assessing the net benefit or cost to the host, the fungus, or both. For example, some species of *Epichloë* are best described as pleiotropic symbionts, being mutualistic pathogens which only rarely cause disease. *Epichloë festucae* A. Leuchtmann C.L. Schardl et M.R. Siegel, symbiotic with *Festuca* L. species (Fig. 1) is an example. The efficient heritability of some *Epichloë* species is an important factor in the evolution of mutualistic e-endophytes. Symbioses of grasses with asexual e-endophytes are generally regarded as mutualistic, though it may sometimes be useful to view an asexual endophyte as a trapped pathogen whose genome and associated functions have been expropriated for the benefit of the host plant (Schardl et al. 1991).

B. Costs and Benefits

Primary producers that fix sunlight or other abiotic energy sources into biological forms ultimately make their fixed energy and nutrients available to a variety of consumers. Symbiotic interactions between primary producers and parasitic microorganisms involve such nutritional exchanges. Characterization of a symbiosis as antagonistic or mutualistic requires assessment of the relative cost and benefit to the host and symbiont. Costs to plant hosts may be relatively apparent, especially if associated with disease, but benefits can be problematic to assess. A plant host confronts numerous parasites and herbivores. These consumers are themselves parasitized or suffer predation, and so on up the food chains. Infection by one symbiont may protect a plant against, or predispose it toward, herbivory or parasitism by other species. Protection may occur, for example, by eliciting physiological changes in the plant in the form of induced systemic resistance to diseases (Freeman and Rodriguez 1993; Neuenschwander et al. 1996), or by virtue of antiherbivore metabolites produced by a plant parasite (Clay 1988b; Chap. 10, Vol. V, Part A).

Costs and benefits of interactions may vary with short and/or long-term environmental or developmental changes in the host (Michalakis et al. 1992). Such variation may influence the evolu-

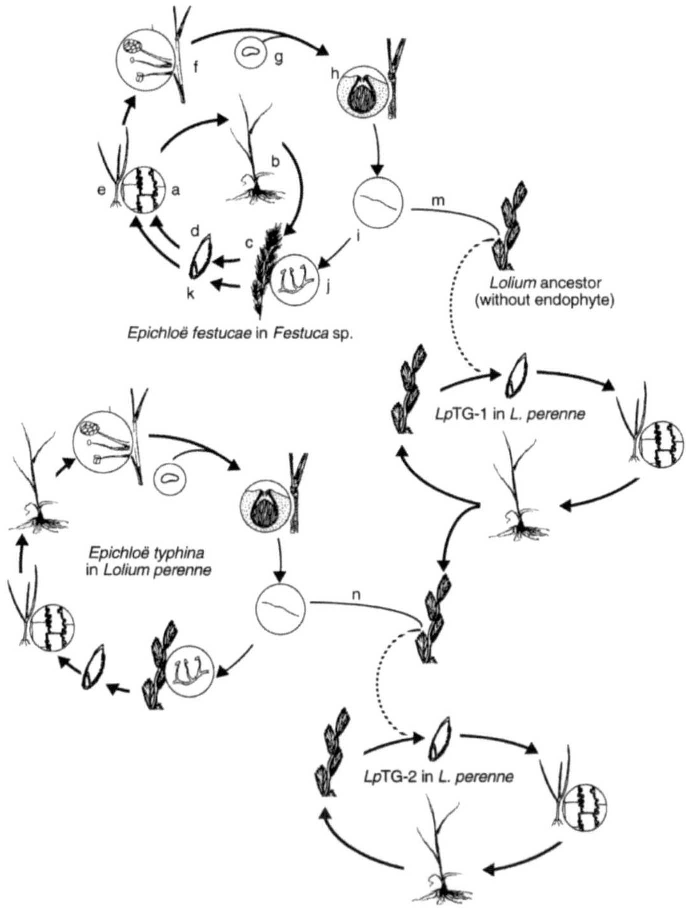

Fig. 1. Possible evolutionary origins of *Neotyphodium lolii* (= LpTG-1) and LpTG-2, endophytes symbiotic with perennial ryegrass. The postulated ancestor of *N. lolii* is *Epichloë festucae* (mating population II), whose alternative sexual and asexual life cycles in association with its hosts (species of genus *Festuca*) are depicted at the *top*. Fungal structures are shown in *circles*. In the asexual cycle, highly convoluted hyphae grow intercellularly in leaf sheaths (*a*), floral meristems (*b*), and in the ovules of the florets (*c*) such that the fungus is transmitted in the seed (*d*). In the sexual cycle, the fungus also grows intercellularly and asymptomatically in vegetative leaf sheaths (*e*), but then produces stromata on flag-leaf sheaths surrounding immature host inflorescences, and floral development is arrested (*f*). Fertilization of the fungus occurs by transfer of spermatia between stromata of opposite mating types (*g*), then perithecia containing asci develop (*h*), and filamentous ascospores are ejected (*i*). Germinating ascospores, initiating cycles of asexual sporulation (conidiation), are thought to cause infection of host florets (*j*) and ultimately of seed (*k*). It is proposed that infection of a floret of a perennial ryegrass ancestor by an *E. festucae* ascospore (*m*) may have given rise to *N. lolii*, which was then maintained in perennial ryegrass by seed transmission. It is further suggested (*n*) that infection of a perennial ryegrass/*N. lolii* symbiotum by *Epichloë typhina* (mating population I), whose life cycle is depicted at the *left*, resulted in co-infection of the ryegrass plant, and that *N. lolii* and *E. typhina* subsequently underwent anastomosis and karyogamy to form the interspecific hybrid ancestral to the endophyte taxon LpTG-2. (Schardl et al. 1994)

tion and coevolution of symbiotic interactions (Thompson 1987). Under certain conditions, evolutionary processes can militate against the negative effects of either or both partners in any interaction (Bull et al. 1991); but sometimes the interactions themselves may reduce the ability of one or both partners to evolve. For example, species of *Danthonia* Lam. et DC infected with the clavicipitaceous fungus *Atkinsonella hypoxylon* (Peck) Diehl are prevented from cross-fertilization with uninfected plants (Clay 1994; Sects. IV.C and V.D). Multitrophic interactions are also highly important in the evolution of grass-endophyte interactions. Thus, there are some well-characterized links between the ecology of these fungi and their evolution.

III. Plant-Endophyte Systems

A. Clavicipitaceous Endophytes of Grasses

A large number of grasses and some sedges are systemically infected by clavicipitaceous endophytes (Clay 1989; see also Chap. 10, Vol. V, Part A). Infections are evident during the flowering season when these fungi produce stromata on either leaves or inflorescences of their hosts. However, grasses infected by asexual e-endophytes never exhibit symptoms and can be identified as infected only by microscopic examination or cultural grow-out tests. Systemic infections are evidenced by the occurrence of fungal hyphae in all tillers and the consistent infection of the same plants from year to year. In seed-transmitted associations, the fungus can be observed microscopically and cultured from seeds (Sampson 1933; Clay 1994). Infections have been reported from a wide variety of host species, geographical regions, and community types (Diehl 1950; Clay 1989).

B. Fungal Characteristics

Members of the Clavicipitaceae produce perithecia that bear filamentous ascospores. The perithecia are embedded in stromata arising either directly on host tissue or from sclerotia. The three tribes of the Clavicipitaceae are Clavicipeae, Balansieae, and Cordycipeae. The Cordycipeae include insect and fungal pathogens and are not discussed here. Clavicipeae (consisting only of genus *Claviceps* Tul.) and Balansieae are parasites of grasses (Poaceae) and sedges (Cyperaceae). *Claviceps* species are specialized parasites of host florets, where they replace host ovules with fungal sclerotia (ergots), the resting and overwintering state that eventually gives rise to stromata in the sexual cycle. The Balansieae (reviewed in Diehl 1950; Clay 1989; Bacon and de Battista 1990) include genera *Balansia* Speg., *Balansiopsis* Höhn, *Myriogenospora* Atk., *Atkinsonella* Diehl, *Echinodothis* Atk. (White 1994b), and *Epichloë* Tul. A number of tropical species are poorly known and need additional research. These include several originally placed in genus *Epichloë*, but which require reassessment (Leuchtmann 1992) and will not be considered here. All known species within the balansioid genera maintain perennial, biotrophic associations with host plants.

Some species of the tribe Balansieae are endophytic (growing between host cells) but generally do not penetrate host cells or produce haustoria (Philipson and Christey 1986; White 1987). In contrast, other species are epiphytic and grow on leaf primordia, within leaf axils and on meristems. Epiphytes include all species of *Atkinsonella*, *Echinodothis*, and *Myriogenospora* and some *Balansia* species (Luttrell and Bacon 1977; Leuchtmann and Clay 1988a; Clay and Frentz 1993). The fungi are triggered to produce sporogenous stromata or sclerotia only at certain developmental stages of their hosts (Figs. 1, 2, and 3). For many species the production of asexual spores, stromata, and/or sclerotia occurs on immature flowering tillers, and maturation of the florets is halted (choke disease). The fungi may then undergo their sexual cycles which, in *Epichloë* species and some others, are known to be heterothallic, and thus obligately outcrossing (White and Bultman 1987; Leuchtmann and Clay 1989c; White and Owens 1992; White 1993, 1994b). Ascospores presumably give rise to the infective stage (Bacon and Hinton 1988), though contagious infections by these fungi have rarely been demonstrated (Diehl 1950; Clay 1990a; Leuchtmann 1992). *Epichloë typhina* was recently observed to infect developing host seeds of uninfected mother plants (Chung 1996).

An important alternative life cycle for some clavicipitaceous endophytes involves seed dissemination without symptom development (Figs. 1, 3; see Leuchtmann and Clay, Chap. 12, this Vol.). Seed dissemination occurs in many *Epichloë* species and e-endophytes, and has also been reported for *A. hypoxylon* (Diehl 1950; Clay 1990a, 1994) and *Balansia obtecta* Diehl (Diehl 1950). In the case of *A. hypoxylon*, specialized basal cleistogamous seeds of host *Danthonia* species are infected, while open-pollinated florets in terminal panicles are aborted by infection (Clay 1984, 1994). In contrast, the frequency with which some *Epichloë* strains and asexual e-endophytes are maternally disseminated in the open-pollinated seeds approaches 100% (Siegel et al. 1984; White 1988). Efficient seed dissemination obviates the need for sporulation, and may predispose *Epichloë* to evolve mutualistic, asexual forms.

C. Host Range and Specificity

Individual species and isolates of clavicipitaceous endophytes can be host-specific (Diehl 1950; Clay 1989; Leuchtmann 1992). The *Epichloë* species are common on pooid grasses. In artificial inoculation experiments, some isolates fail to establish long-term infections of new host species, but in other cases, very wide interspecific host transfers can readily by accomplished (Leuchtmann and Clay 1993). Variation in host specificity is also documented for e-endophytes (Siegel et al. 1990; Leuchtmann 1994a; M.R. Siegel, pers. comm.). Furthermore, molecular phylogenetic evidence and mating studies indicate that some *Epichloë* species are associated with only certain grass genera (*E. festucae* in *Festuca* species), but others are present in more than one host genus (Leuchtmann et al. 1994; Schardl et al. 1994).

Among the other Balansieae, *Atkinsonella texensis* A. Leuchtmann et K. Clay infects the pooid grass species *Stipa leucotricha* Trin. et Rupr., *A. hypoxylon* infects *Danthonia* species (Arundinoideae), *Myriogenospora atramentosa* (Berk. et M.A. Curtis) Diehl infects panicoid grasses (Rykard et al. 1985), *Echinodothis tuberiformis* (B. & Rav.) Atk. (White 1994b) infects a bambusoid grass, and *Balansia* and *Balansiopsis* species infect several subfamilies of the Poaceae, in addition to the Cyperaceae (Diehl 1950; Leuchtmann and Clay 1988a). At least one species of *Balansia* (*B. cyperi* Edg.) infects only members of the sedge genus *Cyperus* L. Host ranges of Balansieae are reviewed elsewhere (Clay 1990b; Leuchtmann 1992; Chap. 12, this Vol.). Some described species with broad host ranges may actually represent a complex of host-specific species, but artificial inoculation experiments have been conducted with only a small number of endophytes.

IV. The Symbiotic Continuum and the Clavicipitaceae

Associations of clavicipitaceous fungi with grasses and sedges evade simple classification as mutualistic or antagonistic interactions. Each interaction entails costs and benefits to each partner, and the relative effects of these costs and benefits are often difficult to assess. We focus here on several examples of the types of associations that occur and indicate what the major costs and benefits are to each partner, without any attempt to sum them. The species discussed form a continuum of fungi that range from more pathogenic to more mutualistic. At one extreme are *Claviceps* species, the causative agents of ergot disease of grasses. *Balansia obtecta* and *A. hypoxylon* are somewhat intermediate with the former being more antagonistic (pathogenic) than the latter. Finally, pleiotropic *Epichloë* species and the related asexual e-endophytes, provide the clearest examples of endophyte/grass mutualism.

A. *Claviceps* Species

The associations of *Claviceps* species with host grasses (reviewed in Tudzynski et al. 1995; Alexopoulos et al. 1996) are more transient than those between grasses and clavicipitaceous endophytes. *Claviceps* species are transmitted to grasses first from sclerotia via ascospores then, secondarily, via conidia to florets of plants. In some species or strains, heterothallic mating (between opposite mating types) is reported, though experiments with some *C. purpurea* (Fr.:Fr.) Tul. strains suggest homothallic (self-) mating (Tudzynski et al. 1982). A thick mycelial growth, perhaps the heterokaryon stage (Spalla 1973), invades and kills the ovule, and eventually gives rise to a sclerotium (ergot), the dormant overwintering stage. The sclerotium germinates to produce stalked stromata bearing fertile perithecia. The sexual spores, products of meiosis, mature within asci in the perithecia, and are forcibly ejected through ostioles to repeat the life cycle.

The associations of grasses with *Claviceps* species appear to be mildly antagonistic. The cost to the plant of *Claviceps* infection is some reduction of seed set, and some nutritional transfer to the fungus. These costs should be proportional to the number of infected florets per plant. In general, only a minority of ovules are destroyed, while most ovules are uninfected and set seed.

There is no evidence that *Claviceps* species can become seed-transmitted, unlike many members of the Balansieae, which infect host ovules without causing their death and then grow systemically in the resulting seedlings. Further, *Claviceps* sclerotia typically arise from single ovules, whereas stromata of *Atkinsonella*, *Epichloë*, and many *Balansia* species typically envelop entire inflorescences. However, Walker (1970) observed a possible intermediate between the *Claviceps* and *Balansia* life cycles in which a fungus described as a *Claviceps* species was systemic within the vegetative plant parts and every ovary was replaced by an individual sclerotium. The evolutionary transition from a transient life cycle to one involving systemic infection may require merely that the fungus proliferate beyond the ovule and persist in the host through nonflowering periods.

Whether *C. purpurea* provides any benefit to the plant is unknown. *Claviceps* sclerotia are rich in anti-insect and anti-mammalian alkaloids, so that grazing mammals may be discouraged from feedings on seed heads with significant ergot frequency, but this possibility has not been investigated empirically. Since, however, there is no evidence of alkaloids occurring outside the sclerotia, leaf material is presumably not protected by *C. purpurea* infection.

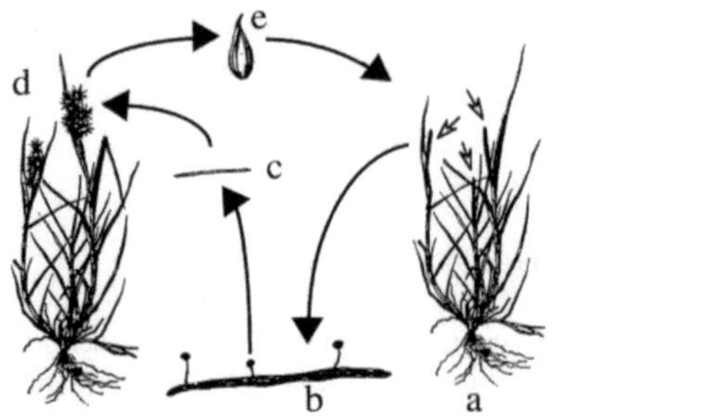

Fig. 2. Life cycle of *Balansia obtecta* infecting *Cenchrus echinatus*. The annual grass *C. echinatus*, when infected by the systemic endophyte *B. obtecta*, bears cylindrical black sclerotia in place of inflorescences (*a*). Sclerotia drop to the ground and overwinter until the following spring when perithecia and ascospores develop (*b*). Ascospores (*c*) are dispersed by wind and may land on exerted stigmata in inflorescences of healthy plants (*d*). The ascospore germinates and grows into the developing ovule, resulting in an infected seed (*e*), which gives rise to a new infected plant in the following season. Additional details are provided in Diehl (1950)

B. *Balansia obtecta*

Among the clavicipitaceous endophytes, most associations are best viewed as antagonistic even when some host benefits are evident. In the interaction between *B. obtecta* and the annual grass, *Cenchrus echinatus* L. (sandspur), the fungus occurs as a systemic endophyte in aerial, vegetative tissues of its host (Fig. 2). Its intercellular growth causes no obvious damage to the vegetative plant, which may be protected from herbivory to a greater extent than uninfected plants, as has been observed for many other grasses and sedges infected with clavicipitaceous endophytes (Clay et al. 1989; Clay 1990a, 1991; Plowman et al. 1990; Marks et al. 1991). Clay et al. (1985) found that the development of lepidopteran larvae was significantly retarded when reared on leaves from infected (vs. uninfected) plants. Ergot alkaloids have been detected in *B. obtecta* culture and infected plants (Powell et al. 1990), and this class of compound is known to be toxic to some insect and mammalian herbivores (Clay and Cheplick 1989; Raisbeck et al. 1991). *Balansia obtecta* most likely produces other biologically active, but as yet unidentified, metabolites in view of what is known of the other Clavicipitaceae. Diehl (1950) reported gigantism of infected plants, also suggesting an ecological benefit from infection.

The most obvious cost to the host of infection by *B. obtecta* is exacted during flowering. Infected plants are effectively sterilized because the fungus produces sclerotia that envelop and mummify host inflorescences, and prevent seed set (Fig. 2). There is some mitigation of this negative effect because infected grasses often become perennial and spread vigorously by vegetative growth (Diehl 1950; K. Clay pers. observ.).

With the exception of its systemic growth, *B. obtecta* is like *Claviceps* in that the sclerotia overwinter separately from the plant and produce perithecia the following spring (Fig. 2). Further, Diehl (1950) reports that contagious spread occurs when ascospores land on the stigmata of uninfected plants and infect developing ovules, again like *Claviceps*, except that the resultant seeds are viable and give rise to systemically infected seedlings.

C. *Atkinsonella hypoxylon*

The interaction between *A. hypoxylon* and *Danthonia* grasses (Fig. 3) is one of the best-studied associations between fungal parasites and wild plants. The fungus infects four *Danthonia* species in eastern North America (Clay 1994). Infection rates range from 0% to over 50% depending on the site and host population (Diehl 1950; K. Clay

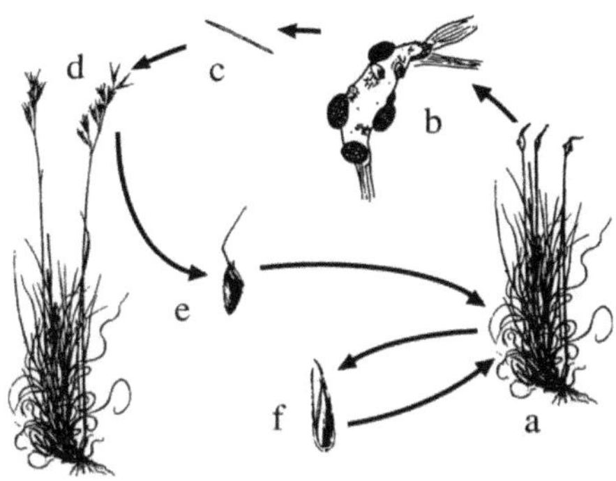

Fig. 3. Life cycle of *Atkinsonella hypoxylon* infecting *Danthonia spicata*. The perennial grass, *D. spicata*, when infected by the systemic epiphytic fungus, *A. hypoxylon*, bears gray stromata in place of the terminal panicles of inflorescences (*a*). Stromata produce perithecia and ascospores if mating (via conidia) with another mating type occurs (*b*). Ascospores (*c*) are dispersed by wind and may land on exerted stigmata in inflorescences of healthy plants (*d*). The ascospore germinates and hyphae are thought to grow into the developing ovule, resulting in an infected seed from the terminal panicle (*e*), which gives rise to a new infected plant the following season (*a*). In addition to terminal panicles, *Danthonia* species produce basal cleistogamous flowers in the lower leaf sheaths (*f*). These flowers are not aborted by infection and bear viable seeds that are infected by the maternal fungus and give rise to new infected plants (*a*). Therefore, contagious spread occurs through terminal flowers while vertical transmission occurs through basal flowers. Additional details are provided in Diehl (1950), Leuchtmann and Clay (1989c), Morgan-Jones and White (1989), and Clay (1994)

1984, pers. observ.). The fungus causes a perennial and systemic infection of tillers and inflorescences of host plants (Diehl 1950; Leuchtmann and Clay 1988a). The mycelium of *Atkinsonella* grows epiphytically in leaf axils of tightly appressed young leaves, on the meristems, and on developing inflorescences, resulting in stunted, aborted terminal inflorescences enveloped by the fungal fruiting body (stroma). Stromata bear both conidia, which act as spermatia, and later ascospores if conidia are transferred between mating types (Leuchtmann and Clay 1989b; Morgan-Jones and White 1989) . Leuchtmann and Clay (1989b) found Mendelian segregation of isozyme alleles within progeny of single asci, as expected in a heterothallic mating system. The mechanism of contagious spread of *Atkinsonella* has not been documented, although field demographic experiments have revealed a low frequency of conversion of uninfected to infected plants (Clay 1984). The finding of mycelium in some seeds from uninfected plants suggests contagious spread through stigmata (Clay 1994).

A more common mechanism of spread is by vertical (maternal) transmission of the fungus through cleistogamous seeds of infected plants (Clay and Jones 1984; Clay 1994). *Danthonia* species produce both wind-pollinated, potentially outcrossed flowers in terminal panicles, and obligately self-pollinated, cleistogamous flowers enclosed in the lower leaf axils of flowering tillers (Clay 1983b, 1994; Cheplick and Clay 1989). Both types of seedlings regularly become established and contribute to the reproductive success of the *Danthonia* species (Clay 1983a). Potentially outcrossing, terminal flowers are aborted by *A. hypoxylon*, but the fungus is clonally transmitted to cleistogamous progeny, resulting in multiple generations of inbred offspring infected by one fungal genotype (Clay 1994). The fungus may gain some advantage by preventing outcrossing, because cleistogamous progeny canot acquire resistance alleles from uninfected plants. Selection for fungi that suppress outcrossing could occur if there was genetic resistance in the host population; preliminary evidence on hypersensitive-like responses suggests such resistance exists (K. Clay, pers. observ.).

Atkinsonella infecting *Danthonia* exhibits aspects of both mutualism (greater host competitive ability, growth, survival, and insect resistance) and antagonism (reduced host fertility) (Clay 1984; Antonovics et al. 1987; Kelley and Clay 1987; Cheplick and Clay 1988; Leuchtmann and Clay 1988b). While vegetative growth is often enhanced, host fecundity is greatly reduced by infection. The effect of infection on host fecundity differs among *Danthonia* species because of differences in the proportions of cleistogamous flowers produced by different host species (Clay 1983b). The impact of infection on host fecundity and infection dynamics is therefore influenced by the variable reproductive breeding system of *Danthonia*.

D. The *Epichloë* Group

The taxonomic relationships of the *Epichloë* species are in flux, but those that have been described recently and in detail grow endophytically in C_3 grasses (mostly subfamily Pooideae) and develop ectophytic plectenchymatous stromata on leaf sheaths surrounding preemergent host inflorescences (White 1994c). *Epichloë* produce only microconidia, whereas *Atkinsonella*, *Balansia*, and *Myriogenospora* can produce *Ephelis* macroconidia. Previously, *Epichloë* strains with these characteristics were grouped together in one morphological species, *Epichloë typhina* (Pers.: Fr.) Tul. sensu lato, but they are now known to comprise several biological species (Schardl and Tsai 1992; White 1993) and are ancestral to most or all of the asexual, clavicipitaceous endophytes (Schardl et al. 1991, 1994). In the late 19th and early 20th centuries, it was common to classify Balansieae as *Epichloë* species. Many of these, such as *Epichloë cinerea* (Mhaskar and Rao 1976), need to be reevaluated (Leuchtmann 1992) and are not considered here.

Among the known plant interactions with Clavicipitaceae, the greatest variation on the symbiotic continuum of mutualism, antagonism, commensalism, and pleiotropy occurs in those involving *Epichloë* species and their asexual relatives. The evolutionary basis for this variation appears to be the ability of many *Epichloë* species to complete either a sexual (thus, somewhat antagonistic) or vegetative (therefore, not antagonistic) life cycle (Fig. 1). However, some species or forms cause complete host sterilization (White 1988; Leuchtmann and Clay, Chap. 12, this Vol.) and are rarely or never transmitted vertically in seeds (C.L. Schardl, pers. observ.), so presumably spread contagiously. Chung (1996; Chung and Schardl 1997) documented ascospore-mediated horizontal infection by *E. typhina* of perennial ryegrass seeds that developed on uninfected mother plants.

Some *Epichloë* species are pleiotropic in causing only some host inflorescences to abort, while infecting developing ovules of other, healthy tillers (Fig. 1). The healthy tillers produce seeds that are systemically infected with the fungus during their development (Philipson and Christey 1986; Clay 1988a). The *Epichloë* life cycle resembles that of *A. hypoxylon*, in which some host flowers are aborted, while some are fertile and produce infected seeds, except that in *Atkinsonella* this occurs on one inflorescence, in *Epichloë* on different inflorescences. This difference is interesting because populations of grasses infected with *Epichloë* could accumulate and redistribute resistance alleles, whereas *A. hypoxylon* may suppress this process (as mentioned in the previous section). However, available evidence suggests that resistance to *Epichloë* infection is uncommon in compatible host species. Typically, seed-disseminated *Epichloë* species infect close to 100% of progeny plants (Saha et al. 1987). As in the *Atkinsonella/Danthonia* interactions, infected seed progeny receive the same benefits from the fungus as the mother plant. If these fungi are not severely pathogenic (i.e., if host sterilization is not so severe as to reduce overall fitness in the face of protective benefits), then the symbiosis is best regarded as mutualistic.

Epichloë has therefore two alternative life cycles, one utilizing sexual recombination and horizontal transmission, and the other employing clonal, vertical transmission through seeds. The latter form of reproduction is otherwise unknown among the Clavicipitaceae except for *Atkinsonella*. Complete asymptomatic infections with seed transmission are known only in pleiotropic *Epichloë* species and anamorphs. The extent of stroma formation versus seed production and vertical transmission in a pleiotropic symbiosis depends to some degree on the environment (Funk et al. 1994), and probably also on host and fungal genotypes (Leuchtmann and Clay 1993). Seed-transmitted *Epichloë* species can be mutualistic, as evidenced, for example, by the extensive infections of *Festuca* species by *E. festucae* (Sampson 1933; Leuchtmann 1994b). Host benefits of *Festuca/E. festucae* symbioses include production of anti-herbivore alkaloids (Siegel et al. 1990; Chap. 10, Vol. V, Part A) and enhanced protection from insects (Saha et al. 1987).

In addition to mutualistic or pleiotropic *Epichloë* species, there are close relatives (e-endophytes) that have no sexual cycle and never, or only rarely, sterilize host flowering tillers. The e-endophytes are confined to clonal propagation and vertical transmission in host seeds and ramets. The linkage of endophyte clones to maternal lineages of grass hosts should cause selection for enhanced mutualism, because increased host survival and fitness necessarily increases dissemination of the endophyte. In fact, extensive comparisons of fitness characteristics of *Festuca arundinacea* Schreb. (tall fescue) and *Lolium perenne* L. (perennial ryegrass) plants with and without e-endophytes indicate numerous benefits of symbiosis. For example, tall fescue exhibits enhanced resistance to insect and mammalian herbivores, nematodes, and drought (reviewed in Clay 1990b), as well as increased resistance to seedling disease caused by *Rhizoctonia zeae* Voorhees

(Gwinn and Gavin 1992), enhanced seed production (Rice et al. 1990), and increased competitiveness (Hill et al. 1991).

A simple hypothesis for evolution of asexual e-endophytes is that they are derived from seed-transmissible *Epichloë* species that are unable to produce stromata due to fungal mutation or suppression by host genotypes (Fig. 1). In some cases, at least, this hypothesis may be correct, but it does not explain the complicated evolutionary histories of many e-endophytes as elucidated by molecular and biochemical genetics. In the next section we discuss our present knowledge regarding evolution of asexual endophytes from more pathogenic sexual endophytes.

V. Phylogenetic Relationships

A. Family Clavicipitaceae, Tribe Balansieae

The relationships of species in order Clavicipitales to other fungal orders were recently investigated by Spatafora and Blackwell (1994) by sequence analysis of the nuclear coding sequences for small subunit rRNAs. The results suggested that the Clavicipitales are monophyletic and closely related to the Hypocreales, another order that includes stroma-producing plant pathogens. Relationships within the Clavicipitaceae have been examined further by comparisons of sequences from additional portions of nuclear rRNA genes (Liu 1993; Kuldau et al. 1997). Parsimony analysis of the large subunit D1 and D2 region sequences appears in Fig. 4. The results do not indicate that Balansieae constitutes a distinct monophyletic tribe. Instead, *Claviceps* may be derived from within Balansieae as presently delineated. However, relevant edges are not strongly supported, so additional sequence information is required. With an expanded data set, representatives of *Epichloë* species and e-endophytes consistently grouped in a monophyletic clade deeply rooted in the Balansieae (Kuldau et al. 1997).

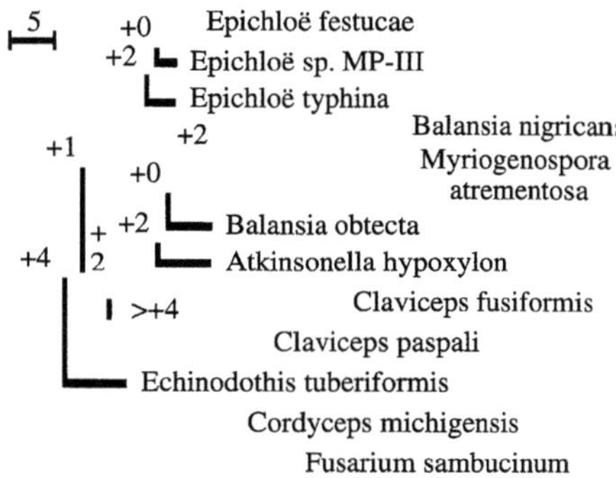

Fig. 4. Gene tree derived by parsimony analysis on variable D1 and D2 regions of the nuclear gene for 26S large subunit rRNA for members of the Clavicipitaceae, with *Fusarium sambucinum* sequence as the outgroup. The *number above each internal edge* indicates the maximum number of characters that can be removed in all permutations while retaining the edge in all most parsimonious trees. Edges with a *decay index of +0* are not present in all equally parsimonious trees derived from all characters. The *bar* indicates a scale of five nucleotide substitutions. *Fusarium sambucinum* sequence is accession number X65476 (K. O'Donnell), *C. fusiformis* sequence is accession number U17402 (S. A. Rehner), and other sequences are from Kuldau et al. 1997. The tree was determined using the branch-and-bound algorithm of PAUP (Swofford 1993), with all transitions and transversions equally weighted, and alignment gaps treated as missing information

B. *Epichloë* Species

Recent investigations of morphology (White 1993), alkaloid profiles (Siegel et al. 1990; TePaske et al. 1994; Chap. 10, Vol. V, Part A), allozyme profiles (Leuchtmann and Clay 1990; Leuchtmann et al. 1994; Chap. 12, this Vol.), sexual interfertility groups, i.e., mating populations (Schardl and Tsai 1992; White 1993; Leuchtmann et al. 1994; Chap. 12, this Vol.), and DNA sequence variation (Schardl et al. 1991, 1994; An et al. 1992) indicate that the former *E. typhina* is a diverse assemblage of different biological species. Within the species complex, several new morphospecies have recently been described (White 1993, 1994a; Leuchtmann et al. 1994), namely *E. typhina* sensu stricto, *E. amarillans* J.F. White, *E. baconii* J.F. White, *E. clarkii* J.F. White, and *E. festucae*. Most, but not all, morphospecies represent populations between which mating is precluded by genetic barriers (Schardl and Tsai 1992; Leuchtmann et al. 1994; Chap. 12, this Vol.; Schardl et al. 1997).

Biological species of *Epichloë* have strong support from molecular and biochemical data. Members of the same species tend to group in clusters based on genetic identity determined from allozyme analysis (Leuchtmann et al. 1994). Likewise, comparison of the rRNA gene (*rrn*) and beta-tubulin gene (*tub2*) trees relating *Epichloë* isolates from several hosts gave a strong indication of speciation (Schardl et al. 1994). Analysis of nine sexual isolates from six grass genera employed noncoding (therefore variable) transcribed portions of the two genes. Both gene trees had the same overall structure, indicating no substantial recombination between different mating populations, and suggesting that most nucleotide substitutions occurred after speciation. Recent, additional surveys coupled with mating tests, allozyme analyses, and molecular phylogenetic analyses indicate at least five more biological species of *Epichloë* that have not yet been described taxonomically (Schardl et al. 1997).

No overall pattern has emerged in molecular phylogenetic analysis of *Epichloë* species that would indicate cospeciation with host grasses, or consistent trends toward (or away from) enhanced mutualism (Liu 1993). There is, nevertheless, a tendency for *Epichloë* strains of similar characteristics and/or related hosts to group in gene trees. If cospeciation occurs, it is most likely in the pleiotropic *Epichloë* species; these might have a balance of vertical and horizontal transmission that requires close coordination with host development, and could be indicative of a long coevolutionary history.

C. Asexual *Epichloë*-Type Endophytes

Various lines of evidence indicate that *Epichloë* and e-endophytes are evolutionary close relatives. Conidial morphology is similar (White and Morgan-Jones 1987) and related or identical alkaloids are produced by both *Epichloë* and e-endophytes (Siegel et al. 1990). Allozyme patterns clearly cluster the *Epichloë* species and their anamorphs (Leuchtmann and Clay 1990). Phylogenetic analyses of DNA sequences further support these relationships. Sequences of the *rrn* internal transcribed spacers (ITS) show great similarity between e-endophytes and *Epichloë* species, but considerable divergence from those of the other Balansieae species *A. hypoxylon*, *B. obtecta* (Schardl et al. 1991), *B. epichloë* (Weese) Diehl, *B. strangulans* (Mont.) Diehl, and *M. atramentosa* (Liu 1993). Despite the evolutionary link between *Epichloë* and e-endophytes, detailed phylogenetic analysis indicates fundamental differences in the evolutionary histories of the sexual and some asexual strains. In particular, interspecific hybridization is a feature of the evolution of many asexual e-endophytes (Schardl et al. 1994; Tsai et al. 1994).

Evidence for hybrid endophytes first emerged in a survey of the *Epichloë* complex by Leuchtmann and Clay (1990). Most of the asexual e-endophytes surveyed exhibited multiband isozyme patterns, suggesting two or more loci for several enzymes. In contrast, ascospore isolates of sexual strains generally had single-band patterns more typical of euascomycetes. These results suggested aneuploidy of many asexual endophytes which, as the authors indicated, is a state more likely to be maintained in a clonally propagated fungus than in a sexual fungus. Apparent aneuploidy is common among the asexual endophytes; 17 genetically distinct endophytes were obtained from nine asymptomatic, endophyte-infected grass species, and 14 exhibited multiband profiles for up to five of the ten isozymes analyzed.

Further evidence of hybridization emerged in investigations of the genetic diversity of endophytes that inhabit single grass species. Christensen et al. (1993) found considerable variation among endophytes of tall fescue and perennial ryegrass in allozyme and alkaloid profiles. The evolutionary basis for this diversity was investigated by molecular phylogenetic analysis of *tub2* and *rrn* noncoding sequences (Schardl et al. 1991; Tsai et al. 1994). The results strongly suggested that most genetic variation between different taxa (as delineated by Christensen et al. 1993) was due to their distinct hybrid ancestries. Three taxa of tall fescue endophytes possessed distinct sets of *tub2* copies, with most isolates containing two or three copies (Table 1) differing substantially in intron sequences. The *tub2* gene tree (Fig. 5) indicated that different copies in the tall fescue endophytes were ultimately derived from different *Epichloë* species. In all, a minimum of three hybridization events are required to account for the distribution of the *tub2* genes among the tall fescue endophytes (Tsai et al. 1994).

Among the perennial ryegrass endophytes, Christensen et al. (1993) identified two distinct taxa, which they designated LpTG-1 (= *Neotyphodium lolii* [G. Latch G. Samuels et M.J. Christensen] Glenn, Bacon et Hanlin) and LpTG-2. Isolates of *N. lolii* and LpTG-2 were analyzed by DNA sequences and polymorphisms and by allozyme profiles (Schardl et al. 1994). Unlike the tall fescue endophytes and LpTG-2, *N. lolii* strains

Table 1. *tub2* copies in e-endophytes and *Epichloë* species

Species or taxon	Isolates	Host species	*tub2* copies				
			1	2	3	4	5
Neotyphodium coenophialum (= FaTG-1)	e19, Tf24, Tf28	*Festuca arundinacea*	–	+	+	+	–
FaTG-2	Tf13	*F. arundinacea*	–	–	–	+	–
FaTG-2	e41, Tf15	*F. arundinacea*	–	+	–	+	–
FaTG-3	Tf18	*F. arundinacea*	–	–	–	+	+
Neotyphodium uncinatum	e166, Fp1, Fp3	*Festuca pratensis*	–	–	+	–	–
Epichloë typhina	E8	*Lolium perenne*	+	–	–	–	–
Epichloë festucae	E28, E32, E189	*Festuca* spp.	–	+	–	–	–
Neotyphodium lolii (= LpTG-1)	Lp5	*Lolium perenne*	–	+	–	–	–
LpTG-2	Lp1, Lp2	*Lolium perenne*	+	+	–	–	–

The *tub2* copies are classified based on similarities of intron sequences. Data are from Tsai et al. (1994) and Schardl et al. (1994).

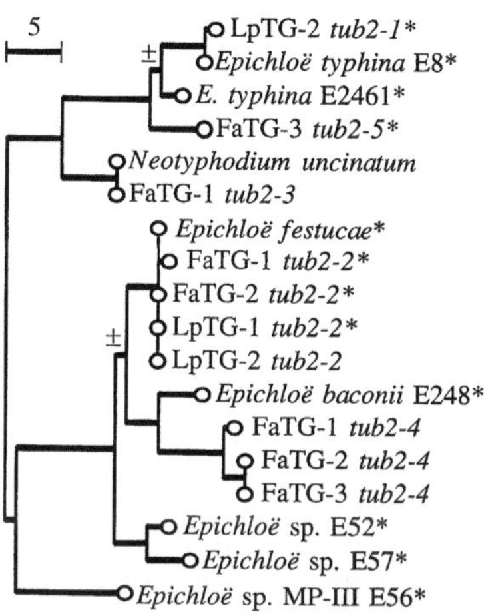

Fig. 5. An unrooted, most parsimonious gene tree relating intron sequences from *tub2* copies in asexual e-endophytes and sexual *Epichloë* species. Taxa, as denoted by Christensen et al. (1993), are: LpTG-1 (= *Neotyphodium lolii*) and LpTG-2 from *Lolium perenne*: FaTG-1 (= *Neotyphodium coenophialum*), FaTG-2 and FaTG-3 from *Festuca arundinacea*, and *Neotyphodium uncinatum* from *Festuca pratensis*. Isolates of e-endophytes and *E. festucae*, and their *tub2* copies, are indicated in Table 1. *Epichloë* species or mating populations (MP) are indicated where known. The tree is the most highly resolved, and edges labeled ± are absent in equally parsimonious trees. *Asterisks* (*) indicate *tub2* gene copies whose phylogenies converge with those of *rrn* ITS from the same isolates. The *bar* indicates a scale of five nucleotide substitutions. (After Tsai et al. 1994; Schardl et al. 1994)

showed no indications of hybrid ancestry. Only single loci were indicated for each isozyme (Christensen et al. 1993), and only a single gene, each, was detected for *tub2*, *rrn*, and *pyr4*. Furthermore, the *tub2* and *rrn* phylogenies converged, both indicating a close relationship to *Epichloë* mating population II (*E. festucae*). In contrast, there were strong indications of hybrid ancestry in LpTG-2, whose likely ancestors were identifiable. LpTG-2 isolates possessed two *tub2* copies (Table 1). Copy *tub2-1* was closely related, by intron sequence, to the single *tub2* copy of an *E. typhina* isolate from perennial ryegrass, whereas *tub2-2* had the same intron sequences as the single *tub2* copy in isolates of *N. lolii* and *E. festucae*. The same relationships were indicated by *pyr4* gene sequences (Collett et al. 1995). The *rrn* ITS sequence was identical to the *E. typhina* sequence. Other molecular and biochemical data were consistent with the postulated origin of LpTG-2 as an interspecific hybrid of *N. lolii* and *E. typhina*. A proposed scheme describing how such a hybridization might occur is shown in Fig. 1.

The identification of hybrid, asexual endophytes is important for several reasons. First, the manner in which they were identified illustrates the importance of analyzing more than one gene tree. In previous studies (Schardl et al. 1991) there were no indications that the tall fescue endophytes were hybrids because, for reasons yet unknown, there was only one apparent sequence of *rrn* ITS in each isolate. Second, the e-endophytes may be the first documented natural interspecific hybrids among the fungi. Their hybrid status contradicts the notion that asexual fungi are simple clones genetically isolated from other fungal lines, and sheds new light on fungal diversity. Finally, as discussed in Section VI.C, hybridization may be important for the long-term persistence of these symbiotic systems.

D. *Atkinsonella* Species

Relationships within the genus *Atkinsonella* have been clarified by a combination of biochemical phylogeny, host specificity studies, and mating studies. An allozyme survey of isolates from four host species indicated that *Atkinsonella* from *S. leucotricha* was genetically distinct from those infecting *Danthonia* species (Leuchtmann and Clay 1989b). This result corresponded to incompatibility of isolates and seedlings from opposite host genera (Leuchtmann and Clay 1989c) and provided the basis of the new species, *A. texensis* (Leuchtmann and Clay 1989a).

Among *Atkinsonella* isolates from *Danthonia* species, allozyme profiles, mitochondrial DNA polymorphisms, and inoculation data indicate that those from *D. compressa* Austin and *D. spicata* (L.) Beauv. are genetically similar and cross-compatible, whereas isolates from *D. epilis* Scribn. and *D. sericea* Nutt. are more divergent (Leuchtmann and Clay 1989b,c; Van Horn and Clay 1995). Therefore, two distinct races appear to exist. The race from *D. compressa* and *D. spicata* occurs throughout eastern North America wherever the hosts occur. In contrast, the race from *D. epilis* and *D. sericea* is restricted to southwestern North Carolina. This is the only place in the ranges of these two host species where they are infected. The two races are cross-fertile (Van Horn and Clay 1995) but may represent a case of incipient speciation. Mitochondrial DNA polymorphisms indicated that the two races were more closely related to each other than either was to *A. texensis*.

Phylogenetic data clearly distinguish the *Atkinsonella* lineage from the *Epichloë* lineage (Sect. V.A). Therefore, the evolution of seed transmission and variation in the degree of mutualism (vs. antagonism) has occurred independently at least twice in the Balansieae. For this reason, comparisons and contrasts of the *Atkinsonella* and *Epichloë* groups may provide special insights into the ecological and genetic factors favoring particular evolutionary pathways.

VI. Underlying Causes and Trends in Evolution of Endophytes

A. Proximate Causes of Mutualism

In the evolution of endophyte symbioses two factors may be especially crucial for emergence of mutualistic associations from pathogenic associations: the coordination of fungal and host life cycles and reproductive systems, and the ability of clavicipitaceous endophytes to confer host resistance to biotic and abiotic stresses.

As described in Section IV, certain Balansieae can infect seeds systemically (vertical transmission) in addition to, or instead of, spore-mediated contagious spread (horizontal transmission). Vertical transmission is well documented for *A. hypoxylon*, e-endophytes, and pleiotropic *Epichloë* species (Sampson 1933; Clay 1994). To the degree that a symbiont depends on vertical transmission, it is advantageous for it to enhance the fitness and fecundity of its host.

The reproductive strategies of most Balansieae inhibit the development of truly mutualistic systems, because the fungal life cycle suppresses sexual reproduction by the host. The loss of genetic recombination ability and seed production is likely to reduce plant fitness in the long run. Thus, clavicipitaceous endophytes that rely on spore transmission as the main or sole means of dissemination are antagonistic. However, there is potential for the evolution of less antagonistic forms with a more benign mode of transmission. The ascospores of these fungi are thought to initiate infections of host florets and give rise to infected seeds (Diehl 1950) which, in turn, give rise to vigorous, systemically infected plants. The maintenance of systemic infections, and inability of the fungi to reproduce until the plant hosts flower would select against highly virulent forms that kill immature seedlings. It may be a relatively small evolutionary change in development from infection of seeds via ascospores to the systemic infection of seeds by vegetative growth of hyphae within the maternal plant. This conjecture seems reasonable, given that both *A. hypoxylon* and some *Epichloë* species can infect seeds systemically, even though these two genera appear no more closely related to each other than to other Balansieae (Fig. 4).

Once vertical transmission is established, the selective advantage of stroma production and contagious spread is reduced. In fact, the two modes of fungal dissemination directly antagonize each other. Nevertheless, each mode probably has certain advantages for the fungus. Seed transmission is highly efficient, while horizontal, spore-mediated infection involves sexual recombination of fungal genomes. Genetic recombination is most likely of some benefit to the fungus (Sect. VI.C), though meiotic spores may be inefficiently disseminated.

The most obvious advantage to infected host plants is protection from biotic and abiotic stresses. Although the biochemical basis for increased drought tolerance and improved growth are unknown, production of alkaloids by the fungi, or by coordinated activities of the fungus and host, results in enhanced resistance to insect and mammalian herbivores (Clay 1991; Chap. 10, Vol. V, Part A). In some instances, this clearly provides a major advantage to the infected grasses over uninfected conspecifics (Clay et al. 1993). Thus, the abilities of Balansieae to contribute to an arsenal of secondary metabolites is almost certainly a major selective factor on the plant hosts for extended symbiosis. The production of alkaloids by endophytes is also conserved in *Claviceps* species but has greater ecological significance to hosts with systemic Balansieae infections. Where the endophytes are vertically transmitted, seeds and seedlings also benefit (Cheplick and Clay 1988; Gwinn and Gavin 1992). In essence, the entire genetic constitution and associated fungal phenotypes, including expression of anti-herbivore metabolites, contribute to the overall characteristics and fitness of the symbiotic systems.

B. Selection in the Symbiotic Systems

Molecular phylogenetic analysis has indicated no overall evolutionary trend in the *Epichloë* group toward or away from mutualism (Liu 1993). More likely, different circumstances cause selection in different directions along the symbiotic continuum. Several factors may select either for or against the antagonistic, stromatal stage of a pleiotropic *Epichloë*. The stromatal stage is required for the sexual cycle of the fungus. In some circumstances, parasitism by the fungus may not be advantageous to the host, so genetic resistance to infection may be selectively favored in host populations. Lack of benefit could occur if the fungi do not produce protective alkaloids, or do not provide specific protection against the particular biotic and abiotic stresses present in the host population. Strong selection for stroma expression and contagious spread may occur in populations in which few potential hosts are infected. In such instances, spore-mediated horizontal transmission would be more efficient than vertical transmission. The reverse is also true: where infection frequencies are already high, contagious spread will be less advantageous if coinfections tend not to occur.

Another possibility is the rise of highly antagonistic relationships by transfer of *Epichloë* genets into host species to which they are poorly adapted. The pleiotropic *Epichloë* grass associations likely represent a highly evolved system of regulatory balances, coordinating the sexual stages of both partners such that each is expressed to a substantial degree. If so, then associations arising from host species transfers may lack these balances and be either entirely asymptomatic or extremely antagonistic (see Leuchtmann and Clay 1993).

Evolution towards increased mutualism may occur with the suppression of the sexual cycle of the fungus for several reasons. First, mutation or genomic arrangement may render the fungus infertile. Second, clones of one mating type may become isolated in a population. Third, the fungus-grass symbiotum may colonize a region devoid of the fly *Phorbia phrenione* Séguy, which facilitates *Epichloë* mating (White and Bultman 1987). Finally, certain ecological circumstances may simply select genotypes of fungus and host that maintain symbiosis entirely by vertical transmission.

A high frequency of infection in host populations provides few potential hosts (it is unknown, however, if superinfections may cause strains to displace one another or interact parasexually). A large number of host species have been described with very high rates of infection (Saha et al. 1987; White 1987; Clay and Leuchtmann 1989). In such symbioses the hosts are probably receiving important benefits. Once potential host populations are saturated, the fungus may gain little benefit from spore dissemination, and if the relationship is essentially mutualistic then seed transmission will benefit both partners. However, the complete absence of sex may exact a significant long-term cost to a clonal fungal symbiont (Sect. VI.C).

Selection against stroma production need not act solely on the fungus. There is most likely selection for host genotypes that suppress the stromatal stage of *Epichloë*. We can use *E. festucae* strains infecting fine fescues and their relatives as examples (Fig. 1). These associations show much variation in the degree of stroma production, versus seed transmission (Sampson 1933; Saha et al. 1987). To what degree host and fungal genotypes contribute to this variation is unknown. Some compatible host species may also suppress stroma production, as suggested in Fig. 1 for the origin of *N. lolii*. As stated earlier, molecular phylogenetic analysis indicates a very close relationship between *E. festucae* and *N. lolii* (Schardl et al.

1994). Inoculation experiments suggest that *E. festucae* isolates are highly compatible with perennial ryegrass (M.J. Christensen, W. Hollin, and M.R. Siegel, pers. comm.). A hypothesis presently under investigation is that *E. festucae* is unable to form stromata on perennial ryegrass.

Another process that may give rise to nonstromatal types is hybridization. Although the hybrid endophytes identified in tall fescue and perennial ryegrass probably arose from in planta associations of nonstromatal endophytes and infecting *Epichloë* genets (Schardl et al. 1994; Tsai et al. 1994), *Epichloë-Epichloë* hybridization may occur in the same way. If the parents differ substantially in karyotype, such hybrids may give rise to stable aneuploids. As in other euascomycetes, aneuploids may exhibit abnormal, variable, and less vigorous growth than haploids (Käfer and Upshall 1973). In fact, endophytes often grow more slowly than their sexual relatives when cultured on sugars commonly present in host tissues (White et al. 1991). Thus, some mutualistic endophytes may actually have been generated by hybridization of sexual *Epichloë* species.

While endophyte systems appear fairly unique in the direct antagonism of vertical versus horizontal transmission, they nevertheless illustrate a choice that is thought critical in the development of mutualism (Yamamura 1993). Production of large numbers of propagules of a parasite generally reduces the fecundity of the host. Vertical transmission reduces the number of propagules that the parasite requires in order to maintain its existence and growth (but in this case, at least, also reduces sexual recombination for the parasite). If the parasite provides no benefit to the host that would compensate for its use of host resources, selection on the host will be to escape infection. However, benefits such as secondary metabolites that protect against herbivory should, if ecologically significant, favor host genotypes that maintain stable, transmissible associations with the parasite. In this way, the parasite gains efficient entry into the next generation of hosts. Furthermore, if a vertically transmitted symbiont provides a selective advantage to its infected host plants over sympatric, noninfected conspecifics, then the host species should evolve to enhanced compatibility with the symbiont, and mutualism should emerge. Several recent theoretical and empirical studies have demonstrated the important role of vertical transmission in the evolution of more mutualistic interactions (Ewald 1983; Bull et al. 1991; Herre 1993). Thus, endophyte associations appear to follow general evolutionary principles evident in a wide range of ecological interactions.

C. Selection on Asymptomatic Associations and Beneficial Endophytes

Protective characteristics of fungal parasites may be a mechanism by which the parasite defends its food source (Wicklow 1988). However, in associations in which the fungus is vertically transmitted, the host not only provides nutrition to the resident endophyte but its progeny represent the food supply of the endophyte's progeny as well. This long-term association may provide even stronger selection on the fungus to benefit the host than in more transient associations. Such long-term associations must be beneficial to the plant if they are to persist. However, there may be long-term costs to the fungus resulting from the loss of the sexual state.

Inherent in the definition of mutualism is the enhancement of each partner's fitness as a result of symbiosis. It is therefore necessary to ask whether the fungal symbiont gains a net benefit from the interaction. While the e-endophytes depend entirely on their hosts for habitat, nutrition, and dissemination, they may be at a disadvantage by the loss of sex. If the e-endophytes are viewed as *Epichloë* lineages that have lost the sexual cycle, they may suffer reduced fitness relative to related, sexual endophytes. Law and Lewis (1983) suggest that sex is not needed for symbiotic inhabitants (like endophytes) because they are protected from the antagonistic biotic interactions that are probably major selective factors favoring sex (Howard and Lively 1994); but another potential benefit of sex, which does apply to symbiotic inhabitants, is to counter the effects of Muller's ratchet. Muller (1964) predicted that clonal entities should relentlessly accumulate deleterious mutations because back mutations or suppressive mutations should occur much less frequently. Consequently, fitness of clonal organisms should decline over successive generations because of the absence of genetic recombination to reduce the load of deleterious mutations. Such recombination normally occurs during the sexual cycle as a result of meiosis. Recent experiments lend support to Muller's ratchet as an important evolutionary phenomenon favoring sex (Howard and Lively

1994). For example, Rice (1994) selected nonrecombined chromosomes of the fruitfly, *Drosophila melanogaster*, through 35 generations, after which they were found to cause significantly reduced fitness relative to homologs that were permitted to recombine in meiosis.

Schardl et al. (1991) suggested that Muller's ratchet was a potential problem for the asexual endophytes and could reduce fitness of both the fungus and the host whenever the host depends on endophyte infection for its survival and reproduction in natural environments. Interspecific hybridization may mitigate the ratchet by providing an influx of entire genomes from sexual relatives. Therefore, although hybridization events may be rare, they may provide such a selective advantage that the products, hybrid endophytes, are common.

Hybridization of maternally transmitted endophytes has other potential advantages for the plant hosts and fungal symbionts. Hybridization would enhance genetic diversity of the endophytes and, thus, diversity of the symbiotic systems. For example, genetic diversity of tall fescue endophytes (measured by allozyme profiles) closely corresponds to the diversity of their profiles of protective alkaloids (Christensen et al. 1993). Most of this diversity apparently results from interspecific hybridization of e-endophytes with *Epichloë* species (Tsai et al. 1994). Another possible benefit is that hybridization with otherwise antagonistic *Epichloë* strains may mitigate their pathogenic effects, as appears to have occurred in the origin of a perennial ryegrass endophyte type LpTG-2 by hybridization of *N. lolii* with *E. typhina* (Schardl et al. 1994; Fig. 1). Disarming the pathogen in this way would contribute to both host and endophyte fitness because the *Epichloë* state reduces host seed production which, in turn, could reduce endophyte dissemination.

VII. Conclusions

Symbiotic associations, it is often suggested, become more mutualistic and less antagonistic over evolutionary time. However, after hundreds of millions of years since complex multicellular eukaryotes originated, extant symbioses still span the continuum from antagonism to mutualism. Insofar as individual organisms are independently subject to selection, evolution may tend toward antagonism under some circumstances, and mutualism under others. Circumstances such as environmental stresses, predation, parasitism, infection status of sympatric hosts, and relative efficiencies of horizontal and vertical transmission may change from year to year, from season to season, or from day to day. The evolutionary histories of the clavicipitaceous endophytes are undoubtedly diverse, and characteristics of their biology – systemic, perennial, asymptomatic infections, heritability (vertical transmission), and the production of protective metabolites – hold the potential for them to evolve towards strict mutualism. Once an endophyte has lost horizontal transmissibility, its evolutionary trajectory must be toward mutualism because survival of seed-transmissible, clonal endophytes depends on survival of their symbiota. Without significant host benefits conferred by the endophytes, the symbiota will most likely be outcompeted by endophyte-free plants. In contrast, more antagonistic endophytes (e.g., *Balansia obtecta*) can persist by contagious spread and, thus, do not depend on long-term fitness of individual host plants.

Phylogenetic evidence does not support the long-term maintenance of clonal endophyte lineages within host populations. In theory, the clonal e-endophytes have extended associations with maternal host lineages and limited potential for genetic variability. However, considerable variation arises from the loss and gain of endophyte strains and interspecific hybridization. The diversity of *Epichloë* species involved in hybridization may be the key to genetic and biochemical diversification of the e-endophytes. If genetic diversity of a grass species enhances its adaptability, then the diversification of its heritable, mutualistic endophytes should also be beneficial. Carroll (1988) points out that short-term symbiotic associations of endophytes in tree leaves may produce spatial and temporal variation in defenses that may enhance the ability of trees to resist pests. Grass endophytes most likely produce similar enhancements in populations of their hosts.

Maintenance of the endophyte-host associations over evolutionary time may require either the occasional completion of the endophyte sexual cycle or some other means of genetic recombination to counteract the effects of Muller's ratchet. A strictly asexual endophyte or group of asexual endophytes in a host grass population may be doomed to extinction as individual lineages accumulate deleterious mutations and die out. Continuance of the systems may therefore depend on the influx of new genes and genomes from sympatric *Epichloë* species; that is, the mutualistic symbioses may ultimately depend on coexisting pathogenic symbioses. It is likely that this dynamic process has existed for a long time, and that many grasses have benefited greatly by their symbiosis with clavicipitaceous endophytes.

Acknowledgments. We thank A. D. Byrd for literary advice in the preparation of the manuscript. This is publication number 94-11-80 of the Kentucky Agricultural Experiment Station, published with the approval of the director.

References

Alexopoulos CJ, Mims CW, Blackwell M (1996) Introductory mycology, 4th edn. John Wiley, New York

An Z-q, Liu J-S, Siegel MR, Bunge G, Schardl CL (1992) Diversity and origins of endophytic fungal symbionts of the North American grass *Festuca arizonica*. Theor Appl Genet 85:366–371

An Z-q, Siegel MR, Hollin W, Tsai H-F, Schmidt D, Schardl CL (1993) Relationships among non-*Acremonium* sp. fungal endophytes in five grass species. Appl Environ Microbiol 59:1540–1548

Antonovics J, Clay K, Schmitt J (1987) The measurement of small-scale environmental heterogeneity using clonal transplants of *Anthoxanthum odoratum* and *Danthonia spicata*. Oecologia 71:601–607

Bacon CW, De Battista J (1990) Endophytic fungi of grasses. In: Arora DK, Rai B, Mukerji KG, Knudsen GR (eds) Handbook of applied mycology, vol 1. Soil and plants. Marcel Dekker, New York, pp 231–256

Bacon CW, Hinton DM (1988) Ascosporic iterative germination in *Epichloë typhina*. Trans Br Mycol Soc 90:563–569

Bull JJ, Molineux IJ, Rice WR (1991) Selection of benevolence in host-parasite system. Evolution 45(4):875–882

Carroll GC (1988) Fungal endophytes in stems and leaves: from latent pathogen to mutualistic symbiont. Ecology 69:2–9

Cheplick GP, Clay K (1988) Acquired chemical defenses of grasses: the role of fungal endophytes. Oikos 52:309–318

Cheplick GP, Clay K (1989) Convergent evolution of cleistogamy and seed heteromorphism in two perennial grasses. Evol Trends Plants 3:127–136

Christensen MJ, Leuchtmann A, Rowan DD, Tapper BA (1993) Taxonomy of *Acremonium* endophytes of tall fescue (*Festuca arundinacea*), meadow fescue (*F. pratensis*), and perennial rye-grass (*Lolium perenne*). Mycol Res 97:1083–1092

Chung K-R (1996) Genetics of the grass symbiont, *Epichloë typhina* (Pers.: Fr.) Tul. PhD Diss, University of Kentucky, Lexington

Chung K-R, Schardl CL (1997) Sexual cycle and horizontal transmission of the grass symbiont, *Epichloë typhina*. Mycol Res 101:295–301

Clay K (1983a) The differential establishment of seedlings from chasmogamous and cleistogamous flowers in natural populations of the grass *Danthonia spicata* (L.) Beauv. Oecologia 57:183–188

Clay K (1983b) Variation in the degree of cleistogamy within and among species of the grass *Danthonia*. Am J Bot 70:835–843

Clay K (1984) The effect of the fungus *Atkinsonella hypoxylon* (Clavicipitaceae) on the reproductive system and demography of the grass *Danthonia spicata*. New Phytol 98:165–175

Clay K (1988a) Clavicipitaceous fungal endophytes of grasses: coevolution and the change from parasitism to mutualism. In: Pirozynski KA, Hawksworth D (eds) Coevolution of fungi with plants and animals. Academic Press, London, pp 79–105

Clay K (1988b) Fungal endophytes of grasses: a defensive mutualism between plants and fungi. Ecology 69:10–16

Clay K (1989) Clavicipitaceous endophytes of grasses: their potential as biocontrol agents. Mycol Res 92:1–12

Clay K (1990a) Comparative demography of three graminoids infected by systemic, clavicipitaceous fungi. Ecology 71:558–570

Clay K (1990b) Fungal endophytes of grasses. Annu Rev Ecol Syst 21:275–297

Clay K (1991) Endophytes as antagonists of plant pests. In: Andrews JH, Hirano SS (eds) Microbial ecology of leaves. Springer Berlin Heidelberg, New York, pp 331–357

Clay K (1994) Hereditary symbiosis in the grass genus *Danthonia*. New Phytol 126:223–231

Clay K, Cheplick GP (1989) Effect of ergot alkaloids from fungal endophyte-infected grasses on fall armyworm (*Spodoptera frugiperda*). J Chem Ecol 15:169–182

Clay K, Frentz IC (1993) *Balansia pilulaeformis*, an epiphytic species. Mycologia 85:527–534

Clay K, Jones JP (1984) Transmission of the fungus *Atkinsonella hypoxylon* (Clavicipitaceae) by cleistogamous seed of *Danthonia spicata* (Gramineae). Can J Bot 62:2893–2898

Clay K, Leuchtmann A (1989) Infection of woodland grasses by fungal endophytes. Mycologia 81:805–811

Clay K, Cheplick GP, Marks S (1989) Impact of the fungus *Balansia henningsiana* on *Panicum agrostoides*: frequency of infection, plant growth and reproduction and resistance to pests. Oecologia 80:374–380

Clay K, Hardy TN, Hammond AM Jr (1985) Fungal endophytes of grasses and their effects on an insect herbivore. Oecologia 66:1–6

Clay K, Marks S, Cheplick GP (1993) Effects of insect herbivory and fungal endophyte infection on competitive interactions among grasses. Ecology 74:1767–1777

Collett MA, Bradshaw RE, Scott DB (1995) A mutualistic fungal symbiont of perennial ryegrass contains two different *pyr4* genes, both expressing orotidine-5′-monophosphate decarboxylase. Gene 158:31–39

Diehl WW (1950) *Balansia* and the Balansiae in America. Agricultural Monograph. United States Department of Agriculture, Washington, DC

Ewald PW (1983) Host-parasite relations, vectors, and the evolution of disease severity. Annu Rev Ecol Syst 14: 465–485

Freeman S, Rodriguez RJ (1993) Genetic conversion of a fungal plant pathogen to a nonpathogenic, endophytic mutualist. Science 260:75–78

Funk CR, Belanger FC, Murphy JA (1994) Role of endophytes in grasses used for turf and soil conservation. In:

Bacon CW, White JF Jr (eds) Biotechnology of endophytic fungi of grasses. CRC Press, Boca Raton, pp 201–209

Glenn AE, Bacon CW, Price R, Hanlin RT (1996) Molecular phylogeny of *Acremonium* and its taxonomic implications. Mycologia 88:369–383

Gwinn KD, Gavin AM (1992) Relationship between endophyte infestation level of tall fescue seed lots and *Rhizoctonia zeae* seedling disease. Plant Dis 76:911–914

Herre EA (1993) Population structure and the evolution of virulence in nematode parasites of fig wasps. Science 259:1442–1445

Hill NS, Belesky DP, Stringer WC (1991) Competitiveness of tall fescue (*Festuca arundinacea* Schreb.) as influenced by endophyte (*Acremonium coenophialum* Morgan-Jones and Gams). Crop Sci 31:185–190

Honegger R (1991) Functional aspects of the lichen symbiosis. Annu Rev Plant Phys Plant Mol Biol 42:553–578

Howard RS, Lively CM (1994) Parasitism, mutation accumulation and the maintenance of sex. Nature 367:554–557

Käfer E, Upshall A (1973) The phenotypes of the eight disomics and trisomics of *Aspergillus nidulans*. J Hered 64:35–38

Kelley SE, Clay K (1987) Interspecific competitive ability and the maintenance of genotypic variation within two perennial grasses. Evolution 41:92–103

Kuldau GA, Liu J-S, White JF, Jr., Siegel MR, Schardl CL (1997) Molecular systematics of Clavicipitaceae supporting monophyly of genus *Epichloë* and form genus *Ephelis*. Mycologia 89:431–441

Latch GCM, Christensen MJ, Samuels GJ (1984) Five endophytes of *Lolium* and *Festuca* in New Zealand. Mycotaxon 20:535–550

Law R, Lewis DH (1983) Biotic environments and the maintenance of sex: some evidence from mutualistic symbioses. Biol J Linn Soc 20:249–276

Leuchtmann A (1992) Systematics, distribution and host specificity of grass endophytes. Nat Toxins 1:150–162

Leuchtmann A (1994a) Isozyme characterization, persistence, and compatibility of fungal and grass mutualists. In: Bacon CW, White JF Jr (eds) Biotechnology of endophytic fungi of grasses. CRC Press, Boca Raton, pp 21–33

Leuchtmann A (1994b) Isozyme relationships of *Acremonium* endophytes from twelve *Festuca* species. Mycol Res 98:25–33

Leuchtmann A, Clay K (1988a) *Atkinsonella hypoxylon* and *Balansia cyperi*, epiphytic members of the Balansiae. Mycologia 80:192–199

Leuchtmann A, Clay K (1988b) Experimental infections of host grasses and sedges with *Atkinsonella hypoxylon* and *Balansia cyperi* (Balansiae, Clavicipitaceae). Mycologia 80:291–297

Leuchtmann A, Clay K (1989a) Experimental evidence for genetic variation in compatibility between the fungus *Atkinsonella hypoxylon* and its three host grasses. Evolution 43:825–834

Leuchtmann A, Clay K (1989b) Isozyme variation in the fungus *Atkinsonella hypoxylon* within and among populations of its host grasses. Can J Bot 67:2600–2607

Leuchtmann A, Clay K (1989c) Morphological, cultural and mating studies in *Atkinsonella*, including *A. texensis*. Mycologia 81:692–701

Leuchtmann A, Clay K (1990) Isozyme variation in the *Acremonium/Epichloë* fungal endophyte complex. Phytopathology 80:1133–1139

Leuchtmann A, Clay K (1993) Nonreciprocal compatibility between *Epichloë* and four host grasses. Mycologia 85:157–163

Leuchtmann A, Schardl CL, Siegel MR (1994) Sexual compatibility and taxonomy of a new species of *Epichloë* symbiotic with fine fescue grasses. Mycologia 86:802–812

Lewis DH (1985) Symbiosis and mutualism: crisp concepts and soggy semantics. In: Boucher DH (eds) The biology of mutualism. Oxford University Press, New York, pp 29–39

Liu J-S (1993) Evolution of the fungal tribe Balansieae (Clavicipitaceae) and related endophytes of grasses. PhD Thesis, University of Kentucky, Lexington

Luttrell ES, Bacon CW (1977) Classification of *Myriogenospora* in the Clavicipitaceae. Can J Bot 55:2090–2097

Marks S, Clay K, Cheplick GP (1991) Effects of fungal endophytes on interspecific and intraspecific competition in the grasses *Festuca arundinacea* and *Lolium perenne*. J Appl Ecol 28:19–204

Meints RH, Schuster AM, Van Etten JL (1985) *Chlorella* viruses. Plant Mol Biol Rep 3:180–187

Mhaskar DN, Rao VG (1976) Development of the ascocarp in *Epichloë cinerea* (Clavicipitaceae). Mycologia 68:994–1001

Michalakis Y, Olivieri I, Renaud F, Raymond M (1992) Pleiotropic action of parasites: how to be good for the host. Trends Ecol Evol 7:59–62

Morgan-Jones G, Gams W (1982) Notes on Hyphomycetes, XLI. An endophyte of *Festuca arundinacea* and the anamorph of *Epichloë typhina*, new taxa in one of two new sections of *Acremonium*. Mycotaxon 15:311–318

Morgan-Jones G, White JF (1989) Concerning *Atkinsonella texensis*, a pathogen of the grass *Stipa leucotricha*: developmental morphology and mating system. Mycotaxon 2:455–467

Muller HJ (1964) The relation of recombination to mutational advance. Mutat Res 1:2–9

Muscatine L, McNeil PL (1989) Endosymbiosis in *Hydra* and the evolution of internal defense systems. Am Zool 29:371–386

Neuenschwander U, Lawton K, Ryals J (1996) Systemic acquired resistance. In: Stacey G, Keen N (eds) Plant-microbe interactions. Chapman & Hall, New York, pp 81–106

Philipson MN, Christey MC (1986) The relationship of host and endophyte during flowering, seed formation, and germination of *Lolium perenne*. N Z J Bot 24:125–134

Plowman TC, Leuchtmann A, Blaney C, Clay K (1990) Significance of the fungus *Balansia cyperi* infecting medicinal species of *Cyperus* (Cyperaceae) from Amazonia. Econ Bot 44:452–462

Powell RG, Plattner RD, Yates SG, Clay K, Leuchtmann A (1990) Ergobalansine, a new ergot-type peptide alkaloid isolated from *Cenchrus echinatus* (sandbur grass) infected with *Balansia obtecta*, and produced in liquid cultures of *B. obtecta* and *Balansia cyperi*. J Nat Prod (Lloydia) 53:1272–1279

Raisbeck MF, Rottinghaus GE, Kendall JD (1991) Effects of naturally occurring mycotoxins on ruminants. In: Smith JE, Henderson RS (eds) Mycotoxins and animal foods. CRC Press, Boca Raton, pp 647–677

Rice WR (1994) Degeneration of a nonrecombining chromosome. Science 263:230–232

Rice JS, Pinkerton BW, Stringer WC, Undersander DJ (1990) Seed production in tall fescue as affected by fungal endophyte. Crop Sci 30:1303–1305

Rowan R, Powers DA (1991) A molecular genetic classification of Zooxanthellae and the evolution of animal-algal symbioses. Science 251:1348–1351

Rykard DM, Bacon CW, Luttrell ES (1985) Host relations of *Myriogenospora atramentosa* and *Balansia epichloë* (Clavicipitaceae). Phytopathology 75:950–956

Saha DC, Johnson-Cicalese JM, Halisky PM, van Heemstra MI, Funk CR (1987) Occurrence and significance of endophytic fungi in fine fescues. Plant Dis 71:1021–1024

Sampson K (1933) The systematic infection of grasses by *Epichloë typhina* (Pers.) Tul. Trans Br Mycol Soc 18: 30–47

Schardl CL, Tsai H-F (1992) Molecular biology and evolution of the grass endophytes. Nat Toxins 1:171–184

Schardl CL, Liu J-S, White JF, Finkel RA, An Z, Siegel MR (1991) Molecular phylogenetic relationships of nonpathogenic grass mycosymbionts and clavicipitaceous plant pathogens. Plant Syst Evol 178:27–41

Schardl CL, Leuchtmann A, Chung K-R, Penny D, Siegel MR (1997) Coevolution by common descent of fungal symbionts (*Epichloë* spp.) and grass hosts. Mol Biol Evol 14:133–143

Schardl CL, Leuchtmann A, Tsai H-F, Collett MA, Watt DM, Scott DB (1994) Origin of a fungal symbiont of perennial ryegrass by interspecific hybridization of a mutualist with the ryegrass choke pathogen, *Epichloë typhina*. Genetics 136:1307–1317

Siegel MR, Johnson MC, Varney DR, Nesmith WC, Buckner RC, Bush LP, Burrus PB II, Jones TA, Boling JA (1984) A fungal endophyte in tall fescue: incidence and dissemination. Phytopathology 74:932–937

Siegel MR, Latch GCM, Bush LP, Fannin FF, Rowan DD, Tapper BA, Bacon CW, Johnson MC (1990) Fungal endophyte-infected grasses: alkaloid accumulation and aphid response. J Chem Ecol 16:3301–3315

Simon L, Bousquet J, Lévesque RC, Lalonde M (1993) Origin and diversification of endomycorrhizal fungi and coincidence with vascular land plants. Nature 363: 67–69

Spalla C (1973) Genetic problems of production of ergot alkaloids in saprophytic and parasitic conditions. In: Vanek Z, Hostlek Z, Cudln J (eds) Genetics of industrial microorganisms. Elsevier, Amsterdam, pp 393–403

Spatafora JW, Blackwell M (1994) Molecular systematics of unitunicate perithecial ascomycetes: the Clavicipitales-Hypocreales connection. Mycologia 85:912–922

Swofford DL (1993) PAUP: phylogenetic analysis using parsimony, version 3.1.1. Illinois Natural History Survey, Champaign

TePaske MR, Powell RG, Clement SL (1993) Analyses of selected endophyte-infected grasses for the presence of loline-type and ergot-type alkaloids. J Agric Food Chem 41:2299–2303

Thompson JN (1987) Symbiont-induced speciation. Biol J Linn Soc 32:385–393

Tsai H-F, Liu J-S, Staben C, Christensen MJ, Latch GCM, Siegel MR, Schardl CL (1994) Evolutionary diversification of fungal endophytes of tall fescue grass by hybridization with *Epichloë* species. Proc Natl Acad Sci USA 91:2542–2546

Tudzynski P, Esser K, Gröshel H (1982) Genetics of the ergot fungus *Claviceps purpurea*. part 2. exchange of genetic material via meiotic recombination. Theor Appl Genet 61:97–100

Tudzynski P, Tenberge KB, Oeser B (1995) Claviceps purpurea. In: Kohmoto K, Singh US, Singh RP (eds) Pathogenesis and host specificity in plant diseases: histopathological, biochemical, genetic and molecular bases. Elsevier, Tarrytown, New York, pp 161–187

Van Horn R, Clay K (1995) Mitochondrial DNA variation in the fungus *Atkinsonella hypoxylon* infecting sympatric *Danthonia* grasses. Evolution 49:360–371

Walker J (1970) Systematic fungal parasite of *Phalaris tuberosa* in Australia. Search 1:81–83

White JF (1987) Widespread distribution of endophytes in the Poaceae. Plant Dis 71:340–342

White JF (1988) Endophyte-host associations in forage grasses. XI. A proposal concerning origin and evolution. Mycologia 80:442–446

White JF Jr (1993) Endophyte-host associations in grasses. XIX. A systematic study of some sympatric species of *Epichloë* in England. Mycologia 85:444–455

White JF Jr (1994a) Endophyte-host associations in grasses. XX. Structural and reproductive studies of *Epichloë amarillans* sp. nov. and comparisons to *E. typhina*. Mycologia 86:571–580

White JF Jr (1994b) Structure and mating system of the graminicolous fungal epiphyte *Echinodothis tuberiformis* (Clavicipitales). Am J Bot 80:1465–1471

White JF Jr (1994c) Taxonomic relationships among the members of the Balansieae (Clavicipitales). In: Bacon CW, White JF Jr (eds) Biotechnology of endophytic fungi of grasses. CRC Press, Boca Raton, pp 3–20

White JF, Bultman TL (1987) Endophyte-host associations in forage grasses. VIII. Heterothallism in *Epichloë typhina*. Am J Bot 74:1716–1721

White JF Jr, Morgan-Jones G (1987) Endophyte-host associations in forage grasses. IX. Concerning *Acremonium typhinum*, the anamorph of *Epichloë typhina*. Mycotaxon 29:489–500

White JF Jr, Owens JR (1992) Stromal development and mating system of *Balansia epichloë*, a leaf-colonizing endophyte of warm-season grasses. Appl Environ Microbiol 58:513–519

White JFJ, Morrow AC, Morgan-Jones G (1990) Endophyte-host associations in forage grasses. XII. A fungal endophyte of *Trichachne insularis* belonging to *Pseudocercosporella*. Mycologia 82:218–226

White JF Jr, Morrow AC, Morgan-Jones G, Chambless DA (1991) Endophyte-host associations in forage grasses. XIV. Primary stromata formation and seed transmission in *Epichloë typhina*: developmental and regulatory aspects. Mycologia 83:72–81

Wicklow DT (1988) Metabolites in the coevolution of fungal chemical defense systems. In: Pirozynski KA, Hawksworth DL (eds) Coevolution of fungi with plants and animals. Academic Press, New York, pp 173–201

Yamamura N (1993) Vertical transmission and evolution of mutualism from parasitism. Theor Popul Biol 44:95–109

15 Mathematical Models of Plant/Fungus Interactions

I. SALONIEMI

CONTENTS

I.	Introduction	239
II.	Concepts and Assumptions	240
A.	The Nature of Genetic Variation	240
B.	Reproductive Characters	241
C.	Natural Selection and Selected Traits	241
D.	Coevolution and Coevolutionary Models	242
E.	Equilibrium and Nonequilibrium Models	242
III.	Ecological Models	243
A.	Epidemiological Models	243
B.	Implicit Genetics	244
IV.	Purely Genetic Models	245
A.	Gene-for-Gene Models	246
B.	Cultivar Mixes	247
C.	Other Genetic Models	247
V.	Explicit Genetics and Epidemiology	248
A.	Gene-for-Gene Models with Ecology	248
B.	Coevolutionary One-Locus Models	248
C.	Coevolutionary Quantitative Genetics Models	249
VI.	Metapopulation Models	250
A.	Ecological Metapopulations	250
B.	Genetics of Subdivided Populations	252
VII.	Conclusions	252
	References	254

I. Introduction

The essence of good model building is to find the right simplifications of reality: what are the most relevant aspects of a system for understanding the problem at hand? Naturally, models should be constructed in accordance with what is known about the biology of a given situation. However, paradigms of what nature is assumed to be, instead of what it is known to be, may receive undue emphasis in models.

Models can be classified as strategic or tactical (May 1974). Tactical models are built to understand a specific system, e.g., the interaction between barley and barley mildew. The aim of these models is often more applied, and the tailored nature of the model allows various empirically measured details to be included. These models may thus give exact predictions about, e.g., yield loss due to pathogenic infections. However, these models may include so many details that the structure of the system, and the effects of slight changes in the model are most difficult to understand. A full (global) analysis of the model is not possible, because only a limited number of parameter values and initial conditions can be studied in computer runs.

Strategic models are not meant to give precise quantitative predictions, but to give a qualitative insight into the system. May (1974) compares strategic model with idealized (like frictionless) systems in physics that are known to be unrealistic, but nevertheless form the backbone of classical physics. For a good strategic plant/pathogen model, we need to concentrate on the most general aspects of such systems.

The questions we want to address should determine the type of the model, although most actual models have both strategic and tactical aspects. Probably the best way to combine these alternatives is to formalize the problem as an analytical (strategic) model, and make use of computers when mathematical tools or skills no longer work. This section concentrates on the general aspects of strategic models.

Traditionally, plant pathology has aimed at understanding how yield of agricultural plants could be improved by introducing new one-locus resistance genes, but studies of nonmanaged plant/pathogen systems have been rare. Practices and theories for agricultural systems have developed quite independently from other biological disciplines.

In ecology/studies on interactions between species have always been in the mainstream. However, animal interactions have been the main target of modeling and data-collecting efforts, and until very recently, interactions between plants and pathogens have been largely ignored. Much

Department of Biology and Kevo Subarctic Research Institute, University of Turku, FIN-20014 Turku, Finland

The Mycota V Part B
Plant Relationships
Carroll/Tudzynski (Eds.)
© Springer-Verlag Berlin Heidelberg 1997

research has been devoted to the specific and unique genetic basis for plant/pathogen interactions, the gene-for-gene system (Flor 1942, 1956), which has seldom been found in interactions between other groups of organisms (Thompson and Burdon 1992). Instead of emphasizing the differences between systems, it is time to acknowledge the similarities in many antagonistic interactions where one species benefits and the other loses. Naturally, I will concentrate on plant/pathogen models, but I am convinced that more progress in this field could be made by integrating research on plant/pathogen systems more closely with related fields such as plant insect studies or animal/parasite studies.

Recent evolutionary biology has tended to merge ecological and genetic aspects of populations. It is now widely acknowledged that both quantity (density) and quality (resistance and virulence) of interacting species influence genetic and ecological characteristics of a given species (Antonovics 1994). The evolutionary process has been called coevolution (Janzen 1980), if the species show reciprocal adaptations. Plants should develop more defenses when attacked by virulent pathogens. In the same way, plants that are harder to attack should cause more aggressive pathogen genotypes to develop if this is evolutionarily possible.

A plant/pathogen interaction may lead to extinction of pathogen population, extinction of both species, or unlimited or limited growth of both populations. If both species persist, genetic variation in one or both species may be maintained or lost. Most plant/pathogen models are equilibrium models, or have concentrated on predicting the outcome of these interactions in a homogeneous environment. In recent years, more realistic assumptions about the environment have produced nonequilibrium models, where the analysis of dynamics is more important than the final outcome.

Plant/pathogen models assume implicitly or explicitly many biological properties of the species, like the sexual system, speed of evolution in host and pathogen populations, or overlap of generations. These assumptions are typically made to allow clearer analysis of the studied problem. However, without further studies, it is extremely difficult to conclude whether the predictions of such models apply only to exceptional (or even impossible) cases.

II. Concepts and Assumptions

A. The Nature of Genetic Variation

Since the ingenious experiments of Flor (1942), the paradigm of plant/pathogen studies has been discrete major-locus variation. If pathogen isolates also differ in their ability to infect plant genotypes, the gene-for-gene hypothesis is routinely invoked. The evidence for gene-for-gene interaction in agricultural systems and some wild relatives of cultivated plants is well established (Alexander 1992), but opinions on the generality of the idea vary widely (Day 1974; Barrett 1985). The importance of these systems has increased with the possibility in the near future for molecular manipulation of major locus resistance genes (Gabriel and Rolfe 1990; de Wit 1992). The number of suggested gene-for-gene interactions is increasing (Thompson and Burdon 1992), but the actual number of proven cases is still somewhere around ten. Without doubt, plants have major locus resistance genes, and discrete variation in the ability of pathogens to attack plant genotypes exists. Most of these pathogens reproduce asexually, so their populations consist of separate lines. However, this does not prove that most plant/pathogen evolution is based on gene-for-gene interactions.

Quantitative resistance may be more common in types of pathogens that have not been intensively studied in agriculture (e.g., nonbiotrophs; Parlevliet 1989; Groth and Crist 1992), but there are good examples available on quantitative variation in agricultural plant/pathogen systems as well (Geiger and Heun 1989). Quantitative variation seems to occur in all studied plants, including the best examples of gene-for-gene system (Sidhu 1988; Parlevliet 1989). Further, genetic interactions between plant and pathogen genotypes are not limited to gene-for-gene systems; but genotypic interactions can be found in quantitative characters as well (De Nooij and Van Damme 1988a,b; Saloniemi 1993b).

Very little can be said about the importance or prevalence of different kinds of resistance. Nearly all models of plant/pathogen interaction are based on gene-for-gene assumptions, although explicit quantitative genetics has been employed in one predator-prey model (Saloniemi 1993a), and assumptions of continuous variation have been included in some models (Frank 1994).

B. Reproductive Characters

Plant/pathogen models normally assume asexual pathogens, and quite often asexual plants as well. This assumption obviously aims at avoiding recombination, which can become a serious problem when several major loci are included in models. Until recently, this was regarded as trivial, because most pathogens were assumed to have a race structure (Burdon 1987), and models dealt with cereals and other mainly self-fertilizing agricultural species. Caten (1987) strongly criticizes the concept of race in plant pathology, and, clearly, there are plenty of examples of pathogens that do not seem to have discrete lines that plant pathologists prefer to call races (Holliday 1989). Accumulating evidence shows that the genetic structure and actual breeding systems of pathogens may be very complicated (McDonald and McDermott 1993). In natural populations, even plant and pathogen species capable of selfing prefer cross-fertilization.

Most, but not all (e.g., Oomycota), pathogens are haploid; models frequently assume haploid host plants as well. In single-locus models this may not prove serious, but if we allow resistance to be of more complicated character, results may be biased. The level of maintained genetic variation and the number of genotypes with the best fitness values are much higher in diploid organisms.

Generations are assumed to be either strictly separate, or totally overlapping. This decision determines the use of two alternative modeling techniques: differential equations for continuous population growth with overlapping generations, and difference equations for separate generations. As with most biological systems, plant/pathogen interactions seem to lie somewhere in between, and both alternatives are oversimplifications. The inclusions of explicit generation times in difference equations is required when pathogen generation time is much shorter than that of the host plant. However, the effect of generation time on selection response may not be as important as previously believed (Rosenheim and Tabashnik 1991).

C. Natural Selection and Selected Traits

Adaptations to the immediate environment are essential to any species and, ever since Darwin, we have known that they are products of natural selection. Natural selection is clearly an ecological process, and multiple aspects of the physical environment as well as the qualities and quantities of coexisting species exert an effect. Inheritance and genetic correlations together with multidimensional selection pressures will change populations in the direction of increasing fitness (e.g., Lande and Arnold 1983).

The first models of classical population genetics assumed the fitness values to be constants. More recently, everyone agrees that this is a simplification for most natural situations, but in practice the paradigm of constant fitnesses is still clearly seen in many models. In plant-pathogen interactions, the fitnesses may depend on the densities (density-dependent selection) and the genetic qualities (frequency-dependent selection) of the species.

Plant/pathogen models reflect the preconceptions of empiricists in the field, and the discipline of plant pathology has developed its own terminology and concepts that make communication with other biologists difficult (Antonovics and Alexander 1989; Shaner et al. 1992). Fitness-related concepts are especially difficult in the plant/pathogen context, because the interest centers on the result of an infection (e.g., disease). For evolutionary studies, we need to separate both the fitness of the plant genotype (see Antonovics and Alexander 1989), and the fitness of the fungal genotype from disease, the specific outcome of infection. Actually, it is possible for both species to show high fitnesses simultaneously, a situation known as tolerance (Clarke 1986).

Shaner et al. (1992) give an outstanding review of the way the concepts fitness, virulence, pathogenicity, and aggressiveness are used in plant pathology. They suggest that virulence be used for the result of the infection. Pathogen fitness (rather than, e.g., virulence or aggressiveness) in the general biological sense should be used for the contribution of the pathogen genotype to the next generation. Pathogen fitness can be divided to different components such as the ability of the pathogen genotype to spread to a new host plant and to penetrate through the plant tissue and overcome plant defenses. If infection takes place, the number of spores or other reproductive propagules that the pathogen produces for the next generation must be taken into account in a proper estimate of fitness.

Furthermore, in natural populations especially we should not exclude the many indirect ways in which the host plant can avoid being infected. The most obvious ones are related to timing: plants may reproduce before, e.g., physical factors are favorable for fungus growth. Burdon (1987) discusses mechanisms like these under the title disease avoidance. However, traits that are involved in indirect or correlational resistance usually have a genetic basis that allows plants to adapt better to their pathogen-infested environment. Thus, evolution may cause disease levels to decrease without clear changes in resistance or virulence genes.

D. Coevolution and Coevolutionary Models

Genetic variability allows species to adapt both to their environment and to other species as an important part of that environment. Change in one species, e.g., a virulence-related mutation in a pathogen, causes a need for adjustment in the host plant population. In some cases, coevolution is very specific and strictly reciprocal between two species, but it may include more diffuse interactions as well.

Plant/pathogen models have traditionally studied two very different subjects that only recently have began to merge. The first involves ecological dynamics and outbreaks of pathogens, while the second deals with genetic variability. Levin and Udovic (1977) give a clear graphic representation of coevolutionary models (Fig. 1). Purely ecological models in the upper part of Fig. 1 specify population growth in isolation (arrows 1), or with the interacting species (arrows 2). These models often consider an elaborate and plausible epidemiological background (close to tactical models), with the aim of understanding changes in population numbers.

The lower part of Fig. 1 represents intraspecific (arrows 3) and interspecific (arrows 4) genetic effects without any ecological complications. The main components of genetic models are the fitnesses of different genotypes present in the population. In basic models, fitnesses are constant for each plant genotype-pathogen genotype combination, but the fate of a certain plant genotype depends on the frequency of the available pathogen races and vice versa. This causes plant and pathogen fitnesses to depend on the frequencies of the other species. The main group of these genetic models is based on gene-for-gene assumptions (Burdon 1987; Thompson and Burdon 1992).

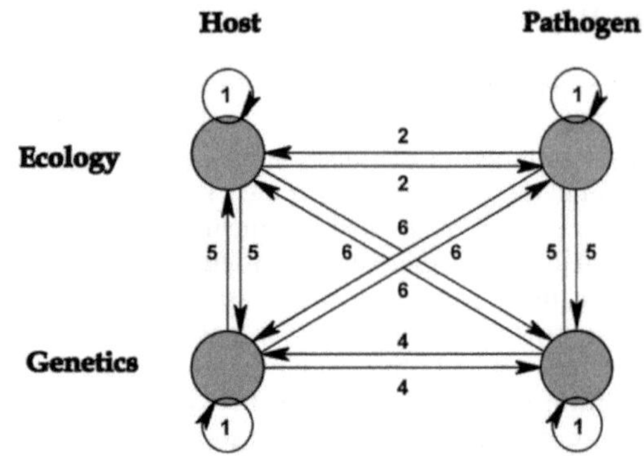

Fig. 1. Graphic representation of coevolution by Levin and Udovic (1977, with permission from American Naturalist and The University of Chicago Press). The *arrows 1* and *2* in the upper part of the graph represent intra- and interpopulation density dynamics. In the lower part, *arrows 3* and *4* show purely genetic intra- and interpopulation effects. Coevolutionary effects by *arrows 5* and *6* connect the genetic and ecological dynamics of the species

Genuine coevolutionary models should include both aspects as shown by arrows 4, 5, and 6 in Fig. 1. Here the density and genetic constitution of the pathogen depends on the genetic characters and numbers of the host and vice versa, i.e., we simply admit a deeper interdependence between ecological and genetic processes. This is clearly more realistic than consideration of only epidemiological or genetic characters in one species or two interacting species. While ecological processes are important for maintenance of genetic variation, genetic variation and evolution may also help to change unstable ecological systems to stable ones.

E. Equilibrium and Nonequilibrium Models

The standard method for analyzing models involves computing the densities and character values (resistance and virulence) that the species would gain after a very long time. Then one checks whether the system tends to remain at this equilibrium or move away from it after a mild perturbation (e.g., changing weather conditions). A stable equilibrium point (Fig. 2a) is found if densities and character values end at a single point regardless of the initial density and character states (or several alternative points when initial

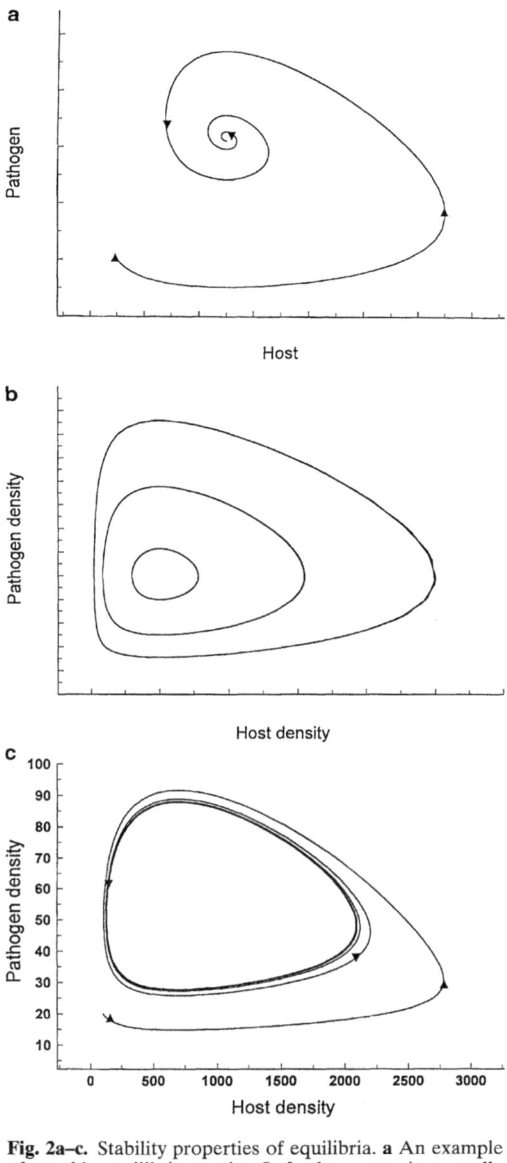

Fig. 2a–c. Stability properties of equilibria. **a** An example of a stable equilibrium point. In **b**, the system is neutrally stable, and **c** shows a stable limit cycle

values are far from each other). It is also possible that there is no single density and genotypic equilibrium point, but that species oscillate around certain values (Fig. 2). Beginning either very near to the equilibrium point, or very far from it, causes the same kind of oscillation without extinction. This means that the populations have another kind of equilibrium: a never-ending cycle around the equilibrium point. In a neutrally stable system the amplitude of the cycle depends on initial densities and trait values, probably an unrealistic situation. In a stable limit cycle the dynamics converge towards a single trajectory when the initial values are quite near to each other.

The procedure explained above is basically valid for both equilibrium and nonequilibrium models, but equilibria may be studied at different levels. Equilibrium models are usually built for a single homogeneous population, while nonequilibrium models may deal with a high number of separate populations. Natural systems are not continuous and homogeneous; rather, species occur in patches of different size and microhabitat. These ideas about metapopulations (Gilpin and Hanski 1991) have influenced both ecological and genetic studies. Stability properties of the system of all populations may be different from those of single populations. Equilibrium models are initial simplifications that can, when necessary, be elaborated by including new factors. Also, more emphasis should be put on studying the dynamic properties of simple models, because the equilibrium point, as such, may be quite irrelevant, e.g., in systems that change because of abiotic factors long before the equilibrium point is reached.

III. Ecological Models

Dynamic models for antagonistic interactions between trophic levels have been built to understand whether biotic interactions could control unlimited growth of populations. For example, the main focus in host-parasite models has been the possible regulation of host population density by a parasite (May and Anderson 1983). Plant/pathogen studies only appear to differ from these more general models, since current models focus mainly on understanding and predicting the conditions for those unlimited pathogen outbreaks that are all too common in intensive agriculture.

A. Epidemiological Models

Epidemiological models are based directly on ecological density models for interacting populations. The effect of plant density (H) and pathogen density (P) on each other when population growth is continuous and generations are overlapping, can be presented as:

$$dH/dt = f(H) - g(H,P) \qquad (1)$$
$$dP/dt = c\,g(H,P) - a\,P.$$

f(H) gives density-dependent growth rate for the plant population. Without density dependence, it is common to assume f(H) = b H, where the constant b is the intrinsic growth rate of the host population. The function g(H,P) gives the decrease in host plant growth when infected by pathogen, proportion c of which can be utilized by pathogen growth, and the constant a is pathogen death rate. Coexistence, equilibrium, and the appearance of cycles depend on functions f(H) and g(H,P). Examples of features that promote stability are density-dependent host growth, aggregation of pathogens, and refuges.

In epidemiological models, emphasis is often on host population dynamics, and information on the pathogen population is included only indirectly. May and Anderson (1983) have developed several animal host-parasite models, where dynamics in the host population has been modeled separately for healthy, diseased, and recovered/immune portions of the host population. Disease-caused immunity is not uncommon in plant pathogen interactions, so Antonovics (1994), for example, applied this model by using two classes

$$X_{t+1} = X_t(1+b) - t(X_t, Y_t) \qquad (2)$$
$$Y_{t+1} = t(X_t, Y_t) + (1-a)Y_t,$$

where X and Y are densities of healthy and diseased individuals, b is growth rate for healthy individuals, and a is death rate of diseased individuals. The function t(X,Y) gives the number of new infections per generation.

The function t(X,Y) will be specified according to the assumptions about the type of pathogen transmission (Thrall et al. 1993; Antonovics 1994). For aerial dissemination transmission is density-dependent such that $t(X,Y) = \alpha X Y$, i.e., the number of new infections depends directly on the number of diseased individuals spreading spores in the population. In sexually transmitted or vector-borne diseases, where vectors actively search for plants, transmission is frequency-dependent, such that $t(X,Y) = \alpha X (Y/N)$, where N is the total number of hosts. A healthy host is more likely to become diseased, if the previous host visited by, e.g., a pollinator, was diseased. Most epidemiological models have assumed N to be constant, and the frequency-dependent model has been applied for any type of disease.

Getz and Pickering (1983) have shown that stability properties depend on the assumed transmission mode. In the sexual transmission mode, a stable coexistence of hosts and pathogens is not possible, but the density-dependent (aerial) transmission model causes neutrally stable cycles. External (independent of disease) density dependency stabilizes the aerial transmission model. To stabilize the sexual model, we need to invoke a stronger density-dependent factor in the healthy than in the diseased class (Thrall et al. 1993; Antonovics 1994).

B. Implicit Genetics

The basic epidemiological models become coevolutionary when individuals are allowed to show genetic differences in contrast to a clonal population structure. However, the simplest models do not specify the genetic basis of inheritance for plant or pathogen traits. There are no genetic complications if only haploid and clonal host plants and pathogens are studied. In this case, populations consist of two or several genetically different types which may represent either clones or species. Each type has different parameter values (birth, death, or transmission rates and virulence). In coevolutionary models, parameter values are assumed to have an unspecified genetic basis. The number of equations increases, because both densities and characteristics (parameter values) of each type involve a new equation.

In a simple model (e.g., Levin and Pimentel 1981; Levin 1983; May and Anderson 1983), a pathogen population consists of lines 1 and 2. Line 1 is more virulent than line 2, so it causes higher mortality (additional death rate due to infection is higher for line 1 compared to line 2, $\alpha_1 > \alpha_2$). Basic birth rates a or death rates b do not differ between the hosts. Both lines are as likely to infect susceptible hosts (with transmission rates β_1 and β_2), and the lines are also able to attack (by rates σ_1 and σ_2) a host previously infected by the other pathogen line. Host populations can now be divided in three: infected by line 1, line 2, or healthy (densities Y_1, Y_2 and X, respectively).

$$\frac{dX}{dt} = a(X + Y_1 + Y_2) - \beta_1 XY_1 - \beta_2 XY_2 - bX$$

$$\frac{dY_1}{dt} = \beta_1 XY_1 - (\sigma_1\beta_1 - \sigma_2\beta_2)\, Y_1Y_2 - (\alpha_1 + b)Y_1$$

$$\frac{dY_2}{dt} = \beta_2 XY_2 - (\sigma_2\beta_2 - \sigma_1\beta_1)\, Y_1Y_2 - (\alpha_2 + b)Y_2.$$

Coexistence of the lines, or competitive exclusion of one of the lines, is possible depending on birth and death rates, transmission rates, and rates of reinfection. Relatively high life expectancy of the line (relative birth rate compared to total death rate) and reinfection rates determine which of the lines will persist. Conditions for coexistence also depend on transmission rates: the more virulent line needs to cause more infections than the less virulent line with lower death rate.

Anderson and May (1981) include recovery rates in their analysis. In plant-pathogen situations recovery could happen in nonsystemic infections when diseased parts of a plant such as tree branches die, or some mechanism (like improving external conditions) stops the spread of a systemic infection. May and Anderson (1983) review their analysis by stating that the winning line must have the highest net reproductive rate. The fitness of a parasite increases by having a high transmission rate, causing low host death rate, and only rarely allowing recovery. The situation in the host is opposite, with highest fitness with low transmission and high recovery, but a low host death rate would favor both the host and the parasite. If the three parameters were not dependent on each other in an actual biological situation, we would expect excess host death rate due to disease to decrease, and transmission and recovery parameters to gain some intermediate values.

Besides showing that coexistence between pathogen genotypes is sometimes possible, they also gave an explanation for the rather common empirical observation (references in May and Anderson 1983) that within endemic host/pathogen systems, there is evolution towards less harmful pathogens, and/or increased resistance in hosts. For example, most epidemics are caused by introduced pests and pathogens (Hokkanen and Pimentel 1983). Unfortunately, most of the classical literature has relied on unpopular group-selection arguments to explain evolution towards less virulent or aggressive strains. In a recent model of the evolution of virulence, Lenski and May (1994) included implicit genetics for host evolution as well, and now their optimality analysis shows that decrease of pathogen virulence and increase in host resistance are the most likely results from a host/pathogen interaction where both species persist. In his review of evolution in animal pathogen virulence, Bull (1994) suggests that transmission mode and tradeoffs between transmission and virulence determine the outcome.

Schaffer and Rosenzweig (1978) obtained a similar result much earlier using a very different approach for animal predator/prey systems. They assume an unknown genetic basis for the parameters of their model and optimizing evolution to change the parameters directly towards the optima. Their analysis finds an optimal parameter value that counterbalances changes in prey and predator characters. Their model causes evolution to favor prey when predators become more effective, and vice versa.

An optimality approach has been used widely in ecology although it is not easy to define criteria for optimality, or make sure that the optimum point can be reached. Including explicit genetics will give a more complete understanding of the dynamics of a coevolutionary system. Even for empirical applications, it will be especially important to check what is the evolutionary outcome of continued coexistence of hosts and pathogens. If more rigorous genetic analyses predict evolution towards less virulent pathogens and more resistant hosts, we may need to change our agricultural practices and breeding programs towards systems that allow coevolution between host and pathogen genotypes.

IV. Purely Genetic Models

Existence and maintenance of genetic variation is still a major unsolved problem in biology. Long before ecologically oriented researchers working with other kinds of predator/prey interactions became interested in genetic variation, models for gene-for-gene interactions in plant/pathogen studies existed (e.g., Mode 1958). Variation in plant resistance and pathogen virulence could be shown empirically in both natural and agricultural species, and it was suggested early on (e.g., Haldane 1949) that host/pathogen interactions could maintain genetic variability. Practically, these studies aimed at understanding the mechanisms and speed of production of new pathogen genotypes that caused breakdowns of resistance in agricultural systems.

Genetic models are built especially to find out whether genetic variation is maintained or generated by interaction. In plant pathology the main questions have been the dynamics of pathogen race buildup and the effects of introducing genetic variation into host plant populations by using cultivar mixes.

A. Gene-for-Gene Models

Models of gene-for-gene interactions have until very recently dealt only with gene frequencies of plant and pathogen populations. Populations sizes were not allowed to vary. Constant fitness values were assigned to plant and pathogen genotypes, and mean fitnesses of the species were calculated according to laws of population genetics by taking into account the frequencies of different genotypes. This causes plant fitness to depend on the pathogen frequencies, and the pathogen fitness to depend on frequencies of plant genotypes. Leonard and Czochor (1980) and Levin (1987) have published good reviews of these gene-for-gene models.

Table 1 (from Leonard and Czochor 1980) shows the four constant fitness values for both pathogens and hosts that Leonard (1977) used in his one-locus model of a diploid host plant and a haploid pathogen. Several earlier models are special cases of this model. These models deal with questions such as whether a virulent pathogen is more or less fit on a susceptible or resistant host genotype (parameters k and a; $k > 0$ indicates a cost of virulence). Some models also allow an avirulent pathogen to reproduce on a resistant plant genotype ($t < 1$). In the same way, resistance in plants may involve costs ($c > 0$). Contrary to previous models, Leonard (1977) also includes a parameter s to take into account disease severity (e.g., in a certain environment). In addition to the cost of resistance, fitness of a host-plant genotype depends on the respective pathogen fitness multiplied by severity of the disease.

Table 1. Relative fitness values of plant and pathogen genotypes in different combinations from Leonard (1977) and Leonard and Czochor (1980). Plant resistance allele R and pathogen virulence allele V are dominant, and the respective alleles r and v are recessive

Fitness of pathogen genotypes in indicated combinations	rr (susceptible)	RR, Rr (resistant)
v (avirulent)	1	$1 - t$
V (virulent)	$1 - k$	$1 - k + a$

Fitness of host genotypes in indicated combinations	rr (susceptible)	RR, Rr (resistant)
v (avirulent)	$1 - s$	$1 - c - s(1 - t)$
V (virulent)	$1 - s(1 - k)$	$1 - c - s(1 - k + a)$

Symbols used in table: k = cost of virulence, c = cost of resistance, t = effectiveness of host resistance, a = effect of plant resistance on a virulent pathogen, s = disease severity, e.g., due to environmental causes. In most analyzed gene-for-gene models $t = 1$, $a = 0$, and $s = 1$.

Analysis of the model shows that plant resistance and pathogen virulence will be fixed unless there are costs for resistance and virulence. With cost, virulence and resistance frequencies show stable oscillation that allow polymorphism in both species. When cost increases, the respective allele frequency of the interacting species increases. Other parameter values have an effect on the gene frequency values that plant and pathogen population gain at equilibrium.

These models assumed diploid (or haploid) hosts with dominant resistance alleles, and haploid pathogens. Levin (1983) points out that even if computer simulations suggest stable cycles or equilibria for these simple models, they seem to be neutrally stable. However, if diploid genetics with heterozygote genotypes different from homozygotes is included in both species, the model stabilizes even without cost of resistance and virulence. A more complicated genetic system allows actual or marginal overdominance to maintain variability. This requires that the heterozygote either always have the highest fitness, or that the gene frequencies of the other species (e.g., pathogen) must allow the heterozygotes (of hosts) on the average to have fitnesses better than either of the homozygotes.

These results on the importance of overdominance or cost of resistance and virulence for stability have had a profound influence on empirical and theoretical work on plant/pathogen studies. Their role and existence in natural populations can only be studied empirically. Despite considerable efforts at finding overdominance in nature, there is very little evidence for it in any biological system. Overdominance is commonly invoked, because genetic variability is ubiquitous in nature, and overdominance is the only mechanism that maintains variability in the simplest genetic models.

Cost of resistance is a more probable explanation, but several empirical studies on costs have found contradictory results (Parker 1992; Simms 1992). Cost could consist, e.g., of the physiological burden of producing an extra chemical or an ecological cost of a disadvantage in competition. Actual gene-for-gene systems seem to consist of several different loci, and the fate of unnecessary loci that do not work any more has caused considerable interest. Van der Plank (1968) suggested that plant varieties with several resistance genes should show lowered fitness. This would lead to selection against unnecessary resistance genes that have already been overcome by pathogens. Cumulative costs would limit the use of several working resistance genes in one single variety. These speculations have also been important in developing cultivar mixes.

B. Cultivar Mixes

Coevolution in the strict sense is not possible in many agricultural systems. In ordinary farms, crop seed does not come from the same field year after year, and thus adaptation to local populations of pathogens would be impossible. Further, modern plant breeding has produced high yielding, self-pollinated, and genetically highly uniform species that could hardly respond to any selection caused by pathogens.

Clearly, in such situations, coevolutionary models are invalid. However, intensive plant breeding constantly seeks new resistance alleles, while pathogen populations change to overcome these defenses. Much of plant/pathogen theory has aimed at understanding whether this human-mediated coevolution could give rise to a stable plant/pathogen system. Finding new resistance alleles and including them to new cultivars is expensive and time-consuming, and it would be useful to find more effective ways of decreasing damage caused by fungal diseases.

Spread of a new pathogen line is fast if the same plant variety is grown on wide areas year after year. Using several cultivars and rotating them between years is not always easy, because one single variety may provide maximum yield and because each of the varieties will be overcome by some pathogen line in the long run.

It has been suggested that growing mixtures of cultivars instead of pure seed could delay buildup of pathogen races (references in Burdon 1987; Marshall 1989; Groth and Christ 1992; Wilhoit 1992). This strategy is necessarily based on a non-equilibrium concept because commercial agriculture demands uniform production characters in cultivars. Thus, the pure seed lots must be grown separately to be blended anew before seed is grown. Otherwise, the proportion of some components would increase, or occasional cross-breeding might produce genetically variable seed.

It has proved very difficult to show in practice that these ideas work (Marshall 1989; Wilhoit 1992). There are several ways of choosing the components, and both long-term and large-scale experiments with several different systems would be needed. Thus, modeling has been an important approach to choosing the most effective strategies for using multilines. In such studies, only gene-for-gene-based systems have been investigated.

Work with multiline models has concentrated on two different problems. First, it has shown that multilines slow the breakdown of resistance and decrease the density of pathogens, but a clear cost of resistance is needed for stabilizing the system. Lower numbers and increased distance between susceptible plants compared with situations with pure lines only largely account for slower growth of epidemics in mixes of varieties. Most modeling effort has been put on finding out whether pathogen populations are likely to develop a superrace, a line that is able to infect all the component plant cultivars. Wilhoit (1992) concludes that, despite 30 years of research, we still do not know what the effect of plant variation is on pathogen variability and superraces. To continue modeling, we should have more information on variation and cost of resistance in actual pathogen populations.

C. Other Genetic Models

Haldane (1949) suggested that infection of a host might be easier for a rare pathogen genotype. This frequency-dependent selection is likely because evolution in the host favors genes that increase resistance towards the most common pathogen genotypes in the population.

Barrett (1988) studied a simple plant pathogen model that included asymmetries in costs of virulence and resistance: certain plant and pathogen genotypes showed higher fitness when grown together. He assumed asexual haploid species. The system oscillated between different plant and pathogen genotypes, but coexistence of the genotypes required mutation or migration. Barrett (1988) concluded that assumptions for genetic basis of virulence and resistance are not relevant for these simple models. Relevant frequency-dependent models should include diploid, sexual species and a more realistic genetic basis. Further, he suggests that the one-locus model also gives a simplistic view of actual gene-for-gene interactions.

Seger (1988) studied models where a host genotype can be infected by only one pathogen genotype and where the effect on plant fitness depends on the genotype of the pathogen that it encounters. A system with two haploid hosts and pathogens was unstable unless mutation was allowed to stabilize the system. Fixation took much longer, or less mutation was needed when several alleles were allowed in the host from the very beginning of the simulation. Linking several loci for host resistance did not have an effect on stability, but it had a considerable effect on the dynamics of the system.

In a later model, Seger (1992) included a more complex genetic structure in his one-locus predator prey models. The main result was more variable dynamics. For example, in the diploid two-allele host pathogen model of Fig. 3, frequencies of the species cycle around the smaller limit cycle, and no mutation is needed for coexistence.

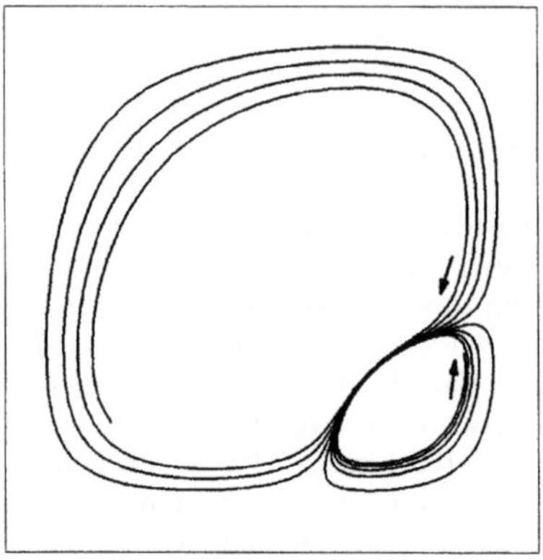

Fig. 3. A diploid host–pathogen model from Seger (1992). Pathogen allele P1 causes better virulence than the alternative allele, and host allele H_1 induces higher resistance than allele H2. Constant fitness values for hosts and pathogens include cost of resistance and virulence (two times higher for homozygotes). Regardless of the initial values, the frequencies of the species cycle around the smaller limit cycle, and no mutation is needed for coexistence

These purely genetic models suggest that studies on very simple asexual or one-locus models may limit our understanding of actual predator/prey interactions regardless of the genetic basis of the system studied. In addition to more analytical approaches to frequency-dependent models, we need studies that include ecological assumptions with density-dependent fitnesses in the models as well.

V. Explicit Genetics and Epidemiology

For unique coevolutionary models, we need to specify explicit inheritance, and the manner in which fitnesses and ecological parameters change with interaction. Epidemiological studies or purely genetic formulations may not be enough for understanding the deeper dependence between ecological and genetic processes. Ecological processes are important for maintenance of genetic variation, but genetic variation and evolution may also help to change unstable ecological systems to stable ones.

A. Gene-for-Gene Models with Ecology

Mode (1961) was the first to include population dynamics in a gene-for-gene model. His general model allows variable population sizes and non-constant fitnesses for plants and pathogens. Stable equilibria exist in the models when frequency-dependent fitnesses are allowed.

Frank (1991, 1993) studied asexual, haploid clones of hosts and pathogens. He includes variable growth rates for different plant and pathogen clones so that population densities vary. His genetic model is loosely reminiscent of the gene-for-gene model, and the questions asked come from this tradition. He shows that demographic parameters like birth and death rates are important for the amount of variation present in a population.

B. Coevolutionary One-Locus Models

A whole family of predator/prey models with single locus genetics dates from Pimentel (1968). His genetic feed-back mechanism consists of prey genetics and predator population growth. Levin

(1972) showed that the system had a stable equilibrium.

Roughgarden (1979) presented a general one-species model with one-locus two-allele genetics and known gene frequencies. Here, ecological dynamics and genetics are connected by using mean Malthusian fitnesses calculated from the genetic model as growth rate of the population. Evolution changes the parameters of the model, and the parameters have a one-locus genetic basis, which is a highly unrealistic assumption. Even then, analytical results are not possible because four equations are needed for a single species, and we would like to include both plant and pathogen dynamics. Based on the models, he claims that a coevolutionary model may be stable only if the underlying ecological model is stable. Thus, genetic variability always emerges as a destabilizing force in these models.

C. Coevolutionary Quantitative Genetics Models

Saloniemi (1993a) presented a general predator-prey model, where both species show explicit quantitative variation, with known heritabilities, in characters relevant for predation (e.g., resistance and virulence, thickness of cuticules and proboscis length, running speeds). Outcomes of random contacts between individuals depend on comparison of traits: a prey individual escapes attack if it has a "better" trait than the predator. Characters must be measured in the same units, such as linear measurements of body size or an enzyme activity needed to detoxify a certain unwanted chemical. Both species are assumed to gain an optimal character value when growing without the other species. When coevolutionary forces tend to change the traits, the final character values are determined both by predation and tendency towards the optimal value.

The dynamics of densities in both the species depends on their traits as well as densities. Without any genetic variation, the species are assumed to show the basic neutrally stable Lotka-Volterra cycles. This model is known to be very unrealistic, because initial values and any minor perturbations will change permanently the basic characteristics of the dynamics. However, any structural change in the model is likely to make the model either stable or unstable. We can study the effect of genetic variation for the dynamics by including heritable traits in the Lotka-Volterra model. I chose a simple model to gain analytical results (Saloniemi 1993a). Later studies (unpubl.) have shown that many other ecological models (including those which are unstable) give the same results.

Analysis of the model shows that genetic variation either stabilizes or destabilizes the system, depending on parameter values such as birth and death rates, as well as heritabilities. Selection for optimal character values, prey population growth rates and predator death rates determine if an equilibrium for densities and traits is possible. If an equilibrium exists, instability of the system depends entirely on the level of genetic variation in the species. When approaching a stable equilibrium point, characters relevant to e.g. plant defense and pathogen virulence gain values that depend on the quality and quantity of both the species. Figure 4 shows a case where variability in both the species leads a neutrally stable ecological model to stability. Parameters and initial values for Fig. 5 are the same, except that the ecological model is unstable to begin with. Genetic variation both in prey and predator stabilizes the coevolutionary system.

Local analysis of the system shows that in all cases stability arises whenever both host and pathogen have high levels of genetic variability; low variability leads either the pathogen or plant population to go extinct. This result suggests that genetic variation in agricultural plant populations may be important for escaping continuous boom and bust cycles with pathogens. Further work along these lines should continue, e.g., by including more specific assumptions about the plant-pathogen interaction in the model.

Frank (1994) has presented a model with densities and frequencies of several genotypes (= phenotypes with heritability = 1) that assumes quantitative inheritance as well. However, his model is less realistic because both hosts and pathogens are haploid, and no recombination is allowed. In a genuine quantitative model, several different genotypes may produce the same phenotype, which makes maintenance of variation more effective, and causes the phenotypic distribution to be continuous in contrast to his model. Also, the quantitative characters he uses, costs of resistance in host and pathogen populations, are not especially explicit or easy to measure. The most interesting result of his model predicts that quantitative variation may decrease dramatically near some equilibria. It would be interesting to include

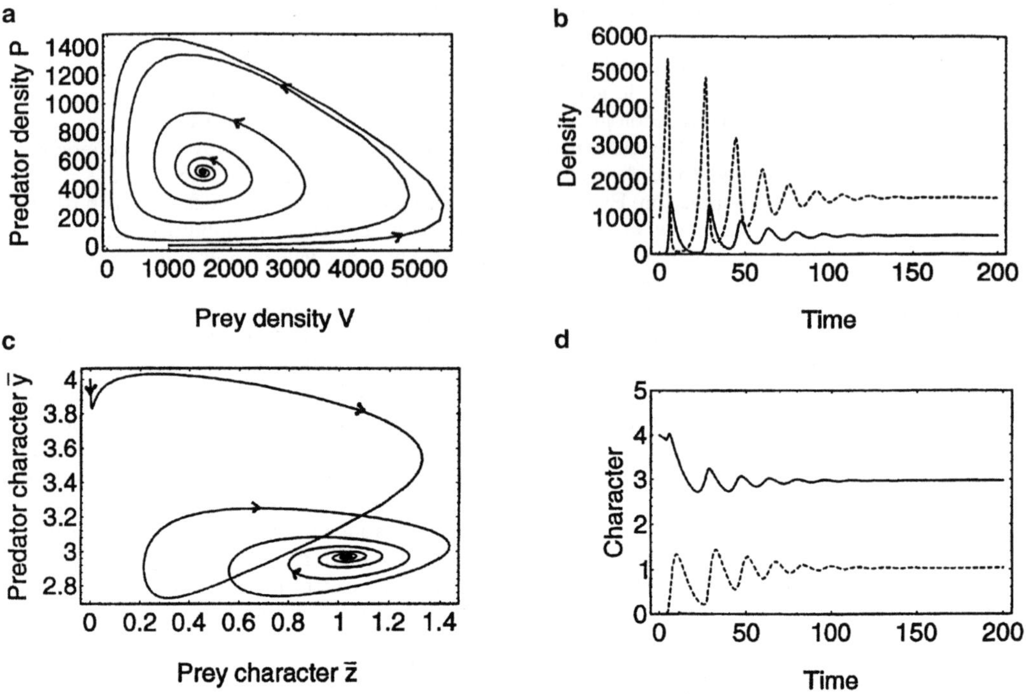

Fig. 4a–d. Stabilization of a predator-prey system by including high enough quantitative variation in prey and predator characters. (Saloniemi 1993a, with permission from American Naturalist and The University of Chicago Press). **a** shows a trajectory of prey and predator densities, and **b** gives a respective trajectory of the characters. **c** shows the change in densities, and **d** the change in character values. *Continuous lines* in **c** and **d** refer to predator, and *dashed lines* to prey values. In this figure additive genetic value for prey is 1 and for predator is 0.5

dynamics of genetic variation in actual quantitative models.

VI. Metapopulation Models

A. Ecological Metapopulations

Natural populations usually consist of a patchwork of more or less suitable habitats (Burdon et al. 1989; Burdon et al. 1989; Burdon 1992). This is especially obvious for pathogens that can only survive in a site where plants are available; but this is true for plants as well: even physical properties of the environment limit the sites where plants can survive.

Rarely are these small local populations isolated; a low level of migration occurs between patches. Extinction of a local population need not have any effect on the rest of the sites. Some patches remain empty for variable times (e.g., plants are free from disease), but migrating propagules will finally get to that site as well. This system of local populations connected by migration is called a metapopulation (Gilpin and Hanski 1991).

Figure 6 shows a prediction from a model by Gotelli (1991) for the proportion of empty patches when new individuals immigrate from other patches while the existing population size is decreasing. A situation like this might apply to a plant pathogen system when effective fungicides or herbicides are used to decrease pathogens while new pathogens are arriving from other plots.

A host and a pathogen may coexist even when local subpopulations are doomed to extinction. Antonovics et al. (1994) showed that a critical factor for coexistence was a delay in pathogen arrival in a newly established host population. If there is no delay in pathogen arrival, and all host populations become diseased, coexistence is still possible

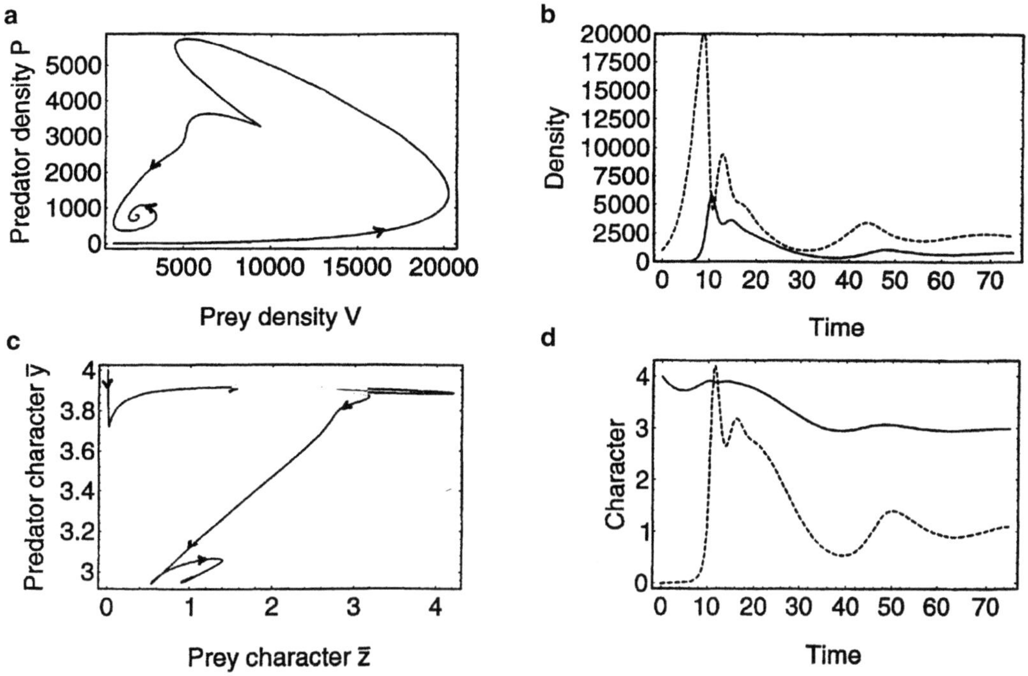

Fig. 5a–d. An unstable ecological system can be stabilized by quantitative genetic variation as well. Additive genetic variance for prey is 3 and for predator is 1. (Saloniemi 1993a, with permission from American Naturalist and The University of Chicago Press). See Fig. 4 for further explanation

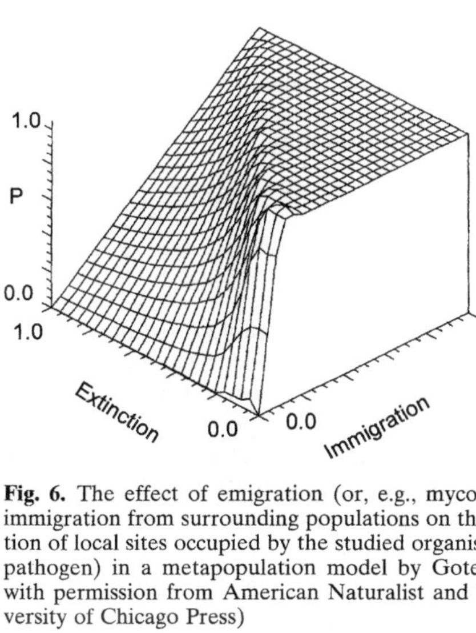

Fig. 6. The effect of emigration (or, e.g., mycocids) and immigration from surrounding populations on the proportion of local sites occupied by the studied organism (like a pathogen) in a metapopulation model by Gotelli (1991, with permission from American Naturalist and The University of Chicago Press)

when it takes several years for the pathogens to arrive.

In a metapopulation, the dynamics of local extinction and colonization from other populations determines the overall dynamics. Each individual population may be unstable (e.g., the pathogen will kill all plants), but the system as a whole may still persist. A metapopulation structure as such does not make an ecological system stable. Taylor (1990) gives a list of basic conditions that must be met for predator/prey metapopulations to persist. The most important requirement both in single species (Gilpin 1990) and predator/prey metapopulation models (Taylor 1988) is asynchronous density fluctuations in different populations (e.g., no fast-spreading epidemics to nearby sites). This happens when, e.g., subpopulations are heterogenous or there are random stochastic fluctuations in demographic parameters, dispersal rates are not very high, and the number of subpopulations is high. Stronger within-population density dependence and asynchronous density-independent migration also enhance metapopulation persistence (Reeve 1988;

Murdoch et al. 1992). Spatial subdivision without any other forces has no effect on stability (Reeve 1988).

Again, it is very difficult to show that a certain population is in an equilibrium state. Data on several generations are needed to separate random and deterministic changes. However, constant densities and character values are not prerequisites for an equilibrium state value: a population may be returning towards equilibrium after a perturbation, or the equilibrium values consists of a whole cycle instead of a single point.

B. Genetics of Subdivided Populations

The effect of population subdivision on genetic variability has been studied thoroughly (Wright 1931; Wade and McCauley 1988). Migration rates and the size of local populations determine differentiation and genetic variability in these models. In the near future, we are likely to see models that combine both demographic and genetic factors for subdivided populations. Frank (1993) has already included extinction and colonization in his haploid genetic models.

Without a doubt, genetic variation in both plant and pathogen characters can be found in natural populations. In a disease-free environment, plants with no resistance would show a higher fitness. If infection levels vary spatially, this mechanism would promote genetic variation (Gillespie 1975; Clarke 1976). Parker (1992) discusses nonequilibrium hypotheses for genetic variation in populations. All genetic variation, like resistance-related characters, may tend towards fixation. However, if new heritable variation is generated fast enough, large amounts of genetic diversity are available at any point. Also, selective pressures may exhibit changing directions and strengths in different areas or seasons.

The genetic consequences of a metapopulation structure are best known for identical finite populations with symmetrical migration rates. Wright (1940) derived a well-known result showing that at least one immigrant every other generation is needed to maintain within-population genetic variability. Slatkin (1977) and Wade and McCauley (1988) included the effects of subpopulation extinction and recolonization in the model. Differences between populations increase if migrants originate from one population; however, if the migrant pool is a mixture from all populations, differences between populations decrease.

These models exclude mutation, an important source of new variability in quantitative characters. Lande (1992) showed that, in a model with a large number of islands, mutation, migration, and drift, within-island equilibrium variance is nearly identical to that in a panmictic population. Total equilibrium genetic variance in the entire metapopulation may be higher than under panmixia. Contrary to Wright's classical result, within-subpopulation variation does not depend on migration rate ($0 < m \ll 1$), because mutation allows subpopulations to diverge. Migration may be a powerful mechanism of maintaining genetic variance in the whole metapopulation.

There are few studies on adaptive evolution in subdivided populations. Via and Lande (1985) studied mainly within-subpopulation patchiness, but they also included a model where matings take place within each of the two subpopulations. The results of these models are basically the same: with time population(s) evolve to the same optimum genotype in both patches, if there is enough genetic variability, and phenotypic values between the two populations are not perfectly opposite. This model will maintain genetic variability only temporarily by genotype-environment interaction in contrast to previous models of disruptive selection. However, this model assumes that means and variances in the different patches will remain the same, and also that the time to reach the same equilibrium in both populations may be very long. Slatkin (1978) studied a different kind of quantitative genetics model with subdivided populations that provided different selective optima. His model predicts that this system maintains increased additive genetic variability.

VII. Conclusions

The factors relevant for durable resistance among crop plants have naturally been subject to intensive research, research which forms the core of plant/pathogen studies. Experiments on a large enough scale to assess the relative importance of numerous factors on plant/pathogen host interactions are extremely expensive, or even impossible, and thus theoretical models are the next-best alternatives. A good model is comparable to solid reasoning: undoubtedly the results from a properly analyzed model are true. However, it is very difficult to understand the meaning and consequences of explicit and implicit assumptions of such models.

Our paradigms of plant/pathogen systems come from agricultural self-fertile grains and their foliar pathogens. We tend to reuse the assumptions made originally for these models, for models of entirely different kinds of systems. It seems unlikely, for example, that a haploid or an asexual model would serve for cross-pollinated diploid or polyploid species.

Some central concepts of plant pathogen (and other) interactions may also limit our imagination. For example, the models include the one-dimensional characters resistance and virulence as the only determinants of plant and pathogen fitnesses. It is extremely unlikely that these characters act independently of other characters related to physiology, life history, or morphology of the species; thus, genetic correlation between characters should be taken into account. We should either broaden the meaning of resistance and virulence to multidimensional phenotypes, or try to deal directly with the fitnesses as such.

No single model will answer the various questions that are relevant for various plant/pathogen systems. For example, in many self-fertilizing cultivated plants, it is natural to ignore plant genetic variation. In modern agriculture, cultivated species are not allowed to evolve with their pathogens. Single farms do not save their own seed stocks, as was common in the time of landraces. Here, coevolution takes place only between efforts of plant breeders and pathogens in a wider area.

Studies of plant/pathogen interactions in natural populations are amazingly few compared to those in managed populations. It is generally believed that natural plant/pathogen populations are more stable than those in managed systems. Although epidemics are common in intensive agriculture, resistance has lasted for a long time in some cases (Lamberti et al. 1983), and many traditional crops and landraces have been grown for hundreds of years without any major pest or pathogen problem (Ball 1983; Van der Graaff 1983). Very commonly, pests and pathogens are not regarded as pests in their native ecosystems (Pimentel 1988; Anagnostakis 1992), or they are minor diseases (Eskes 1983; Van der Graaff 1983).

Unfortunately, there are so many differences between managed and natural systems that we still have very little understanding of the factors that make natural systems more stable. There are, however, some promising candidate factors that could be studied theoretically as well as empirically in more detail.

Ecological factors promoting stability and coexistence of plants and pathogens in natural populations are best known. Studies comparing cultivar mixes to monocultures have shown that ecological factors such as distance between susceptible host plants promote stability. Natural populations are more likely to meet these conditions. Another important ecological research direction, the metapopulation approach, has found conditions for persistence of hosts and pathogens. It will be interesting to learn the effect of including genetic variability in these models.

Most epidemics are caused by introduced pests and pathogens, and epidemics are rare between endemic plants and their pathogens (Hokkanen and Pimentel 1983). Plant/pathogen systems need not evolve towards stability (Frank 1992; May and Anderson 1983), but even new theoretical evidence (Lenski and May 1994) suggests that virulence decreases with time when e.g., plants and pathogens are allowed to evolve together. Anagnostakis (1992) provides some evidence for this in plant/pathogen systems as well. Jarosz and Davelos (1995) claim that only pathogens causing systemic plant diseases have potential for evolution towards decreasing virulence. The mechanism of coevolution towards less virulent types is unclear in all systems (Bull 1994), but in addition to directional selection on plant and pathogen characters, selection may act on their interaction as well. This might cause tolerance (Clarke 1986), plasticity (Via 1994), or genotype × environment interactions (Falconer 1981) between the species. It would be very interesting to include such factors in plant-pathogen models.

One of the main differences between natural and intensely managed plant-pathogen systems is found in the amount of genetic variability (Barrett 1981). Plant breeders consciously seek single-locus resistance, which is easy to backcross to existing lines without the need to maintain high population numbers. However, all studied plant/pathogen systems seem to show quantitative variation as well. Besides the level of variance, the genetic basis of variation may turn out to be very important. High mutation rates for quantitative characters and recombination in sexual diploid species may maintain quantitative genetic variation (Lande 1976) at levels that allow response to many kinds of selection pressures.

Besides broadening out understanding of the factors influencing host/pathogen coexistence in the wild, modeling may help us find new ways of breeding more durable resistance in agricultural plants as well.

Acknowledgments. I am grateful to Seppo Neuvonen, Pekka Vakkari, and George Carroll for support and helpful comments. This study was supported by Academy of Finland grants 7638 and 2990.

References

Alexander HM (1992) Evolution of disease resistance in natural plant populations. In: Fritz RS, Simms EL (eds) Plant resistance to herbivores and pathogens: ecology, evolution and genetics. University of Chicago Press, Chicago, pp 326–344

Anagnostakis SL (1992) Diversity within populations of fungal pathogens on perennial parts of perennial plants. In: Carroll GC, Wicklow DT (eds) The fungal community, 2nd edn. Marcel Dekker, New York, pp 183–192

Anderson JB, Kohn LM, Leslie JF (1992) Genetic mechanisms of fungal adaptation. In: Carroll GC, Wicklow DT (eds) The fungal community, 2nd edn. Marcel Dekker, New York, pp 73–98

Anderson RM, May RM (1981) The population dynamics of microparasites and their invertebrate hosts. Philos Trans R Soc Lond B 291:451–524

Antonovics J (1994) The interplay of numerical and gene-frequency dynamics in host- pathogen systems. In: Real LA (ed) Ecological genetics. Princeton University Press, Princeton, pp 129–145

Antonovics J, Alexander HM (1989) The concept of fitness in plant-fungal pathogen systems. In: Leonard KJ, Fry WE (eds) Plant disease epidemiology, genetics, resistance and management, vol 2. McGraw-Hill, New York, pp 185–214

Antonovics J, Thrall P, Jarosz A, Stratton D (1994) Ecological genetics of metapopulations: the *Silene-Ustilago* plant-pathogen system. In: Real LA (ed) Ecological genetics. Princeton University Press, Princeton, pp 146–170

Ball S (1983) Pathogenic variability of downy mildew (*Sclerospora graminicola*) on pearl millet. I. Host cultivar reactions to infection by different pathogen isolates. Ann Appl Biol 102:257–264

Barrett JA (1981) The evolutionary consequences of monoculture. In: Bishop JA, Cook LM (eds) Genetic consequences of man-made change. Academic Press, London, pp 209–248

Barrett J (1985) The gene-for-gene hypothesis: parable or paradigm. In: Rollison D, Anderson RM (eds) Ecology and genetics of host-parasite interactions. Academic Press, London, pp 215–225

Barrett JA (1988) Frequency-dependent selection in plant-fungal interactions. Philos Trans R Soc Lond B 319: 473–483

Bull JJ (1994) Virulence. Evolution 48:1423–1437

Burdon JJ (1987) Disease and plant population biology. Cambridge University Press, Cambridge

Burdon JJ (1992) Host population subdivision and the genetic structure of natural pathogen populations. Adv Plant Pathol 8:81–94

Burdon JJ, Jarosz AM, Kirby GC (1989) Pattern and patchiness in plant-pathogen interactions – causes and consequences. Annu Rev Ecol Syst 20:119–136

Caten CE (1987) The concept of race in plant pathology. In: Wolfe MS, Caten CE (eds) Populations of plant pathogens: their dynamics and genetics. Blackwell, Oxford, pp 21–38

Clarke B (1976) The ecological genetics of host-parasite relationships. In: Taylor AER, Muller R (eds) Genetic aspects of host-parasite relationships. Blackwell, Oxford, pp 87–103

Clarke DD (1986) Tolerance of parasites and disease in plants and its significance in host-parasite interactions. Adv Plant Pathol 5:161–197

Day PR (1974) Genetics of host-parasite interaction. WH Freeman, San Francisco

De Nooij MP, Van Damme JMM (1988a) Variation in host susceptibility among and within populations of *Plantago lanceolata* L infected by the fungus *Phomopsis subordinaria* (Desm). Evolution 42:1166–1171

De Nooij MP, Van Damme JMM (1988b) Variation in pathogenicity among and within populations of the fungus *Phomopsis subordinaria* infecting *Plantago lanceolata*. Evolution 42:1166–1171

De Wit PJGM (1992) Molecular characterization of gene-for-gene systems in plant-fungus interactions and the application of avirulence genes in control of plant pathogens. Annu Rev Phytopathol 30:391–418

Eskes AB (1983) Incomplete resistance to coffee leaf rust. In: Lamberti F, Waller JM, Van der Graaff NA (eds) Durable resistance in crops. Plenum Press, New York, pp 291–316

Falconer DS (1981) Introduction to quantitative genetics, 2nd edn. Longman, London

Flor HH (1942) Inheritance of pathogenicity in *Melampsora lini*. Phytopathology 32:653–659

Flor HH (1956) The complementary genic system in flaxx and flax rust. Adv Genet 8:29–54

Frank SA (1991) Ecological and genetic models of host-pathogen coevolution. Heredity 67:73–83

Frank SA (1992) Models of plant-pathogen coevolution. Trends Genet 8:213–219

Frank SA (1993) Coevolutionary genetics of plants and pathogens. Evol Ecol 7:45–75

Frank SA (1994) Coevolutionary genetics of hosts and parasites with quantitative inheritance. Evol Ecol 8:74–94

Gabriel DW, Rolfe BG (1990) Working models of specific recognition in plant-microbe interactions. Annu Rev Phytopathol 28:365–391

Geiger HH, Heun M (1989) Genetics of quantitative resistance to fungal diseases. Annu Rev Phytopathol 27: 311–341

Getz WM, Pickering J (1983) Epidemic models: thresholds and population regulation. Am Nat 121:892–898

Gillespie JH (1975) Natural selection for resistance to epidemics. Ecology 56:493–495

Gilpin M, Hanski I (1991) Metapopulation dynamics: empirical and theoretical investigations. Academic Press, New York

Gotelli JN (1991) Metapopulation models: The rescue effect, the propagule rain, and the core-satellite hypothesis. Am Nat 38:768–776

Groth JV, Christ BJ (1992) Diversity in plant pathogenic fungi as influenced by host characteristics. In: Fritz RS, Simms EL (eds) Plant resistance to herbivores and pathogens: ecology, evolution and genetics. University of Chicago Press, Chicago, pp 172–191

Haldane JBS (1949) Disease and evolution. Ric Sci 19 Suppl 68–76

Hokkanen H, Pimentel D (1983) New approach for selecting biological control agents. Can Entomol 116:1109–1121

Holliday P (1989) A dictionary of plant pathology. Cambridge University Press, Cambridge

Janzen DH (1980) When is it coevolution? Evolution 34:611–612

Jarosz AM, Davelos AL (1995) Effects of disease in wild populations and the evolution of pathogen aggressiveness. New Phytol 129:371–387

Kiyosawa S (1982) Genetics and epidemiological modeling of breakdown of plant disease resistance. Annu Rev Phytopathol 20:93–117

Lamberti F, Waller JM, Van Der Graaff NA (1983) Durable resistance in crops. Plenum Press, New York

Lande R (1976) The maintenance of genetic variability by mutation in a polygenic character with linked loci. Genet Res 26:221–235

Lande R (1992) Neutral theory of quantitative genetic variance in an island model with local extinction and colonization. Evolution 46:381–389

Lande R, Arnold J (1983) The measurement of selection on correlated characters. Evolution 37:1210–1226

Lenski RE, May RM (1994) The evolution of virulence in parasites and pathogens: reconciliation between two competing hypotheses. J Theor Biol 169:253–265

Leonard KJ (1977) Selection pressure and plant pathogens. Ann NY Acad Sci 287:207–222

Leonard KJ, Czochor RJ (1980) The theory of genetic interactions among populations of plants and their pathogens. Annu Rev Phytopathol 18:237–258

Levin SA (1972) A mathematical analysis of the genetic feedback mechanism. Am Nat 106:145–164

Levin SA (1983) Some approaches to the modeling of coevolutionary interactions. In: Nitecki MH (ed) Coevolution. University of Chicago Press, Chicago, pp 21–65

Levin SA, Pimentel D (1981) Selection of intermediate rates in parasite-host systems. Am Nat 117:308–315

Levin S, Udovic JD (1977) A mathematical model of evolving populations. Am Nat 111:657–675

Marshall DR (1989) Modeling the effect of multiline varieties on the population genetics of plant pathogens. In: Leonard KJ, Fry WE (eds) Plant disease epidemiology, genetics, resistance and management, vol 2. McGraw-Hill, New York, pp 284–318

May RM (1974) Stability and complexity in model ecosystems, 2nd edn. Princeton University Press, Princeton, 265 pp

May RM, Anderson RM (1983) Parasite-host coevolution. In: Futuyma DJ, Slatkin M (eds) Coevolution. Sinauer, Sunderland, pp 186–206

McDonald BA, McDermott JM (1993) Population genetics of plant pathogenic fungi. BioScience 43:311–319

Mode CJ (1958) A mathematical model for the coevolution of obligate parasites and their hosts. Evolution 12:158–165

Mode CJ (1961) A generalized model of a host-pathogen system. Biometrics 17:386–404

Murdoch WW, Briggs RM, Nisbet W, Gurney WSC, Stewart-Oaten A (1992) Aggregation and stability in metapopulation models. Am Nat 140:41–58

Parker MA (1992) Disease and plant population genetic structure. In: Fritz RS, Simms EL (eds) Plant resistance to herbivores and pathogens: ecology, evolution and genetics. University of Chicago Press, Chicago, pp 345–362

Parlevliet JE (1989) Identification and evaluation of quantitative resistance. In: Leonard KJ, Fry WE (eds) Plant disease epidemiology, vol 2. Genetics, resistance and management. McGraw-Hill, New York, pp 215–248

Parlevliet JE, Zadoks JC (1977) The integrated concept of disease resistance; a new view including horizontal and vertical resistance in plants. Euphytica 26:5–21

Pimentel D (1968) Population regulation and genetic feedback. Science 159:1432–1437

Pimentel D (1988) Herbivore population feeding pressure on plant hosts: feedback evolution and host conservation. Oikos 53:289–302

Reeve JD (1988) Environmental variability, migration, and persistence in host-parasitoid systems. Am Nat 132: 810–836

Rosenheim JA, Tabashnik BE (1991) Influence of generation time on the rate of response to selection. Am Nat 137:527–541

Roughgarden J (1979) The theory of population genetics and evolutionary biology: an introduction. Macmillan, New York

Saloniemi I (1993a) A coevolutionary predator-prey model with quantitative characters. Am Nat 141:880–896

Saloniemi I (1993b) Coevolution in predator-prey systems and plant-pathogen interactions. PhD Thesis, University of Oregon, Eugene

Schaffer WM, Rosenzweig ML (1978) Homage to the red queen. I. Coevolution of predators and their victims. Theor Popul Biol 14:135–157

Seger J (1988) Dynamics of some simple host-parasite models with more than two genotypes in each species. Philos Trans R Soc Lond B 319:541–555

Seger J (1992) Evolution of exploiter-victim relationships. In: Crawley MJ (ed) Natural enemies, the population biology of predators, parasites and diseases. Blackwell, London

Shaner G, Stromberg EL, Lacy GH, Barker KR, Pirone TP (1992) Nomenclature and concepts of pathogenicity and virulence. Annu Rev Phytopathol 30:47–66

Sidhu GS (1988) Genetics of plant pathogenic fungi. Advances in plant pathology, vol 6. Academic Press, London

Simms EL (1992) Cost of plant resistance to herbivory. In: Fritz RS, Simms EL (eds) Plant resistance to herbivores and pathogens: ecology, evolution and genetics. University of Chicago Press, Chicago, pp 392–425

Simms EL (1996) The evolutionary genetics of plant-pathogen systems. BioScience 46:136–145

Simms EL, Triplett J (1994) Costs and benefits of plant response to disease: resistance and tolerance. Evolution 48:1973–1985

Slatkin M (1977) Gene flow and genetic drift in a species subject to frequent local extinctions. Theor Popul Biol 12:253–262

Slatkin M (1978) Spatial patterns in the distribution of polygenic characters. J Theor Biol 70:213–228

Taylor AD (1988) Parasitoid competition and dynamics of host-parasitoid models. Am Nat 132:417–436

Thompson JN, Burdon JJ (1992) Gene-for-gene coevolution between plants and parasites. Nature 360:121–125

Thrall PH, Jarosz AM (1994) Host-pathogen dynamics in experimental populations of *Silene alba* and *Ustilago violacea*. II. Experimental tests of theoretical models. J Ecol 82:561–570

Thrall PH, Antonovics J, Hall DW (1993) Host and pathogen coexistence in sexually transmitted and vector-borne diseases characterized by frequency-dependent disease transmission. Am Nat 142:543–552

Van der Graaff NA (1983) Resistance in perennial crops. In: Lamberti F, Waller JM, Van der Graaff NA (eds) Durable resistance in crops. Plenum Press, New York, pp 263–276

Van der Plank JE (1968) Disease resistance in plants. Academic Press, New York

Via S (1994) The evolution of phenotypic plasticity: what do we really know? In: Real LA (ed) Ecological genetics. Princeton University Press, Princeton, pp 35–57

Via S, Lande R (1985) Genotype-environment interaction and the evolution of phenotypic plasticity. Evolution 39:505–523

Wade MJ, McCauley (1988) Extinction and recolonization: their effect on the genetic differentiation of local populations. Evolution 42:995–1005

Wilhoit LR (1992) Evolution of herbivore virulence to plant resistance: influence of variety mixtures. In: Fritz RS, Simms EL (eds) Plant resistance to herbivores and pathogens: Ecology, evolution and genetics. University of Chicago Press, Chicago, pp 91–139

Wright S (1931) Evolution in Mendelian populations. Genetics 16:97–169

Wright S (1940) Breeding structure of populations in relation to speciation. Am Nat 74:232–248

Wright S (1977) Evolution and the genetics of populations, vol 3. Experimental results and evolutionary deductions. University of Chicago Press, Chicago

16 Evolution of Plant Resistance and Susceptibility to Fungal Parasites

M.C. Heath

CONTENTS

I. Introduction ... 257
II. An Overview of Host-Parasite Specificity in Extant Plants and Fungi ... 257
A. Nature and Genetic Basis of Resistance Mechanisms ... 257
B. Intraspecific Variability in Resistance Mechanisms ... 259
C. Fungus Specificity, Parasite Specificity, and Parasite Selectivity of Resistance and Resistance Mechanisms ... 259
D. The Concept of Basic Resistance ... 259
E. Nature and Genetic Basis of Pathogenicity Factors in Extant Fungi ... 260
F. The Host Specificity of Pathogenicity Factors ... 261
G. The "Matching Factors" Concept of Host Specificity ... 261
H. The Concept of Basic Compatibility ... 261
I. Gene-for-Gene Interactions ... 262
1. Lack of a Universal Mechanism in Gene-for-Gene Interactions ... 262
2. Prevalence of Gene-for-Gene Interactions ... 263
III. Evolution of Resistance and Susceptibility in Vascular Plants ... 264
A. Role of Parasite Selective Pressure in the Evolution of Basic Resistance ... 264
B. Evolution of Basic Compatibility ... 265
C. Evolution of Resistance Within Plant Species ... 266
D. Coevolutionary Arms Races Between Interacting Parasite and Plant Populations ... 268
E. The Evolutionary Paths of Plant-Parasite Interactions ... 269
1. Long-Term Changes in Host Specificity ... 270
2. Evolutionary Changes in the Form of Plant-Parasite Relationships ... 270
F. Mutualism and the Enigma of Widespread Susceptibility to Vesicular-Arbuscular Mycorrhizal (VAM) Fungi ... 272
IV. Conclusions ... 272
References ... 273

Botany Department, University of Toronto, Toronto, Ontario, Canada M5S 1A1

I. Introduction

Like other parasites that have intimate associations with their hosts, fungal parasites of plants show host specificity. Although the complete host range of a single fungal parasite is rarely known for certain (Heath 1985), it is generally assumed that any one species can successfully attack relatively few of the many species of extant flowering plants, and may even be restricted to only certain genotypes within otherwise susceptible species. Resistance to parasitism, therefore, is the norm and, as a corollary, susceptibility to fungal infection is the more unusual condition. This chapter discusses the possible ways in which both of these conditions may evolve, but first, it reviews our current knowledge of the bases of resistance and susceptibility, since it is my contention that the mechanisms that determine these conditions impose constraints upon their past and future evolution (Heath 1991b). No attempt has been made to separate data obtained from studies of true fungi and oomycota, even though the latter are no longer considered to be closely related to the former. The considerable similarity in the ways in which parasitic members of both of these groups interact with higher plants provides no reason to believe that the evolutionary principles derived from one group will differ from those derived from the other.

II. An Overview of Host-Parasite Specificity in Extant Plants and Fungi

A. Nature and Genetic Basis of Resistance Mechanisms

Although there are very few situations in which the mechanism of resistance towards a given fungus has been unequivocally proven, it is unargu-

The Mycota V Part B
Plant Relationships
Carroll/Tudzynski (Eds.)
© Springer-Verlag Berlin Heidelberg 1997

able that every flowering plant has a battery of potential antifungal defenses (Table 1). Many of these are constitutive and passive, such as features of the plant's life history that may favor disease avoidance (Alexander 1990), structural features that prevent fungal ingress into the tissue (Parker 1988), or the presence of antifungal compounds (Esele et al. 1993; Bennett and Wallsgrove 1994). In addition, many, if not all, plants contain at least one constitutive enzyme capable of degrading some fungal phytotoxins (Meeley et al. 1992), and it is possible that certain plants owe their resistance to particular fungi because of the lack of specific molecules that may act as (1) receptors for fungal products (Chap. 5, Vol. V, Part A; Chap. 1, this Vol.), (2) signals of activities essential for pathogenesis (Chap. 5, Vol. V, Part A), or (3) essential nutrients (Ride 1985). Other potential defenses are inducible by attempted fungal invasion, such as the production of active oxygen species (which also may act as elicitors of additional defenses; Levine et al. 1994; Chap. 5, Vol. V, Part A), the accumulation of antimicrobial phytoalexins (Dixon and Lamb 1990; Smith 1996; Chap. 6, Vol. V, Part A), structural changes in the plant cell wall (Bradley et al. 1992), the synthesis of "defense" or "pathogenesis-related" proteins such as hydrolytic enzymes capable of degrading fungal walls (Dixon and Lamb 1990; Chap. 6, Vol. V, Part A), and the death of plant cells. The latter, in addition to directly inhibiting biotrophic fungi that require living cells for their nutrition, may trigger additional defenses in adjacent cells (Mansfield 1984). The importance of these inducible defenses is illustrated by the facts that plant resistance commonly is reduced by treatments that interfere with metabolism and that many fungal saprophytes and parasites will grow rapidly through nonhost tissue that is dead but structurally intact, even though they cannot do so when the tissue is alive (Fernandez and Heath 1989). Although many potential defenses may be common to many, or all, plants, various anatomical features or specific toxic chemicals are likely to be species-specific (Heath 1991b). Moreover, to account for the host species specificity of fungal parasites, and for the fact that fungi with hosts in common do not necessarily share the same host range, some aspect or combination of the defenses of each plant species must be unique.

Table 1. Some of the potential defense mechanisms that comprise the basic resistance of extant plants, and pathogenicity factors demonstrated to exist in some fungal parasites

Plant defense mechanisms	Fungal pathogenicity factors
Constitutive	Adhesives
Life history characteristics	Infection structures
Structural and developmental features	Cell wall-degrading enzymes
Toxic chemicals	Factors to aid nutrient acquisition
Enzymes that degrade fungal pathogenicity factors such as phytotoxins	Enzymes that detoxify phytoalexins and other defense compounds
Lack of fungus-required:	Phytotoxins
receptors	Suppressors of defense responses
signals	
nutrients	Lack of defense inducers
Induced	
Active oxygen species	
Phytoalexins and other toxic nonproteinaceous molecules	
Cell wall modifications	
Defense-related enzymes and other proteins	
Cell death	

It is important to remember, however, that **the defensive potential, and effectiveness, of any plant feature depends on the attributes and activities of the potential pathogen**. The epidermal cell wall, for example, is a defense only against "wound pathogens" that lack the ability to breach this wall when intact. The absence of the correct topographical signal required by rust fungi to form infection structures (Allen et al. 1991; Chaps. 2, 3, Vol. V, Part A; Chap. 5, this Vol.) is irrelevant to a fungus that does not require such a signal. Therefore, the specific features that account for resistance of a single plant to one fungus may not be exactly the same as those that protect the same plant against another.

Many potential resistance mechanisms are complex traits involving morphological features and/or complex biochemical pathways that are likely to be polygenically controlled. For constitutive defenses, the genes involved have only rarely been identified (e.g., Esele et al. 1993). In contrast, many of the genes that produce inducible "defense proteins", such as pathogenesis-related proteins or enzymes involved in phytoalexin synthesis, have been well studied and their expression commonly is elicited by a variety of fungi, fungal and plant products (elicitors), or other stresses (Dixon and Lamb 1990; Smith 1996; Chap. 6, Vol. V, Part A). Several belong to gene families, other members of which control processes that have roles to play in the plant other than defense, and the pattern and levels of gene expression may vary with the mode of elicitation (Dixon and Lamb 1990). In addition to these nonspecific modes of elicitation, it has been suggested that inducible defenses also may be triggered in a parasite-specific manner as discussed later.

B. Intraspecific Variability in Resistance Mechanisms

Evidence suggests that individual plants within a single species may differ quantitatively and qualitatively in their potential resistance mechanisms against fungal invaders. For example, the expression of defense genes induced by fungal products may vary with plant genotype (Ellis et al. 1989), although the significance of this in plant-parasite interactions is unknown. However, in sorghum grain, resistance to several fungi has been shown to be affected by levels of phenolic materials in the caryopsis that are controlled by the additive effects of several dominant genes (Esele et al. 1993). Similarly, the existence of quantitatively inherited "field resistance" in wild plants (Crute 1990) also suggests that individuals may differ qualitatively and quantitatively in defense expression. Nevertheless, polymorphisms in defense mechanisms may not necessarily affect overall resistance, as exemplified by the fact that there can be intraspecific variation in expression of specific defenses against rust fungi in nonhost plant species (Elmhirst and Heath 1989; M.C. Heath, unpubl.).

C. Fungus Specificity, Parasite Specificity, and Parasite Selectivity of Resistance and Resistance Mechanisms

Because of differences between fungi in life history and other attributes, resistance mechanisms often are **parasite-selective** (i.e., selectively effective with respect to different fungal parasites), and a few of the mechanisms listed in Table 1 that are not inherently defensive (e.g., lack of a required compound) may be **parasite-specific** if they are effective against only one fungal species. Nevertheless, plant features that have a primarily defensive role generally seem not to be **fungus-specific** (i.e., effective only against fungal parasites). Even for certain pathogenesis-related proteins that can degrade fungal walls, it is still equivocal whether this is their sole function, since some forms of these enzymes are not highly antifungal (Sela-Buurlage et al. 1993). Similarly, inducible toxins (Krischik 1991) and oxidative responses (Bi and Felton 1995) may be involved in defense against herbivores as well as other pathogens, and wall modifications may be nonspecific responses to, and protect against, various forms of damage (Heath 1991b). This nonspecificity towards fungal parasites is even more obvious for many constitutive defenses, such as the epidermal wall and cuticle, which also protect the plant against a variety of biotic and abiotic stresses (Heath 1991b).

Plant resistance to successful fungal invasion, like its underlying mechanisms, is commonly parasite-selective but, unlike the defenses on which it is assumed to be based, resistance expressed by certain plant genotypes within an otherwise susceptible host species is often parasite-specific. This latter resistance may also be **race-** (or **pathotype-**) **specific** when it is effective against only specific genotypes of the parasite. Such specificity is typical of gene-for-gene interactions in which fungus-nonspecific defense mechanisms are thought to be triggered by the recognition of pathotype-specific, fungal features (see Sect. II.I).

D. The Concept of Basic Resistance

The fact that every plant appears to have a battery of constitutive and inducible features that can potentially protect it against fungal invasion, and the fact that most of these defenses are not specific for particular fungi, or for fungi in general, has led to the concept that all plants have non-specific, **basic resistance** against potential fungal parasites (Heath 1987a). It is this resistance that accounts for **nonhost resistance**, i.e., the resistance shown by a plant species towards a fungal species for which it is not considered a host (Heath 1985). Data from studies with rust fungi show that nonhost resistance usually is the result of the sequential effects of several constitutive and inducible defenses (Heath 1985). Comparable data have been more difficult to obtain for other fungal pathogens, but the large number of potential defenses in each plant and the fact that nonhost resistance seems difficult to overcome, judging by the infrequency with which fungal parasites have changed their host range over recorded history, suggests that most examples of nonhost resistance depend on more than one plant defense (Heath

1985). Although there is considerable evidence that inducible defenses can be triggered by a variety of products and activities common to most fungi, it has been suggested from studies with bacterial pathogens that the induction of certain defenses in nonhost plants may depend on the presence of parasite-specific resistance genes that control the specific recognition of each pathogen (Dixon and Lamb 1990). While it is possible that each plant may have relic resistance genes against fungi related to parasites of its ancestors (Heath 1991a; also see Sect. III.B, E), if **all** induced defenses relied on such specific recognition, each plant would have to have a different resistance gene to control the recognition of every fungal parasite for which it is a nonhost. No evidence for this unlikely situation yet exists for plant-fungal interactions, and it has been suggested that even for bacterial pathogens, the evidence on which this hypothesis is based may have other interpretations (Heath 1991a).

E. Nature and Genetic Basis of Pathogenicity Factors in Extant Fungi

It is usually accepted that parasitism represents a specialized ecological condition and that fungal parasites have special attributes, herein called pathogenicity factors, that facilitate the invasion and colonization of living plant tissue. The search for such factors has been rather cursory until recently, but the list (Table 1) now includes such features as adhesives that attach spores (Chap. 1, Vol. V. Part A), germ tubes, and other infection structures to the plant surface (Howard et al. 1991; Kunoh et al. 1991; Chap. 2, Vol. V, Part A), the development of infection structures (Allen et al. 1991; Chaps. 2, 3, Vol. V, Part A), and the secretion of enzymes and other molecules required to degrade plant cell walls and facilitate nutrient acquisition. (Chap. 4, Vol V, Part A).

If one accepts that all plants have a variety of potential defenses against fungal attack, then at least some pathogenicity factors must be involved in specifically negating these defenses. Identified examples include enzymes that detoxify phytoalexins (Miao and VanEtten 1992), phytotoxins that kill or narcotize cells and prevent active defenses (Kohmoto et al. 1989; Chaps. 7, 8, Vol. V, Part A), and suppressors of these defenses (Toyoda et al. 1992; Chap. 1, this Vol.). It also seems likely that some fungi may lack molecules that might otherwise trigger defensive responses in their host plants, and this phenomenon can also be considered a pathogenicity factor. For example, chitin, a fungal wall component that triggers defensive responses in many plants, is absent in walls of young intracellular haustoria of rust fungi, thereby possibly preventing an adverse response of the invaded cell (Harder and Chong 1984). It must be remembered that fungi may have attributes that are essential for successful infection of one host, but not another. Thus, just as effective plant defenses can only be recognized in terms of a particular pathogen, **a fungal attribute is an effective pathogenicity factor only when proven to aid successful infection of a particular host.**

Intuitively, it seems that each fungus must have a number of pathogenicity factors to be a parasite. This assumption is supported by genetic studies of the roles of phytoalexin detoxification and tolerance in fungal virulence (Miao and VanEtten 1992) and the demonstration that transforming a fungal saprophyte with a gene to detoxify the phytoalexin of pea did not turn it into a pea parasite (Schäfer et al. 1989). In contrast, a maize parasite transformed with the same gene was able to grow in pea, presumably because it already had an array of additional, suitable pathogenicity factors (Schäfer et al. 1989). Nevertheless, the differences in the number of potential pathogenicity factors between some saprophytes and parasites may not be large. Although some saprophytes lack features associated with invasiveness (Fernandez and Heath 1989), they may possess other potentially pathogenic attributes, such as the ability to produce cell wall-degrading enzymes. Conversely, some wound pathogens may resemble saprophytes in their lack of pathogenicity factors needed to penetrate intact plant surfaces.

The genetic basis of most pathogenicity factors in unknown, although enzymes and other proteins are likely to be products of single genes. More complex traits, such as those involving parasite morphogenesis, are probably polygenically controlled, although the production of peptide phytotoxins, which involves several enzymatic steps, can be inherited as a single-gene trait if the toxins are synthesized by large, multifunctional, enzymes (Walton and Panaccione 1993). In *Nectria haematococca*, genes for phytoalexin detoxification appear be on an unstable chromosome that carries other genes involved in pathogenicity; the instability of this chromosome means that all genes on the chromosome will be coinherited in crosses with genotypes lacking this chromosome (Miao and VanEtten 1992).

F. The Host Specificity of Pathogenicity Factors

Many pathogenicity factors (Table 1) are not host-specific, and many parasitic fungi will germinate on, adhere to, form infection structures on, and attempt to enter nonhost species (Fernandez and Heath 1989) or even inert surfaces (Howard et al. 1991). Other factors may have limited specificity. For example, germ tubes of rust fungi may efficiently locate and recognize stomata on host and nonhost leaves that have a similar surface topography (Heath 1977). Still other factors are highly plant species- or plant genotype-specific, such as host-selective phytotoxins and the enzymes and suppressors that prevent the accumulation of specific phytoalexins. Presumably, it is these plant-selective factors that play a prime role in the host specificity of the fungus.

G. The "Matching Factors" Concept of Host Specificity

If plants have a battery of potential features to deter fungal parasitism, and fungal parasites have a battery of potential pathogenicity factors, it seems obvious that the susceptibility of a given plant to a given fungus depends on whether the pathogenicity factors of the latter "match" the features of the plant (Fig. 1). Nonhost resistance may be the result of a considerable degree of mismatch between plant and fungal attributes (Fig. 1A), since it often appears to be based on more than one plant defense (see Sect. II.D). Many examples of resistance within an otherwise susceptible host species also can be attributed to such a mismatch, although the number of mismatches may be fewer (Heath 1985, 1994; Fig. 1C).

H. The Concept of Basic Compatibility

It has been suggested that when a parasite has sufficient pathogenicity factors to successfully parasitize a plant **species** (Fig. 1B), it has achieved a **basic compatibility** with the latter (see Heath 1991b). As seen later, this term is conceptually useful in discussions of host-parasite evolution. However, it also is important to realize that resistance mechanisms may not be totally negated in susceptible plants. Recent studies of trangenic tobacco plants with reduced, constitutive, levels of phenylpropanoids show them to have increased susceptibility to the tobacco pathogen, *Cercospora nicotianae*, suggesting that constitutive defenses may have "residual" effects even when a parasite has a basically compatible relationship with its susceptible host (Maher et al. 1994).

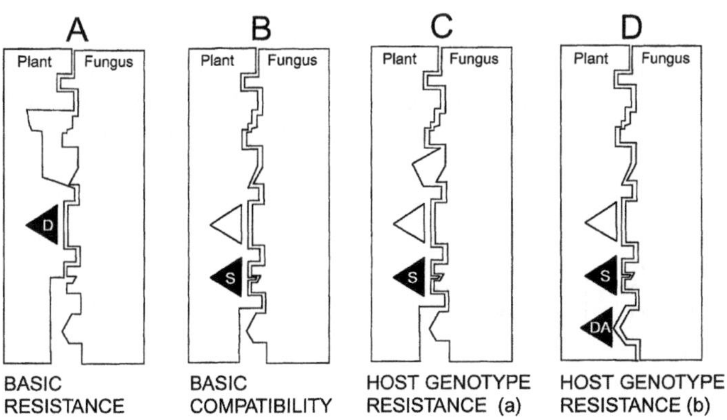

Fig. 1A–D. Diagrammatic representation of the different interactions that may take place between plants and parasitic fungi. **A** Nonhost plants exhibit basic resistance due to the mismatch between plant attributes and fungal pathogenicity factors. Fungal products or activities commonly elicit inducible defenses (*D*). **B** Matching attributes result in a basic compatibility between fungus and plant species, commonly involving the suppression (*S*) or negation of inducible defenses. **C**, **D** Genotypes within a host species may be resistant due to a mismatch of plant and fungal attributes or through the evolution of a parasite-specific recognition system (*DA*) that elicits defense responses

I. Gene-for-Gene Interactions

For some fungal diseases of crop plants, a single plant may be resistant to some genotypes of a given fungal parasite and susceptible to others (i.e., race- or pathotype-specific resistance, see Sect. II.C). Following pioneering studies of the genetic basis of this type of host-parasite specificity in flax rust, Flor (1955) discovered that resistance in the plant was controlled by single resistance genes, each one of which was effective only if the fungus had a "corresponding" gene for avirulence. The current molecular interpretation of such gene-for-gene interactions is that the interaction between the avirulence gene product (elicitor) and the product of the resistance gene (receptor) results in the activation of defense responses in the plant (Figs. 1D, 2; Gabriel and Rolfe 1990; Chap. 1, this Vol.). The less popular alternative explanation is that the interaction between the avirulence and resistance gene products directly inhibit the fungus (Fig. 2), in which case the observed defensive responses of the plant are irrelevant (Ellingboe 1982; Crute and Norwood 1986). Since, in the absence of a gene for resistance, the plant is susceptible to the parasite, one must assume that gene-for-gene interactions are superimposed on the basic compatibility (see Sect. II.H) between the fungus and its host species, and that the original function of the avirulence gene product was not related to evoking plant resistance (see Heath 1991b). The former conclusion is supported by the fact that race-specific resistance to rust fungi is usually expressed later (and involves different defense mechanisms) than nonhost resistance and, therefore, cannot be effective unless the latter has been overcome (Heath 1974, also see refs. in Heath 1991b).

The recent cloning of some fungal avirulence genes (Van den Ackerveken et al. 1992; Rohe et al. 1995; Sweigard et al. 1995 Chap. 1, this Vol.) and plant resistance genes (Lawrence et al. 1995; Staskawicz et al. 1995) suggests that we shall soon have a clearer understanding of the nature and role of their products. Avirulence genes typically are not allelic, suggesting that they are diverse in function, a conclusion supported by the lack of similarity among those characterized so far. In contrast, resistane genes in the plant are often clustered on specific chromosomes and genetic assortment within related clusters may produce new resistance specificities (Richter et al. 1995). The cloned resistance genes against fungal, bacterial, and viral pathogens have significant homology (Lawrence et al. 1995), suggesting that parasite-specific resistance employs similar parasite-recognition systems regardless of the type of organism producing the eliciting signal. The structure of these resistance genes currently suggest that they are involved in signal transduction processes (Staskawicz et al. 1995; Chap. 5, Vol. V, Part A). However, the interactions between avirulence and resistance gene products may be quite complex, since avirulence gene products of *Cladosporium fulvum* are seemingly recognized by both resistant and susceptible host plants, although the ultimate response may differ (Wubber et al. 1986; Chap. 1, this Vol.).

1. Lack of a Universal Mechanism in Gene-for-Gene Interactions

If the popular molecular interpretation is correct for at least some gene-for-gene systems, then these systems depart from the "matching factors" concept of host specificity described in Section

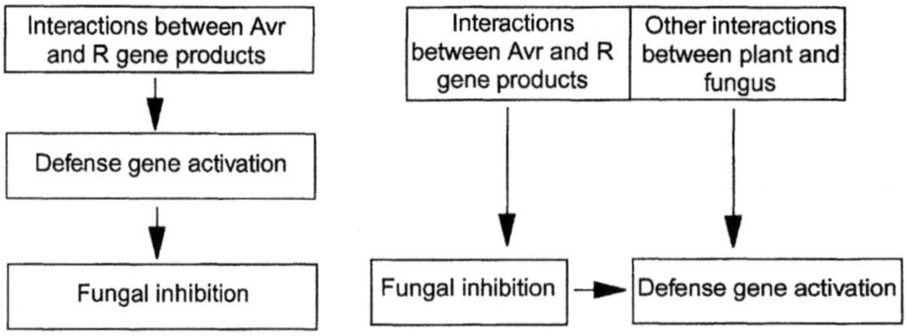

Fig. 2. Diagrammatic representation of the two alternative ways in which interactions between the products of fungal avirulence (*Avr*) genes and plant resistance (*R*) genes can result in the inhibition of fungal growth

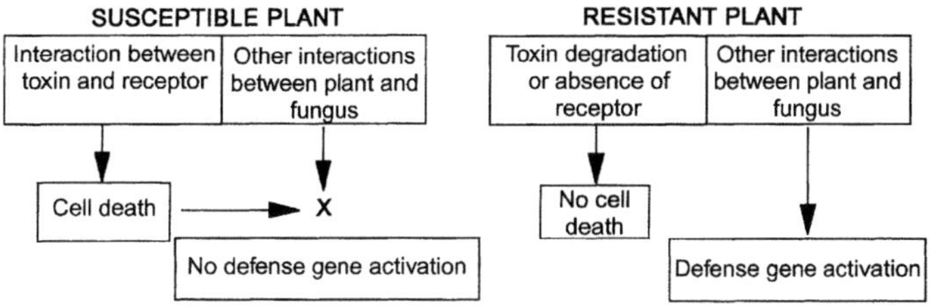

Fig. 3. Diagrammatic representation of the interactions that lead to intraspecific susceptibility or resistance against fungi that rely on the action of host-selective toxins for their host specificity

II.G that suggests that some host, and most nonhost resistance is based on fungus-nonspecific constitutive features, and/or the nonspecific triggering of active defenses, which are not overcome in some way by fungal pathogenicity factors. It is important to realize that the "matching factors" basis of specificity also can produce a one-gene-for-one gene situation. For example, a single-gene-controlled pathogenicity factor such as a phytotoxin may interact with a single-gene-controlled resistance mechanism such as the absence of a receptor or the presence of a toxin-degrading enzyme. In these situations, resistance is not the result of specific recognition coupled to defense activation, but is the consequence of nonspecific induction of defenses by some fungal product or activity, which normally are prevented by the action of the phytotoxin (Fig. 3; Heath 1985, 1994a). It is evident, therefore, that the number of genes involved in a plant-fungus interaction does not define the mechanism of resistance or susceptibility or identity gene-for-gene interactions involving elicitor-receptor interactions. Even within gene-for-gene systems involving biotrophic fungi and crop plants, resistance and avirulence genes show enough variability, in terms of dominance and influence by other genes (Christ et al. 1987; Gates and Loegering 1991) to suggest that they are not mechanistically identical.

2. Prevalence of Gene-for-Gene Interactions

Only about 12 fungal systems involving crop plants have been proven by genetic studies **of both fungus and plant** to involve gene-for-gene interactions similar to those described for flax rust (Christ and Groth 1982; Thompson and Burdon 1992; Van den Ackerveken et al. 1992). In addition, about 20 more have been assumed to be gene-for-gene systems on the basis of single-gene resistance in the plant and the existence of races with different plant genotype specificities in the fungus (Thompson and Burdon 1992). Most proven and assumed gene-for-gene systems involve biotrophic parasites, and it is possible that the molecular intimacy of these parasites and their hosts makes it more likely that the elicitor/receptor type of gene-for-gene interactions evolve (Heath 1994). Since most studies of fungal parasites of wild plants have involved biotrophic fungi that are related to crop pathogens for which gene-for-gene interactions have been demonstrated, the assumed existence of elicitor/receptor gene-for-gene interactions in natural ecosystems (Burdon 1987) may be valid. However, biotrophs often account for less than 10% of the known fungal pathogens of some uncultivated plants (Farr et al. 1989) and the existence and basis of host genotype resistance to the remaining fungal parasites is completely unknown. Such resistance may not involve elicitor/receptor gene-for-gene interactions and could be the result of intraspecific variability in resistance mechanisms (see Sect. II.B) and the resulting mismatch between fungal and plant attributes.

The commonly assumed universality of gene-for-gene interactions among other plant-organism combinations also is becoming more tenuous with increasing scrutiny. Apart from the wheat-Hessian fly system, no other plant-insect combination has been found to show gene-for-gene complementarity (Thompson 1990), and molecular genetics has revealed that in many plant-bacterial pathogen combinations, the presence of a single avirulence gene in the bacterium may be responsible for triggering resistance controlled by more than one plant resistance gene, and vice versa (De Feyter et al. 1993; Whalen et al. 1993). Although, as discussed above, there seems to be some homology between resistance genes active against fungal and bacterial parasites, it still remains to be determined how much mechanistic homology there is between fungal and bacterial gene-for-gene systems.

III. Evolution of Resistance and Susceptibility in Vascular Plants

A. Role of Parasite Selective Pressure in the Evolution of Basic Resistance

The components of basic resistance in extant plants (Table 1) can be divided into three categories: (1) those that are inducible and appear to have a primarily defensive role although they may not be specific for fungal parasites (e.g., phytoalexin accumulation); (2) those that are constitutive and also seem to be primarily involved in broad-spectrum defense (e.g., toxic secondary metabolites); and (3) those that are usually constitutive and whose defensive properties appear to be secondary (even incidental) to other roles in the plant (e.g., the plant cell wall or the surface topography of the leaf). In theory, there are four ways in which any of these components could have evolved (Fig. 4): (1) as the summation of numerous coevolutionary arms races between each lineage of plant and each lineage of fungus with which it came into contact; (2) as a result of diffuse selection (Janzen 1980; Fox 1981) by an array of potentially parasitic and predatory organisms that generate selective pressure as groups; (3) as the incidental result of adaptive changes in the plant that evolved for reasons other than any selective effects of parasites or predators, and (4) as the result of nonadaptive changes in the plant that were established because of genetic or functional (sensu Gould and Lewontin 1979) linkage to other adaptive traits.

All of these processes appear to have been involved in the evolution of basic resistance. Defenses that are parasite-selective and related to anatomical features of the plant or to the lack of particular signals or molecules that certain parasites require, may well be the result of the third or fourth evolutionary paths, although it cannot be ruled out that these traits may be maintained in some extant plants by parasite pressure. For features that are primarily defensive, the nonspecificity of their action and/or induction suggests that they have arisen primarily through diffuse selection in response to the collective selective pressure exerted by a number of parasites, herbivores, and abiotic stresses. Jones and Firn (1991) have suggested that the evolution of chemical defenses may have proceeded independently of predator or parasite adaptation once the fundamental trait of continual production of a diversity of secondary metabolites had been established. Nevertheless, the fact that inducible defenses (e.g., phytoalexin accumulation) can be triggered by a variety of fungal products as well as other stimuli suggests

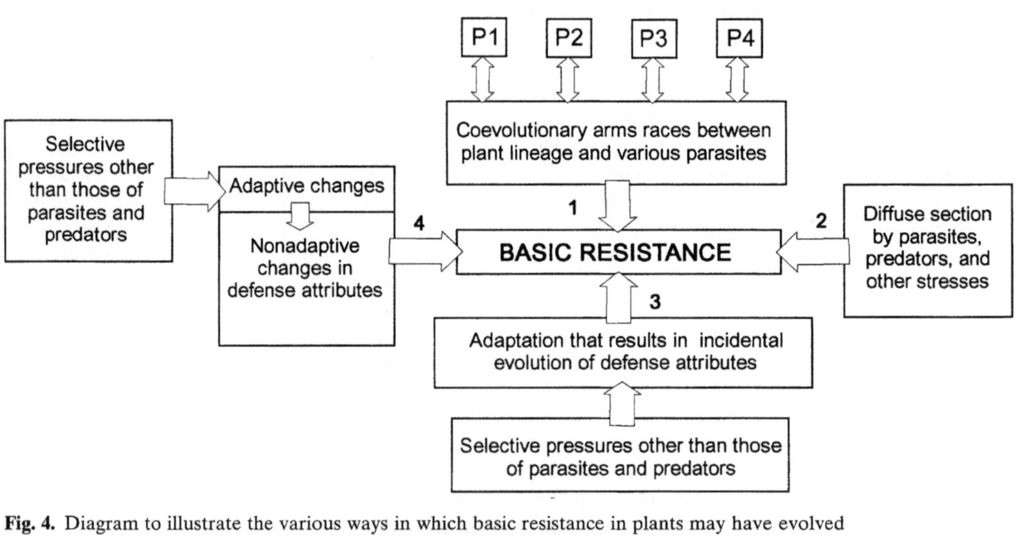

Fig. 4. Diagram to illustrate the various ways in which basic resistance in plants may have evolved

that at some time during the evolution of vascular plants, fungi as a group have had a significant selective influence on these defenses (Heath 1991b). This ability to respond to fungal products may predate the evolution of specific phytoalexins, since the signal system seems to be conserved between plants that produce phytoalexins by different biosynthetic pathways (Stermer et al. 1990). Data also suggest that the signal pathway for wound- and fungal elicitor-induced expression of chitinase (a pathogenesis-related protein) evolved before the divergence of mono- and dicotyledonous plants (Zhu et al. 1993).

Coevolutionary arms races, involving reciprocal adaptations of interacting plant and fungus lineages, have been suggested to explain the origin of parasite-specific gene-for-gene relationships (see Heath 1991b and Sect. III.D), but their role in the evolution of basic resistance is uncertain. To have played a major role in this process, coevolutionary arms races between ancient plants and parasites must have resulted in plant defenses with broad-spectrum affects, rather than parasite-specific recognition systems since, as argued in Section II.D, the latter are unlikely to be involved in most modern examples of nonhost resistance. Nevertheless, there may be some situations in which plant species have evolved to become nonhosts for members of a fungal parasite lineage to which their ancestors were susceptible, and as a result have retained some form of specific recognition for some of the extant fungal parasites (Elmhirst and Heath 1989). The existence of this recognition does not necessarily mean that it is important for the nonhost resistance currently expressed against the fungus (Heath 1991a), and it is unlikely that this situation exists between plant and fungal lineages that have had no contact during their evolutionary history. Currently, the HC-toxin reductase enzyme, whose sole function has been suggested to be to protect maize against the toxin of *Cochliobolus carbonum* and which may be widespread in other plant species (Meeley et al. 1992), is the only known defense that may be maintained in nonhost plants by the selective pressure of a single, or small group, of fungal parasites.

As discussed in detail elsewhere (Heath 1991b), many constitutive and inducible defensive features in extant plants are related to, or deployed in, activities that have functions other than defense against other organisms. These features often can be traced to the very early history of green plants, and it is possible that some of these traits were originally established because of their nondefensive functions and only more recently have defense-specific forms of some enzymes and their corresponding genes evolved. For example, the stepwise evolution of many parts of the phenylpropanoid pathway, which is currently involved in the synthesis of a great diversity of defensive and nondefensive compounds such as phytoalexins, lignin, and flower pigments, may be the result of a multitude of selective pressures, all of which may have varied in relative importance with time (Heath 1991b).

B. Evolution of Basic Compatibility

If one assumes that the first terrestrial fungi were saprophytic (Luttrell 1974; Heath 1987b), then the first terrestrial parasites must have arisen from saprophytes. However, over the last 500 million years, plant parasitic fungi have arisen from a variety of saprophytic or parasitic ancestors (Berbee and Taylor 1993) and in the latter case, their current host range may represent association by descent (Mitter and Brooks 1983) or "jumps" from other host species (Heath 1985).

On the assumption that basic compatibility depends on a fungus having pathogenicity factors to match the features of its host, there are two ways in which a new parasite-plant combination may arise; (1) through the acquisition of new pathogenicity factors by the fungus and (2) by a change in the plant that makes it vulnerable to the potential pathogenicity factors that a particular fungus already has. Emerging molecular data suggests that the second method is particularly common in the case of crop plants and fungi such as *Cochliobolus* species that produce host-selective toxins. These fungi can grow saprophytically, often secrete a variety of potentially toxic compounds, and appear to have sufficient broad-spectrum pathogenicity factors to allow them to invade a variety of senescing plants (Scheffer 1991). They become significant pathogens only when one of their secreted products is sufficiently toxic to a particular plant species to allow the fungus to invade and grow in nonsenescent tissue. For two *Cochliobolus* parasites of maize, toxin-sensitive plant genotypes appear to have arisen fortuitously by nuclear mutation (see Meeley et al. 1992) or by genetic rearrangements in the mitochondrial genome (Levings 1990; Chap. 8, Vol. V, Part A). Whether plant species susceptibility arises by similar means in natural ecosystems is uncertain, however, since the susceptible genotypes of maize were deliberately selected for reasons other than their response to the pathogen. It might be expected that toxin-sensitive individuals would be rapidly eliminated from a wild plant population if they arose in the presence of the parasite and if successful parasite infection resulted in a significant loss of plant fitness. In the absence of the fungus, toxin sensitivity would have to be a pleiotropic trait or be tightly linked with some other factor to account for its establishment and maintenance in the plant species.

In other situations, it appears that basic compatibility has resulted from a series of highly specific adaptations by the parasite to its host species. This is particularly evident with obligately biotrophic fungi such as the rust fungi, that require a series of plant signals for continued development (Heath 1995b) and form a complex metabolic relationship with the living plant cell that seems to require the specific negation of certain active plant defenses which they might otherwise trigger (Heath 1981). Since these fungi cannot grow in the absence of the living plant, it seems likely that the pathogenicity factors of each fungal lineage primarily have evolved as the result of coevolutionary interactions with specific plant lineages and that much (but not all, as discussed later) of the current host range is the result of association by descent.

Association by descent, or even the fortuitous independent evolution of matching fungus and plant attributes, cannot easily explain the host range of all extant fungi. Even ecologically obligate biotrophs such as the rust fungi appear to have made frequent jumps to ecologically associated hosts during their evolution (Baum and Savile 1985). The suggestion that this occurs when a plant taxon is evolutionarily young and "genetically plastic" (Savile 1955) seems unlikely if nonspecific defenses evolved early in the history of green plants and have been consistently maintained in both "young" and "old" taxa by a variety of selective forces. Only when we know how many genetic and phenotypic changes were required for specific jumps will we be able to speculate knowledgeably on how they evolved.

It is important to realize that in related groups of parasites or plants, the susceptibility of a plant species to a particular parasite may not be controlled solely by fungal pathogenicity factors which can match the nonspecific defensive features of the plant. For example, *Magnaporthe grisea* (Valent and Chumley 1987; Sweigard et al. 1995; Chap. 3, this Vol.), and some powdery mildew fungi (Tosa 1992; Chap. 4, this Vol.) exist as intraspecific pathotypes in which single genes can control which of a related group of host species are successfully attacked. For the powdery mildew fungi, it has been suggested than once basic compatibility was achieved with an ancestral host, plant speciation was accompanied by the evolution of different parasite-specific resistance genes in different plant species, each triggered by the product of a different fungal gene. The different fungal pathotypes (forma speciales) then arose as coevolutionary adaptations to the presence of these genes (Fig. 5; Tosa 1992). In this system, therefore, gene-for-gene interactions, superimposed on basic compatibility, control both race-cultivar and forma specialis-host species specificity (Tosa 1992). However, there are other groups of related fungal parasites in which the ability of different pathotypes to be pathogens of different plant species seems based more on the "matching factors" principle of basic compatibility than the ability to negate parasite-specific resistance genes. In these systems, the correlation between host range and intersterility groupings in the fungi are most likely explained by the need to maintain the more complex gene combinations required for host species specialization (Brasier 1987; Otrosina et al. 1993).

C. Evolution of Resistance Within Plant Species

While there is considerable controversy over the degree to which parasites shape the character of plant communities (Harper 1990), it is commonly assumed that they alter the genetic structure of host populations by selecting for more resistant phenotypes. However, although intraspecific variation in susceptibility to a given fungal parasite is common (Burdon 1987), this does not necessarily imply that resistance in the population is established or maintained by parasite pressure.

Once a parasite species has achieved a basic compatibility with its host species, intraspecific variation in plant resistance or susceptibility theoretically may arise in one of at least two ways (Heath 1987a, 1991b). The first is as a result of the polymorphism (see Sect. II.B) in some components of basic resistance within the plant species that allows the pathogenicity factors of the fungus

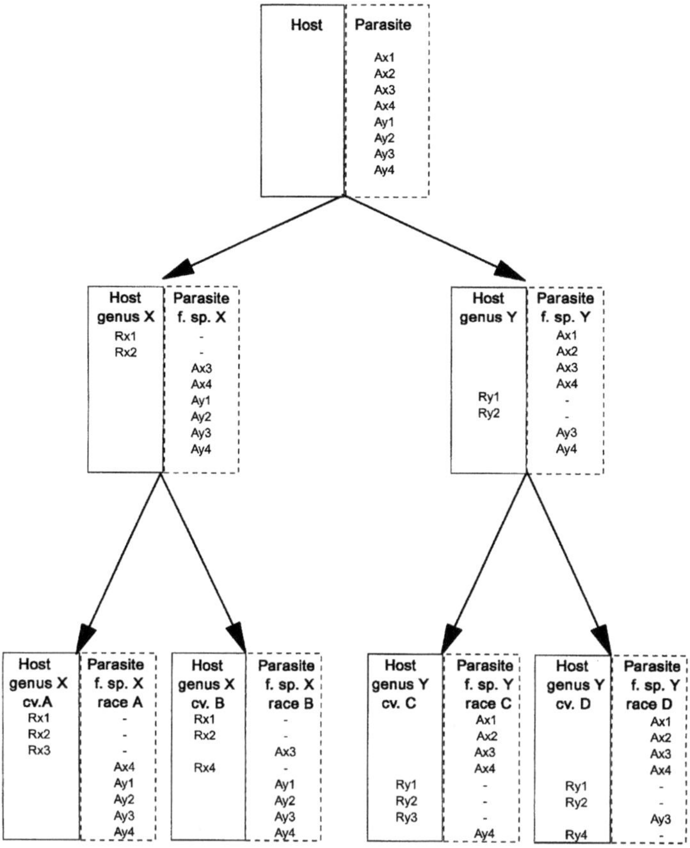

Fig. 5. Hypothetical model of the evolution of formae speciales and races of some powdery mildew fungi. *Rx* and *Ry* are resistance genes that arose in plant genera *X* and *Y*, respectively. *Ax* and *Ay* are avirulence genes corresponding to Rx and Ry, respectively. (Tosa 1992)

to be "matched" in some individuals but not others. The resistance that results from any "mismatch" between plant and fungal attributes may vary in degree, depending on the number of defensive attributes involved and their roles in preventing fungal growth. Such mismatches may account for quantitatively inherited "field resistance" such as that associated with low infection efficiency and small lesion size (Crute 1990), as well as more compete resistance.

In theory, there are no restrictions on the components of basic resistance that may be polymorphic, or the number of genes that control them. However, polymorphisms in the relative proportions and levels of potentially toxic chemicals may be particularly common since alterations in enzyme levels can have dramatic effects on the nature and quantity of these chemicals through changes in the partitioning of intermediates between branch pathways (e.g., Vögeli and Chappell 1988). Given the multiple roles of many components of basic resistance and the possibility that predators and parasites have different and opposing selective effects (Simms and Rausher 1993), polymorphisms in basic resistance mechanisms may be maintained by opposing selective pressures of a variety of factors of which a single fungal parasite may be a minor component. Therefore, a single fungal parasite may not exert sufficient selective pressure to increase the frequency of resistant individuals in the plant population, depending on the frequency of plant infection, the effect of infection on plant survival and fecundity, and to what degree resistant plants are selected against by other selective forces.

The second way of generating resistant plant genotypes within a host species is by their evolution from previously susceptible ancestors. Unless resistance is linked to some other adaptive feature, the establishment and maintenance of this resistance might be expected to depend heavily on the selective pressure exerted by a specific parasite. Such resistance is likely to be parasite-specific and controlled by relatively few genes, since mutation is likely to generate such defenses faster than those that are polygenically controlled (Heath 1987a). Therefore, this mode of evolution of resistance is the one most likely to initiate elicitor/receptor gene-for-gene interactions. If the fungus can mutate so that it can overcome the new form of resistance, the scene also is set for a coevolutionary arms race between fungus and plant.

D. Coevolutionary Arms Races Between Interacting Parasite and Plant Populations

The often rapid evolution of new fungal races following the introduction of new resistant crop cultivars has led to the common assumption that the latter have a strong selective effect on the parasite population. This conclusion is further supported by data suggesting that identical virulences (i.e., ability to overcome specific forms of resistance in the host population) may arise from different genetic lineages within a fungal species (Goodwin et al. 1992; Elias et al. 1993; Levy et al. 1993). These observations are commonly extrapolated to imply that once a fungal parasite has achieved a basic compatibility with its host species, reciprocal selective pressures in natural ecosystems will result in an arms race between the two organisms, with new forms of resistance arising in the plant and new forms of virulence arising in the fungus to overcome each form of resistance. Gene-for-gene coevolution is a specific version of this arms race in which resistance and virulence are controlled by single genes. However, despite the widespread acceptance of the existence of coevolutionary arms races in natural ecosystems (Thompson 1990), it is still equivocal whether, and to what degree, reciprocal selection between plant and parasite occurs.

Considerable polymorphisms for virulence and resistance have been found in several, but not all (Crute 1990; Heath 1992), studies of biotrophic fungal parasites of wild plants (Burdon 1987; Clarke et al. 1990; Burdon and Jarosz 1992). If fungal polymorphism in virulence is balanced rather than transient, as has been suggested for some pathogens (Burdon and Jarosz 1992), there must be selection against unneeded virulence genes (Vanderplank 1968; Leonard and Czochor 1980). However, for known or assumed gene-for-gene systems involving crop plants, changes in virulence gene frequencies may depend both on direct selection by matching host resistance genes and indirect selection associated with gametic disequilibrium (Hovmoller et al. 1993). Moreover, some populations of fungal parasites have large amounts of "unnecessary" virulences (Goodwin et al. 1992; Andrivon 1994), and individual virulences may or may not be associated with fitness (Kolmer 1993). Therefore, selection for virulence to resistance genes in crop plants may be balanced by other selective forces in the parasite population (Kolmer 1993). It also has been suggested (Jarosz and Burdon 1991; Burdon and Jarosz 1992) that the lack of close correspondence between the race composition of *Melampsora lini* and resistance genes in natural populations of *Linum marginale* is influenced by genetic drift and migration in fungal populations, leading to the conclusion that the evolution of the pathogen may be best studied at the level of the metapopulation rather than individual demes (Burdon 1993). Overall, therefore, data suggest that the frequency of specific pathotypes (races) in the fungal population can be influenced by factors other than the presence of resistant host genotypes.

One of the more intriguing hypotheses concerning the role of parasites in natural ecosystems is that sexual reproduction in plants is an adaptation to parasite pressure that, by promoting the rapid recombination of defenses, allows resistance to keep pace with rapid changes in virulence in the parasite (Hamilton et al. 1990). However, there are data indicating that such virulence changes do not always take place even in the face of host resistance (Parker 1991; also see Sect. III.E), and this hypothesis requires a strong selective advantage for resistance to certain parasites. The degree to which fungal parasites select and maintain pathotype-specific resistance genes in the plant, what type of advantages these genes confer, and the fitness cost of resistance, are controversial. Studies of the fitness cost of resistance are rare and, in crop plants, have yielded variable results (Burdon 1987). Nevertheless, the fact that resistance genes often are more frequent in plant populations growing in areas favorable for disease development (Burdon 1987) is generally interpreted as evidence of parasite-driven selection. However, pathogen attack does not always promote genetic diversity among hosts (Parker 1989) and it also has been argued that inbred species may accumulate resistance genes that are largely ineffective against the prevalent races of the parasite (Jarosz and Burdon 1990), or may show parasite-unrelated changes in resistance frequency due to genetic hitchhiking with alleles influencing other traits that are subject to natural selection (Parker 1991). There are also indications that plants may have resistance genes against normally unencountered parasites, possibly as relic genes that protected their ancestors against related fungi (Kinlock 1992).

It is significant that several studies of pathotype-specific resistance in natural ecosystems have shown that high numbers or frequencies of pathotype-specific resistance genes in the plant population rarely preclude the presence of the pathogen (Burdon 1993). Therefore, rather than protecting individual plants against disease, the main selective advantage of pathotype-specific resistance may, given the effective dispersal systems of many plant pathogens, lie in the postponement of the reestablishment of pathogen demes following local extinction (Burdon 1993). It also is conceivable that pathotype-specific resistance may be selected for because of the induction of systemic resistance to other parasites elicited by the necrosis that accompanies many forms of this resistance (Martinelli et al. 1993; Heath 1995a). The role of induced resistance in natural ecosystems is still uncertain (Heath 1995a) but in *Ipomoea purpurea*, lines with higher baseline levels of resistance to *Colletotrichum dematium* were more fit than those dependent on induced resistance (Simms and Vision 1995).

From studies of populations of *Senecio vulgaris* and *Erysiphe fischeri*, in which no plant lines were resistant to all local pathotypes of the powdery mildew fungus, and each fungal pathotype could infect a large proportion of plants in each host population, Clarke et al. (1990) conclude that pathotype-specific resistance may be less important in the survival strategy of the plant than tolerance of parasitism that maximizes reproductive output despite infection. Similarly, De Nooij et al. (1995) have shown that populations of *Tussilago farfara* with higher levels of resistance to the rust fungus, *Coleosporium tussilaginis*, may show greater rust-induced reduction in growth than populations with lower levels of resistance. Further studies of inbred plant lines of *S. vulgaris*, however, revealed considerable diversity in resistance phenotypes, such as complete, partial, age-dependent, and temperature-dependent resistance (Bevan et al. 1993). The authors raise previous suggestions that environmentally sensitive resistance genes may result in more stable resistance because of their lack of continuous selection on the pathogen population, and suggest that partial resistance may be more important than pathotype-specific complete resistance in minimizing the effects of disease. They conclude that the defense strategy in this pathosystem is based on a complex mosaic of interacting factors, suggesting that the reciprocal selective pressures exerted by plant and fungus are equally diverse and complex.

Despite this potential complexity of plant-fungal parasite interactions in natural ecosystems, relatively simple mathematical models based on gene-for-gene coevolution can predict some of the features seen in natural and agricultural pathosystems, such as various dynamics of polymorphisms in resistance or virulence genes (May 1990; Frank 1993; Leonard 1994) and, by including multilocus genetics and local extinctions and migration, weak association of frequencies of matching resistance and virulence alleles (Frank 1993). These models also suggest that rather low costs of resistance and virulence are required for coevolution to explain polymorphisms in these traits in natural populations (Frank 1993). Nevertheless, given the data described above, it seems that coevolutionary arms races represent just one part of a complex tapestry of interacting factors that affect plant and parasite populations.

E. The Evolutionary Paths of Plant Parasite Interactions

The current uncertainty of the role of coevolutionary arms races in natural ecosystems is not surprising, since on a relatively short time scale, any particular fungal parasite species may show dramatic fluctuations in population size, genetic diversity, and spatial distribution brought about by environmental factors, life-history traits, as well as interactions with its host (Burdon 1993). To complicate matters further, the effect of a fungus on a susceptible plant may depend on the timing of infection during plant development (Jarosz and Burdon 1992), its trophic relationship with the plant (Shtienberg 1992), whether it castrates, kills, or debilitates the plant (Burdon 1993), whether it is systemic (Burdon 1993), and whether the presence of other organisms reduces (e.g., Duchesne et al. 1989) or enhances infection and/or damage (De Nooij and Paul 1992; Hatcher et al. 1994). Moreover, even susceptible plants may exhibit resistance, apparently as a result of changes in components of nonspecific basic resistance (Heath 1995a), to a normally successful fungal pathogen if they have been subject to previous infection (Kuc 1985) or other treatments (Gottstein and Kuc 1989; Kehlenbeck and Schonbeck 1995). Obviously, the degree of selective pressure exerted on a host by the fungus, and vice versa, may vary considerably in space and time.

The possible evolutionary paths followed by single interacting lineages of plant and fungal parasite are summarized in Fig. 6. I have argued (Heath 1991b) that these paths are governed not just by selective forces but also by the genetic capabilities of the two organisms and the mechanistic bases of their interactions. This argument parallels the broader concept of genostasis in which evolution is limited by lack of **appropriate** genetic variation (Bradshaw 1991). Thus, a fungus will not become a parasite of a particular plant, regardless of how long the two lineages coexist, if it does not have the genetic capability to generate the particular pathogenicity factors dictated by specific features of the plant. Similarly, a plant species may never evolve effective defense mechanisms against fungi that have modes of parasitism that are difficult to negate. Equally, new races or pathotypes of a fungus may never arise if the latter cannot devise a method of overcoming the resistance exhibited by resistant host genotypes. In support of these conclusions, there are data to indicate that some forms of resistance have few or no obvious means of being negated by a particular parasite (Parker 1988) and that the genetic background of a fungal lineage within a parasite species defines the types of new virulences that may evolve within the lineage (Levy et al. 1993). As discussed before (Heath 1985), fungal pathogenicity factors and plant defenses with high molecular specificity, such as new toxins, suppressors of active defenses, or plant receptors, may be more difficult to generate by random mutation than those that represent the absence of a particular product or activity. These constraints on future evolution implied by specific mechanisms of resistance or susceptibility appear to be ignored by many theoretical discourses on host-parasite evolution in which it is assumed that resistance may arise or be overcome with equal ease in all plant-parasite interactions.

1. Long-Term Changes in Host Specificity

In the long term, the parasitic interaction between any given fungus and plant lineage may continue in some form, or it may disappear. In the latter case, either or both partners may become extinct because of a variety of factors not necessarily associated with the plant-parasite interaction, or the entire plant species may become resistant to successful infection. In theory, such resistance could be the result of certain parasite-specific resistance genes becoming fixed in the plant population, coupled with a lack of adaptation to the resulting resistance by the fungus. In this case, plants may have, in addition to nonspecific basic resistance, genes for specific resistance against fungal lineages that once were parasites of the plant's ancestors (Heath 1991a). Alternatively, or in addition, selective forces unrelated to parasitism may cause the evolution of plant features that destroy the basic compatibility between a particular fungal and plant lineage by producing a mismatch between fungal and plant attributes (Heath 1987a, 1991b). Data suggest that within the *Phaseolus-Vigna* complex, extant species that are now nonhosts to certain rust fungi may have followed either of these paths as they evolved from susceptible ancestors (Elmhirst and Heath 1989).

2. Evolutionary Changes in the Form of Plant Parasite Relationships

The existence of differing forms of parasitism within certain fungal taxa (e.g., the Peronosporales and Clavicipitaceae; Chaps. 12, 14, this Vol.) suggests that the type of relationship between plant and fungus may also be subject to evolution. Whether biotrophic parasites evolved from necrotrophs, or vice versa, is a controversial issue (Cooke and Whipps 1987; Heath 1987b), although experimental systems with phage parasites of bacteria suggest that the evolution of more benevolent interactions may be expected in systems with high partner fidelity (Bull et al. 1991). For fungal taxa that have parasitic and mutualistic members, such as the Clavicipitaceae, it is generally assumed that the latter evolved from the former following selection for fungal genotypes that induced few pathogenic symptoms and afforded some fitness advantage to their host (Clay 1988; Chaps. 12, 14,

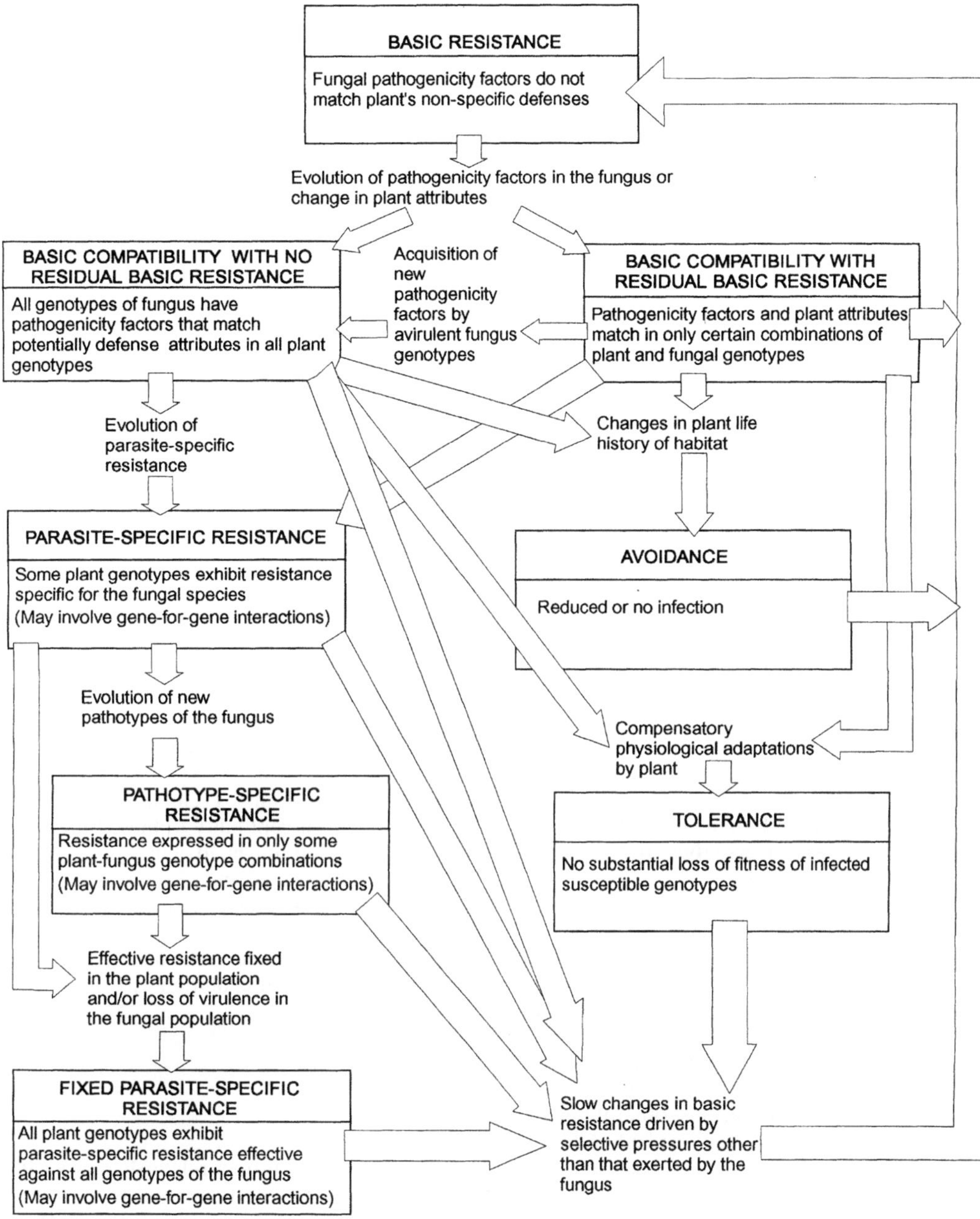

Fig. 6. The possible evolutionary paths that could be followed by a single interacting lineage of plant and fungal parasite. Changes in the nature of the association (i.e., the evolution of mutualism) and the loss of the association through extinction of either plant or fungus have been omitted

this Vol.). However, claims of the conversion of a fungal pathogen to an endophytic mutualist by a single-locus mutation (Freeman and Rodriguez 1993) are misleading because no fitness advantage to the plant was demonstrated, only the lack of seedling death in response to infection. A better interpretation of these results is that the mutant fungus had lost an attribute important in the development of disease symptoms but not mandatory for parasitism.

As long as infected plants have a fitness advantage, complete resistance to fungal invasion would have no selective advantage and stabilizing selection might be expected to select for plant-fungus combinations in which the parasitic cost of the fungus did not outweigh the fitness benefit to the plant, and the successful dissemination of the fungus was maintained. In these situations, host species specificity should be based on the same matching factors concept applicable to fungal parasites, and accordingly, mutualistic fungi often show host specificity similar to their parasitic relatives. In this respect, these mutualists differ strikingly from the vesicular-arbuscular mycorrhizal (VAM) fungi that have enormously large and diverse host ranges while having cellular interactions with their hosts that closely resemble those of biotrophic fungal parasites with much narrower host ranges (Bonfante-Fasolo and Perotto 1991).

F. Mutualism and the Enigma of Widespread Susceptibility to Vesicular-Arbuscular Mycorrhizal (VAM) Fungi

Since the fossil record suggests that VAM fungi evolved with the first land plants (Simon et al. 1993), it can be assumed that the current wide host range of VAM fungi (Chap. 7, this Vol.) is the result of association by descent. Nevertheless, one still must explain how each fungal species can mechanistically cope with the various forms of basic resistance, effective against other fungi, that have evolved in each lineage of plant. Although plant roots and shoots may have been subject to different selective pressures because of the potential for soil fungi to provide mineral nutrients to subterranean plant parts (Heath 1987b), roots do not lack defense mechanisms, and they are capable of exhibiting resistance to soil-borne fungi or to VAM fungi in nonmycorrizal species (Allen et al. 1989; Chap. 11, this Vol.). The suggestion that VAM fungi do not trigger inducible defenses (Codignola et al. 1989; Wyss et al. 1991) because of their low level of hydrolytic enzyme production (Bonfante-Fasolo and Perotto 1991) may explain their biotrophic way of life, but it does not explain their wide host range, since this suggestion has also been used to explain the lack of active defenses elicited in host species by biotrophic fungal parasites (Cooke and Whipps 1987) with relatively narrow host ranges. Although there are reports that VAM fungi **can** trigger plant defenses (Dumas-Gaudot et al. 1992; Harrison and Dixon 1993), and that VAM fungi are more tolerant than pathogens of certain defense response (Vierheilig et al. 1993), there also are data to suggest that the synthesis of a variety of defense proteins is suppressed by VAM formation, possibly by alterations in the hormonal balance of the root (Lambais and Mehdy 1993). Such a mode of inducing susceptibility could be effective in a wide range of plants, but it is an interesting question why fungal parasites have not evolved the same approach, given that many of them can produce plant hormones.

Perhaps the solution to this enigma lies in reports that mycorrhizal dependence in wheat is an inherited trait, possibly independent of mycorrhizal colonization (Hetrick et al. 1993), and that single, recessive genes in pea not only prevent both root nodulation and mycorrhiza formation, but also allow the plant to respond in a resistant manner to the VAM fungus (Gollotte et al. 1993). The widespread susceptibility to these fungi may depend not only on fungal attributes to negate or tolerate basic resistance, but also on "symbiosis" genes in the plant that foster susceptibility to, and dependence on, mycorrhizal fungi. These genes may have been conserved in the majority of plant lineages following their evolution early in the history of land plants.

IV. Conclusions

All plants have a battery of relatively nonspecific constitutive or inducible features that can act as defenses against fungal parasites, many of which have multiple defensive and nondefensive roles in the plant and are probably subject to a variety of selective forces. Consequently, susceptibility to a fungal parasite must depend on the latter possessing pathogenic attributes that "match" the features of the host plants with which they interact.

Nonhost resistance, therefore, results from a mismatch between fungal pathogenicity factors and plant attributes, which allows some components of the plant's basic resistance to be effective. Resistance within a host species may arise because of intraspecific polymorphisms in certain components of basic resistance maintained by selective pressures of which a single fungus may be only a minor component, or by parasite-driven evolution of resistance in populations of previously susceptible plants. The latter situation may result in gene-for-gene interactions, some of which have been suggested to depart from the "matching factors" concept of host-parasite specificity and involve specific recognition of parasite products that subsequently leads to pathotype-specific defense activation. However, the prevalence of gene-for-gene interactions in natural ecosystems is controversial.

Available data from natural ecosystems suggest that polymorphisms in resistance and virulence within interacting plant and parasite species may not be entirely due to coevolutionary arms races, and raise questions about the selective advantage of pathotype-specific resistance. It is argued that the evolutionary paths followed by interacting plant and fungal lineages are constrained by their genetic capabilities and the mechanisms of resistance and susceptibility, as well as by the selective pressures acting on each lineage. These interactions may persist over time or may eventually disappear due to (1) extinction of plant or fungus, (2) effective parasite-specific resistance becoming fixed in the plant population, or (3) a change in the plant's basic resistance in response to selective pressures other than that applied by the parasite. The form of each plant-fungus association may also be subject to evolution, with some forms of parasitism becoming mutualistic as a result of reduced plant damage and a fitness advantage afforded to infected plants. Susceptibility to these mutualists presumably is based on the same matching factors principle as susceptibility to their parasitic relatives. In contrast, the unusually wide host range of the vesicular-arbuscular mycorrhizal fungi seems to be based not only on the ability of these fungi to tolerate or negate the basic resistance of their hosts, but also on the existence of "symbiosis genes" in the latter that foster susceptibility and mycorrhizal dependence, and which probably evolved early in the history of land plants.

References

Alexander HM (1990) Dynamics of plant-pathogen interactions in natural plant communities. In: Burdon JJ, Leather SR (eds) Pests, pathogens and plant communities. Blackwell, Oxford, pp 31–45

Allen EA, Hoch HC, Stavely JR, Steadman JR (1991) Uniformity among races of *Uromyces appendiculatus* in response to topographic signaling for appressorium formation. Phytopathology 81:883–887

Allen MF, Allen EB, Friese CF (1989) Responses of the non-mycotrophic plant *Salsola kali* to invasion by vesicular-arbuscular mycorrhizal fungi. New Phytol 111:45–49

Andrivon D (1994) Race structure and dynamics in populations of *Phytophthora infestans*. Can J Bot 72:1681–1687

Baum BR, Savile DBO (1985) Rusts (Uredinales) of Triticeae: evolution and extent of coevolution, a cladistic analysis. Bot J Linn Soc 91:367–394

Bennett RN, Wallsgrove RM (1994) Secondary metabolites in plant defence mechanisms. New Phytol 127:617–633

Berbee ML, Taylor JW (1993) Dating the evolutionary radiations of the true fungi. Can J Bot 71:1114–1127

Bevan JR, Crute IR, Clarke DD (1993) Diversity and variation in expression of resistance to *Erysiphe fischeri* in *Senecio vulgaris*. Plant Pathol 42:647–653

Bi JL, Felton GW (1995) Foliar oxidative stress and insect herbivory: primary compounds, secondary metabolites, and reactive oxygen species as components of induced resistance. J Chem Ecol 21:1511–1530

Bonfante-Fasolo P, Perotto S (1991) Mycorrhizal and pathogenic fungi: do they share any features? In: Mendgen K, Lesemann D-E (eds) Electron microscopy of plant pathogens. Springer, Berlin Heidelberg New York, pp 265–275

Bradshaw AD (1991) Genostasis and the limits to evolution. Philos Trans R Soc Lond B 333:289–305

Brasier CM (1987) The dynamics of fungal speciation. In: Rayner ADM, Brasier CM, Moore D (eds) Evolutionary biology of the fungi. Cambridge University Press, Cambridge, pp 231–260

Bull JJ, Molineux IJ, Rice WR (1991) Selection of benevolence in a host-parasite system. Evolution 45:875–882

Burdon JJ (1987) Disease and plant population biology. Cambridge University Press, Cambridge

Burdon JJ (1993) The structure of pathogen populations in natural plant communities. Annu Rev Phytopathol 31: 305–323

Burdon JJ, Jarosz AM (1992) Temporal variation in the racial structure of flax rust (*Melampsora lini*) populations growing on natural stands of wild flax (*Linum marginale*): local versus metapopulation dynamics. Plant Pathol 41:165–179

Bradley DJ, Kjellbom P, Lamb CJ (1992) Elicitor- and wound-induced oxidative cross-linking of a proline-rich plant cell wall protein: a novel, rapid defense response. Cell 70:21–30

Christ BJ, Groth JV (1982) Inheritance of resistance in three cultivars of beans to the bean rust pathogen and the interaction of virulence and resistance genes. Phytopathology 72:771–773

Christ BJ, Person CO, Pope DD (1987) The genetic determination of variation in pathogenicity. In: Wolfe MS, Caten CE (eds) Populations of plant pathogens: their dynamics and genetics. Blackwell, Oxford, pp 7–19

Clarke DD, Campbell FS, Bevan JR (1990) Genetic interactions between *Senecio vulgaris* and the powdery mildew fungus *Erysiphe fischeri*. In: Burdon JJ, Leather SR (eds) Pests, pathogens and plant communities. Blackwell, Oxford, pp 189–201

Clay K (1988) Clavicipitaceous fungal endophytes of grasses: coevolution and the change from parasitism to mutualism. In: Pirozynski KA, Hawksworth DL (eds) Coevolution of fungi with plants and animals. Academic Press, London, pp 79–105

Codignola A, Verotta L, Spanu P, Maffei M, Scannerini S, Bonfante-Fasolo (1989) Cell wall-bound phenols in roots of vesicular-arbuscular mycorrhizal plants. New Phytol 112:221–228

Cooke RC, Whipps JM (1987) Saprotrophy, stress and symbiosis. In: Rayner ADM, Brasier CM, Moore D (eds) Evolutionary biology of the fungi. Cambridge University Press, Cambridge, pp 137–148

Crute IR (1990) Resistance to *Bremia lactucae* (downy mildew) in British populations of *Lactuca serriola* (prickly lettuce). In: Burdon JJ, Leather SR (eds) Pests, pathogens and plant communities. Blackwell, Oxford, pp 203–217

Crute IR, Norwood JM (1986) Gene-dosage effects on the relationship between *Bremia lactucae* (downy mildew) and *Lactuca sativa* (lettuce): the relevance to a mechanistic understanding of host-parasite specificity. Physiol Mol Plant Pathol 29:133–145

De Feyter R, Yang Y, Gabriel DW (1993) Gene-for-genes interactions between cotton R genes and *Xanthomonas campestris* pv. *malvacearum* avr genes. Mol Plant-Microbe Interact 6:225–237

De Nooij MP, Paul ND (1992) Invasion of rust (*Puccinia poarum*) pycnia and aecia on coltsfoot (*Tussilago farfara*) by secondary pathogens: death of host leaves. Mycol Res 96:309–312

De Nooij MP, Paul ND, Ayres PG (1995) Variation in susceptibility and tolerance within and between populations of *Tussilago farfara* L. infected by *Coleosporium tussilaginis* (Pers.) Berk. New Phytol 129:117–123

Dixon RA, Lamb CJ (1990) Molecular communication in interactions between plants and microbial pathogens. Annu Rev Plant Physiol Plant Mol Biol 41:339–367

Duchesne LC, Ellis BE, Peterson RL (1989) Disease suppression by the ectomycorrhizal fungus *Paxillus involutus*: contribution of oxalic acid. Can J Bot 67:2726–2730

Dumas-Gaudot E, Grenier J, Furlan V, Asselin A (1992) Chitinase, chitosanase and β-1,3-glucanase activities in *Allium* and *Pisum* roots colonized by *Glomus* species. Plant Sci 84:17–24

Elias KS, Zamir D, Lichtman-Pleban T, Katan T (1993) Population structure of *Fusarium oxysporum* f. sp. *lycopersici*: restriction fragment length polymorphisms provide genetic evidence that vegetative compatibility group is an indicator of evolutionary origin. Mol Plant-Microbe Interact 6:565–572

Ellingboe AH (1982) Genetical aspects of active defence. In: Wood RKS (ed) Active defense mechanisms in plants. Plenum Press, New York, pp 179–192

Ellis JS, Jennings AC, Edwards LA, Mavandad M, Lamb CJ, Dixon RA (1989) Defense gene expression in elicitor-treated cell suspension cultures of French bean cv. Imuna. Plant Cell Rep 8:504–507

Elmhirst JF, Heath MC (1989) Interactions of the bean and cowpea rust fungi with species of the *Phaseolus-Vigna* plant complex. II. Histological responses to infection in heat-treated and untreated leaves. Can J Bot 67:58–72

Esele JP, Frederiksen RA, Miller FR (1993) The association of genes controlling caryopsis traits with grain mold resistance in sorghum. Phytopathology 83:490–495

Farr DF, Bills GF, Chamuris GP, Rossman AY (1989) Fungi on plants and plant products in the United States. APS Press, St Paul

Fernandez MR, Heath MC (1989) Interactions of the nonhost French bean plant (*Phaseolus vulgaris*) with parasitic and saprophytic fungi. I. Fungal development on and in killed, untreated, heat-treated or Blasticidin S-treated leaves. Can J Bot 67:661–669

Flor HH (1955) Host-parasite interactions in flax rust – its genetics and other implications. Phytopathology 45:680–685

Fox LR (1981) Defense and dynamics in plant-herbivore systems. Am Zool 21:853–864

Frank SA (1993) Coevolutionary genetics of plants and pathogens. Evol Ecol 7:45–75

Freeman S, Rodriguez RJ (1993) Genetic conversion of a fungal plant pathogen to a nonpathogenic, endophytic mutualist. Science 260:75–78

Gabriel DW, Rolfe BG (1990) Working models of specific recognition in plant-microbe interactions. Annu Rev Phytopathol 28:365–391

Gates JE, Loegering WQ (1991) Mutation to wider virulence in *Puccinia graminis* f. sp. *tritici*: evidence for the existence of loci which allow the fungus to overcome several host stem rust resistance genes simultaneously. Appl Environ Microbiol 57:2332–2336

Gollotte A, Gianinazzi-Pearson V, Giovannetti M, Sbrana C, Avio L, Gianinazzi S (1993) Cellular localization and cytochemical probing of resistance reactions to arbuscular mycorrhizal fungi in a "locus a" myc$^-$ mutant of *Pisum sativum* L. Planta 191:112–122

Goodwin SB, Allard RW, Hardy SA, Webster RK (1992) Hierarchical structure of pathogenic variation among *Rhynchosporium secalis* populations in Idaho and Oregon. Can J Bot 70:810–817

Gottstein HD, Kuc J (1989) Induction of systemic resistance to anthracnose in cucumber by phosphates. Phytopathology 79:176–179

Gould SJ, Lewontin RC (1979) The spandrels of San Marco and the Panglossian paradigm: a critique of the adaptationist programme. Proc R Soc Lond B 205:581–598

Hamilton WD, Axelrod R, Tanese R (1990) Sexual reproduction as an adaptation to resist parasites. Proc Natl Acad Sci USA 87:3566–3573

Harder DE, Chong J (1984) Structure and physiology of haustoria. In: Bushnell WR, Roelfs AP (eds) The cereal rusts. Academic Press, Orlando, pp 431–476

Harper JL (1990) Pests, pathogens and plant communities: an introduction. In: Burdon JJ, Leather SR (eds) Pests, pathogens and plant communities. Blackwell, Oxford, pp 3–14

Harrison MJ, Dixon RA (1993) Isoflavonoid accumulation and expression of defense gene transcripts during the establishment of vesicular-arbuscular mycorrhizal associations in roots of *Medicago truncatula*. Mol Plant-Microbe Interact 6:643–654

Hatcher PE, Paul ND, Ayres PG, Whittaker JB (1994) Interactions between *Rumex* spp., herbivores and a rust fungus: *Gasterophysa viridula* grazing reduces subsequent infection by *Uromyces rumicis*. Funct Ecol 8:265–272

Heath MC (1974) Light and electron microscope studies of the interactions of host and nonhost plants with cowpea

rust – *Uromyces phaseoli* var. *vignae*. Physiol Plant Pathol 4:403–414
Heath MC (1977) A comparative study of non-host interactions with rust fungi. Physiol Plant Pathol 10:73–88
Heath MC (1981) The suppression of the development of silicon-containing deposits in French bean leaves by exudates of the bean rust fungus and extracts from bean rust-infected tissue. Physiol Plant Pathol 18:149–155
Heath MC (1985) Implications of non-host resistance for understanding host-parasite interactions. In: Groth JV, Bushnell WR (eds) Genetic basis of biochemical mechanisms of disease. APS Press, St Paul, pp 25–42
Heath MC (1987a) Evolution of plant resistance and susceptibility to fungal invaders. Can J Plant Pathol 9:389–397
Heath MC (1987b) Evolution of parasitism in the fungi. In: Rayner ADM, Brasier CM, Moore D (eds) Evolutionary biology of the fungi. Cambridge University Press, Cambridge, pp 149–160
Heath MC (1991a) The role of gene-for-gene interactions in the determination of host species specificity. Phytopathology 81:127–130
Heath MC (1991b) Evolution of resistance to fungal parasitism in natural ecosystems. New Phytol 119:331–343
Heath MC (1992) Host species specificity of the goldenrod rust fungus and the existence of rust resistance within some goldenrod species. Can J Bot 70:2461–2466
Heath MC (1994) Current concepts of the determinants of plant-fungal specificity. In: Kohmoto K, Yoder OC (eds) Host-specific toxin: biosynthesis, receptor and molecular biology. Tottori University, Tottori, Japan, pp 3–21
Heath MC (1995a) Thoughts on the role and evolution of induced resistance in natural ecosystems, and its relationship to other types of plant defenses against disease. In: Hammerschmidt R, Kuc J (eds) Induced resistance to disease. Kluwer, Dordrecht, pp 141–151
Heath MC (1995b) Signal exchange between higher plants and rust fungi. Can J Bot 73 (Suppl 1):S616–S623
Hetrick BAD, Wilson GWT, Cox TS (1993) Mycorrhizal dependence of modern wheat cultivars and ancestors: a synthesis. Can J Bot 71:512–518
Hovmoller MS, Munk L, Ostergård H (1993) Observed and predicted changes in virulence gene frequencies at 11 loci in a local barley powdery mildew population. Phytopathology 83:253–260
Howard R, Bourett TM, Ferrari MA (1991) Infection by Magnaporthe: an in vitro analysis. In: Mendgen K, Lesemann D-E (eds) Electron microscopy of plant pathogens. Springer, Berlin Heidelberg New York, pp 251–264
Janzen DH (1980) When is it coevolution? Evolution 34:611–612
Jarosz AM, Burdon JJ (1990) Predominance of a single major gene for resistance to *Phakopsora pachyrhizi* in a population of *Glycine argyrea*. Heredity 64:347–353
Jarosz AM, Burdon JJ (1991) Host-pathogen interactions in natural populations of *Linum marginale* and *Melampsora lini*: II. Local and regional variation in patterns of resistance and racial structure. Evolution 45: 1618–1627
Jarosz AM, Burdon JJ (1992) Host-pathogen interactions in natural populations of *Linum marginale* and *Melampsora lini*: III. Influence of pathogen epidemics on host survivorship and flower production. Oecologia 89:53–61
Jones CG, Firn RD (1991) On the evolution of plant secondary chemical diversity. Philos Trans R Soc Lond B 333:273–280
Kehlenbeck H, Schonbeck F (1995) Effects of induced resistance on disease severity/yield in mildewed barley. J Phytopathol 143:561–567
Kinlock BB Jr (1992) Distribution and frequency of a gene for resistance to white pine blister rust in natural populations of sugar pine. Can J Bot 70:1319–1323
Kohmoto K, Otani H, Kodama M, Nishimura S (1989) Host recognition: can accessibility to fungal invasion be induced by host-specific toxins without necessitating necrotic cell death? In: Graniti A, Durbin RD, Ballio A (eds) Phytotoxins and plant pathogens. Springer, Berlin Heidelberg New York, pp 249–265
Kolmer JA (1993) Selection in a heterogeneous population of *Puccinia recondita* f. sp. *tritici*. Phytopathology 83: 909–914
Krischik VA (1991) Specific or generalized plant defense: reciprocal interactions between herbivores and pathogens. In: Barbosa P, Krischik VA, Jones CG (eds) Microbial mediation of plant-herbivore interactions. John Wiley, New York, pp 309–340
Kuć J (1985) Expression of latent genetic information for disease resistance in plants. In: Key JL, Kosuge T (eds) Cellular and molecular biology of plant stress. Alan R Liss, New York, pp 303–318
Kunoh H, Nicholson RL, Kobayashi I (1991) Extracellular materials of fungal structures: their significance at prepenetration stages of infection. In: Mendgen K, Lesemann D-E (eds) Electron microscopy of plant pathogens. Springer, Berlin Heidelberg New York, pp 223–234
Lambais MR, Mehdy MC (1993) Suppression of endochitinase, β-1,3-endoglucanase, and chalcone isomerase expression in bean vesicular-arbuscular mycorrhizal roots under different soil phosphate conditions. Mol Plant-Microbe Interact 6:75–83
Lawrence GJ, Finnegan EJ, Ayliffe MA, Ellis JG (1995) The *L6* gene for flax rust resistance is related to the *Arabidopsis* bacterial resistance gene *RPS2* and the tobacco viral resistance gene *N*. Plant Cell 7:1195–1206
Leonard KJ (1994) Stability of equilibria in a gene-for-gene coevolution model of host-parasite interactions. Phytopathology 84:70–77
Leonard KJ, Czochor RJ (1980) Theory of genetic interactions among populations of plants and their pathogens. Annu Rev Phytopathol 18:237–258
Levine A, Tenhaken R, Dixon R, Lamb C (1994) H_2O_2 from the oxidative burst orchestrates the plant hypersensitive disease resistance response. Cell 79:583–593
Levings III CS (1990) The Texas cytoplasm of maize: cytoplasmic male sterility and disease susceptibility. Science 250:942–947
Levy M, Correa-Victoria FJ, Zeigler RS, Xu S, Hamer JE (1993) Genetic diversity of the rice blast fungus in a disease nursery in Colombia. Phytopathology 83:1427–1433
Luttrell ES (1974) Parasitism of fungi on vascular plants. Mycologia 66:1–15
Maher EA, Bate NJ, Mi W, Elkind Y, Dixon RA, Lamb CJ (1994) Increased disease susceptibility of transgenic tobacco plants with suppressed levels of preformed phenylpropanoid products. Proc Natl Acad Aci USA 91: 7802–7806
Mansfield JW (1984) Plant cell death during infection by fungi. In: Davies I, Sigee DC (eds) Cell ageing and cell

death. Cambridge University Press, Cambridge, pp 323–345

Martinelli JA, Brown JKM, Wolfe MS (1993) Effects of barley genotype on induced resistance to powdery mildew. Plant Pathol 42:195–202

May RM (1990) Population biology and population genetics of plant-pathogen associations. In: Burdon JJ, Leather SR (eds) Pests, pathogens and plant communities. Blackwell, Oxford, pp 309–325

Meeley RB, Johal GS, Briggs SP, Walton JD (1992) A biochemical phenotype for a disease resistance gene of maize. Plant Cell 4:71–77

Miao VPW, VanEtten HD (1992) Genetic analysis of the role of phytoalexin detoxification in virulence of the fungus *Nectria haematococca* on chickpea (*Cicer arietinum*). Appl Environ Microbiol 58:809–814

Mitter C, Brooks DR (1983) Phylogenetic aspects of coevolution. In: Futuyma DJ, Slatkin M (eds) Coevolution. Sinauer, Sunderland

Otrosina WJ, Chase TE, Cobb Jr FW, Korhonen K (1993) Population structure of *Heterobasidion annosum* from North America and Europe. Can J Bot 71:1064–1071

Parker MA (1988) Genetic uniformity and disease resistance in a clonal plant. Am Nat 132:538–549

Parker MA (1989) Disease impact and local genetic diversity in the clonal plant *Podophyllum peltatum*. Evolution 43:540–547

Parker MA (1991) Nonadaptive evolution of disease resistance in an annual legume. Evolution 45:1209–1217

Richter TE, Pryor TJ, Bennetzen JL, Hulbert SH (1995) New rust specificities associated with recombination in the *Rp1 complex* in maize. Genetics 141:373–381

Ride JP (1985) Non-host resistance to fungi. In: Fraser RSS (ed) Mechanisms of resistance to plant diseases, Martinus Nijhoff/Dr W Junk, Dordrecht, pp 29–61

Rohe M, Gierlich A, Hermann H, Hahn M, Schmidt B, Rosahl S, Knogge W (1995) The race-specific elicitor, NIP1, from the barley pathogen, *Rhynchosporium secalis*, determines avirulence on host plants of the Rrs1 resistance genotype. EMBO J 14:4168–4177

Savile DBO (1955) A phylogeny of the basidiomycetes. Can J Bot 33:60–103

Schäfer W, Straney D, Ciuffetti L, Van Etten HD, Yoder OC (1989) One enzyme makes a fungal pathogen, but not a saprophyte, virulent on a new host plant. Science 246:247–249

Scheffer RP (1991) Role of toxins in evolution and ecology of plant pathogenic fungi. Experientia 47:804–811

Sela-Buurlage MB, Ponstein AS, Bres-Vloemans SA, Melchers LS, van den Elzen PJM, Cornelissen BJC (1993) Only specific tobacco (*Nicotiana tabacum*) chitinases and β-1,3-glucanases exhibit antifungal activity. Plant Physiol 101:857–863

Shtienberg D (1992) Effects of foliar diseases on gas exchange processes: a comparative study. Phytopathology 82:760–765

Simms EL, Rausher MD (1993) Patterns of selection on phytophage resistance in *Ipomoea purpurea*. Evolution 47:970–976

Simms EL, Vision TJ (1995) Pathogen-induced systemic resistance in *Ipomoea purpurea*. Oecologia 102:494–500

Simon L, Bousquet J, Lévesque RC, Lalonde M (1993) Origin and diversification of endomycorrhizal fungi and coincidence with vascular land plants. Nature 363:67–69

Smith CJ (1996) Accumulation of phytoalexins: defence mechanism and stimulus response system. New Phytol 132:1–45

Staskawicz BJ, Ausubel FM, Baker BJ, Ellis JG, Jones JDG (1995) Molecular genetics of plant disease resistance. Science 268:661–667

Stermer BA, Schmid J, Lamb CJ, Dixon RA (1990) Infection and stress activation of bean chalcone synthase promoters in transgenic tobacco. Mol Plant-Microbe Interact 3:381–388

Sweigard JA, Carroll AM, Kang S, Farrall L, Chumley FG, Valent B (1995) Identification, cloning, and characterization of *PWL2*, a gene for host species specificity in the rice blast fungus. Plant Cell 7:1221–1233

Thompson JN (1990) Coevolution and the evolutionary genetics of interactions among plants and insects and pathogens. In: Burdon JJ, Leather SR (eds) Pests, pathogens and plant communities. Blackwell, Oxford, pp 249–271

Thompson JN, Burdon JJ (1992) Gene-for-gene coevolution between plants and parasites. Nature 360:121–125

Tosa Y (1992) A model for the evolution of formae speciales and races. Phytopathology 82:728–730

Toyoda K, Shiraishi T, Yoshioka H, Yamada T, Ishinose Y, Oku H (1992) Regulation of polyphosphoinositide metabolism in pea plasma membranes by elicitor and suppressor from a pea pathogen, *Mycosphaerella pinodes*. Plant Cell Physiol 33:445–452

Valent B, Chumley FG (1987) Genetic analysis of host species specificity in *Magnaporthe grisea*. In: Arntzen CJ, Ryan C (eds) Molecular strategies for crop protection. Alan R Liss, New York, pp 83–93

Van den Ackerveken GFJM, Van Kan JAL, De Wit PJGM (1992) Molecular analysis of the avirulence gene *avr9* of the fungal tomato pathogen *Cladosporium fulvum* supports the gene-for-gene hypothesis. Plant J 2:359–366

Vanderplank JE (1968) Disease resistance in plants. Academic Press, London

Vierheilig H, Alt M, Neuhaus J-M, Boller T, Wiemken A (1993) Colonization of transgenic *Nicotiana sylvestris* plants, expressing different forms of *Nicotiana tabacum* chitinase, by the root pathogen *Rhizoctonia solani* and by the mycorrhizal symbiont *Glomus mosseae*. Mol Plant-Microbe Interact 6:261–264

Vögeli U, Chappell J (1988) Induction of sesquiterpene cyclase and suppression of squalene synthetase activities in plant cell cultures treated with fungal elicitor. Plant Physiol 88:1291–1296

Walton JD, Panaccione DG (1993) Host-selective toxins and disease specificity: perspectives and progress. Annu Rev Phytopathol 31:275–303

Whalen MC, Wang JF, Carland FM, Heiskell ME, Dahlbeck D, Minsavage GV, Jones JB, Scott JW, Stall RE, Staskawicz BJ (1993) Avirulence gene *avrRxv* from *Xanthomonas campestris* pv. *vesicatoria* specifies resistance on tomato line Hawaii 7998. Mol Plant-Microbe Interact 6:616–627

Wubber JP, Lawrence CB, de Wit PJGM (1996) Differential induction of chitinase and 1,3-β-glucanse gene expression in tomato by *Cladosporium fulvum* and its race-specific elicitors. Physiol Mol Plant Pathol 48:105–116

Wyss P, Boller T, Wiemken A (1991) Phytoalexin response is elicited by a pathogen (*Rhizoctonia solani*) but not by a mycorrhizal fungus (*Glomus mosseae*) in soybean roots. Experientia 47:395–399

Zhu Q, Doerner PW, Lamb CJ (1993) Stress induction and developmental regulation of a rice chitinase promoter in transgenic tobacco. Plant J 3:203–212

Biosystematic Index

Abies, 97
– *alba*, 137, 139
– *balsamea*, 139
– *grandis*, 130
Absidia glauca, 210, 217, 219
Acacia, 127
– *pulchella*, 118, 128
Acaulospora
– *elegans*, 175
– *gerdemannii*, 175
Acaulosporaceae, 117
Achlya ambisexualis, 27
Acremonium, 185–202, 221, 231, 236–238
– *coenophialum*, 192, 195, 200–202, 231, 237
– *implicatum*, 88
– *lolii*, 187, 202, 223, 230, 231, 233, 235
Actinomycetes, 122, 127, 128
Aecidium, 81
– *berberidis*, 90
Aeschynomene, 159, 161
– *virginica*, 159, 161
Agaricales, 97, 113
Agrobacterium, 216, 219
– *rhizogenes*, 214
– *tumefaciens*, 214
Agropyron caninum, 191, 192
Agrostis, 185, 191, 196
– *hiemalis*, 191, 202
– *perennans*, 191
– *tenuis*, 189, 200
Albuginaceae, 18
Albugo, 18
– *candida*, 18
Alfalfa, 151, 152, 155
Allium, 274
Almond, 164, 166
Alnus, 136, 139, 172, 179
Alpova, 172
American chestnut, 144
Ammophila, 169, 175
Angiosperms, 131
Antelope, 173
Anthomyiid fly, 187
Anthoxanthum odoratum, 191, 236
Apiognomonia errabunda, 132, 139, 140
Apple, 160–163, 165–168
Arabidopsis, 34, 76, 91
– *thaliana*, 18, 34
Artemesia tridentata, 182
Arthrosporiella, 210, 216
Arthuriomyces peckianus, 93
Arundinoideae, 195
Ascomycotina, 185
Aspergillus, 112, 114, 219
– *nidulans*, 27, 35, 50, 51, 102, 107, 110, 111, 210, 216, 218, 237
– *niger*, 215, 217
– *terrus*, 52
Atkinsonella, 185, 189–193, 195, 201, 224, 226–228, 232, 237
– *hypoxylon*, 191, 192, 194, 195, 200–202, 224–228, 230, 232, 236–238
– *texensis*, 201, 225, 232, 237
Avena sativa, 92
Avocado, 159, 162, 166, 167

Bacillus, 88
– *subtilis*, 60, 92
Bacteria, 119, 122, 124, 126–128
Badger, 172, 176
Balansia, 185, 189, 190, 195–197, 200, 224–226, 228, 236–238
– *cyperi*, 200–202, 225
– *epichloë*, 201, 202, 229, 230
– *henningsiana*, 200, 236
– *obtecta*, 188, 196, 224–226, 230, 235, 237
– *pilulaeformis*, 200, 236
– *strangulans*, 230
Balansieae, 185, 222, 224, 225, 228–230, 232, 233, 237, 238
Balansiopsis, 224, 225
Bambusoideae, 195
Banana, 144, 161, 164, 168
Banksia, 127
Barberry (see also *Ribes*), 80, 87, 92
Barley (see also *Hordeum*), 4, 11, 15, 16, 55–78
Basidiobolus ranarum, 208, 218
Beech (see also *Fagus*), 132, 133, 135–140
Bell pepper, 149, 156
Betula, 97
– *pendula*, 115
Birch (see also *Betula*), 132, 139
Birds, 118
Blackberry (see also *Rubus*), 88
Blumeria graminis, 55
Boletus, 98, 111
Botrytis, 145, 155
– *cinerea*, 145, 147
Bouteloua gracilis, 125, 126, 182
Brachiaria, 48
Brachyelytrum, 194
– *erectum*, 191, 193, 194
Brachypodium sylvaticum, 189, 191, 192, 194, 200
Brassica
– *napus*, 206, 214, 215, 217
– *nigra*, 215
Brassicaceae, 18, 179
Bremia lactucae, 18, 25–27, 34, 274
Bromus
– *benekenii*, 191, 192
– *erectus*, 191
– *ramosus*, 191, 192

Cajanus cajan, 127
Calamagrostis villosa, 191
Calocedrus, 127
Candida
– *albicans*, 26
– *parapsilopsis*, 217
Cantharellus, 111
Capsicum, 81, 90, 91
Carica papaya, 166
Carthamus tinctorius, 88
Carya, 179
Castanea sativa, 97
Casuarina equisetifolia, 138
Cenangium ferruginosum, 131
Cenchrus echinatus, 188, 196, 226, 237
Cenococcum geophilum, 107
Centaurea
– *diffusa*, 88, 92
– *maculosa*, 88
Cercospora, 145
– *nicotianae*, 261
Cereals, 241
Cerotelium, 81, 92
Chamaecyparis thyoides, 138
Chlorella, 221, 237
Chondrilla juncea, 88, 93
Chrysolepis sempervirens, 175
Cicada, 173
Cicer arietinum, 276
Citrus, 163–167
– *jambhiri*, 153
– *sinensis*, 163
Cladosporium, 12
– *fulvum*, 3–16, 26, 33, 35, 36, 58, 75, 207, 212, 216, 218, 219, 262, 276
Claviceps, 224–226, 229, 233, 238
– *fusiformis*, 229
– *purpurea*, 225, 226
Clavicipitaceae, 129, 221–226, 228, 229, 236–238, 270

Clavicipitales, 229, 238
Clover (see also *Trifolium*), 149, 153, 156
Coastal redwood (see also *Sequoia*), 133, 138, 139
Cochliobolus, 265
– *carbonum*, 265
Coffea arabica, 82
Coffee, 144
Coleosporium tussilaginis, 269, 274
Colletotrichum, 157–168
– *acutatum*, 163, 164, 166, 167
– *aurantifolia*, 163
– *capsici*, 157, 164
– *caudatum*, 164
– *cingulata*, 160
– *coccodes*, 164
– *dematium*, 157, 164, 166, 269
– *destructivum*, 159, 164, 167
– *falcatum*, 159
– *fragariae*, 163, 164
– *gloeosporioides*, 157, 159–165, 210, 211, 216, 218
– *gloeosporioides aeschynomene*, 158, 159, 161
– *gossypii*, 159
– *graminicola*, 157, 159–161, 164, 166–168
– *lagenarium*, 166, 168
– *lindemuthianum*, 159, 161, 164, 165
– *magna*, 164, 167
– *malvarum*, 164
– *musae*, 134, 159, 161, 164, 167, 168
– *orbiculare*, 159, 164, 163, 166–168
– *trifolii*, 164, 167
– *truncatum*, 164
Conifer, 81, 130, 172
Coprinus cinereus, 99, 111, 211, 216
Coral, 221
Cortinariaceae, 97
Cotton, 148, 155
Crabgrass, 38
Cronartium ribicola, 83, 148, 154
Cronartium, 84
Cryptosphaeria cubensis, 139
Cryptosporiopsis, 137–139
Cucurbits, 164, 166
Cyclaneusma minus, 131
Cylindrocladium crotalariae, 149, 155
Cyperaceae, 185, 195
Cyperus, 48, 225, 237
– *esculentus*, 88, 89, 93
– *rotundus*, 196, 200, 202
– *virens*, 196, 200
Cytospora eucalypticola, 138

Dactylis glomerata, 188, 189, 191, 202
Danthonia, 189–194, 200–202, 223–228, 232, 236, 238
– *compressa*, 190 193, 194, 232
– *epilis*, 190, 191, 193, 194, 232
– *sericea*, 190, 191, 193, 194, 232
– *spicata*, 190, 191, 193–195, 200, 227, 232, 236
Daucus carota, 120
Deer mouse, 173
Deer, 173
Diaporthales, 37
Dictyostelium discoideum, 32
Digitaria, 41, 42, 47, 48, 49
– *sanguinalis*, 38
Discula
– *quercina*, 140
– *umbrinella*, 132–136, 139, 140
Douglas fir, 133–136, 139, 140, 172, 182
Drosophila melanogaster, 235
Dryas
– *integrifolia*, 97, 114
– *octopetala*, 97, 101, 113

Earthworm, 118
Echinodothis, 224
– *tuberiformis*, 225, 238
Eleusine, 41, 42, 47, 48, 51
– *coracana*, 38
Elk, 173, 174
Elm, 144
Elymus, 193, 194
– *europaeus*, 191, 192
– *villosus*, 191, 194
– *virginicus*, 187, 191, 194
Endocronartium harknessii, 91
Endogonales, 179, 183
Endophyllum, 81
– *euphorbiae-sylvaticae*, 81
– *pampeanum*, 90, 91
Endothia eugeniae, 139
Eocronartium muscicola, 84, 89
Ephelis, 185, 228
Epichloë, 140, 185–197, 200–202, 221–238
– *amarillans*, 229
– *baconii*, 229
– *clarkii*, 187, 229
– *festucae*, 223, 225, 228, 229, 231, 233, 234
– *typhina*, 139, 185, 187–189, 192–194, 196, 200–202, 221, 223, 228–231, 235, 238
Eragrostis curvula, 38
Erwinia, 88
– *herbicola*, 91
Erysiphe
– *fischeri*, 269, 273, 274
– *graminis*, 55–78
– *graminis agropyri*, 58
– *graminis avenae*, 58
– *graminis bromi*, 58
– *graminis hordei*, 56–78
– *graminis poae*, 58
– *graminis secalis*, 58
– *graminis tritici*, 56, 58
– *pisi*, 65, 75, 77
Escherichia coli, 111, 114, 214
Eucalyptus, 110, 113, 127, 173
– *nitens*, 138
Euphorbia lathyris, 150, 151
Euterpe oleracea, 139

Fagus, 97
– *sylvatica*, 102, 113, 132, 133, 137–140
Fern, 81
Festuca, 188, 191, 194, 201, 202, 222, 223, 225, 228, 231, 237
– *altissima*, 191
– *arizonica*, 200, 236
– *arundinacea*, 189–198, 200, 201, 228, 231, 236, 237
– *gigantea*, 191, 192
– *glauca*, 191
– *heterophylla*, 191
– *longifolia*, 191
– *obtusa*, 191
– *ovina*, 191
– *pratensis*, 191, 195, 200, 231, 236
– *pulchella*, 191
– *rubra*, 188, 191, 195
– *valesiaca*, 191
– *versuta*, 202
Festucae, 222, 223
Finger millet (see also *Eleusine*), 38, 47, 54
Fireweed, 176
Flax (see also *Linum*), 4
Fraxinus, 97
Fucus, 36
Fulvia fulva, 15, 16, 217
Funaria hygometrica, 127
Fusarium, 145, 146, 210, 212, 213, 216–219
– *culmorum*, 91
– *moniliforme*, 212
– *oxysporum*, 167, 210, 217, 50
– *oxysporum lycopersici*, 216, 217, 274
– *oxysporum pisi*, 219
– *oxysporum vasinfectum*, 148
– *sambucinum*, 229
– *solani pisi*, 26
– *subglutinans pini*, 216
Fusicoccum amygdali, 11

Garden cress, 151
Gasterophysa viridula, 274
Gibberella fujikuroi, 212, 217, 219
Gigaspora
– *gigantea*, 126, 127
– *margarita*, 119, 128, 175, 180
Gigasporaceae, 117
Gloeosporium, 165
Glomaceae, 117, 127
Glomales, 117, 178, 183
Glomerella, 157, 159–161, 165–168
– *cingulata*, 158–168
– *glycines*, 161, 167
– *graminicola*, 168
– *lindemuthianum*, 161
– *lycopersici*, 166
– *magna*, 166
– *musae*, 161, 167
– *tucumanensis*, 167
Glomus, 128, 175, 176, 274, 276
– *albidum*, 175
– *deserticola*, 175
– *epigaeum*, 126
– *fasciculatum*, 119, 127
– *intraradices*, 118, 120
– *mosseae*, 126, 127, 169, 276
– *occultum*, 175

– *versiforme*, 118, 127
Glyceria striata, 191, 193
Glycine argyrea, 275
Gopher, 171, 172, 176, 179, 182
Grape, 18, 144, 145, 147, 155
Grass, 55, 58, 71, 74, 77, 171–174, 182, 221–228, 230, 233, 236–238
Grasshoppers, 173, 183
Grizzly bear, 173
Gymnosperm, 131
Gymnosporangium
– *fuscum*, 92
– *juniperi-virginianae*, 92

Hansenula polymorpha, 218
Harvester ant, 172, 182
Hebeloma, 97, 98, 102–105, 113–115
– *alpinum*, 97, 101, 113
– *circinans*, 107
– *cylindrosporum*, 95–112, 183
– *edurum*, 107, 112
– *hiemale*, 113, 115
– *marginatulum*, 97, 101, 113
– *sinapizans*, 107
Hedysarum, 173
Hemileia vastatrix, 82
Heterobasidion annosum, 276
Holcus lanatus, 187, 191
Hordeum
– *murinum*, 58, 73
– *spontaneum*, 74
– *vulgare*, 73–76
Hormonema dematiodes, 139
Hydra, 221, 237
Hyphochytridiomycota, 18
Hypocreales, 229, 238
Hyponectria, 51
Hypoxylon, 133, 137
– *fragiforme*, 133, 134, 137, 138
Hystrix, 193, 194
– *patula*, 191, 194

Inocybe, 175
Ipomoea purpurea, 269, 276

Jack rabbit, 174
Johnsongrass, 164
Juncus, 138
– *bufonius*, 135
– *effusus*, 135

Knapweed (see also *Centaurea*), 88

Labyrinthulomycota, 18
Laccaria, 98, 102, 113
– *altaica*, 101
– *bicolor*, 97, 101, 104–107, 113–115, 179, 183
– *laccata*, 97, 98, 101, 107, 111–113
Lactarius hysginus, 114
Lactuca
– *sativa*, 274
– *serriola*, 274
Larix, 134
– *laricina*, 97
Leersia, 48
Legume, 159, 161, 167, 173
Lepidium sativum, 154
Leptosphaeria, 212, 213
– *maculans*, 206, 207, 211, 213, 215–219
– *nodorum*, 210
Leptostroma, 136, 139
Lettuce (see also *Lactuca*), 18, 34
Lichen, 221
Lilium longiflorum, 36
Linum
– *marginale*, 268, 273, 275
– *usitatissimum*, 207, 92
Lolium, 188, 194, 202
– *perenne*, 187, 189–191, 195, 196, 200, 201, 228, 231, 236, 237
Lophodermium
– *piceae*, 131
– *pinastri*, 136, 140
– *seditiosum*, 131
Lotus corniculatus, 155
Lupine, 176
Lycopersicon esculentum, 17, 207, 5

Macrophomina phaseolina, 150, 151, 156
Madrone, 172, 182
Magnaporthe, 51
– *grisea*, 4, 26, 29, 33, 37–54, 131, 133, 140, 168, 209, 211, 218, 219, 266, 276
Maize (see also *Zea*), 144, 152, 156, 161, 164, 168, 260, 265, 266, 275, 276
Mango, 162, 165
Medicago truncatula, 274
Melampsora lini, 4, 15, 79, 84, 85, 89, 92, 207, 216, 254, 268, 273, 275
Melanconium, 136, 139
– *apiocarpum*, 139
– *marginale*, 132, 139
Melica ciliata, 191, 193
Meloidogyne, 12, 201
Meria, 134, 135
laricis, 134, 139
Moose, 173
Morchella, 183
Mucor, 210, 215
– *circinelloides*, 210, 218
Mycosphaerella
– *graminicola*, 212
– *pinodes*, 276
Myriogenospora, 185, 189, 190, 195, 196, 224, 228, 237
– *atramentosa*, 202, 225, 230, 238

Naemacyclus minor, 131, 138
Nectria haematococca, 133, 140, 209, 260, 276
Neottiella, 219
– *rutilans*, 208
Neurospora, 211, 216, 218
– *crassa*, 9, 15, 27, 39, 102, 107, 110, 164, 208, 211, 218
Nicotiana
– *sylvestris*, 276
– *tabacum*, 276
Northern jointvetch, 161, 162, 166
Oak (see also *Quercus*), 175
Onion (see also *Allium*), 169
Oomycota, 18, 20, 257
Ophiostoma ulmi, 210, 216
Orchid, 151, 155
Oryza sativa, 37, 41, 53
Oryzoideae, 195
Oxalis pes-caprae, 127

Panicoideae, 195
Panicum, 48
– *agrostoides*, 200, 236
Papaya (see also *Carica*), 162
Paramecium, 221
Parasitella parasitica, 210, 211, 213, 215, 216
Parsley, 28, 32, 34, 35
Paspalum, 48
Paxillus involutus, 111, 112, 274
Pea (see also *Pisum*), 65, 75, 260, 272, 276
Peanut, 145, 149, 155
Pecan, 158, 160–162, 165
Pennisetum, 41, 42, 48, 49
Perameles, 183
Perennial ryegrass (see also *Lolium*), 223, 228, 230, 231, 234–236, 238
Peronospora
– *parasitica*, 18, 26, 28, 34
– *tabacina*, 149
Peronosporaceae, 18
Peronosporales, 18, 35, 270
Pezicula, 139
Phakopsora, 81–84, 92
– *pachyrhizi*, 83, 91, 92, 275
Phalaris tuberosa, 238
Phaseolus, 90, 270, 274
– *vulgaris*, 84, 92, 274
Phoma lingam, 206, 211, 216–218
Phomopsis subordinaria, 254
Phorbia, 200
– *phrenione*, 187, 233
Phragmidium violaceum, 88
Phyllosticta, 136, 139
– *abietis*, 130
Phylloxera, 144
Physopella, 81, 92
Phytophthora, 17, 18, 20, 21, 25–28, 32–36, 154–156
– *cactorum*, 149
– *capsici*, 26, 28, 149
– *cinnamomi*, 18, 26, 128
– *citricola*, 26
– *infestans*, 17–36, 143, 148, 155, 208, 216, 217, 219, 273
– *megasperma glycinea*, 18
– *megasperma medicaginis*, 151
– *nicotianae*, 149
– *palmivora*, 153, 156
– *parasitica*, 153, 156
– *sojae*, 18, 27, 28, 145, 156
Picea, 139, 97
– *abies*, 131, 139
– *excelsa*, 113, 114
– *mariana*, 139
Pichia angusta, 218
Pinaceae, 179, 182

Pine, 87, 132, 139, 172, 173, 175
Red pine, 136
Pineapple, 149
Pinus, 97, 98, 113, 115
– *banksiana*, 97, 114
– *monticola*, 154
– *pinaster*, 95–99, 101–104, 108, 109, 112–115
– *resinosa*, 140
– *sylvestris*, 98, 131, 138, 139
– *taeda*, 200
– *virginiana*, 97
Pisolithus, 102, 110, 113
– *arhizus*, 98, 114
– *tinctorius*, 107, 112, 175, 176, 178, 183
Pisum, 274
– *sativum*, 107, 113, 274
Plantago lanceolata, 254
Plasmodiophora, 146, 214
– *brassicae*, 214, 216
Plasmopara viticola, 18
Pleurophomopsis, 137
Pleuroplaconema, 138
Poa
– *autumnalis*, 191
– *sylvestris*, 191
– *trivialis*, 191
– *wolfii*, 191
Poaceae, 37, 129, 186, 195, 201, 202
Podophyllum peltatum, 276
Pooideae, 189, 190, 195, 227
Populus, 97, 179
Potato (see also *Solanum*), 17–36, 143, 144, 147, 148, 154, 155, 208, 216
Powdery mildew fungi (see also *Erysiphe*), 55–78, 266, 267
Prochimys, 173, 182
Protista, 27
Pseudocercosporella, 238
Pseudomonas syringae, 14
Pseudotsuga menziesii, 106, 114, 113
Puccinia, 81, 90
– *acroptili*, 92
– *arachidis*, 81
– *asarina*, 79
– *canaliculata*, 88, 89, 93
– *carthami*, 92
– *centaureae*, 92
– *chondrillina*, 88
– *coronata*, 88, 92
– *coronata avenae*, 91
– *graminis*, 16, 79, 80, 87–91, 93, 202, 211, 216
– *graminis avenae*, 93
– *graminis tritici*, 85, 91, 93, 148, 274
– *hordei*, 84
– *jaceae*, 88, 92
– *lagenophorae*, 92
– *melanicephala*, 82
– *menthae*, 91
– *pampeana*, 81, 90, 91
– *poarum*, 93, 274
– *recondita*, 84, 91, 93, 156
– *recondita tritici*, 275
– *sorghi*, 84, 89, 92, 152
– *xanthii*, 92
Pyrenomycete, 37
Pyrenophora tritici-repentis, 153, 154
Pyricularia, 51, 54
– *grisea*, 37, 38, 51–54
– *oryzae*, 37, 51–54, 217
Pythiaceae, 18
Pythium, 18, 25, 145, 152, 155
– *irregulare*, 151, 154

Quercus, 97
– *englemannii*, 175
– *garryana*, 140
– *petraea*, 139
– *robur*, 138

Rattus, 183
Rhabdocline parkeri, 130, 133–136, 140
Rhizobium, 69, 107, 138
Rhizoctonia, 145, 146, 155, 201
– *solani*, 140, 276
– *zeae*, 228, 237
Rhizopogon, 114
– *luteolus*, 98
Rhynchosporium secalis, 5, 11, 15, 16, 274, 276
Ribes
– *lacustre*, 154
– *viscosissimus*, 154
Rice (see also *Oryza*), 4, 37–54, 59, 131, 140, 144, 145, 153, 156
Rodents, 118, 173, 182
Rubus, 88
Rumex, 274
Russula, 98
– *emetica*, 175
Rust fungi, 79–93, 258–262, 266, 270, 274, 275

Saccharomyces cerevisiae, 27, 52, 160, 165, 209, 211, 216, 217
Safflower (see also *Carthamnus*), 88
Sagebrush, 171, 175, 179, 181
Salix, 79, 97
– *reticulata*, 97
– *retusa*, 97
Salsola kali, 179, 182, 273
Schizachyrium scoparium, 183
Schizophyllum commune, 99, 115
Schizosaccharomyces pombe, 39, 52, 210, 216
Sclerospora graminicola, 254
Sclerotinia sclerotiorum, 212, 216
Sclerotium cepivorum, 139
Sedge, 221, 224–226, 237
Senecio vulgaris, 92, 269, 273, 274
Septoria
– *nodorum*, 153, 154, 210, 216
– *tritici*, 210, 212, 218
Sequoia, 127
– *sempervirens*, 133, 135
Sequoiadendron, 127
Setaria, 48
Silene, 254
– *alba*, 200, 256
Sinapis alba, 214
Skeleton weed (see also *Chondrilla*), 88
Solanaceae, 17, 34, 91
Solanum, 22, 35
– *demissum*, 34
– *tuberosum tuberosum*, 22
– *tuberosum*, 17, 208
Sordariaceae, 207, 217
Sorghum, 164, 167, 168, 259, 274
Soybean (see also *Glycine*), 18, 28, 32, 145
Sphacelia, 221
Sphenopholis
– *nitida*, 191
– *pallens*, 191
Sphenospora saphena, 151, 155
Spinach, 164, 166
Spiny cocklebur (see also *Xanthium*), 164
Spodoptera frugiperda, 236
Sporotrichiella, 210, 216
Squirrels, 173, 175, 182
Stachybotrys chartarum, 128
Stagonospora innumerosa, 135
Stemphylium loti, 155
Stipa, 189, 190
– *leucotricha*, 190, 191, 225, 232, 237
– *robusta*, 191
Stramenopila, 18
Strawberry, 149, 163–167
Streptomyces, 128
Stylosanthes, 162, 165, 167, 216
Suillus, 98, 101, 102, 115
– *bellinii*, 107
– *bovinus*, 113
– *collinitus*, 98
– *granulatus*, 105, 112, 113
– *luteus*, 113
– *tomentosus*, 115

Talaromyces flavus, 147, 155
Tall fescue (see also *Festuca*), 228, 230, 231, 234–238
Telimenella, 51
Thelephora, 175
– *terrestris*, 98
Thomomys talpoides, 182
Thuja, 127
Tilletia
– *caries*, 218
– *controversa*, 218
Tilletiopsis pallescens, 75
Tobacco (see also *Nicotiana*), 10, 13, 28, 32, 144, 149, 154, 261, 275, 276
Tolumnia variegata, 151, 155
Tomato (see also *Lycopersicon*), 3–16, 17, 20, 25, 35, 36, 160, 165
Trichachne insularis, 238
Trichoderma, 88
– *harzianum*, 147, 155, 90
– *longibrachiatum*, 210, 217
– *reesei*, 218
Trifolium
– *pratense*, 155
– *repens*, 127

Triticeae, 59
Triticum aestivum, 76
Trypanosoma brucei, 219
Trypanosomes, 207
Tuber, 111
– *maculatum*, 98
Tussilago farfara, 93, 269, 274

Uromyces, 80, 82, 83, 86, 89–91, 93
– *appendiculatus*, 83–85, 89–93, 273
– *phaseoli*, 92
– *phaseoli typica*, 91
– *phaseoli vignae*, 91, 275
– *rumicis*, 274
– *viciae-fabae*, 80, 82, 83, 86, 89–91
– *vignae*, 81, 84, 93
Ustilago, 254
– *hordei*, 209, 218
– *maydis*, 26, 27, 52, 209, 217
– *violacea*, 200, 256

Vaccinium myrtillus, 138
Venturia inaequalis, 146
Verticillium, 88
– *lecanii*, 90
– *psalliotae*, 92
Vicia faba, 80, 82, 113
Vigna, 270, 274
– *sinensis*, 84, 90, 93
Vertebrates, 197, 199

Wapati, 173, 176
Weeping love-grass, 38
Wheat (see also *Triticum*), 41, 42, 48, 53, 57–59, 73–78, 148, 153, 154, 156, 263, 272, 275
Wheatgrass (see also *Elymus*), 58

Xanthium
– *occidentale*, 92
– *spinosum*, 164
Xanthomonas campestris
– *campestris malvacearum*, 274
– *campestris vesicatoria*, 276
Xylaria, 136, 137–139
– *cubensis*, 136, 139
Xylariaceae, 133, 137, 139, 140

Yarrowia lipolytica, 218

Zygomycotina, 117, 210, 211

Subject Index

Aegricorpus, 86, 87
Amino acids
  Arginine, 110
  Cysteine, 7, 8
  Glutamine/Glutamate, 102–105, 109, 113, 115
  Histidine, 110
  Methionine, 110
  Tryptophane, 108, 111
  Tyrosine, 8
Avirulence, 5–10, 55, 58, 73, 79, 81, 89, 246

Biological pest control, 146–148
Biological weed control, 88, 157

Chromosomes, 260, 261
  Aneuploidy, 207, 209, 230
  B chromosomes, 38, 49, 209
  Chromatin structure, 30
  Deletions, 49, 50, 209, 210
  Insertions, 209
  Foldback structures, 211
  Karyotype patterns, 211
  Length polymorphisms, 38, 209, 216–218
  Maps, 40, 57
  Minichromosomes, 38, 39, 162, 209
  Molecular karyotype, 208, 216
  Nucleosome structure of chromosomal DNA, 214
  Pachytene centromeres, 84
  Translocations, 209–211
  Transpositions, 207
Competition,
  Competitive ability, 124, 196, 198
  Competitive exclusion, 245
  Population displacement, 24–25
Compatibility, 20, 29, 33
  Basic, 4, 5
  Non-reciprocal, 194

Demography, 197, 200, 248, 251
  Birth rates, 244
  Death rates, 244, 245, 248, 249
  Extinction, 240, 243, 250–252, 255, 256
  Continuous growth, 222
  Population growth, 241
  Generation times, 241
  Fecundity, 227, 267
  Life expectancy, 245
  Overlap of generations, 240, 241
  Residual survival, 170, 173
  Separate generations, 241
  Survival, 146–148
Disease, 241, 242, 244–246, 250, 252, 254–256
  Avoidance, 242, 258
  Contagiousness, 148
  Disease cycle, 146
  Distribution in host population, 148
  Effects on host plant, 69–79, 143–147
  Etiology, 146
  Foci, 149
  Functional types, 145
  Infection chain, 146
  Phases of development, 151, 158
  Progress curves, 147, 149, 155
  Severity, 246
  Transmission, 146, 151
Diseases
  Anthracnoses, 157
  Apple scab, 146, 155
  Banana leaf spot, 144
  Bitter-rot of apple, 163
  Black shank, 28
  Blue mold, 144, 149, 155
  Chestnut blight, 144
  Choke disease, 188, 201, 202
  Clubroot disease, 214
  Coffee leaf rust, 82, 144
  Corn smut, 209
  Cotton wilt, 148
  Cowpea rust, 84, 93
  Crown gall, 214
  Downy mildew pathogens, 18, 144
  Dry rot, 29
  Dutch elm disease, 144, 210
  Ergot disease, 144, 225
  Flax rust, 4, 15, 79, 85, 87, 90–92, 262, 263, 273–275
  Foliar blight, 18
  Grape bunch rot, 145, 147
  Leaf spot necrosis, 145
  Leather rot of strawberry, 149
  Panama disease of bananas, 144
  Pea wilt, 209
  Peanut rust, 81, 93
  Pineapple heart rot, 149
  Potato late blight, 17, 19, 21, 33, 35, 143, 144, 147, 148, 208, 216
  Rice blast, 37, 38, 41, 42, 46, 47, 51–54, 144, 209, 219
  Root rot, 18
  Rusts, 80–81, 87, 148
  Seedling damping-off, 18
  Seedling disease, 228, 237
  Soft rot, 18, 210
  Southern leaf blight of maize, 144
  Stem rot, 28
  Sugar cane rust, 82
  Wheat blast, 42
  Wheat stem rust, 79, 91, 148
Disease-tolerance, 241, 253–255
Disturbances to habitats, 169–174, 179, 182, 183
DNA elements,
  Dissociator (Ds) elements, 13
  Fosbury or MAGGY, 50
  Gene promoters, 26, 27
  Grasshopper, 50, 51
  Gypsy family of retrotransposons, 50
  Hexanucleotide sequence TAGATA, 9
  Interspersed repetitive DNA segments, 211
  Introns, 26
  Intron sequences, 230, 231
  ITS1 region of the ribsomal DNA, 131
  ITS1, 162, 167
  ITS2, 164
  Maize activator, 12
  Mitochondrial DNA (MtDNA), 162–166, 190, 192, 194, 21
  Polymorphic DNA markers, 21
  Pot2 family of inverted repeat transposons, 50
  Promoters, 26, 27
  Repeat units for ribosomal DNA, 214
  Repetitive DNA, 57, 58, 211
  Retroelement, 50, 51
  Retrotransposons, 50, 57
  RFLP and ITS sequences, 41
  RFLPs, 12, 15, 21, 39, 41, 50, 56, 57, 59, 73, 74, 76, 78, 84, 89, 132, 136, 162–164, 192, 209
  RIP-like phenomena, 211
  Rrn internal transcribed spacers (ITS), 230
  Short Interspersed Nuclear Elements (SINE), 57
  Small subunit rRNA sequences, 229
  Telomeric 1.5 kb DNA fragment, 48
  Transposable elements, 49, 50
Ectomycorhizins, 110
  Ectomycorrhizae, 95–115
  Enzymatic solubilization of nutrients, 96

Ectomycorhizins,
Hartig net, 96, 101, 102, 108, 109, 112
Mantle, 102, 112
Nitrogen nutrition, 96, 102–106
Water uptake, 95
Elicitors, 17, 20, 26–29, 33–35, 58, 62, 75, 87, 91, 130, 258, 263, 268, 276
Arachidonic acid (AA), 28, 35
AVR4 protein, 7–9
AVR9 elicitor, 6–8, 10, 12, 13
Capsicein, 28
Elicitins, 28, 35
Eicosapentaenoic acid (EPA), 28
NIP1 protein, 11
Oligoglucoside elicitor, 28
Parasiticein, 28, 35
Race-specific, 4, 6, 9–11, 13, 15, 28
Endophytes, 129–140, 185–202, 221–238
Bark endophytes, 129
Effect on host drought tolerance, 233
E-endophytes, 221, 222, 224, 225, 228–232, 234, 235
Evolution, 221–238
Grass endophytes, 185–202, 221–238
Hybrid, 222, 230, 234, 235, 238
Host incompatibility, 232
On perennial plants, 130, 138, 185
P-endophytes, 130
Substrate utilization by, 136, 139
Symptomless, 186
Endophyte phylogenies, 190
Xylotropic, 129
Enzymes
1,3-β-glucanases, 6, 15, 16, 29
α-glucosidase, 65, 73
Acidic cellulases, 84
Adenosine deaminase, 65
Amylase, 132
Alternative oxidase, 69, 74
Anthranilate synthetase, 108
ATPase, 11, 12, 16
β-glucanase, 70, 71
CAMP-dependent protein kinase, 46, 53
Chitinase, 6, 15, 16, 60, 65, 70, 71, 77, 195, 201, 265, 274, 276
Cinnamyl alcohol dehydrogenase, 68
Cutinase, 29, 36, 45, 53, 61, 76, 83, 134, 159, 166–168
Cytochrome oxidase, 69
Endocellulases, 29
Endoxylanase, 46
Esterases, 61, 62, 83
Extracellular enzymes, 136
Galactanases, 29
Germin-like oxalate oxidase, 70, 78
Glutamate dehydrogenase, 102–105, 115
Glutathione-S-transferase, 70
H^+-ATPase, 11
H_2O_2-catalases, 10
HC-toxin reductase enzyme, 265
Hygromycin phosphotransferase gene, 111
Housekeeping enzymes, 26
Invertases, 5
Isopenicillin synthase, 214
Lipoxygenases, 10
Neutral zinc metallopeptidases, 48
Nitrate reductase, 102–106, 110, 113, 115
Pectin methyl esterase, 84
Pectin-esterases, 29
Pectinase, 132
Permeases, 65
Peroxidase, 60, 65, 71, 73, 76, 77, 86
Phosphatase, 102
Phospholipase A2, 28
Phenylalanine ammonia lyase, 68, 70, 71, 73, 77
Phytase, 102
Polygalacturonases, 29
Polygalacturonate lyase, 84, 90
Protease, 48
Protein kinase, 11, 16
Rubisco, 70
β-glucosidases, 29
Tryptophane synthase, 108
Xanthine oxidase, 87, 92
Evolution, 205–276
Coevolutionary arms races, 268, 269
Coevolution, 240, 242, 245, 247, 253–256
Convergent evolution, 42
Cospeciation, 230
Evolution of virulence, 245, 255
Gene-for-gene coevolution, 268, 269, 275, 276
Host-parasite evolution, 270
Epidemiological parameters
Density dependent host growth, 244
Density of susceptible hosts, 244
Disease-caused immunity, 244
Germination efficiency, 147
Infectivity, 103, 104, 106
Inoculum potential, 118, 123
Pathogen aggressiveness, 240, 241, 255
Pathogen death rate, 244
Pathogen transmission, 244
Persistence, 170, 183, 194
Rates of reinfection, 245
Spore density, 148
Epidemics (epiphytotics), 143–156, 169–183, 242
Analysis, 149
– Multivariate statistical approaches, 147
– Temporal analysis, 149
– Two-dimensional distance analysis, 149
– Spatial-temporal autocorrelation, 149
Boom and boost cycles, 249
Development, 19, 25, 143, 144, 146, 148, 151–154, 246, 247
Focal, 25
Source areas, 173, 176

Functional groups of fungi
Biotrophs, 263, 270
Endophytes, 129–140, 185–202, 221–238
Epigeous fungi, 174, 175
Epiphytes, 129, 137, 185, 186, 190, 200, 201, 224, 227, 236, 237
Hemibiotrophs, 17, 25
Hyperparasites, 88, 120
Latent pathogens, 129, 130, 136, 137
Mycoparasites, 211, 213
Necrotrophs, 270
Pioneer species, 98
Saprophytes, 258, 260, 265
Seral mycorrhizal fungi, 179
Soilborne pathogenic fungi, 149, 150
Symbionts, 169, 177, 183, 185, 186, 200
Wood decomposers, 129
Wound pathogens, 258, 260
Fungal attachment, 133, 134, 139, 140, 158, 167
Adhesion pad, 83
Adhesives, 258, 260
Glucose/mannose-specific lectins, 133
Hydrophobicity, 45, 46, 52
IPI proteins (suspected role in adhesion), 31–33
Mucilage, 122, 133, 159, 167
Spore tip mucilage, 43
Fungal dispersal and establishment, 143–156, 169, 170, 173, 176–183, 251, 252, 255
Abiotic factors, 146–150, 155, 221, 232, 233, 259, 264
Aerial dissemination, 18, 19, 82, 150, 151, 174–176, 182, 244
Animal dispersal of spores (VAM), 172, 176
Artificial inoculation, 173, 176
Cyclic dispersal, 148
Contagious spread (endophytes), 187–190, 192, 193, 195, 197–199
Dispersal by humans, 82
Entrainment (wind), 174, 181
Environmental heterogeneity, 180
Genetic constraints, 178
Horizontal transmission, 187, 190
Root colonization, 118, 123, 126, 272
Seed transmission, 186, 190–195, 198, 199, 202, 224
Spatial constraints, 179, 180
Spore deposition, 158, 170, 174–177, 179, 180
Vectors, 169, 173–176, 179, 182, 183
Vertical (maternal) transmission, 185, 188, 189, 221, 227, 228
Fungal fruiting bodies, 174
Acervuli, 159, 168

Aecia, 80, 81
Basidiocarps, 98, 101
Cleistothecia, 56, 57
Conidiomata, 157
Pycnia, 79, 80, 81
Sori, 80, 81, 88
Spermogonia, 79, 80
Sporangia, 18, 19, 35
Telia, 80
Uredia, 80
Fungal infection, 131–135, 139, 146, 147, 150, 151, 155, 158, 159, 166 168, 258, 260, 261
Appressorium, 18, 29, 31, 61, 75, 39, 44–46, 51–54, 82–84, 91–93, 121, 134, 135, 158, 159
Dikaryotic infection, 85, 86
Infection fans, 120
Monokaryotic infection, 85
Mycorrhizal infection, 171, 178, 182
Penetration hyphae, 18, 84, 86, 121, 134, 135
Penetration pegs, 44, 46, 62, 64, 121
Quiescent infection, 129, 157
Running hyphae, 85
Systemic infection, 224, 232, 245
Fungal intersterility groups, 266
Fungal/plant interface
Arbuscules, 117–121, 125, 214
Extracellular matrix, 133
Extracellular proteins, 5
Extracellular sheath, 134, 140
Extrahaustorial matrix, 32, 33, 62, 65–68, 78, 86
Extrahaustorial membrane, 62, 65, 67, 69, 78
Haustoria, 18, 56, 61, 64–68, 73–77, 80, 85, 87, 90, 91, 93, 135, 260
Hartig net, 96, 101, 102, 108, 109, 112
Intercellular hyphae, 121
Fungal spores
Aeciospores, 80–85
Ascospores, 147, 223–227, 232
Basidiospores, 80–84, 90, 92, 93
Conidia, 192
Chlamydospores, 118, 126
Oospores, 19–21, 24, 25, 34
Pycniospores, 80, 84, 85
Spermatia, 80, 186, 187, 192, 200, 223, 227
Teliospores, 79–81, 86, 87, 92
Urediospores, 80–87, 89, 93
Zoospores, 18, 214
Zygospores, 179
Fungal spore germination, 118–121, 130, 132–135, 138, 139, 158, 159, 167, 168
Eclosion, 133, 138
Germ pore, 79
Germ tubes, 79, 83
Iterative germination of ascospores, 187
Maternal supressor, 61
Primary germ tube, 56, 61, 62, 64, 67, 71, 76, 78
Proline rich proteins, 159
Role of CO_2, 119, 120, 122, 126
Self-inhibiting dormancy factors, 120
Root exudates, 118–120, 125, 126
Volatile compounds, 119, 125
Fungal toxins, 29, 258, 260, 261, 275
Fusicoccin, 11, 16
Host selective toxins, 265
Peptide phytotoxins, 11, 16, 260
Sirodesmins, 206, 216
Tenuazonic acid, 46, 52
Fungicides, 250
Anilide group, 107
Azole fungicides, 87
Benomyl, 107, 108, 49
Benodanil, 107
Carboxine, 107
Dithiocarbamates, 87
Ethirimol, 65, 74
Ergosterol inhibitors, 87
Indar, 87
Metalaxyl, 24, 25, 34
Oxycarboxine, 107

Gene cloning, 12–13
Map-based cloning, 12
Transposon tagging, 12–13
Gene expression
Differential, 9–10, 29–33
Genes
14-3-3 (defense-related gene, barley), 71
Avirulence genes, 4, 5, 7, 15, 18, 20, 26, 28, 36, 38, 55–59, 72, 75, 207, 210
AVR2–YAMO (*Magnaporthe* resistance, rice), 43, 48–51
BUF1 (melanin biosynthesis, *Magnaporthe*), 49
CalA (calmodulin synthesis, *Phytophthora*), 27, 30, 31
CPKA (protein kinase subunit, *Magnaporthe*), 46
CUT1 (cutinase, *Magnaporthe*), 45
Defense genes, 4, 12, 14
ecp genes (extracellular proteins, *Cladosporium*), 5, 6
Fungal avirulence genes, 262
HPG (hygromycin resistance, bacteria), 26
HSP90 (defense-related gene, barley), 71
INF genes (infection structures, rusts), 85, 93
Ipi genes (cell wall proteins?, *Phytophthora*), 27, 31–33, 35
L^6 (resistance to rust, flax) , 87
Lineage-specific resistance genes, 43
Mat1 (mating type, many fungi), 37
Ml genes (resistance to *Erysiphe*, barley), 59
MPG1 (hydrophobin, *Magnaporthe*), 46, 53
Nar genes (signal transduction, barley), 59, 74
Nia genes (nitrate reductase, *Phytophthora*, *Aspergillus*, (26, 50)
NPTII (resistance to G-418, bacteria), 26
Pathogenicity genes, 26, 29, 30, 33, 35, 152
Pathotype-specific resistance genes, 268, 269
Ph1, (*Phytophthora* resistance, Tomato), 20
Pm genes (*Erysiphe* resistance, wheat), 58
PWL genes (host specificity genes, *Magnaporthe*), 47–51, 53
Pyr genes (orotidine-5′-mono-phosphate decarboxylase, *Epichloë*), 231, 236
Recessive housekeeping genes, 210
Resistance genes, 56, 58–60, 69, 72–74
R3 (resistance to *Phytophthora*, potato), 20
rRNA genes (ribosomal RNA, universal), 229
SMO1 (appressoria differentiation, *Magnaporthe*), 45, 49
Sr genes (rust resistance, wheat), 88
Symbiosis genes, 272
β-tubulin genes (*Neurospora*, *Aspergillus*), 107, 108
Trp1 (tryptophane biosynthesis, *Phytophthora*), 26, 35
tub2 (β-tubulin gene, *Epichloë*), 230, 231
Ubi3R (ubiquitin, *Phytophthora*), 30, 31
Unnecessary resistance genes, 247
Unneeded virulence genes, 268
Genetic instability, 49–51
Gene-transfer systems, 110
Autolytic processes, 214
DNA transformation, 26
Extrachromosomal inheritance, 21
Horizontal, 213, 214, 217
Parasexual, 49, 159, 165, 193, 199, 208
Sexual reproduction, 25, 268, 274
Genomic organization (see also chromosomes)
Heterokaryosis, 159, 166, 168, 193, 215
Heterozygosis, 246
Homozygosis, 243, 246, 248
Linkage groups, 39, 209
Glycoproteins, 84, 86, 89, 93, 132, 134, 159, 167
Green islands, 87
Group species, 157, 162, 167

Habitats for fungal colonization
Agricultural systems, 143, 144, 153, 154, 170, 239, 240, 246, 247
Alder thickets, 175
Alpine, 97

Habitats for fungal colonization
Ant mounds, 172
Arid ecosystems, 179
Badger mounds, 172, 176
Beach fronts, 175
Canopy edge, 175
Chaparral, 172
Dunes, 98, 169, 175, 183
Endemic plants, 245, 253
Exotic habitats, 178
Forest edges, 181
Forests, 171, 173
Gaps, 175
Glacial moraine, 171
Glaciers, 175
Gopher mounds, 172, 176, 182
Grapevine canopies, 147
Mt. St. Helens, 171
Mycorrhizosphere, 122, 124, 125, 127
Native ecosystems, 171
Native plant communities, 169, 170, 173, 181
Plant canopy, 176
Plant communities, 266, 273, 274, 276
Plant patches, 176, 180
Potato refuse piles, 25
Refuges, 244
Rhizosphere, 150, 155
Savannas, 175
Shrub rhizosphere, 176
Shrub-steppe, 173–175, 179, 182
Shrublands, 171
Strip mines, 172, 173
Subalpine and alpine meadows, 173
Tropical seasonal forests, 172
Unstable ecological systems, 242, 248
Woody hosts, 130
Herbicides, 171, 250
Herbivores, 172, 196, 197, 200–222, 226, 228, 233, 259, 264, 274, 275
Herbivory, 171, 172, 197, 222, 226, 234, 236
Host plant defenses (see also Hypersensitive response, Pathogenesis-related proteins, Phytoalexins, 10, 249, 258
Callose deposition, 6, 62, 64, 74, 76
Cell wall lignification, 10
Papilla formation, 60, 62, 69, 71, 74–78, 87
Halos, 135
Inducible defenses, 258–261, 264, 272
Mechanisms of defense, 205–207, 215, 258
Silicon compounds, 61, 62, 64, 75, 86
Wound plugs, 62
Host specificity, 192–195, 201, 257, 258, 261–263, 266, 270, 272, 275, 276
Hydrophobins, 46
Hypersensitive response (HR), 4, 20, 36, 195
Active oxygen species, 10, 258
Mitochondrial degeneration, 68
Necrosis, 10, 13–16, 65, 68, 69, 87, 130
Oxidative burst, 10
Superoxide, 10
Tonoplast breakdown, 68

Induced susceptibility, 60, 75
Interactions,
Antagonistic, 240, 243
Compatible, 6, 7
Incompatible, 20,
Gene-for-gene, 3, 4, 18, 20, 55, 58, 59, 79, 90, 161, 207, 240, 254 259, 261–263, 266, 268, 272, 273, 275
Genotype/environment, 103, 105
Genotypic, 240
Plant/parasite, 259, 269, 270
Plant/pathogen, 240, 241, 244, 253
Insect herbivory, 129, 137, 199, 200

Karyotypes (see also chromosomes), 208–211

Life cycles
Autoecious, 79, 80, 90
Colletotrichum, 157–158
Demicyclic, 80
Dikaryons, 81–83, 85, 86, 91, 99, 101, 102, 104–108, 110, 111, 114
Erysiphe, 56
Epichloë, 186, 223
Hebeloma, 98–101
Heteroecious, 79–82, 90
Karyogamy, 179
Magnaporthe, 39
Macrocyclic, 80, 81
Meiosis, 179
Microcyclic, 80
Monokaryons, 82, 85, 86, 91, 99–102, 105, 107, 108, 110, 114, 115, 179
Neotyphodium, 223
Phytophthora, 18–20
Rusts, 79–81
VAM, 118

Mannitol, 5
Mating systems, 17, 19–25, 34, 240
Bifactorial, 101, 160
Female-sterile strains, 38
Hermaphrodites, 37, 38
Heterothallic, 79, 160, 161, 187, 192, 207, 225, 227
Homothallic (self) mating, 160, 161, 225
Migration, 21, 34, 36, 173, 174, 176, 177, 180, 182, 248, 250–252, 255, 268, 269
Models and modeling, 151, 239–256
Coevolutionary, 242, 244, 247, 248
Ecological, 242, 243–245
Epidemiological, 243, 244
Equilibrium, 240, 242, 243
Genetic, 248
Host/parasite, 243, 244
Logistic, 149
Multiline, 247
Non-equilibrium, 243
Monomolecular model, 149
Old-field model of succession, 181
Plant/pathogen, 240–242, 253
Predator/prey, 240, 248, 249
Predictive, 147, 239
Single locus, 241
Spatial, 176
Strategic, 239
Tactical, 239, 242
Molecular genetic markers
Allozymes, 21, 36, 229, 230, 232, 235
Amplified fragment length polymorphism (AFLP), 12
DNA fingerprints, 21, 22, 23, 25, 34, 162–164, 212–219
DNA sequences, 162
Electrophoretic karyotyping, 162
Electrophoretic profiles, 136
Isozymes, 132, 136
Heterologous transposon tagging, 12
Mitochondrial DNA polymorphisms, 41, 232
Random amplified polymorphic DNAs (RAPDs), 12, 57, 78, 132, 136, 139, 162–164, 212, 217, 218
Restriction fragment length polymorphism (RFLP)s, 8, 12, 212, 217
Mutants
Nitrate non-utilizing mutants (*Hebeloma*), 110
Nitrate non-utilizing mutants (*Colletotrichum*), 159
Buf$^-$ (melanin biosynthesis, *Magnaporthe*), 49
Con mutants (conidiation-defective, *Magnaporthe*), 46
Fluorindole resistant (FIR) mutants(*Hebeloma*), 108, 113
Fungicide resistant mutant, 107
IAA-overproducer mutants (*Hebeloma*), 108, 109
Mpg1$^-$ (appressorial development, *Magnaporthe*), 46
Smo$^-$, (abnormal spore morphology, *Magnaporthe*), 45–47
Mutation, 8, 13, 58, 59, 74, 77, 193, 242, 243, 248, 252–255
Mutational load, 24, 234, 235
Muller's ratchet, 234, 235
Somatic, 210
Mutualistic symbioses, 169, 221–225, 227, 228, 230, 232–238
Benign parasitism, 222
Costs, 222, 225, 234
Endophytes, 129–140, 185–202, 221–238
Mycorrhizal dependence, 272, 273, 275
Mycorrhizae, 95–128, 169–183
Protective, 129
Zooxanthellate algae, 221

Mycorrhizae, 95–128, 169–183
Colonization, 272
Formation, 272
Mycorrhization helper bacteria, 96
Mycorrhizal propagules, 172

Nitrogen metabolism, 5, 9, 15, 16, 70, 76, 77, 95, 96, 102, 110, 114, 117, 123, 126, 127
Ammonium assimilation, 102
Glutamate dehydrogenase, 102, 105–106
Nitrate uptake, 70
Nitrogen fixation, 124, 221
Nitrate reductase, 96, 102–106, 109, 110, 112–115
Organic nitrogenous molecules, 109

Pathogenesis-related proteins (PRP), 6, 60, 70, 76
Chitinase, 6, 15, 16, 60, 65, 70, 71, 77, 195, 201, 265, 274, 276
Thaumatin-like protein, 70
Thionins, 71, 73
Pathogenicity factors, 4, 15, 33, 258–266, 270, 273
Peptides, 6–9, 11, 13, 15, 16
Amino terminal signal peptide, 31, 32
AVR9 elicitor peptide, 7
GRGDSP hexapeptide, 32
Peptide phytotoxins, 260
RGD tripeptide motif, 32
Pesticides, 171
Pests, 196, 199, 200
Phosphorus metabolism and nutrition
Phosphate, 60, 65, 70, 77, 78, 106, 96, 123–124
Phospholipid, 122
Phosphorus, 122, 123, 126–128
Polyphosphate metabolism, 70, 102
Phylogeny, 229–232
Phytoalexins, 86, 87, 92, 123, 258, 260, 261, 265, 276
Phenylpropanoids, 261
Diterpenoid rice phytoalexins, 46
Pisatin, 209
Plant developmental and physiological processes
Apoplastic flow, 86
CN-insensitive respiration, 69
Dark respiration, 69
Lignification, 87, 124, 125, 195
Ripening (of fruit), 129
Suberization, 121
Transpiration, 123
Plant growth regulators, 123, 196
Auxins, 87, 92, 96, 108, 112, 113
Cytokinins, 87, 90, 96
Ethylene, 10, 87, 92, 96
Gibberellins, 96
Indolacetic Acid (IAA), 103–105, 108, 109, 111–113
Jasmonic acid, 28, 71, 76
Plant morphology – organs, tissues and cells
Apoplast, 5, 6, 9, 15
Basal leaf meristems, 186
Bulbil, 196
Caryopsis, 259, 274
Casparian strip, 86, 91
Cleistogamous flowers, 190, 227, 236
Cleistogamous seeds, 224, 227
Cotyledons, 10, 12, 28
Crop seed, 247
Crotches of trees, 175
Culms, 186, 187, 189, 190
Cuticle, 131, 134, 259
Epidermis, 60, 61, 65
Epidermal cells, 18, 134, 135
Endodermis, 109
Flowers, 187
Guard cells, 18, 83
Hypocotyls, 145
Inflorescences, 196, 223, 224, 226–228
Leaves, 186, 187, 196, 200
Evergreen foliage, 130
Leaf axils, 224, 227
Leaf mesophyll, 5
Leaf primordia, 224
Petioles, 206
Lenticels, 19
Mesophyll, 5, 9, 10, 18, 206
Nodules, 124
Nucellus tissues, 186
Outer cortex, 109
Ovule, 32, 186, 188
Phloem, 5, 9
Pollen, 32
Rhizomes, 169, 189, 190
Root cortex, 117, 118
Root endodermal cells, 86
Root systems, 172
Seed, 185–190, 192–202
Seed chambers, 172
Seed heads, 226
Seedlings, 145, 151, 154, 155
Seedling radicles, 145
Sheaths, 186, 190
Shoot meristems, 186, 189
Stems, 186
Stigmata, 188
Stolons, 186, 20
Stomata, 5, 18, 261
Subterranean plant parts, 272
Tillers, 186, 188, 189, 194, 195, 224, 227, 228
Tropical fruits, 157–159, 162, 165–167
Tubers, 189
Viviparous plantlets, 189, 190, 196
Xylem vessels, 206
Xylem, 5, 9
Plant toxins and other products (see also plant growth regulators, phytoalexins, and polysaccharides
Alkaloids, 197, 226, 228–230
Anti-herbivore metabolites, 222, 233
Ergot alkaloids, 226, 236, 238
Fatty alcohols, 159
Flavonoids, 119
Gramine, 71, 77
Leaf essential oils, 133
Lignin, 265
Monolignol glucosides, 133, 138
Phenylpropanoids, 261
Polyphenols, 62, 68, 159
Wax, 159, 167
Polysaccharides, 86, 132, 133
3-β-glucans, 6
Glucans, 65, 84
Pectin, 84, 86, 90
Chitin, 133, 6, 12, 15, 46, 65, 84, 90, 260
Starch, 70, 74
Populations, 185–202, 240–256
Population genetics, 17, 21
Constant fitnesses, 241, 248
Density-dependent selection, 241
Diploid genetics, 87, 246
Diffuse selection, 264
Epistasis, 103
Frequency dependent fitness, 241, 244
Frequency dependent selection, 241
Gene flow, 136
Genetic differentiation, 190, 193, 200
Genetic drift, 252, 256, 268
Genetic hitchhiking, 269
Group-selection, 245
Immigration (see also migration), 170, 173
Host-specialized populations, 157, 158
Mating populations, 229–231
Multidimensional selection pressures, 241
Overdominance, 246
Plant genotype, 241, 242, 246
Plant reproductive fitness, 151
Polymorphism, 246, 266, 268
Sexual reproduction, 25
Virulence diversity, 23, 24
Population structure
Aggregation, 149, 151, 244, 255
Asynchronous density, 251
Clonal structure, 244
Cultivar mixes, 246, 247, 253
Density, 240–244, 247, 248, 251
Metapopulations, 243, 250–252, 254, 268, 273
Local extinction, 181, 269
Pathogen density, 243
Patches, 169–172, 180, 182, 243, 250, 252
Panmixis, 252
Plant density, 243
Safe sites, 181
Within-subpopulation patchiness, 252
Seed bank, 151

Quantitative genetics, 98–108, 240, 249, 252, 254, 255
Heritability, 103
Genetic correlations, 241
Genotype/environment interactions, 103, 105

Receptors and binding-proteins (see also signals), 11, 13, 30, 32, 58, 72
Na$^+$-sensitive receptor, 133
Guanine nucleotide (GDP or GTP)-binding regulatory proteins, 11
Integrins, 32, 34
NIT2 protein (transcriptional factor), 9
Ubiquitin (regulator of gene expression) , 30, 31, 35
Recombination, 25, 159, 162
Activators of, 210
Parasexual, 49, 159, 165, 193, 199, 208
Resistance, 240–249, 252–256
Basic, 195
Correlational, 242
Costs of, 243, 246–248
Field, 259, 267
Fungicide, 21
Host genotypes, 152
Host genotype-specific, 195
Hygromycin, 215
Local induced, 60
Nonhost, 259, 261, 262, 265, 273
One-locus, 239
Papilla-based, 60
Quantitative, 240, 254, 255
Race-specific, 18, 20
Systemic, 222
Temperature-dependent, 269

Sclerotia, 157, 224–226
Signals and signal transduction, 4, 10–12, 14, 59, 61, 159, 262, 265
14-3-3 protein kinase regulator-like protein, 70
Amino terminal signal peptide, 31, 32
Calcium ion fluxes, 30, 31
Calmodulin, 30, 31, 35
Cutin monomers, 61, 134
Cyclic AMP (CAMP), 46, 52, 53
CAMP-dependent protein kinase, 46, 53
Dephosphorylation, 11, 16
Elicitor perception, 4, 10, 11
Endogenous signal, 10
G proteins, 11, 15
H$^+$-ATPase, 11
H_2O_2, 10
Linolenic acid, 28
Lipid peroxidation, 10, 16
Lipoxygenases, 10
Peroxidases, 60, 65, 71, 73, 76, 77, 86
Protein kinase, 11, 16
Salicylic acid, 10, 14, 15
Serine/threonine protein kinases, 14
Stromata, 186, 187, 189, 190, 194, 196, 202, 223–227, 229, 234, 238
Succession, 132, 169–172, 175, 177, 181, 182
Bootstrapping, 172, 183
Sugars
Fructose, 5
Glucose, 5
Sucrose, 5, 65, 70, 74
Suppression of host defense gene expression, 20, 60, 70, 75, 87, 270
Symbiosis, 169, 177, 183, 185, 186, 200
Commensalisms, 222
Multipartner, 96
Mutualistic, 169, 221–225, 227, 228, 230, 232–238
Pleiotropic, 222
Symbiotic continuum, 221, 222, 225, 228, 233

Variance
Additive variance, 103
Genotypic variance (VG), 103
Environmental variance, 103, 105
Phenotypic, 103
Within-island equilibrium variance, 252
Variation
Continuous, 240
Genetic, 23–25, 207–208, 266
Glutamate dehydrogenase activity, 105–106
IAA-synthesizing activity, 105, 106
Intraspecific, 106
Major-locus, 240
Vegetative compatibility groups, 159, 164
Vesicular-arbuscular mycorrhizae (VAM), 107, 117–127, 171–183
Auxiliary cells, 118
Arbuscules, 117–121, 125, 214
Role in host drought tolerance, 123, 124, 127
Host osmoregulation, 123, 125
Hyphal coils, 118, 121, 214
Hyphal exudates, 122
Interplant nutrient transfer, 123
Role in host mineral nutrition, 123
Role in host seedling establishment, 123
Soil aggregation, 122, 123
Vesicles, 117–121, 125, 126, 214
Virulence, 20–25, 33, 35, 36, 130, 241–249, 253–256